BIOLOGY

The Unity and Diversity of Life

STARR

TAGGART

EVERS

STARR

Volume 1
Cell Biology and Genetics

CENGAGE
Learning

14TH EDITION

Australia • Brazil • Mexico • Singapore • United Kingdom • United States

CENGAGE
Learning®

Cell Biology and Genetics
Biology: The Unity and Diversity of Life,
Fourteenth Edition
Cecie Starr, Ralph Taggart, Christine Evers,
Lisa Starr

Product Director: Mary Finch

Senior Product Team Manager: Yolanda Cossio

Senior Product Manager: Peggy Williams

Associate Content Developers: Kellie Petruzzelli,
   Casey Lozier

Product Assistant: Victor Luu

Media Developer: Lauren Oliveira

Senior Market Development Manager:
   Tom Ziolkowski

Content Project Manager: Harold Humphrey

Senior Art Directors: John Walker, Bethany Casey

Manufacturing Planner: Karen Hunt

Production Service: Grace Davidson & Associates

Photo Researcher: Cheryl DuBois, PreMedia Global

Text Researcher: Kristine Janssens,
   PreMedia Global

Copy Editor: Anita Wagner Heuftle

Illustrators: Lisa Starr, Gary Head,
   ScEYEnce Studios

Text Designer: Lisa Starr

Cover Designer: Bethany Casey

Cover and Title Page Image:
   © Pete Oxford/Minden Pictures

   Butterflies sip the tears of a yellow-spotted river turtle sunning itself in Yasuní National Park,
   Ecuador. Turtle tears supply the butterflies with
   sodium, an essential nutrient missing from their
   flower nectar diet in the Amazon rainforest.
   Butterflies are almost never observed sipping
   turtle tears outside of this small region, which
   is famous for having one of the most diverse
   assortments of species in the world. Currently,
   oil drilling operations are destroying the forest
   and wildlife in the park.

Compositor: Lachina Publishing Services

For product information and technology assistance, contact us at
Cengage Learning Customer & Sales Support, 1-800-354-9706.
For permission to use material from this text or product,
submit all requests online at **www.cengage.com/permissions**.
Further permissions questions can be emailed to
permissionrequest@cengage.com.

Library of Congress Control Number: 2014944586

ISBN-13: 978-1-305-25124-3

ISBN-10: 1-305-25124-5

Cengage Learning
20 Channel Center Street
Boston, MA 02210
USA

Cengage Learning is a leading provider of customized learning solutions with office
locations around the globe, including Singapore, the United Kingdom, Australia, Mexico,
Brazil, and Japan. Locate your local office at:
www.cengage.com/global.

Cengage Learning products are represented in Canada by Nelson Education, Ltd.

To learn more about Cengage Learning Solutions, visit **www.cengage.com.**

Purchase any of our products at your local college store or at our preferred online store
www.cengagebrain.com.

Printed in Canada
Print Number: 01 Print Year: 2014

Contents in Brief

Highlighted chapters are not included in *Cell Biology and Genetics*.

INTRODUCTION

1   Invitation to Biology

UNIT I   PRINCIPLES OF CELLULAR LIFE

2   Life's Chemical Basis
3   Molecules of Life
4   Cell Structure
5   Ground Rules of Metabolism
6   Where It Starts—Photosynthesis
7   How Cells Release Chemical Energy

UNIT II   GENETICS

8   DNA Structure and Function
9   From DNA to Protein
10 Control of Gene Expression
11 How Cells Reproduce
12 Meiosis and Sexual Reproduction
13 Observing Patterns in Inherited Traits
14 Chromosomes and Human Inheritance
15 Studying and Manipulating Genomes

UNIT III   PRINCIPLES OF EVOLUTION

16 Evidence of Evolution
17 Processes of Evolution
18 Organizing Information About Species
19 Life's Origin and Early Evolution

UNIT IV   EVOLUTION AND BIODIVERSITY

20 Viruses, Bacteria, and Archaea
21 Protists—The Simplest Eukaryotes
22 The Land Plants
23 Fungi
24 Animal Evolution—The Invertebrates
25 Animal Evolution—The Chordates
26 Human Evolution

UNIT V   HOW PLANTS WORK

27 Plant Tissues
28 Plant Nutrition and Transport
29 Life Cycles of Flowering Plants
30 Communication Strategies in Plants

UNIT VI   HOW ANIMALS WORK

31 Animal Tissues and Organ Systems
32 Neural Control
33 Sensory Perception
34 Endocrine Control
35 Structural Support and Movement
36 Circulation
37 Immunity
38 Respiration
39 Digestion and Nutrition
40 Maintaining the Internal Environment
41 Animal Reproductive Systems
42 Animal Development
43 Animal Behavior

UNIT VII   PRINCIPLES OF ECOLOGY

44 Population Ecology
45 Community Ecology
46 Ecosystems
47 The Biosphere
48 Human Impacts on the Biosphere

Detailed Contents

INTRODUCTION

1 Invitation to Biology

1.1 The Secret Life of Earth 3

1.2 Life Is More Than the Sum of Its Parts 4

1.3 How Living Things Are Alike 6
Organisms Require Energy and Nutrients 6
Organisms Sense and Respond to Change 7
Organisms Use DNA 7

1.4 How Living Things Differ 8

1.5 Organizing Information About Species 10
A Rose by Any Other Name . . . 10

1.6 The Science of Nature 12
Thinking About Thinking 12
The Scientific Method 12

1.7 Examples of Experiments in Biology 14
Potato Chips and Stomachaches 14
Butterflies and Birds 14

1.8 Analyzing Experimental Results 16
Sampling Error 16
Bias in Interpreting Results 17

1.9 The Nature of Science 18
The Limits of Science 18

The Secret Life of Earth (revisited) 19

UNIT I   PRINCIPLES OF CELLULAR LIFE

2 Life's Chemical Basis

2.1 Mercury Rising 23

2.2 Start With Atoms 24
Isotopes and Radioisotopes 24

iv

2.3 Why Electrons Matter **26**
About Vacancies **27**

2.4 Chemical Bonds: From Atoms to Molecules **28**
Ionic Bonds **28**
Covalent Bonds **28**

2.5 Hydrogen Bonds and Water **30**
Hydrogen Bonding in Water **30**
Water's Special Properties **30**

2.6 Acids and Bases **32**

Mercury Rising (revisited) **33**

3 Molecules of Life

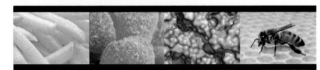

3.1 Fear of Frying **37**

3.2 Organic Molecules **38**
Carbon: The Stuff of Life **38**
Modeling Organic Molecules **38**

3.3 Molecules of Life—From Structure to Function **40**
Functional Groups **40**
What Cells Do to Organic Compounds **40**

3.4 Carbohydrates **42**
Carbohydrates in Biological Systems **42**

3.5 Lipids **44**
Lipids in Biological Systems **44**

3.6 Proteins **46**

3.7 Why Is Protein Structure So Important? **48**

3.8 Nucleic Acids **49**

Fear of Frying (revisited) **49**

4 Cell Structure

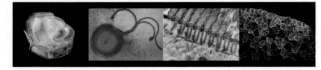

4.1 Food for Thought **53**

4.2 What Is a Cell **54**
Cell Theory **54**
Components of All Cells **54**
Constraints on Cell Size **55**

4.3 How Do We See Cells? **56**

4.4 Introducing Prokaryotes **58**
Biofilms **59**

4.5 Introducing Eukaryotic Cells **60**

4.6 The Nucleus **62**
Chromatin **62**
The Nuclear Envelope **62**
The Nucleolus **63**

4.7 The Endomembrane System **64**

4.8 Mitochondria **66**

4.9 Chloroplasts and Other Plastids **67**

4.10 The Cytoskeleton **68**

4.11 Cell Surface Specializations **70**
Cell Matrices **70**
Cell Junctions **70**

4.12 The Nature of Life **72**

Food for Thought (revisited) **73**

5 Ground Rules of Metabolism

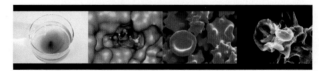

5.1 A Toast to Alcohol Dehydrogenase **77**

5.2 Energy in the World of Life **78**
Energy Disperses **78**
Energy's One-Way Flow **78**

5.3 Energy in the Molecules of Life **80**
Chemical Bond Energy **80**
Why Earth Does Not Go Up in Flames **80**
Energy In, Energy Out **81**

5.4 How Enzymes Work **82**
The Need For Speed **82**
The Transition State **82**
Enzyme Activity **83**

Detailed Contents (continued)

5.5 Metabolism—Organized, Enzyme-Mediated
Reactions **84**
Controls Over Metabolism **84**
Electron Transfers **84**

5.6 Cofactors in Metabolic Pathways **86**
ATP—A Special Coenzyme **86**

5.7 A Closer Look at Cell Membranes **88**
The Fluid Mosaic Model **88**
Proteins Add Function **88**

5.8 Diffusion and Membranes **90**
Semipermeable Membranes **90**
Turgor **91**

5.9 Membrane Transport Mechanisms **92**
Transport Protein Specificity **92**
Facilitated Diffusion **92**
Active Transport **92**

5.10 Membrane Trafficking **94**
Endocytosis and Exocytosis **94**
Recycling Membrane **95**

A Toast to Alcohol Dehydrogenase (revisited) **96**

6 Where It Starts—Photosynthesis

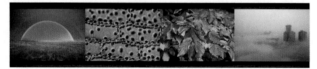

6.1 Biofuels **101**

6.2 Sunlight as an Energy Source **102**
Properties of Light **102**
Pigments: The Rainbow Catchers **102**

6.3 Exploring the Rainbow **104**

6.4 Overview of Photosynthesis **105**

6.5 Light-Dependent Reactions **106**
The Noncyclic Pathway **106**
The Cyclic Pathway **107**

6.6 The Light-Independent Reactions **108**
Energy Flow in Photosynthesis **108**
Light-Independent Reactions **108**

6.7 Adaptations: Alternative Carbon-Fixing Pathways **110**

Biofuels (revisited) **112**

7 How Cells Release Chemical Energy

7.1 Risky Business **117**

7.2 Overview of Carbohydrate Breakdown Pathways **118**

7.3 Glycolysis—Sugar Breakdown Begins **120**

7.4 Second Stage of Aerobic Respiration **122**
Acetyl–CoA Formation **122**
The Krebs Cycle **122**

7.5 Aerobic Respiration's Big Energy Payoff **124**

7.6 Fermentation **126**

7.7 Alternative Energy Sources in Food **128**
Energy From Dietary Molecules **128**

Risky Business (revisited) **129**

UNIT II GENETICS

8 DNA Structure and Function

8.1 A Hero Dog's Golden Clones **133**

8.2 The Discovery of DNA's Function **134**

8.3 The Discovery of DNA's Structure **136**
Building Blocks of DNA **136**
DNA's Base Sequence **137**

8.4 Eukaryotic Chromosomes **138**
Chromosome Number and Type **139**

8.5 DNA Replication **140**

8.6 Mutations: Cause and Effect **142**
Replication Errors **142**
Agents of DNA Damage **142**
Rosalind Franklin, X-Rays, and Cancer **143**

8.7 Cloning Adult Animals **144**

A Hero Dog's Golden Clones (revisited) **145**

9 From DNA to Protein

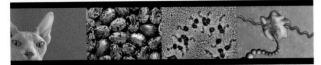

9.1 Ricin, RIP 149

9.2 DNA, RNA, and Gene Expression 150
DNA to RNA 150
RNA to Protein 150

9.3 Transcription: DNA to RNA 152
Post-Transcriptional Modifications 153

9.4 RNA and the Genetic Code 154

9.5 Translation: RNA to Protein 156

9.6 Mutated Genes and Their Protein Products 158

Ricin, RIP (revisited) 160

10 Control of Gene Expression

10.1 Between You and Eternity 163

10.2 Switching Genes On and Off 164
Gene Expression Control 164

10.3 Master Genes 166
Homeotic Genes 166

10.4 Examples of Gene Control in Eukaryotes 168
X Marks the Spot 168
Male Sex Determination in Humans 168
Flower Formation 169

10.5 Examples of Gene Control in Prokaryotes 170
The *lac* Operon 170
Lactose Intolerance 171
Riboswitches 171

10.6 Epigenetics 172

Between You and Eternity (revisited) 173

Detailed Contents (continued)

11 How Cells Reproduce

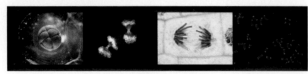

11.1   Henrietta's Immortal Cells   177

11.2   Multiplication by Division   178
Controls Over the Cell Cycle   179

11.3   A Closer Look at Mitosis   180

11.4   Cytokinesis: Division of Cytoplasm   182

11.5   Marking Time With Telomeres   183

11.6   When Mitosis Becomes Pathological   184
Cancer   184

Henrietta's Immortal Cells (revisited)   186

12 Meiosis and Sexual Reproduction

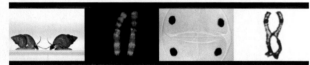

12.1   Why Sex?   189

12.2   Meiosis in Sexual Reproduction   190
Introducing Alleles   190
Meiosis Halves the Chromosome Number   190
Fertilization Restores Chromosome Number   191

12.3   Visual Tour of Meiosis   192

12.4   How Meiosis Introduces Variations in Traits   194
Crossing Over   194
Chromosome Segregation   194

12.5   Mitosis and Meiosis—An Ancestral Connection?   196

Why Sex? (revisited)   197

13 Observing Patterns in Inherited Traits

13.1   Menacing Mucus   201

13.2   Mendel, Pea Plants, and Inheritance Patterns   202
Mendel's Experiments   202
Inheritance in Modern Terms   202

13.3   Mendel's Law of Segregation   204

13.4   Mendel's Law of Independent Assortment   206
The Contribution of Crossovers   206

13.5   Beyond Simple Dominance   208
Codominance   208
Incomplete Dominance   208
Epistasis   209
Pleiotropy   209

13.6   Nature and Nurture   210
Some Environmental Effects   210

13.7   Complex Variations in Traits   212
Continuous Variation   212

Menacing Mucus (revisited)   213

14 Chromosomes and Human Inheritance

14.1   Shades of Skin   217

14.2   Human Chromosomes   218
Types of Genetic Variation   218

14.3   Examples of Autosomal Inheritance Patterns   220
The Autosomal Dominant Pattern   220
The Autosomal Recessive Pattern   221

14.4   Examples of X-Linked Inheritance Patterns   222
The X-Linked Recessive Pattern   222

14.5   Heritable Changes in Chromosome Structure   224
Types of Chromosomal Change   224
Chromosome Changes in Evolution   225

14.6   Heritable Changes in Chromosome Number   226
Autosomal Aneuploidy and Down Syndrome   226
Sex Chromosome Aneuploidy   226

14.7   Genetic Screening   228

Shades of Skin (revisited)   229

15 Studying and Manipulating Genomes

15.1 Personal Genetic Testing 233

15.2 Cloning DNA 234
Cut and Paste 234
cDNA Cloning 235

15.3 Isolating Genes 236
DNA Libraries 236
PCR 236

15.4 DNA Sequencing 238
The Human Genome Project 239

15.5 Genomics 240
DNA Profiling 240

15.6 Genetic Engineering 242

15.7 Designer Plants 242

15.8 Biotech Barnyards 244
Knockouts and Organ Factories 245

15.9 Safety Issues 245

15.10 Genetically Modified Humans 246
Gene Therapy 246
Eugenics 247

Personal Genetic Testing (revisited) 247

Appendix I   Periodic Table of the Elements
Appendix II   Amino Acids
Appendix III Closer Look at Some Major Metabolic Pathways
Appendix IV   A Plain English Map of the Human Chromosomes
Appendix VI   Units of Measure
Appendix VII Answers to Self-Quizzes and Genetics Problems

Preface

This edition of *Biology: The Unity and Diversity of Life* includes a wealth of new information reflecting recent discoveries in biology (details can be found in the *Power Bibliography*, which lists journal articles and other references used in the revision process; available upon request). Descriptions of current research, along with photos and videos of scientists who carry it out, underscore the concept that science is an ongoing endeavor carried out by a diverse community of people. Discussions include not only what was discovered, but also how the discoveries were made, how our understanding has changed over time, and what remains to be discovered. These discussions are provided in the context of a thorough, accessible introduction to well-established concepts and principles that underpin modern biology. Every topic is examined from an evolutionary perspective, emphasizing the connections between all forms of life.

Throughout the book, text and art have been revised to help students grasp difficult concepts. This edition also continues to focus on real world applications pertaining to the field of biology, including social issues arising from new research and developments. This edition covers in detail the many ways in which human activities are continuing to alter the environment and threaten both human health and Earth's biodiversity.

Changes to this Edition

Here are a few highlights of the revisions to this edition.

1 Invitation to Biology Renewed and updated emphasis on the relevance of new species discovery and the process of science.

2 Life's Chemical Basis New graphics illustrate elements and radioactive decay.

3 Molecules of Life New figure illustrates protein domains.

4 Cell Structure and Function New table summarizing cell theory; new photos of prokaryotes. Comparison of microscopy techniques updated using *Paramecium*. New figure shows food vacuoles in *Nassula*.

5 Ground Rules of Metabolism Temperature-dependent enzyme activity now illustrated with polymerases. New art and photos illustrate coenzymes, adhesion proteins, membrane trafficking, and energy transfer in redox reactions.

6 Where It Starts—Photosynthesis New photos illustrate phycobilins, stomata, adaptations of C4 plants, ice core sampling, smog in China. Light-dependent reactions art simplified.

7 How Cells Release Chemical Energy New photos illustrate mitochondrial disease and aerobic respiration.

8 DNA Structure and Function Concepts and illustrations of DNA hybridization and primers added to replication section. New photo of mutations caused by radiation at Chernobyl; new illustration of mutation.

9 From DNA to Protein Expanded material on the effects of mutation includes discussion of hairlessness in cats and a new micrograph of a sickled blood cell.

10 Gene Control New photos show transcription factors, X chromosome inactivation; new material explains evolution of lactose tolerance. New critical thinking question requires understanding of the effects of floral identity gene mutations.

11 How Cells Reproduce New photos illustrate mitosis, the mitotic spindle, and telomeres.

12 Meiosis and Sexual Reproduction New material on asexuality in mud snails and bdelloid rotifers. New micrograph shows multiple crossovers.

13 Observing Patterns in Inherited Traits New material about environmental effects on hemoglobin gene expression in *Daphnia*. New photos illustrate continuous variation.

14 Chromosomes and Human Inheritance Material on Tay-Sachs has been moved to this chapter as an illustration of autosomal recessive inheritance.

15 Studying and Manipulating Genomes Coverage of personal genetic testing updated with new medical applications, including the social impact of Angelina Jolie's response to her test. New photos of genetically modified animals. New "who's the daddy" critical thinking question offers students an opportunity to analyze a paternity test based on SNPs.

16 Evidence of Evolution New MRI showing coccyx illustrates a vestigial structure. Photos of 19th century naturalists added to emphasize the process of science that led to natural selection theory. Expanded coverage of fossil formation includes how banded iron formations provide evidence of the evolution of photosynthesis.

17 Processes of Evolution New opening essay on resistance to antibiotics as an outcome of agricultural overuse (warfarin material moved to illustrate directional selection). New art illustrates founder effect, and hypothetical example in text replaced with reduced diversity of *ABO* alleles in Native Americans. New art illustrates stasis in coelacanths.

18 Organizing Information About Species New material on DNA barcoding added to biochemical comparisons section. Data analysis activity revised to incorporate new data on honeycreeper ancestry.

19 Life's Origin and Early Evolution Added material about new discovery of 3.4-billion-year old fossil bacteria. New graphic illustrates endosymbiotic origin of mitochondria and chloroplasts.

20 Viruses, Bacteria, and Archaea Added information about Ebola and West Nile viruses, and newly discovered giant viruses.

21 Protists—The Simplest Eukaryotes New graphic depicts primary and secondary endosymbiosis. Added information about diatoms as a source of oil.

22 The Land Plants New essay about seed banks and the importance of sustain plant biodiversity.

23 Fungi More extensive coverage of fungal ecology; added information about white nose syndrome, a fungal disease of bats.

24 Animal Evolution—Invertebrates Updated information of medicines from invertebrates. New photos of terrestrial flatworm, plant-infecting roundworm.

25 Animal Evolution—Vertebrates Improved discussion of transition to land, with new illustration. Reorganized coverage of mammal evolution and diversity.

26 Human Evolution Updated to include latest discoveries about *Australopithecus sediba*, Denisovans, and Neanderthals.

27 Plant Tissues Carbon sequestration essay revised to include new data on wood production by old-growth redwoods. Reorganized to consolidate primary growth into its own section. Many new photos illustrate stem, leaf, and root structure. Material on fire scars added to section on dendroclimatology.

28 Plant Nutrition and Transport Illustration of Casparian strip integrated with new micrograph. Revisited section discusses phytoremediation at Ford's Rouge Center.

29 Life Cycles of Flowering Plants Updated material reflects current research on bee pollination behavior and colony collapse. New photos illustrate pollinators, fruit classification, asexual reproduction.

30 Communication Strategies in Plants Updates reflect ongoing major breakthroughs in the field of plant hormone function. New photos show apical dominance, effect of gibberellin, and abscission.

31 Animal Tissues and Organ Systems Added information about tissue regeneration in nonhuman animals; updated information about use of human and embryonic stem cells. Added information about blubber as a specialized adipose tissue.

32 Neural Control New opening essay about the effects of concussion on the brain. Reorganized coverage of psychoactive drugs. Added information about epidural anesthesia. Updated, improved coverage of memory.

33 Sensory Perception New opening essay about cochlear implants; revisited section discusses retinal implants, artificial limbs. Updated information about human sense of taste.

34 Endocrine Control Updated discussion of endocrine disruptors. New examples of pituitary gigantisms, dwarfism. Added information about role of melatonin in seasonal coat color changes.

35 Structural Support and Movement Added information about myostatin polymorphism in race horses to opening essay. New section discusses principles of animal location. Added information about boneless muscular organs such as the tongue.

36 Circulation More extensive coverage of plasma components. Discussion of genetics of blood types deleted. Improved coverage of and illustration of capillary exchange. Added information about blood pressure and jugular vein valves in giraffes.

37 Immunity Updated material on HIV/AIDS treatment strategies. New photos show T cell/APC interaction, skin as a surface barrier, a cytotoxic T cell killing a cancer cell, contact allergy, and victims of HIV.

38 Respiration Improved comparison of water and air as respiratory media with accompanying figure. Revised figure depicting first aid for choking victims to reflect latest guidelines. Discussion of human adaptation to high altitude now compares mechanisms in Tibetan and Andean populations.

39 Digestion and Nutrition New graphic depicting functional variations in animal dentition. New figure showing arrangement of organs that empty into the small intestine. Improved discussion of vitamin and mineral functions. New MRI illustrates how abdominal fat compresses internal organs. Added information about basal metabolic rate.

40 Maintaining the Internal Environment New subsection about climate-related adaptations in human populations.

41 Animal Reproductive Systems Coverage of intersex conditions dropped. Opening essay now discusses reproductive technology (IVF, egg banking); Revisited section discusses sperm banks. New section discusses location of animal gonads and the general mechanism of gamete formation. Reproductive function of human females now discussed before that of males; improved figure depicting the ovarian cycle.

42 Animal Development New opening essay about human birth defects, with a focus on cleft lip and palate. Improved photos illustrating apoptosis in digit development. Reorganized coverage of early human development. Added information about surgical delivery (cesarean section).

43 Animal Behavior Opening essay about effects of noise pollution on animal communication moved here and updated to reflect recent research. Revised discussion of the possible benefits of grouping.

44 Population Ecology Improved presentation of effects of predation on guppy life history. Revised, updated graphics.

45 Community Ecology Added information about and a photo of a brood parasite of ants. Added photo of the keystone species Pisaster.

46 Ecosystems More extensive discussion of aquifer depletion, salination; added information about ecological effects of over-allocation of river water. Updated discussion of the rise in atmospheric CO_2.

47 The Biosphere New opening essay about how winds and ocean currents distributed and are distributing material from the 2011 earthquake and tidal wave that affected Japan. Discussion of El Nino now a subsection within the chapter.

48 Human Impacts on the Biosphere New graphics of extinct animals: mastadon and dodo. Added information about and photo of endangered Florida lichen; added information about the Great Pacific Garbage Patch. Updated coverage of ozone depletion and effects of global climate change.

Student and Instructor Resources

Cengage Learning Testing Powered by Cognero is a flexible, online system that allows you to:

• author, edit, and manage test bank content from multiple Cengage Learning solutions
• create multiple test versions in an instant
• deliver tests from your LMS, your classroom or wherever you want

Instructor Companion Site Everything you need for your course in one place! This collection of book-specific lecture and class tools is available online via www.cengage.com/login. Access and download PowerPoint presentations, images, instructor's manual, videos, and more

Cooperative Learning Cooperative Learning: Making Connections in General Biology, 2nd Edition, authored by Mimi Bres and Arnold Weisshaar, is a collection of separate, ready-to-use, short cooperative activities that have broad application for first year biology courses. They fit perfectly with any style of instruction, whether in large lecture halls or flipped classrooms. The activities are designed to address a range of learning objectives such as reinforcing basic concepts, making connections between various chapters and topics, data analysis and graphing, developing problem solving skills, and mastering terminology. Since each activity is designed to stand alone, this collection can be used in a variety of courses and with any text.

MindTap A personalized, fully online digital learning platform of authoritative content, assignments, and services that engages students with interactivity while also offering instructors their choice in the configuration of coursework and enhancement of the curriculum via web-apps known as MindApps. MindApps range from ReadSpeaker (which reads the text out loud to students), to Kaltura (allowing you to insert inline video and audio into your curriculum). MindTap is well beyond an eBook, a homework solution or digital supplement, a resource center website, a course delivery platform, or a Learning Management System. It is the first in a new category —the Personal Learning Experience.

New for this edition! MindTap has an integrated Study Guide, expanded quizzing and application activities, and an integrated Test Bank.

Aplia for Biology The Aplia system helps students learn key concepts via Aplia's focused assignments and active learning opportunities that include randomized, automatically graded questions, exceptional text/art integration, and immediate feedback. Aplia has a full course management system that can be used independently or in conjunction with other course management systems such as MindTap, D2L, or Blackboard.

Acknowledgments

Writing, revising, and illustrating a biology textbook is a major undertaking for two full-time authors, but our efforts constitute only a small part of what is required to produce and distribute this one. We are truly fortunate to be part of a huge team of very talented people who are as committed as we are to creating and disseminating an exceptional science education product.

Biology is not dogma; paradigm shifts are a common outcome of the fantastic amount of research in the field. Ideas about what material should be taught and how best to present that material to students changes even from one year to the next. It is only with the ongoing input of our many academic reviewers and advisors (see opposite page) that we can continue to tailor this book to the needs of instructors and students while integrating new information and models. We continue to learn from and be inspired by these dedicated educators. A special thanks goes to Jose Panero for his extensive and detailed review for this edition.

On the production side of our team, the indispensable Grace Davidson orchestrated a continuous flow of files, photos, and illustrations while managing schedules, budgets, and whatever else happened to be on fire at the time. Grace, thank you as always for your patience and dedication. Thank you also to Cheryl DuBois, John Sarantakis, and Christine Myaskovsky for your help with photoresearch. Copyeditor Anita Hueftle and proofreader Kathy Dragolich, your valuable suggestions kept our text clear and concise.

Yolanda Cossio, thank you for continuing to support us and for encouraging our efforts to innovate and improve. Peggy Williams, we are as always grateful for your enthusiastic, thoughtful guidance, and for your many travels (and travails) on behalf of our books.

Thanks to Hal Humphrey our Cengage Production Manager, Tom Ziolkowski our Marketing Manager, Lauren Oliveira who creates our exciting technology package, Associate Content Developers Casey Lozier and Kellie Petruzzelli, and Product Assistant Victor Luu.

Lisa Starr and Christine Evers, May 2014

Influential Class Testers and Reviewers

Brenda Alston-Mills
North Carolina State University

Kevin Anderson
Arkansas State University - Beebe

Norris Armstrong
University of Georgia

Tasneem Ashraf
Coshise College

Dave Bachoon
Georgia College & State University

Neil R. Baker
The Ohio State University

Andrew Baldwin
Mesa Community College

David Bass
University of Central Oklahoma

Lisa Lynn Boggs
Southwestern Oklahoma State University

Gail Breen
University of Texas at Dallas

Marguerite "Peggy" Brickman
University of Georgia

David Brooks
East Central College

David William Bryan
Cincinnati State College

Lisa Bryant
Arkansas State University - Beebe

Katherine Buhrer
Tidewater Community College

Uriel Buitrago-Suarez
Harper College

Sharon King Bullock
Virginia Commonwealth University

John Capehart
University of Houston - Downtown

Daniel Ceccoli
American InterContinental University

Tom Clark
Indiana University South Bend

Heather Collins
Greenville Technical College

Deborah Dardis
Southeastern Louisiana University

Cynthia Lynn Dassler
The Ohio State University

Carole Davis
Kellogg Community College

Lewis E. Deaton
University of Louisiana - Lafayette

Jean Swaim DeSaix
University of North Carolina - Chapel Hill

(Joan) Lee Edwards
Greenville Technical College

Hamid M. Elhag
Clayton State University

Patrick Enderle
East Carolina University

Daniel J. Fairbanks
Brigham Young University

Amy Fenster
Virginia Western Community College

Kathy E. Ferrell
Greenville Technical College

Rosa Gambier
Suffok Community College - Ammerman

Tim D. Gaskin
Cuyahoga Community College - Metropolitan

Stephen J. Gould
Johns Hopkins University

Laine Gurley
Harper College

Marcella Hackney
Baton Rouge Community College

Gale R. Haigh
McNeese State University

John Hamilton
Gainesville State

Richard Hanke
Rose State Community College

Chris Haynes
Shelton St. Community College

Kendra M. Hill
South Dakota State University

Juliana Guillory Hinton
McNeese State University

W. Wyatt Hoback
University of Nebraska, Kearney

Kelly Hogan
University of North Carolina

Norma Hollebeke
Sinclair Community College

Robert Hunter
Trident Technical College

John Ireland
Jackson Community College

Thomas M. Justice
McLennan College

Timothy Owen Koneval
Laredo Community College

Sherry Krayesky
University of Louisiana - Lafayette

Dubear Kroening
University of Wisconsin - Fox Valley

Jerome Krueger
South Dakota State University

Jim Krupa
University of Kentucky

Mary Lynn LaMantia
Golden West College

Dale Lambert
Tarrant County College

Kevin T. Lampe
Bucks County Community College

Susanne W. Lindgren
Sacramento State University

Madeline Love
New River Community College

Dr. Kevin C. McGarry
Kaiser College - Melbourne

Ashley McGee
Alamo College

Jeanne Mitchell
Truman State University

Alice J. Monroe
St. Petersburg College - Clearwater

Brenda Moore
Truman State University

Erin L. G. Morrey
Georgia Perimeter College

Rajkumar "Raj" Nathaniel
Nicholls State University

Francine Natalie Norflus
Clayton State University

Harold Olivey
Indiana University Northwest

Alexander E. Olvido
Virginia State University

John C. Osterman
University of Nebraska, Lincoln

Jose L. Panero
University of Texas

Bob Patterson
North Carolina State University

Shelley Penrod
North Harris College

Carla Perry
Community College of Philadelphia

Mary A. (Molly) Perry
Kaiser College - Corporate

John S. Peters
College of Charleston

Carlie Phipps
SUNY IT

Michael Plotkin
Mt. San Jacinto College

Ron Porter
Penn State University

Karen Raines
Colorado State University

Larry A. Reichard
Metropolitan Community College - Maplewood

Jill D. Reid
Virginia Commonwealth University

Robert Reinswold
University of Northern Colorado

Ashley E. Rhodes
Kansas State University

David Rintoul
Kansas State University

Darryl Ritter
Northwest Florida State College

Amy Wolf Rollins
Clayton State University

Sydha Salihu
West Virginia University

Jon W. Sandridge
University of Nebraska

Robin Searles-Adenegan
Morgan State University

Erica Sharar
IVC; National University

Julie Shepker
Kaiser College - Melbourne

Rainy Shorey
Illinois Central College

Eric Sikorski
University of South Florida

Phoebe Smith
Suffolk County Community College

Robert (Bob) Speed
Wallace Junior College

Tony Stancampiano
Oklahoma City Community College

Jon R. Stoltzfus
Michigan State University

Peter Svensson
West Valley College

Jeffrey L. Travis
University at Albany

Nels H. Troelstrup, Jr.
South Dakota State University

Allen Adair Tubbs
Troy University

Will Unsell
University of Central Oklahoma

Rani Vajravelu
University of Central Florida

Jack Waber
West Chester University of Pennsylvania

Kathy Webb
Bucks County Community College

Amy Stinnett White
Virginia Western Community College

Virginia White
Riverside Community College

Robert S. Whyte
California University of Pennsylvania

Kathleen Lucy Wilsenn
University of Northern Colorado

Penni Jo Wilson
Cleveland State Community College

Robert Wise
University of Wisconsin Oshkosh

Michael L. Womack
Macon State College

Maury Wrightson
Germanna Community College

Mark L. Wygoda
McNeese State University

Lan Xu
South Dakota State University

Poksyn ("Grace") Yoon
Johnson and Wales University

Muriel Zimmermann
Chaffey College

1 Invitation to Biology

LEARNING ROADMAP

Whether or not you have studied biology, you already have an intuitive understanding of life on Earth because you are part of it. Every one of your experiences with the natural world—from the warmth of the sun on your skin to the love of your pet—contributes to that understanding.

THE SCIENCE OF NATURE

We can understand life by studying it at many levels, starting with atoms that are components of all matter, and extending to interactions of organisms with their environment.

LIFE'S UNITY

All living things require ongoing inputs of energy and raw materials; all sense and respond to change; and all have DNA that guides their functioning.

LIFE'S DIVERSITY

Observable characteristics vary tremendously among organisms. Various classification systems help us keep track of the differences.

THE NATURE OF SCIENCE

Carefully designing experiments helps researchers unravel cause-and-effect relationships in complex natural systems.

WHAT SCIENCE IS (AND WHAT IT IS NOT)

Science addresses only testable ideas about observable events and processes. It does not address the untestable, including beliefs and opinions.

This book parallels nature's levels of organization, from atoms to the biosphere. Learning about the structure and function of atoms and molecules will prime you to understand how living cells work. Learning about processes that keep a single cell alive can help you understand how multicelled organisms survive. Knowing what it takes for organisms to survive can help you see why and how they interact with one another and their environment.

1.1 The Secret Life of Earth

In this era of detailed satellite imagery and cell phone global positioning systems, could there possibly be any places left on Earth that humans have not yet explored? Actually, there are plenty of them. In 2005, for example, helicopters dropped a team of scientists into the middle of a vast and otherwise inaccessible cloud forest atop New Guinea's Foja Mountains. Within a few minutes, the explorers realized that their landing site, a dripping, moss-covered swamp, had been untouched by humans. Team member Bruce Beehler remarked, "Everywhere we looked, we saw amazing things we had never seen before. I was shouting. This trip was a once-in-a-lifetime series of shouting experiences."

How did the explorers know they had landed in uncharted territory? For one thing, the forest was filled with plants and animals previously unknown even to native peoples that have long inhabited other parts of the region. During the next month, the team members discovered many new species, including a rhododendron plant with flowers the size of a plate and a frog the size of a pea. They also came across hundreds of species that are on the brink of extinction in other parts of the world, and some that supposedly had been extinct for decades. The animals had never learned to be afraid of humans, so they could easily be approached. A few were discovered as they casually wandered through campsites (**FIGURE 1.1**).

New species are discovered all the time, often in places much more mundane than Indonesian cloud forests. How do we know what species a particular organism belongs to? What is a species, anyway, and why

should discovering a new one matter to anyone other than a scientist? You will find the answers to such questions in this book. They are part of the scientific study of life, **biology**, which is one of many ways we humans try to make sense of the world around us.

Trying to understand the immense scope of life on Earth gives us some perspective on where we fit into it. For example, hundreds of new species are discovered every year, but about 20 species become extinct every minute in rain forests alone—and those are only the ones we know about. The current rate of extinctions is about 1,000 times faster than normal, and human activities are responsible for the acceleration. At this rate, we will never know about most of the species that are alive on Earth today. Does that matter? Biologists think so. Whether or not we are aware of it, humans are intimately connected with the world around us. Our activities are profoundly changing the entire fabric of life on Earth. These changes are, in turn, affecting us in ways we are only beginning to understand.

Ironically, the more we learn about the natural world, the more we realize we have yet to learn. But don't take our word for it. Find out what biologists know, and what they do not, and you will have a solid foundation upon which to base your own opinions about how humans fit into this world. By reading this book, you are choosing to learn about the human connection—your connection—with all life on Earth.

---

biology The scientific study of life.

FIGURE 1.1 Explorers found hundreds of rare species and dozens of new ones during recent survey expeditions to the Foja Mountain cloud forest (left). Right, Paul Oliver discovered this tree frog (*Litoria*) perched on a sack of rice during a particularly rainy campsite lunch. The explorers dubbed the new species "Pinocchio frog" after the Disney character because the male frog's long nose inflates and points upward during times of excitement.

✔ Biologists study life by thinking about it at different levels of organization.

✔ The quality of life emerges at the level of the cell.

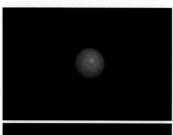

❶ Atoms
Atoms are fundamental units of all substances, living or not. This image shows a model of a single atom.

❷ Molecule
Atoms join other atoms in molecules. This is a model of a water molecule. The molecules special to life are much larger and more complex than water.

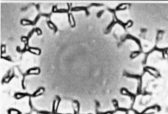

❸ Cell
The cell is the smallest unit of life. Some, like this plant cell, live and reproduce as part of a multicelled organism; others do so on their own.

❹ Tissue
Organized array of cells that interact in a collective task. This is epidermal tissue on the outer surface of a flower petal.

❺ Organ
Structural unit of interacting tissues. Flowers are the reproductive organs of many plants.

❻ Organ system
A set of interacting organs. The shoot system of this poppy plant includes its aboveground parts: leaves, flowers, and stems.

FIGURE 1.2 ▶Animated Levels of life's organization.

What, exactly, is the property we call "life"? We may never actually come up with a good definition, because living things are too diverse, and they consist of the same basic components as nonliving things. When we try to define life, we end up with a list of properties that differentiate living from nonliving things. These properties often emerge from the interactions of basic components. To understand how that works, take a look at these groups of squares:

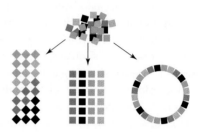

The property of "roundness" emerges when the component squares are organized one way, but not other ways. Characteristics of a system that do not appear in any of the system's components are called **emergent properties**. The idea that structures with emergent properties can be assembled from the same basic building blocks is a recurring theme in our world—and also in biology.

Life has successive levels of organization, with new emergent properties appearing at each level (**FIGURE 1.2**). This organization begins with interactions between **atoms**, which are fundamental building blocks of all substances ❶. Atoms bond together to form **molecules** ❷. There are no atoms unique to living things, but there are unique molecules. In today's natural world, only living things make the "molecules of life," which are lipids, proteins, DNA, RNA, and complex carbohydrates. The emergent property of "life" appears at the next level, when many molecules of life become organized as a cell ❸. A **cell** is the smallest unit of life. Cells survive and reproduce themselves using energy, raw materials, and information in their DNA.

Some cells live and reproduce independently. Others do so as part of a multicelled organism. An **organism** is an individual that consists of one or more cells. A poppy plant is an example of a multicelled organism ❼. In most multicelled organisms, cells are organized as tissues ❹. A **tissue** consists of specific types of cells organized in a particular pattern. The arrangement allows the cells to collectively perform a special function such as protection from injury (dermal tissue) or movement (muscle tissue). An **organ** is an organized array of tissues that collectively carry out a particular task or set of tasks ❺. For example, a flower is an organ of reproduction in plants; a heart, an organ that pumps blood in animals. An **organ system** is a set of organs and tissues that interact to keep the individual's body working properly ❻. Examples of organ systems include the aboveground parts of a plant (the shoot system), and the heart and blood vessels of an animal (the circulatory system).

A **population** is a group of interbreeding individuals of the same type, or species, living in a given area ❽. An example may be all California poppies living in California's Antelope Valley Poppy Reserve. At the next level, a **community** consists of all populations of all species in a given area. The Antelope Valley Reserve community includes California poppies and all other plants, animals, microorganisms, and so on ❾. Communities may be large or small, depending on the area defined.

The next level of organization is the **ecosystem**, which is a community interacting with its environment ❿. The most inclusive level, the **biosphere**, encompasses all regions of Earth's crust, waters, and atmosphere in which organisms live ⓫.

atom Fundamental building block of all matter.
biosphere All regions of Earth where organisms live.
cell Smallest unit of life.
community All populations of all species in a given area.
ecosystem A community interacting with its environment.
emergent property A characteristic of a system that does not appear in any of the system's component parts.
molecule Two or more atoms bonded together.
organ In multicelled organisms, a grouping of tissues engaged in a collective task.
organism Individual that consists of one or more cells.
organ system In multicelled organisms, set of organs engaged in a collective task that keeps the body functioning properly.
population Group of interbreeding individuals of the same species that live in a given area.
tissue In multicelled organisms, specialized cells organized in a pattern that allows them to perform a collective function.

TAKE-HOME MESSAGE 1.2
How do living things differ from nonliving things?

✔ All things, living or not, consist of the same building blocks: atoms. Atoms join as molecules.

✔ In today's natural world, only living things make lipids, proteins, DNA, RNA, and complex carbohydrates. The unique properties of life emerge as these molecules become organized into cells.

✔ Higher levels of life's organization include multicelled organisms, populations, communities, ecosystems, and the biosphere.

✔ Emergent properties occur at each successive level of life's organization.

❼ Multicelled organism
Individual that consists of more than one cell. Cells of this California poppy plant are part of its two organ systems: aboveground shoots and belowground roots.

❽ Population
Group of single-celled or multicelled individuals of a species in a given area. This population of California poppy plants is in California's Antelope Valley Poppy Reserve.

❾ Community
All populations of all species in a specified area. These plants are part of the Antelope Valley Poppy Reserve community.

❿ Ecosystem
A community interacting with its physical environment through the transfer of energy and materials. Sunlight and water sustain the natural community in the Antelope Valley.

⓫ Biosphere
The sum of all ecosystems: every region of Earth's waters, crust, and atmosphere in which organisms live. No ecosystem in the biosphere is truly isolated from any other.

CREDITS: (2) #7: Michael Szoenyi/Science Source; #8: Exactostock/SuperStock; #9: © Sergei Krupnov, www.flickr.com/photos/7969319@N03; #10: © Mark Koberg Photography; #11: NASA.

1.3 How Living Things Are Alike

✔ Continual inputs of energy and the cycling of materials maintain life's complex organization.

✔ Organisms sense and respond to change.

✔ All organisms use information in the DNA they inherited from their parent or parents to function.

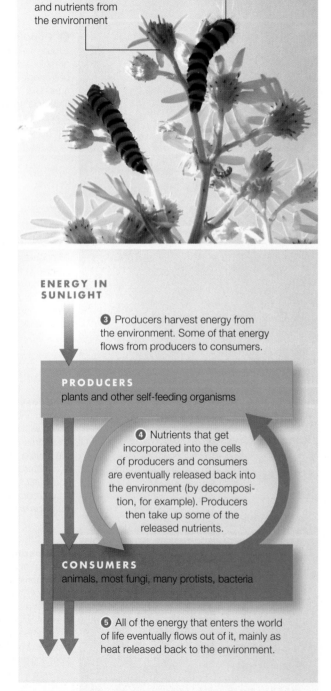

❶ producer acquiring energy and nutrients from the environment

❷ consumer acquiring energy and nutrients by eating a producer

ENERGY IN SUNLIGHT

❸ Producers harvest energy from the environment. Some of that energy flows from producers to consumers.

PRODUCERS
plants and other self-feeding organisms

❹ Nutrients that get incorporated into the cells of producers and consumers are eventually released back into the environment (by decomposition, for example). Producers then take up some of the released nutrients.

CONSUMERS
animals, most fungi, many protists, bacteria

❺ All of the energy that enters the world of life eventually flows out of it, mainly as heat released back to the environment.

FIGURE 1.3 ▶Animated The one-way flow of energy and cycling of materials in the world of life.

Even though we cannot precisely define "life," we can intuitively understand what it means because all living things share a set of key features. All require ongoing inputs of energy and raw materials; all sense and respond to change; and all pass DNA to offspring.

Organisms Require Energy and Nutrients

Not all living things eat, but all require energy and nutrients on an ongoing basis. Both are essential to maintain the functioning of individual organisms and the organization of life. A **nutrient** is a substance that an organism needs for growth and survival but cannot make for itself.

Organisms spend a lot of time acquiring energy and nutrients (**FIGURE 1.3**). However, the source of energy and the type of nutrients acquired differ among organisms. These differences allow us to classify all living things into two categories: producers and consumers. **Producers** make their own food using energy and simple raw materials they obtain from nonbiological sources ❶. Plants are producers that use the energy of sunlight to make sugars from water and carbon dioxide (a gas in air), a process called **photosynthesis**. By contrast, **consumers** cannot make their own food. They obtain energy and nutrients by feeding on other organisms ❷. Animals are consumers. So are decomposers, which feed on the wastes or remains of other organisms. The leftovers from consumers' meals end up in the environment, where they serve as nutrients for producers. Said another way, nutrients cycle between producers and consumers.

Unlike nutrients, energy is not cycled. It flows through the world of life in one direction: from the environment ❸, through organisms ❹, and back to

consumer Organism that gets energy and nutrients by feeding on tissues, wastes, or remains of other organisms.
development Multistep process by which the first cell of a new multicelled organism gives rise to an adult.
DNA Deoxyribonucleic acid; carries hereditary information that guides development and other activities.
growth In multicelled species, an increase in the number, size, and volume of cells.
homeostasis Process in which an organism keeps its internal conditions within tolerable ranges by sensing and responding to change.
inheritance Transmission of DNA to offspring.
nutrient Substance that an organism needs for growth and survival but cannot make for itself.
photosynthesis Process by which producers use light energy to make sugars from carbon dioxide and water.
producer Organism that makes its own food using energy and nonbiological raw materials from the environment.
reproduction Processes by which parents produce offspring.

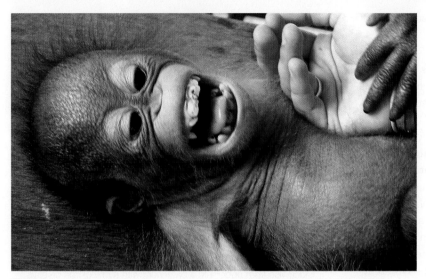

FIGURE 1.4 Organisms sense and respond to stimulation. This baby orangutan is laughing in response to being tickled. Apes and humans make different sounds when being tickled, but the airflow patterns are so similar that we can say apes really do laugh.

the environment ❺. This flow maintains the organization of every living cell and body, and it also influences how individuals interact with one another and their environment. The energy flow is one-way, because with each transfer, some energy escapes as heat, and cells cannot use heat as an energy source. Thus, energy that enters the world of life eventually leaves it (we return to this topic in Chapter 5).

Organisms Sense and Respond to Change

An organism cannot survive for very long in a changing environment unless it adapts to the changes. Thus, every living thing has the ability to sense and respond to change both inside and outside of itself (**FIGURE 1.4**). For example, after you eat, the sugars from your meal enter your bloodstream. The added sugars set in motion a series of events that causes cells throughout the body to take up sugar faster, so the sugar level in your blood quickly falls. This response keeps your blood sugar level within a certain range, which in turn helps keep your cells alive and your body functioning.

The fluid portion of your blood is a component of your internal environment, which is all of the body fluids outside of cells. That internal environment must be kept within certain ranges of temperature and other conditions, or the cells that make up your body will die. By sensing and adjusting to change, you and all other organisms keep conditions in the internal environment within a range that favors survival.

Homeostasis is the name for this process, and it is one of the defining features of life.

Organisms Use DNA

With little variation, the same types of molecules perform the same basic functions in every organism. For example, information in an organism's **DNA** (deoxyribonucleic acid) guides ongoing functions that sustain the individual through its lifetime. Such functions include **development**: the process by which the first cell of a new individual gives rise to a multicelled adult; **growth**: increases in cell number, size, and volume; and **reproduction**: processes by which individuals produce offspring.

Individuals of every natural population are alike in certain aspects of their body form and behavior because their DNA is very similar: Orangutans look like orangutans and not like caterpillars because they inherited orangutan DNA, which differs from caterpillar DNA in the information it carries. **Inheritance** refers to the transmission of DNA to offspring. All organisms inherit their DNA from one or two parents.

DNA is the basis of similarities in form and function among organisms. However, the details of DNA molecules differ, and herein lies the source of life's diversity. Small variations in the details of DNA's structure give rise to differences among individuals, and also among types of organisms. As you will see in later chapters, these differences are the raw material of evolutionary processes.

> **TAKE-HOME MESSAGE 1.3**
> How are all living things alike?
>
> ✔ A one-way flow of energy and a cycling of nutrients sustain life's organization.
>
> ✔ Organisms sense and respond to conditions inside and outside themselves. They make adjustments that keep conditions in their internal environment within a range that favors cell survival, a process called homeostasis.
>
> ✔ All organisms use information in the DNA they inherited from their parent or parents to develop, grow, and reproduce. DNA is the basis of similarities and differences in form and function among organisms.

✔ There is great variation in the details of appearance and other observable characteristics of living things.

Living things differ tremendously in their observable characteristics. Various classification schemes help us organize what we understand about the scope of this variation, which we call Earth's **biodiversity**.

For example, organisms can be grouped on the basis of whether they have a nucleus, which is a sac with two membranes that encloses and protects a cell's DNA. **Bacteria** (singular, bacterium) and **archaea** (singular, archaeon) are organisms whose DNA is *not* contained within a nucleus. All bacteria and archaea are single-celled, which means each organism consists of one cell (**FIGURE 1.5A,B**). Collectively, these organisms are the most diverse representatives of life. Different kinds are producers or consumers in nearly all regions of Earth. Some inhabit such extreme environments as frozen desert rocks, boiling sulfurous lakes, and nuclear reactor waste. The first cells on Earth may have faced similarly hostile environments.

Traditionally, organisms without a nucleus have been called **prokaryotes**, but this designation is now used only informally. This is because, despite the similar appearance of bacteria and archaea, the two types of cells are less related to one another than we once thought. Archaea turned out to be more closely related to **eukaryotes**, which are organisms whose DNA is contained within a nucleus. Some eukaryotes live as individual cells; others are multicelled (**FIGURE 1.5C**). Eukaryotic cells are typically larger and more complex than bacteria or archaea.

Structurally, **protists** are the simplest eukaryotes, but as a group they vary dramatically, from single-celled consumers to giant, multicelled producers.

Fungi (singular, fungus) are eukaryotic consumers that secrete substances to break down food externally, then absorb nutrients released by this process. Many fungi are decomposers. Most fungi, including those that form mushrooms, are multicellular. Fungi that live as single cells are called yeasts.

Plants are multicelled eukaryotes; the majority are photosynthetic producers that live on land. Besides feeding themselves, plants also serve as food for most other land-based organisms.

Animals are multicelled consumers that ingest tissues or juices of other organisms. Unlike fungi, animals break down food inside their body. They also develop through a series of stages that lead to the adult form. All animals actively move about during at least part of their lives.

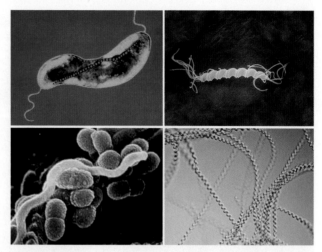

A Bacteria are the most numerous organisms on Earth. Clockwise from upper left, a bacterium with a row of iron crystals that acts like a tiny compass; a common resident of cat and dog stomachs; spiral cyanobacteria; types found in dental plaque.

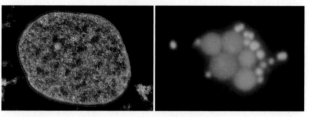

B Archaea resemble bacteria, but are more closely related to eukaryotes. Left, an archaeon that grows in sulfur hot springs. Right, two types of archaea from a seafloor hydrothermal vent.

FIGURE 1.5 ►Animated A few representatives of life's diversity.

animal Multicelled consumer that develops through a series of stages and moves about during part or all of its life.
archaea Group of single-celled organisms that lack a nucleus but are more closely related to eukaryotes than to bacteria.
bacteria The most diverse and well-known group of single-celled organisms that lack a nucleus.
biodiversity Scope of variation among living organisms.
eukaryote Organism whose cells characteristically have a nucleus.
fungus Single-celled or multicelled eukaryotic consumer that breaks down material outside itself, then absorbs nutrients released from the breakdown.
plant A multicelled, typically photosynthetic producer.
prokaryote Single-celled organism without a nucleus.
protist Member of a diverse group of simple eukaryotes.

TAKE-HOME MESSAGE 1.4
How do organisms differ from one another?

✔ Organisms differ in their details; they show tremendous variation in observable characteristics.

✔ We divide Earth's biodiversity into broad groups based on traits such as having a nucleus or being multicellular.

CREDITS: (5A) top left, Dr. Richard Frankel; top right, Science Source; bottom left, www.zahnarzt-stuttgart.com; bottom right, © Susan Barnes; (5B) left, Eye of Science/Science Source; right, © Dr. Harald Huber, Dr. Michael Hohn, Prof. Dr. K.O. Stetter, University of Regensburg, Germany.

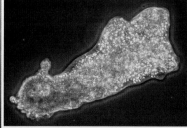

Protists are a group of extremely diverse eukaryotes that range from giant multicelled seaweeds to microscopic single cells.

Plants are multicelled eukaryotes, most of which are photosynthetic. Nearly all have roots, stems, and leaves.

Fungi are eukaryotic consumers that secrete substances to break down food outside their body. Most are multicelled (left), but some are single-celled (right).

Animals are multicelled eukaryotes that ingest tissues or juices of other organisms. All actively move about during at least part of their life.

C **Eukaryotes** are single-celled or multicelled organisms whose DNA is contained within a nucleus.

✔ Each type of organism, or species, is given a unique name.

✔ We define and group species based on shared traits.

Each time we discover a new **species**, or unique kind of organism, we name it. **Taxonomy**, a system of naming and classifying species, began thousands of years ago, but naming species in a consistent way did not become a priority until the eighteenth century. At that time, European explorers who were just discovering the scope of life's diversity started having more and more trouble communicating with one another because species often had multiple names. For example, the dog rose (a plant native to Europe, Africa, and Asia) was alternately known as briar rose, witch's briar, herb patience, sweet briar, wild briar, dog briar, dog berry, briar hip, eglantine gall, hep tree, hip fruit, hip rose, hip tree, hop fruit, and hogseed—and those are only the English names! Species often had multiple scientific names too, in Latin that was descriptive but often cumbersome. The scientific name of the dog rose was *Rosa sylvestris inodora seu canina* (odorless woodland dog rose), and also *Rosa sylvestris alba cum rubore, folio glabro* (pinkish white woodland rose with smooth leaves).

An eighteenth-century naturalist, Carolus Linnaeus, standardized a naming system that we still use. By the Linnaean system, every species is given a unique two-part scientific name. The first part is the name of the **genus** (plural, genera), a group of species that share a unique set of features. The second part is the **specific epithet**. Together, the genus name and the specific epithet designate one species. Thus, the dog rose now has one official name, *Rosa canina*, that is recognized worldwide.

Genus and species names are always italicized. For example, *Panthera* is a genus of big cats. Lions belong to the species *Panthera leo*. Tigers belong to a different species in the same genus (*Panthera tigris*), and so do leopards (*P. pardus*). Note how the genus name may be abbreviated after it has been spelled out.

A Rose by Any Other Name . . .

The individuals of a species share a unique set of inherited characteristics, or **traits**. For example, giraffes normally have very long necks, brown spots on white coats, and so on. These are morphological traits (*morpho*– means form). Individuals of a species also share biochemical traits (they make and use the same molecules) and behavioral traits (they respond the same way to certain stimuli, as when hungry giraffes feed on tree leaves).

We can rank species into ever more inclusive categories based on traits. Each rank, or **taxon** (plural, taxa), is a group of organisms that share a unique set of traits. Each category above species—genus, family, order, class, phylum (plural, phyla), kingdom, and

	wild carrot	marijuana	apple	prickly rose	dog rose
domain	Eukarya	Eukarya	Eukarya	Eukarya	Eukarya
kingdom	Plantae	Plantae	Plantae	Plantae	Plantae
phylum	Magnoliophyta	Magnoliophyta	Magnoliophyta	Magnoliophyta	Magnoliophyta
class	Magnoliopsida	Magnoliopsida	Magnoliopsida	Magnoliopsida	Magnoliopsida
order	Apiales	Rosales	Rosales	Rosales	Rosales
family	Apiaceae	Cannabaceae	Rosaceae	Rosaceae	Rosaceae
genus	*Daucus*	*Cannabis*	*Malus*	*Rosa*	*Rosa*
species	*carota*	*sativa*	*domestica*	*acicularis*	*canina*

FIGURE 1.6 Taxonomic classification of five species that are related at different levels. Each species has been assigned to ever more inclusive groups, or taxa: in this case, from genus to domain.

Which of the plants shown here are in the same order?

Answer: Marijuana, apple, prickly rose, and dog rose

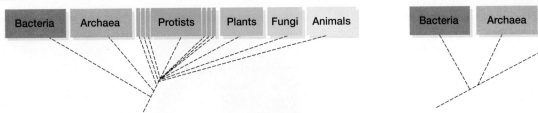

A Six-kingdom classification system. The protist kingdom includes the most ancient multicelled and all single-celled eukaryotes.

B Three-domain classification system. The Eukarya domain includes protists, plants, fungi, and animals.

FIGURE 1.7 ▶**Animated** Two ways to see the big picture of life. Lines in such diagrams indicate evolutionary connections. Compare **FIGURE 1.6**.

domain—consists of a group of the next lower taxon (**FIGURE 1.6**). Using this system, all life can be sorted into categories (**FIGURE 1.7** and **TABLE 1.1**).

It is easy to tell that orangutans and caterpillars are different species because they appear very different. Distinguishing between species that are more closely related may be much more challenging (**FIGURE 1.8**). In addition, traits shared by members of a species often vary a bit among individuals, as eye color does among people. How do we decide whether similar-looking organisms belong to the same species? The short answer to that question is that we rely on whatever information we have. Early naturalists studied anatomy and distribution—essentially the only methods available at the time—so species were named and classified according to what they looked like and where they lived. Today's biologists are able to compare traits that the early naturalists did not even know about, including biochemical ones.

The discovery of new information sometimes changes the way we distinguish a particular species or how we group it with others. For example, Linnaeus grouped plants by the number and arrangement of reproductive parts, a scheme that resulted in odd pairings such as castor-oil plants with pine trees. Having more information today, we place these plants in separate phyla.

Evolutionary biologist Ernst Mayr defined a species as one or more groups of individuals that potentially can interbreed, produce fertile offspring, and do not interbreed with other groups. This "biological species concept" is useful in many cases, but it is not universally applicable. For example, we may never know whether two widely separated populations could interbreed if they got together. As another example, populations often continue to interbreed even as they diverge, so the exact moment at which two populations become two species is often impossible to pinpoint. We return to speciation and how it occurs in Chapter 17, but for now it is important to remember that a "species" is a convenient but artificial construct of the human mind.

Table 1.1	All of Life in Three Domains
Bacteria	Single cells, no nucleus. Most ancient lineage.
Archaea	Single cells, no nucleus. Evolutionarily closer to eukaryotes than bacteria.
Eukarya	Eukaryotic cells (with a nucleus). Single-celled and multicelled species of protists, plants, fungi, and animals.

FIGURE 1.8 Four butterflies, two species: Which are which?

The top row shows two forms of the butterfly species *Heliconius melpomene*; the bottom row, two forms of *H. erato*.

H. melpomene and *H. erato* never cross-breed. Their alternate but similar patterns of coloration evolved as a shared warning signal to predatory birds that these butterflies taste terrible.

genus A group of species that share a unique set of traits; also the first part of a species name.
species Unique type of organism.
specific epithet Second part of a species name.
taxon A rank of organisms that share a unique set of traits.
taxonomy The science of naming and classifying species.
trait An inherited characteristic of an organism or species.

TAKE-HOME MESSAGE 1.5
How do we keep track of known species?

✔ Each type of organism, or species, is given a unique, two-part scientific name.

✔ Classification systems group species on the basis of shared, inherited traits.

CREDITS: (7, Table 1.1) © Cengage Learning; (8) © 2006 Axel Meyer, "Repeating Patterns of Mimicry." *PLoS Biology* Vol. 4, No. 10, e341 doi:10.1371/journal.pbio.0040341. Used with Permission.

✔ Judging the quality of information before accepting it is an active process called critical thinking.

✔ Researchers practice critical thinking by testing predictions about how the natural world works.

Most of us assume that we do our own thinking, but do we, really? You might be surprised to find out how often we let others think for us. Consider how a school's job (which is to impart as much information to students as quickly as possible) meshes perfectly with a student's job (which is to acquire as much knowledge as quickly as possible). In this rapid-fire exchange of information, it can be very easy to forget about the quality of what is being exchanged. Anytime you accept information without questioning it, you let someone else think for you.

Thinking About Thinking

Critical thinking is the deliberate process of judging the quality of information before accepting it. "Critical" comes from the Greek *kriticos* (discerning judgment). When you use critical thinking, you move beyond the content of new information to consider supporting evidence, bias, and alternative interpretations. How does the busy student manage this? Critical thinking does not necessarily require extra time, just a bit of extra awareness. There are many ways to do it. For example, you might ask yourself some of the following questions while you are learning something new:

> What message am I being asked to accept?
> Is the message based on facts or opinion?
> Is there a different way to interpret the facts?
> What biases might the presenter have?
> How do my own biases affect what I'm learning?

Such questions are a way of being conscious about learning. They can help you decide whether to allow new information to guide your beliefs and actions.

The Scientific Method

Critical thinking is a big part of **science**, the systematic study of the observable world and how it works (**FIGURE 1.9**). A scientific line of inquiry usually begins with curiosity about something observable, such as, say, a decrease in the number of birds in a particular area. Typically, a scientist will read about what others have discovered before making a **hypothesis**, a testable explanation for a natural phenomenon. An example of a hypothesis would be, "The number of birds is decreasing because the number of cats is increasing." Making a hypothesis this way is an

A Devising a vaccine for cancer.

FIGURE 1.9 Examples of research in the field of biology.

example of **inductive reasoning**, which means arriving at a conclusion based on one's observations. Inductive reasoning is the way we come up with new ideas about groups of objects or events.

A **prediction**, or statement of some condition that should exist if the hypothesis is correct, comes next. Making predictions is called the if–then process, in which the "if" part is the hypothesis, and the "then" part is the prediction: *If* the number of birds is decreasing because the number of cats is increasing, *then* reducing the number of cats should stop the decline. Using a hypothesis to make a prediction is a form of **deductive reasoning**, the logical process of using a general premise to draw a conclusion about a specific case.

---

control group Group of individuals identical to an experimental group except for the independent variable under investigation.
critical thinking Evaluating information before accepting it.
data Experimental results.
deductive reasoning Using a general idea to make a conclusion about a specific case.
dependent variable In an experiment, a variable that is presumably affected by an independent variable being tested.
experiment A test designed to support or falsify a prediction.
experimental group In an experiment, a group of individuals who have a certain characteristic or receive a certain treatment.
hypothesis Testable explanation of a natural phenomenon.
independent variable Variable that is controlled by an experimenter in order to explore its relationship to a dependent variable.
inductive reasoning Drawing a conclusion based on observation.
model Analogous system used for testing hypotheses.
prediction Statement, based on a hypothesis, about a condition that should exist if the hypothesis is correct.
science Systematic study of the observable world.
scientific method Making, testing, and evaluating hypotheses.
variable In an experiment, a characteristic or event that differs among individuals or over time.

B Improving efficiency of biofuel production from agricultural waste.

C Studying the ecological benefits of weedy buffer zones on farms.

D Discovering medically active natural products made by marine animals.

Next, a researcher will test the prediction. Tests may be performed on a **model**, or analogous system, if working with an object or event directly is not possible. For example, animal diseases are often used as models of similar human diseases. Careful observations are one way to test predictions that flow from a hypothesis. So are **experiments**: tests designed to support or falsify a prediction. A typical experiment explores a cause-and-effect relationship using **variables**, which are characteristics or events that can differ among individuals or over time. An **independent variable** is defined or controlled by the person doing the experiment. A **dependent variable** is an observed result that is supposed to be influenced by the independent variable. For example, in an investigation of our hypothetical bird–cat relationship, an independent variable may be the presence or absence of cats. The dependent variable in this test would be the number of birds.

Biological systems are typically complex, with many interdependent variables. It can be difficult to study one variable separately from the rest. Thus, biology researchers often test two groups of individuals simultaneously. An **experimental group** is a set of individuals that have a certain characteristic or receive a certain treatment. This group is tested side by side with a **control group**, which is identical to the experimental group except for one independent variable: the characteristic or the treatment being tested. Any differences in experimental results between the two groups is likely to be an effect of changing the variable.

Test results—**data**—that are consistent with the prediction are evidence in support of the hypothesis. Data inconsistent with the prediction are evidence that the hypothesis is flawed and should be revised.

Table 1.2 The Scientific Method

1. Observe some aspect of nature.

2. Think of an explanation for your observation (in other words, form a hypothesis).

3. Test the hypothesis.

    a. Make a prediction based on the hypothesis.

    b. Test the prediction using experiments or surveys.

    c. Analyze the results of the tests (data).

4. Decide whether the results of the tests support your hypothesis or not (form a conclusion).

5. Report your results to the scientific community.

A necessary part of science is reporting one's results and conclusions in a standard way, such as in a peer-reviewed journal article. The communication gives other scientists an opportunity to evaluate the information for themselves, both by checking the conclusions drawn and by repeating the experiments. Forming a hypothesis based on observation, and then systematically testing and evaluating the hypothesis, are collectively called the **scientific method** (**TABLE 1.2**).

TAKE-HOME MESSAGE 1.6
How does science work?

✔ The scientific method consists of making, testing, and evaluating hypotheses. It is a way of critical thinking—systematically judging the quality of information before allowing it to guide one's beliefs and actions.

✔ Experiments measure how changing an independent variable affects a dependent variable.

CREDITS: (9B) © Roger W. Winstead, NC State University; (9C) Photo by Scott Bauer, USDA/ARS; (9D) Courtesy of Susanna López-Legentil; (Table 1.2) © Cengage Learning.

✔ Researchers unravel cause-and-effect relationships in complex natural processes by changing one variable at a time.

There are many different ways to do research, particularly in biology. Some biologists survey, which means they observe without making hypotheses. Some make hypotheses and leave the experimentation to others. When it comes to scientific experimentation, however, consistency is the rule. Researchers try to change one independent variable at a time, and see what happens to a dependent variable. To give you a sense of how biology experiments work, we summarize two published studies here.

Potato Chips and Stomachaches

In 1996 the U.S. Food and Drug Administration (FDA) approved Olestra®, a fat replacement manufactured from sugar and vegetable oil, as a food additive. Potato chips were the first Olestra-containing food product

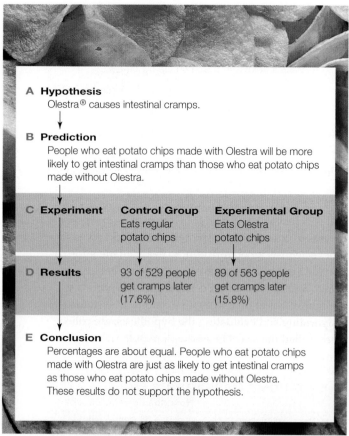

A Hypothesis
Olestra® causes intestinal cramps.

B Prediction
People who eat potato chips made with Olestra will be more likely to get intestinal cramps than those who eat potato chips made without Olestra.

C Experiment	Control Group Eats regular potato chips	Experimental Group Eats Olestra potato chips
D Results	93 of 529 people get cramps later (17.6%)	89 of 563 people get cramps later (15.8%)

E Conclusion
Percentages are about equal. People who eat potato chips made with Olestra are just as likely to get intestinal cramps as those who eat potato chips made without Olestra. These results do not support the hypothesis.

FIGURE 1.10 The steps in a scientific experiment to determine whether Olestra causes cramps. A report of this study was published in the *Journal of the American Medical Association* in January 1998.

What was the dependent variable in this experiment?

Answer: Whether or not a person got cramps

to be sold in the United States. Controversy about the chip additive soon raged. Many people complained of intestinal problems after eating the chips, and thought that the Olestra was at fault. Two years later, researchers at the Johns Hopkins University School of Medicine designed an experiment to test whether Olestra causes cramps.

The researchers predicted *if* Olestra causes cramps, *then* people who eat Olestra should be more likely to get cramps than people who do not eat it. To test the prediction, they used a Chicago theater as a "laboratory." They asked 1,100 people between the ages of thirteen and thirty-eight to watch a movie and eat their fill of potato chips. Each person received an unmarked bag containing 13 ounces of chips.

In this experiment, the individuals who received Olestra-laden potato chips constituted the experimental group, and individuals who received regular chips were the control group. The independent variable was the presence or absence of Olestra in the chips.

A few days after the movie, the researchers contacted all of the people and collected any reports of post-movie gastrointestinal problems. Of 563 people making up the experimental group, 89 (15.8 percent) complained about cramps. However, so did 93 of the 529 people (17.6 percent) making up the control group—who had eaten the regular chips.

People were about as likely to get cramps whether or not they ate chips made with Olestra. These results did not support the prediction, so the researchers concluded that eating Olestra does not cause cramps (**FIGURE 1.10**).

Butterflies and Birds

The peacock butterfly is a winged insect named for the large, colorful spots on its wings. In 2005, researchers reported the results of experiments investigating whether certain behaviors help peacock butterflies defend themselves against insect-eating birds. The study began with the observation that a resting peacock butterfly sits motionless, wings folded (**FIGURE 1.11A**). The dark underside of the wings provide appropriate camouflage. However, when the butterfly sees a predator approaching, it repeatedly flicks its wings open, exposing brilliant spots (**FIGURE 1.11B**). At the same time, it moves the hindwings in a way that produces a hissing sound and a series of clicks. Typically, a colorful, moving, noisy insect is very attractive to predatory birds, so the researchers were curious about why the peacock butterfly moves and makes noises only in the *presence* of a predator. After review-

A With wings folded, a resting peacock butterfly resembles a dead leaf.

B When a bird approaches, a butterfly repeatedly flicks its wings open. This behavior exposes brilliant spots and also produces hissing and clicking sounds.

C Researchers tested whether peacock butterfly wing flicking and hissing reduce predation by blue tits.

FIGURE 1.11 Testing the defensive value of peacock butterfly behaviors.

Researchers painted out the spots of some butterflies, cut the sound-making part of the wings on others, and did both to a third group; then exposed each butterfly to a hungry blue tit.

Results, listed below in **TABLE 1.3**, support the hypotheses that peacock butterfly spots and sounds can deter predatory birds.

What was the dependent variable in this series of experiments?

Answer: Getting eaten

Table 1.3	Results of Peacock Butterfly Experiment*			
Wing Spots	**Wing Sound**	**Total Number of Butterflies**	**Number Eaten**	**Number Survived**
Spots	Sound	9	0	9 (100%)
No spots	Sound	10	5	5 (50%)
Spots	No sound	8	0	8 (100%)
No spots	No sound	10	8	2 (20%)

* Proceedings of the Royal Society of London, Series B (2005) 272: 1203–1207.

ing earlier studies, the scientists made two hypotheses to explain these behaviors:

1. The wing flicking may startle predatory birds because the peacock butterfly's wing spots resemble owl eyes, and anything that looks like owl eyes is known to startle birds.

2. The hissing and clicking sounds produced when the peacock butterfly moves its hindwings may be an additional defense that startles predatory birds.

The researchers used these hypotheses to make the following predictions:

1. If peacock butterflies startle predatory birds by exposing their brilliant wing spots, then individuals having wing spots will be less likely to get eaten by predatory birds than those lacking wing spots.

2. If peacock butterfly hisses and clicks deter predatory birds, then sound-producing individuals will be less likely to get eaten by predatory birds than silent individuals.

The next step was the experiment. The researchers used a marker to paint the wing spots of some butterflies black, and scissors to cut off the sound-making

part of the hindwings of others. A third group had their wing spots painted and their hindwings cut. The researchers then put each butterfly into a large cage with a hungry blue tit (**FIGURE 1.11C**) and watched the pair for thirty minutes.

TABLE 1.3 lists the results of the experiment. All of the butterflies with unmodified wing spots survived, regardless of whether they made sounds. By contrast, only half of the butterflies that had spots painted out but could make sounds survived. Most of the silenced butterflies with painted-out spots were eaten quickly. The test results confirmed both predictions, so they support both hypotheses. Predatory birds are indeed deterred by peacock butterfly sounds, and even more so by wing spots.

TAKE-HOME MESSAGE 1.7

Why do biologists perform experiments?

✔ Natural processes are often very complex and influenced by many interacting variables.

✔ Experiments help researchers unravel causes of complex natural processes by focusing on the effects of changing a single variable.

✔ Checks and balances inherent in the scientific process help researchers to be objective about their observations.

✔ Science is, ideally, a self-correcting process because scientists present their work in a way that allows others to check it.

Sampling Error

Researchers can rarely observe all individuals of a group. For example, the explorers you read about in Section 1.1 did not—and could not—survey every uninhabited part of the Foja Mountains. The cloud forest itself cloaks more than 2 million acres, so surveying all of it would take unrealistic amounts of time and effort.

When researchers cannot directly observe all individuals of a population, all instances of an event, or some other aspect of nature, they may test or survey a subset. Results from the subset are then used to make generalizations about the whole. However, generalizing from a subset is risky because subsets are not necessarily representative of the whole. Consider the golden-mantled tree kangaroo, which was first discovered in 1993 on a single forested mountaintop in New Guinea. For more than a decade, the species was never seen outside of that habitat, which is getting smaller every year because of human activities. Thus, the golden-mantled tree kangaroo was considered to be one of the most endangered animals on the planet. Then, in 2005, the New Guinea explorers discovered

FIGURE 1.12 Kris Helgen holds a golden-mantled tree kangaroo he found during the 2005 Foja Mountains survey. This kangaroo species is extremely rare in other areas, so it was thought to be critically endangered prior to the expedition.

that this kangaroo species is fairly common in the Foja Mountain cloud forest (**FIGURE 1.12**). As a result, biologists now believe its future is secure, at least for the moment.

Sampling error is a difference between results obtained from a subset, and results from the whole (**FIGURE 1.13A**). Sampling error may be unavoidable, but knowing how it can occur helps researchers design their experiments to minimize it. For example, sampling error can be a substantial problem with a

A Natalie chooses a random jelly bean from a jar. She is blindfolded, so she does not know that the jar contains 120 green and 280 black jelly beans.

The jar is hidden from Natalie's view before she removes her blindfold. She sees one green jelly bean in her hand and assumes that the jar must hold only green jelly beans. This assumption is incorrect: 30 percent of the jelly beans in the jar are green, and 70 percent are black. The small sample size has resulted in sampling error.

B Still blindfolded, Natalie randomly chooses 50 jelly beans from the jar. She ends up choosing 10 green and 40 black ones.

The larger sample leads Natalie to assume that one-fifth of the jar's jelly beans are green (20 percent) and four-fifths are black (80 percent). The larger sample more closely approximates the jar's actual green-to-black ratio of 30 percent to 70 percent.

The more times Natalie repeats the sampling, the greater the chance she has of guessing the actual ratio.

FIGURE 1.13 ▶Animated How sample size affects sampling error.

small subset, so experimenters try to start with a relatively large sample, and they repeat their experiments (**FIGURE 1.13B**). To understand why these practices reduce the risk of sampling error, think about flipping a coin. There are two possible outcomes of each flip: The coin lands heads up, or it lands tails up. Thus, the chance that the coin will land heads up is one in two (1/2), or 50 percent. However, when you flip a coin repeatedly, it often lands heads up, or tails up, several times in a row. With just 3 flips, the proportion of times that heads actually land up may not even be close to 50 percent. With 1,000 flips, however, the overall proportion of times the coin lands heads up is much more likely to approach 50 percent.

In cases such as flipping a coin, it is possible to calculate **probability**, which is the measure, expressed as a percentage, of the chance that a particular outcome will occur. That chance depends on the total number of possible outcomes. For instance, if 10 million people enter a drawing, each has the same probability of winning: 1 in 10 million, or (an extremely improbable) 0.00001 percent.

Analysis of experimental data often includes probability calculations. If a result is very unlikely to have occurred by chance alone, it is said to be **statistically significant**. In this context, the word "significant" does not refer to the result's importance. Rather, it means that a rigorous statistical analysis has shown a very low probability (usually 5 percent or less) of the result being skewed by sampling error.

Variation in data is often shown as error bars on a graph (**FIGURE 1.14**). Depending on the graph, error bars may indicate variation around an average for one sample set, or the difference between two sample sets.

Bias in Interpreting Results

Particularly when studying humans, changing a single variable apart from all others is not often possible. For example, remember that the people who participated in the Olestra experiment were chosen randomly. Thus, the study was not controlled for gender, age, weight, medications taken, and so on. Such variables may have influenced the results.

---

probability The chance that a particular outcome of an event will occur; depends on the total number of outcomes possible.
sampling error Difference between results derived from testing an entire group of events or individuals, and results derived from testing a subset of the group.
statistically significant Refers to a result that is statistically unlikely to have occurred by chance alone.

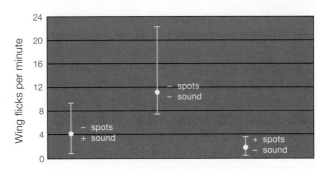

FIGURE 1.14 Example of error bars in a graph. This graph was adapted from the peacock butterfly research described in Section 1.7. The researchers recorded the number of times each butterfly flicked its wings in response to an attack by a bird.

The squares represent average frequency of wing flicking for each sample set of butterflies. The error bars that extend above and below the dots indicate the range of values—the sampling error.

What was the fastest rate at which a butterfly with no spots or sound flicked its wings?

Answer: 22 times per minute

Human beings are by nature subjective, and scientists are no exception. Researchers risk interpreting their results in terms of what they want to find out. That is why they often design experiments to yield quantitative results, which are counts or some other data that can be measured or gathered objectively. Such results minimize the potential for bias, and also give other scientists an opportunity to repeat the experiments and check the conclusions drawn from them.

This last point gets us back to the role of critical thinking in science. Scientists expect one another to recognize and put aside bias in order to test their hypotheses in ways that may prove them wrong. If one scientist does not, then others will, because exposing errors is just as useful as applauding insights. The scientific community consists of critically thinking people trying to poke holes in one another's ideas. Their collective efforts make science a self-correcting endeavor.

TAKE-HOME MESSAGE 1.8
How do scientists avoid potential pitfalls of sampling error and bias when doing research?

✔ Researchers minimize sampling error by using large sample sizes and by repeating their experiments.

✔ Probability calculations can show whether a result is likely to have occurred by chance alone.

✔ Science is a self-correcting process because it is carried out by a community of people systematically checking one another's work and conclusions.

✔ Scientific theories are our best descriptions of reality.

✔ Science is limited to the observable.

Suppose a hypothesis stands even after years of tests. It is consistent with all data ever gathered, and it has helped us make successful predictions about other phenomena. When a hypothesis meets these criteria, it is considered to be a **scientific theory** (TABLE 1.4).

To give an example, all observations to date have been consistent with the hypothesis that matter consists of atoms. Scientists no longer spend time testing this hypothesis for the compelling reason that, since we started looking 200 years ago, no one has discovered matter that doesn't consist of atoms. Thus, researchers use the hypothesis, now called atomic theory, to make other hypotheses about matter and the way it behaves.

Scientific theories are our best descriptions of reality. However, they can never be proven absolutely, because to do so would necessitate testing under every possible circumstance. For example, in order to prove atomic theory, the atomic composition of all matter in the universe would have to be checked—an impossible task even if someone wanted to try.

Like all hypotheses, a scientific theory can be disproven by a single observation or result that is inconsistent with it. For example, if someone discovers a form of matter that does not consist of atoms, atomic theory would have to be revised. The potentially falsifiable nature of scientific theories means that science has a built-in system of checks and balances. A theory is revised until no one can prove it to be incorrect. For

example, the theory of evolution, which states that change occurs in a line of descent over time, still holds after a century of observations and testing. As with all other scientific theories, no one can be absolutely sure that it will hold under all possible conditions, but it has a very high probability of not being wrong. Few other theories have withstood as much scrutiny.

You may hear people apply the word "theory" to a speculative idea, as in the phrase "It's just a theory." This everyday usage of the word differs from the way it is used in science. Speculation is an opinion, belief, or personal conviction that is not necessarily supported by evidence. A scientific theory is different. By definition, it is supported by a large body of evidence, and it is consistent with all known data.

A scientific theory also differs from a **law of nature**, which describes a phenomenon that has been observed to occur in every circumstance without fail, but the scientific explanation for it is incomplete. The laws of thermodynamics, which describe energy, are examples. We know how energy behaves, but not exactly why it behaves the way it does.

The Limits of Science

Science helps us be objective about our observations in part because of its limitations. For example, science does not address many questions, such as "Why do I exist?" Answers to questions like this can only come from within, as an integration of the personal experiences and mental connections that shape our consciousness. This is not to say subjective answers have

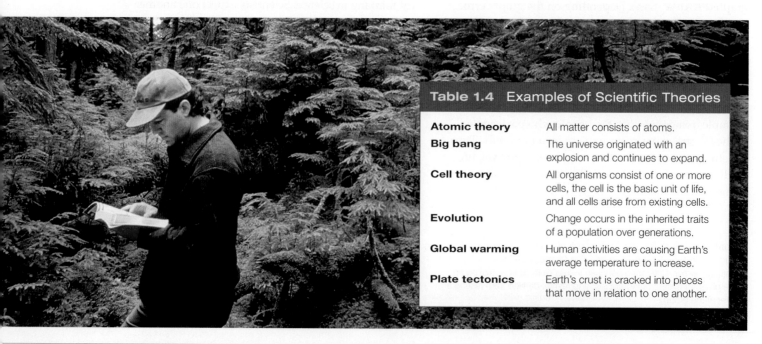

Table 1.4 Examples of Scientific Theories

Atomic theory	All matter consists of atoms.
Big bang	The universe originated with an explosion and continues to expand.
Cell theory	All organisms consist of one or more cells, the cell is the basic unit of life, and all cells arise from existing cells.
Evolution	Change occurs in the inherited traits of a population over generations.
Global warming	Human activities are causing Earth's average temperature to increase.
Plate tectonics	Earth's crust is cracked into pieces that move in relation to one another.

The Secret Life of Earth (revisited)

Of an estimated 100 billion species that have ever lived, at least 100 million are still with us. That number is only an estimate because we are still discovering them. For example, a mouse-sized opossum and a cat-sized rat turned up on a return trip to the Foja Mountains. Other recent surveys have revealed new species of dolphin, gecko, skink, and frog in Australia; a giant fish in Brazil; legless lizards in Southern California; a giant crayfish in Tennessee; a rat-eating plant in the Philippines; a sausage-sized millipede in Tanzania; and scores of plants and single-celled organisms. Most were discovered by biologists simply trying to find out what lives where.

Biologists discover thousands of new species per year (**FIGURE 1.15**). Each is a reminder that we do not yet know all of the organisms living on our own planet. We don't even know how many to look for. The vast information about the 1.8 million species we do know about changes so quickly that collating it has been impossible—until recently. A website titled

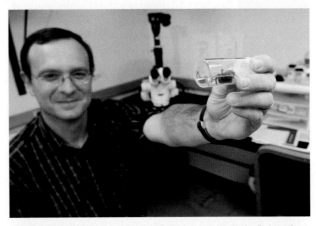

FIGURE 1.15 The discoverer of a new species typically has the honor of naming it. Dr. Jason Bond holds a new species of spider he discovered in California in 2008. Bond named the spider *Aptostichus stephencolberti*, after TV personality Stephen Colbert.

the Encyclopedia of Life is intended to be an online reference source and database of species information maintained by collaborative effort. See its progress at www.eol.org.

---

no value, because no human society can function for long unless its individuals share standards for making judgments, even if they are subjective. Moral, aesthetic, and philosophical standards vary from one society to the next, but all help people decide what is important and good. All give meaning to our lives.

Neither does science address the supernatural, which is anything "beyond nature." Science neither assumes nor denies the existence of supernatural phenomena, but controversy may arise when researchers discover a natural explanation for something that was thought to have none. Such controversy often occurs when a society's moral standards are interwoven with its understanding of nature. Consider Nicolaus Copernicus, who concluded in 1540 that Earth orbits the sun. Today that idea is generally accepted, but during Copernicus's time the prevailing belief system had Earth as the immovable center of the universe. In 1610, astronomer Galileo Galilei published evidence for the Copernican model of the solar system, an act that resulted in his imprisonment. He was publicly forced to recant his work, spent the rest of his life under house arrest, and was never allowed to publish again.

As Galileo's story illustrates, exploring a traditional view of the natural world from a scientific perspective is often misinterpreted as a violation of morality. As a group, scientists are no less moral than anyone else. However, they follow a particular set of rules that do not necessarily apply to others: Their work concerns only the natural world, and their ideas must be testable in ways others can repeat.

Science helps us communicate our experiences without bias. As such, it may be as close as we can get to a universal language. We are fairly sure, for example, that the law of gravity applies everywhere in the universe. Intelligent beings on a distant planet would likely understand the concept of gravity. We might well use gravity or another scientific concept to communicate with them, or anyone, anywhere. The point of science, however, is not to communicate with aliens. It is to find common ground here on Earth.

TAKE-HOME MESSAGE 1.9

Why does science work?

✔ Science is concerned only with testable ideas about observable aspects of the natural world.

✔ Because a scientific theory is thoroughly tested and revised until no one can prove it wrong, it is our best way of describing reality.

---

law of nature Generalization that describes a consistent natural phenomenon for which there is incomplete scientific explanation.
scientific theory Hypothesis that has not been disproven after many years of rigorous testing.

Section 1.1 **Biology** is the scientific study of life. We know about only a fraction of the organisms that live on Earth, in part because we have explored only a fraction of its inhabited regions.

Section 1.2 Biologists think about life at different levels of organization, with **emergent properties** appearing at successive levels. All matter consists of **atoms**, which combine as **molecules**. **Organisms** are individuals that consist of one or more **cells**, the organizational level at which life emerges. Cells of larger multicelled organisms are organized as **tissues**, then **organs**, and **organ systems**. A **population** is a group of interbreeding individuals of a species in a given area; a **community** is all populations of all species in a given area. An **ecosystem** is a community interacting with its environment. The **biosphere** includes all regions of Earth that hold life.

Section 1.3 Life has underlying unity in that all living things have similar characteristics: (1) All organisms require energy and **nutrients** to sustain themselves. **Producers** harvest energy from the environment to make their own food by processes such as **photosynthesis**; **consumers** ingest other organisms, their wastes, or remains. (2) Organisms keep the conditions in their internal environment within ranges that their cells tolerate—a process called **homeostasis**. (3) **DNA** contains information that guides an organism's **growth**, **development**, and **reproduction**. The passage of DNA from parents to offspring is called **inheritance**.

Section 1.4 The many types of organisms that currently exist on Earth differ greatly in the details of their body form and function. **Biodiversity** is the sum of differences among living things. **Bacteria** and **archaea** are both **prokaryotes**, single-celled organisms whose DNA is not contained within a nucleus. The DNA of single-celled or multicelled **eukaryotes** (**protists**, **plants**, **fungi**, and **animals**) is contained within a nucleus.

Section 1.5 Each **species** is given a two-part name. The first part is the **genus** name. When combined with the **specific epithet**, it designates the particular species. With **taxonomy**, species are ranked into ever more inclusive **taxa** (genus, family, order, class, phylum, kingdom, domain) on the basis of shared inherited **traits**.

Section 1.6 **Critical thinking**, the self-directed act of judging the quality of information as one learns, is an important part of **science**. Generally, a researcher observes something in nature, uses **inductive reasoning** to form a **hypothesis** (testable explanation) for it, then uses **deductive reasoning** to make a testable **prediction** about what might occur if the hypothesis is correct. **Experiments** with

variables may be performed on an **experimental group** as compared with a **control group**, and sometimes on **models**. A researcher typically changes an **independent variable**, then observes the effects of the change on a **dependent variable**. Conclusions are drawn from the resulting **data**. The **scientific method** consists of making, testing, and evaluating hypotheses, and sharing results with the scientific community.

Section 1.7 Biological systems are typically influenced by many interacting variables. Research approaches differ, but experiments are designed in a consistent way, in order to study a single cause-and-effect relationship in a complex natural system.

Section 1.8 Small sample size increases the potential for **sampling error** in experimental results. In such cases, a subset may be tested that is not representative of the whole. Researchers design experiments carefully to minimize sampling error and bias, and they use **probability** rules to check the **statistical significance** of their results. Science is ideally a self-correcting process because scientists check and test one another's ideas.

Section 1.9 Science helps us be objective about our observations because it is only concerned with testable ideas about observable aspects of nature. Opinion and belief have value in human culture, but they are not addressed by science. A **scientific theory** is a long-standing hypothesis that is useful for making predictions about other phenomena. It is our best way of describing reality. A **law of nature** is a phenomenon that occurs without fail, but has an incomplete scientific explanation.

self-quiz

1. _____ are fundamental building blocks of all matter.
   a. Atoms   c. Cells
   b. Molecules   d. Organisms

2. The smallest unit of life is the _____ .
   a. atom   c. cell
   b. molecule   d. organism

3. Organisms require _____ and _____ to maintain themselves, grow, and reproduce.
   a. DNA; energy   c. nutrients; energy
   b. food; sunlight   d. DNA; cells

4. By sensing and responding to change, organisms keep conditions in the internal environment within ranges that cells can tolerate. This process is called _____ .

5. DNA _____ .
   a. guides form   c. is transmitted from
      and function   parents to offspring
   b. is the basis of traits   d. all of the above

Peacock Butterfly Predator Defenses The photographs below represent the experimental and control groups used in the peacock butterfly experiment discussed in Section 1.7. See if you can identify the experimental groups, and match them up with the relevant control group(s). *Hint:* Identify which variable is being tested in each group (each variable has a control).

A Wing spots painted out   **B** Wing spots visible; wings silenced   **C** Wing spots painted out; wings silenced   **D** Wings painted but spots visible   **E** Wings cut but not silenced   **F** Wings painted, spots visible; wings cut, not silenced

6. A process by which an organism produces offspring is called _____ .
   - a. reproduction
   - b. development
   - c. homeostasis
   - d. inheritance

7. _____ is the transmission of DNA to offspring.
   - a. Reproduction
   - b. Development
   - c. Homeostasis
   - d. Inheritance

8. A butterfly is a(n) _____ (choose all that apply).
   - a. organism
   - b. domain
   - c. species
   - d. eukaryote
   - e. consumer
   - f. producer
   - g. prokaryote
   - h. trait

9. _____ move around for at least part of their life.

10. A bacterium is _____ (choose all that apply).
   - a. an organism
   - b. single-celled
   - c. an animal
   - d. a eukaryote

11. Bacteria, Archaea, and Eukarya are three _____ .

12. A control group is _____ .
   - a. a set of individuals that have a certain characteristic or receive a certain treatment
   - b. the standard against which an experimental group is compared
   - c. the experiment that gives conclusive results

13. Fifteen randomly selected students are found to be taller than 6 feet. The researchers concluded that the average height of a student is greater than 6 feet. This is an example of _____ .
   - a. experimental error
   - b. sampling error
   - c. a subjective opinion
   - d. experimental bias

14. Science only addresses that which is _____ .
   - a. alive
   - b. observable
   - c. variable
   - d. indisputable

15. Match the terms with the most suitable description.
   - ___ life
   - ___ probability
   - ___ species
   - ___ hypothesis
   - ___ prediction
   - ___ producer
   - a. if–then statement
   - b. unique type of organism
   - c. emerges with cells
   - d. testable explanation
   - e. measure of chance
   - f. makes its own food

critical thinking

1. A person is declared dead upon the irreversible ceasing of spontaneous body functions: brain activity, blood circulation, and respiration. Only about 1% of a body's cells have to die in order for all of these things to happen. How can a person be dead when 99% of his or her cells are alive?

2. Explain the difference between a one-celled organism and a single cell of a multicelled organism.

3. Why would you think twice about ordering from a restaurant menu that lists the specific epithet but not the genus name of its offerings? *Hint:* Look up *Homarus americanus, Ursus americanus, Bufo americanus, Lepus americanus, Necator americanus, Lysichiton americanus, Leucoagaricus americanus,* and *Nicrophorus americanus.*

4. Once there was a highly intelligent turkey that had nothing to do but reflect on the world's regularities. Morning always started out with the sky turning light, followed by the master's footsteps, which were always followed by the appearance of food. Other things varied, but food always followed footsteps. The sequence of events was so predictable that it eventually became the basis of the turkey's theory about the goodness of the world. One morning, after more than 100 confirmations of this theory, the turkey listened for the master's footsteps, heard them, and had its head chopped off. Any scientific theory is modified or discarded upon discovery of contradictory evidence. The absence of absolute certainty has led some people to conclude that "theories are irrelevant because they can change." If that is so, should we stop doing scientific research? Why or why not?

5. In 2005, researcher Woo-suk Hwang reported that he had made immortal stem cells from human patients. His research was hailed as a breakthrough for people affected by degenerative diseases, because stem cells may be used to repair a person's own damaged tissues. Hwang published his results in a peer-reviewed journal. In 2006, the journal retracted his paper after other scientists discovered that Hwang's group had faked their data.

   Does the incident show that results of scientific studies cannot be trusted? Or does it confirm the usefulness of a scientific approach, because other scientists discovered and exposed the fraud?

CREDIT: Scientific Paper; Adrian Vallin, Sven Jakobsson, Johan Lind and Christer Wiklund, *Proc. R. Soc. B* (2005 272, 1203, 1207). Used with permission of The Royal Society and the author.

2 Life's Chemical Basis

LEARNING ROADMAP

In this chapter, you will explore the first level of life's organization—atoms—as you encounter an example of how the same building blocks, arranged different ways, form different products (Section 1.2). You will also see one aspect of homeostasis, the process by which organisms keep themselves in a state that favors cell survival (Section 1.3).

ATOMS AND ELEMENTS

Atoms, the building blocks of all matter, differ in their numbers of protons, neutrons, and electrons. Atoms of an element have the same number of protons.

WHY ELECTRONS MATTER

Whether an atom interacts with other atoms depends on the number of electrons it has. An atom with an unequal number of electrons and protons is an ion.

ATOMS BOND

Atoms of many elements interact by acquiring, sharing, and giving up electrons. Interacting atoms may form ionic, covalent, or hydrogen bonds.

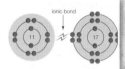

WATER

Hydrogen bonding among individual molecules gives water properties that make life possible: temperature stabilization, cohesion, and the ability to dissolve many other substances.

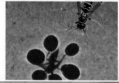

HYDROGEN POWER

Most of the chemistry of life occurs in a narrow range of pH, so most fluids inside organisms are buffered to stay within that range.

Electrons will come up again as you learn how energy drives metabolism, especially in photosynthesis (Chapter 6) and respiration (Chapter 7). Hydrogen bonding is critical for the molecules of life (3.4, 3.6–3.8); the properties of water, for membranes (Section 4.2 and Chapter 5), plant nutrition and transport (28.4), and temperature regulation (40.9). You will also see how radioisotopes are used to date rocks and fossils (16.6), and explore the dangers of free radicals (8.6) and acid rain (48.5).

Actor Jeremy Piven, best known for his Emmy-winning role on the television series *Entourage*, began starring in a Broadway play in 2008. He quit suddenly after two shows, citing medical problems. Piven explained that he was suffering from mercury poisoning caused by eating too much sushi. The play's producers and his co-actors were skeptical, and the playwright ridiculed Piven, saying he was leaving to pursue a career as a thermometer. However, mercury poisoning is no laughing matter.

Mercury is a naturally occurring toxic metal. Most of it is safely locked away in rocky minerals, but volcanic activity and other geologic processes release it into the atmosphere. So do human activities, especially burning coal (**FIGURE 2.1**). Airborne mercury can drift long distances before settling to Earth's surface, where microbes combine it with carbon to form a substance called methylmercury.

Unlike mercury alone, methylmercury easily crosses skin and mucous membranes. In water, it ends up in the tissues of aquatic organisms. All fish and shellfish contain it. Humans contain it too, mainly as a result of eating seafood.

When mercury enters the body, it damages the nervous system, brain, kidneys, and other organs. An average-sized adult who ingests as little as 200 micrograms of methylmercury may experience blurred vision, tremors, itching or burning sensations, and loss of coordination. Exposure to larger amounts can result in thought and memory impairment, coma, and death. Methylmercury in a pregnant woman's blood passes to her unborn child, along with a legacy of permanent developmental problems.

It takes months or even years for mercury to be cleared from the body, so the toxin can build up to high levels if even small amounts are ingested on a regular basis. That is why large predatory fish have a lot of mercury in their tissues. It is also why the U.S. Environmental Protection Agency recommends that adult humans ingest less than 0.1 microgram of mercury per kilogram of body weight per day. For an average-sized person, that limit works out to be about 7 micrograms per day, which is not a big amount if you eat seafood. A typical 6-ounce can of albacore tuna contains about 60 micrograms of mercury, and the occasional can has many times that amount. It does not matter if the fish is canned, grilled, or raw, because methylmercury is unaffected by cooking. Eat a medium-sized tuna steak, and you could be getting more than 700 micrograms of mercury along with it.

With this chapter, we turn to the first of life's levels of organization: atoms. Interactions between atoms make the molecules that sustain life, and also some that destroy it.

FIGURE 2.1 The Navajo Generating Station, a coal-fired power plant in Arizona. Emissions from such plants are by far the largest source of mercury pollution worldwide.

Mercury can accumulate to toxic levels in the tissues of tuna and other large predatory fish that we harvest for food.

2.2 Start With Atoms

✔ Atomic structure gives rise to chemical properties of atoms.
✔ The number of protons in the atomic nucleus defines the element, and the number of neutrons defines the isotope.

Even though atoms are about 20 million times smaller than a grain of sand, they consist of even smaller subatomic particles. Positively charged **protons** (p^+) and uncharged **neutrons** occur in an atom's core, or **nucleus**. Negatively charged **electrons** (e^-) move around the nucleus (**FIGURE 2.2A**). **Charge** is an electrical property: Opposite charges attract, and like charges repel.

A typical atom has about the same number of electrons and protons. The negative charge of an electron is the same magnitude as the positive charge of a proton, so the two charges cancel one another. Thus, an atom with exactly the same number of electrons and protons carries no charge.

All atoms have protons. The number of protons in the nucleus is called the **atomic number**, and it determines the type of atom, or element. **Elements** are pure substances, each consisting only of atoms

with the same number of protons in their nucleus (**FIGURE 2.2B**). For example, the element carbon has an atomic number of 6. All atoms with six protons in their nucleus are carbon atoms, no matter how many electrons or neutrons they have. Elemental carbon (the substance) consists only of carbon atoms, and all of those atoms have six protons.

Knowing the numbers of electrons, protons, and neutrons in atoms helps us predict how elements will behave. In 1869, chemist Dmitry Mendeleyev arranged the elements known at the time by their chemical properties. The arrangement, which he called the **periodic table**, turned out to be by atomic number, even though subatomic particles would not be discovered until the early 1900s. In the periodic table, each element is represented by a symbol that is typically an abbreviation of the element's Latin or Greek name (**FIGURE 2.2C**). For example, the symbol for lead, Pb, is short for its Latin name: *plumbum*. The word "plumbing" is related (ancient Romans made their water pipes with lead). Carbon's symbol, C, is from *carbo*, the Latin word for coal, which is mostly carbon.

Isotopes and Radioisotopes

All atoms of an element have the same number of protons, but they can differ in the number of other subatomic particles. Those that differ in the number of neutrons are called **isotopes**. The total number of neutrons and protons in the nucleus of an isotope is its **mass number**. Mass number is written as a superscript to the left of the element's symbol. For example, the most common isotope of hydrogen has one proton

A Atoms consist of electrons moving around a nucleus of protons and neutrons. Models such as this one do not show what atoms look like. Electrons move in defined, three-dimensional spaces about 10,000 times bigger than the nucleus.

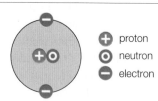

⊕ proton
⊙ neutron
⊖ electron

B Example of an element.

atomic number ⎯⎯ 6
element symbol ⎯⎯ C
mass number ⎯⎯ 12
elemental substance
element name
carbon

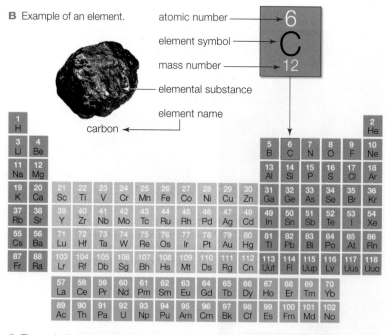

C The periodic table of the elements.

FIGURE 2.2 ▶Animated Atoms and elements.

atomic number Number of protons in the atomic nucleus; determines the element.
charge Electrical property. Opposite charges attract, and like charges repel.
electron Negatively charged subatomic particle.
element A pure substance that consists only of atoms with the same number of protons.
isotopes Forms of an element that differ in the number of neutrons their atoms carry.
mass number Of an isotope, the total number of protons and neutrons in the atomic nucleus.
neutron Uncharged subatomic particle in the atomic nucleus.
nucleus Core of an atom; occupied by protons and neutrons.
periodic table Tabular arrangement of all known elements by their atomic number.
proton Positively charged subatomic particle that occurs in the nucleus of all atoms.
radioactive decay Process by which atoms of a radioisotope emit energy and/or subatomic particles when their nucleus spontaneously breaks up.
radioisotope Isotope with an unstable nucleus.
tracer A substance that can be traced via its detectable component.

and no neutrons, so it is designated ^{1}H. Other isotopes include deuterium (^{2}H, one proton and one neutron), and tritium (^{3}H, one proton and two neutrons).

The most common isotope of carbon has six protons and six neutrons (^{12}C). Another naturally occurring carbon isotope has six protons and eight neutrons (^{14}C). Carbon 14 is an example of a **radioisotope**, or radioactive isotope. Atoms of a radioisotope have an unstable nucleus that breaks up spontaneously. As a nucleus breaks up, it emits radiation (subatomic particles, energy, or both), a process called **radioactive decay**. The atomic nucleus cannot be altered by ordinary means, so radioactive decay is unaffected by external factors such as temperature, pressure, or whether the atoms are part of molecules.

Each radioisotope decays at a predictable rate into predictable products. For example, when carbon 14 decays, one of its neutrons splits into a proton and an electron. The nucleus emits the electron as radiation. Thus, a carbon atom with eight neutrons and six protons (^{14}C) becomes a nitrogen atom, with seven neutrons and seven protons (^{14}N):

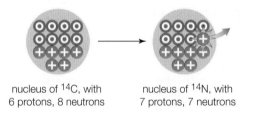

nucleus of ^{14}C, with   nucleus of ^{14}N, with
6 protons, 8 neutrons   7 protons, 7 neutrons

This process is so predictable that we can say with certainty that about half of the atoms in any sample of ^{14}C will be ^{14}N atoms after 5,730 years. The predictable rate of radioactive decay makes it possible for scientists to estimate the age of a rock or fossil by measuring its isotope content (we return to this topic in Section 16.6).

All isotopes of an element generally have the same chemical properties regardless of the number of neutrons in their atoms. This consistency means that the atoms of one isotope behave the same way inside organisms as atoms of another isotope. Thus, radioisotopes can be used as or in **tracers**, which are substances with a detectable component. For example, a molecule in which an atom (such as ^{12}C) has been replaced with a radioisotope (such as ^{14}C) can be used as a radioactive tracer. When delivered into a biological system such as a cell, body, or ecosystem, this tracer may be followed as it moves through the system with instruments that detect radiation.

Radioactive tracers are widely used in research. A famous example is a series of experiments carried out

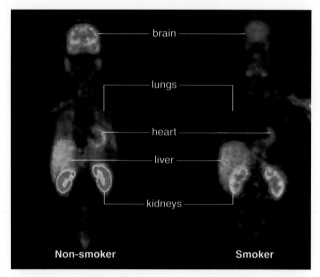

FIGURE 2.3 ▸Animated PET scans use radioactive tracers to form a digital image of a process in the body's interior. These two PET scans reveal the activity of a molecule called MAO-B in the body of a nonsmoker (left) and a smoker (right). The activity is color-coded from red (highest activity) to purple (lowest). Low MAO-B activity is associated with violence, impulsiveness, and other behavioral problems.

by Melvin Calvin and Andrew Benson. These researchers synthesized carbon dioxide with ^{14}C, then let green algae take up the radioactive gas. Using instruments that detect electrons emitted by the radioactive decay of ^{14}C, they tracked carbon through steps by which the algae—and all plants—make sugars.

Radioisotopes have medical applications as well. For example, PET (short for positron-emission tomography) helps us "see" a functional process inside the body. By this procedure, a radioactive sugar or other tracer is injected into a patient. Inside the patient's body, cells with differing rates of activity take up the tracer at different rates. A scanner detects radioactive decay wherever the tracer is, then translates that data into an image (**FIGURE 2.3**).

TAKE-HOME MESSAGE 2.2

What are the basic building blocks of all matter?

✔ All matter consists of atoms, tiny particles that in turn consist of electrons moving around a nucleus (core). Protons and neutrons are components of the atomic nucleus.

✔ An elemental substance consists only of atoms with the same number of protons. Isotopes are forms of an element that have different numbers of neutrons.

✔ Unstable nuclei of radioisotopes emit radiation as they spontaneously break down (decay). Radioisotopes decay at a predictable rate to form predictable products.

2.3 Why Electrons Matter

✔ Whether an atom will interact with other atoms depends on how many electrons it has.

The more we learn about electrons, the weirder they seem. Consider that an electron has mass but no size, and its position in space is described as more of a smudge than a point. It carries energy, but only in incremental amounts (this concept will be important to remember when you learn how cells harvest and release energy). An electron gains energy only by absorbing the precise amount needed to boost it to the next energy level. Likewise, it loses energy only by emitting the exact difference between two energy levels.

A lot of electrons may be occupying the same atom. However, despite moving very fast—almost the speed of light—they never collide. Why not? For one reason, electrons in an atom occupy different orbitals, which are defined volumes of space around the atomic nucleus. To understand how orbitals work, imagine that an atom is a multilevel apartment building, with the nucleus in the basement. Each "floor" of the building corresponds to a certain energy level, and each has a certain number of "rooms" (orbitals) available for rent. Two electrons can occupy each room. Pairs of electrons populate rooms from the ground floor up; in other words, they fill orbitals from lower to higher energy levels. The farther an electron is from the nucleus in the basement, the greater its energy. An electron can move to a room on a higher floor if an energy input gives it a boost, but it immediately emits the extra energy and moves back down.

A **shell model** helps us visualize how electrons populate atoms (**FIGURE 2.4**). In this model, nested "shells" correspond to successively higher energy levels. Thus, each shell includes all of the rooms (orbitals) on one floor (energy level) of our atomic apartment building.

We draw a shell model of an atom by filling it with electrons (represented as balls or dots), from the innermost shell out, until there are as many electrons as the atom has protons. There is only one room on the first floor, one orbital at the lowest energy level. It fills up first. In hydrogen, the simplest atom, a single electron occupies that room (**FIGURE 2.4A**). Helium, with two protons, has two electrons that fill the room—and the first shell. In larger atoms, more electrons rent the second-floor rooms (**FIGURE 2.4B**). When the second floor fills, more electrons rent third-floor rooms (**FIGURE 2.4C**), and so on.

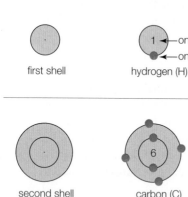 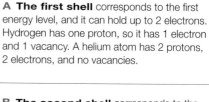

A **The first shell** corresponds to the first energy level, and it can hold up to 2 electrons. Hydrogen has one proton, so it has 1 electron and 1 vacancy. A helium atom has 2 protons, 2 electrons, and no vacancies.

first shell   hydrogen (H)   helium (He)

one proton
one electron

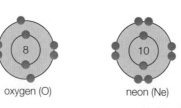

second shell   carbon (C)   oxygen (O)   neon (Ne)

B **The second shell** corresponds to the second energy level, and it can hold up to 8 electrons. Carbon has 6 electrons, so its first shell is full. Its second shell has 4 electrons and four vacancies. Oxygen has 8 electrons and two vacancies. Neon has 10 electrons and no vacancies.

 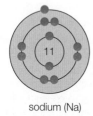 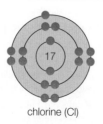 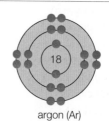

third shell   sodium (Na)   chlorine (Cl)   argon (Ar)

C **The third shell** corresponds to the third energy level, and it can hold up to 8 electrons. A sodium atom has 11 electrons, so its first two shells are full; the third shell has one electron. Thus, sodium has seven vacancies. Chlorine has 17 electrons and one vacancy. Argon has 18 electrons and no vacancies.

FIGURE 2.4 ▶Animated Shell models. Each circle (shell) represents one energy level. To make these models, we fill the shells with electrons from the innermost shell out, until there are as many electrons as the atom has protons. The number of protons in each model is indicated.

FIGURE IT OUT Which of these models have unpaired electrons in their outer shell?

Answer: Hydrogen, carbon, oxygen, sodium, and chlorine

About Vacancies

When an atom's outermost shell is filled with electrons, we say that it has no vacancies, and it is in its most stable state. Helium, neon, and argon are examples of elements with no vacancies. Atoms of these elements are chemically stable, which means they have very

little tendency to interact with other atoms. Thus, these elements occur most frequently in nature as solitary atoms.

By contrast, when an atom's outermost shell has room for another electron, it has a vacancy. Atoms with vacancies tend to get rid of them by interacting with other atoms; in other words, they are chemically active. For example, the sodium atom (Na) depicted in FIGURE 2.4C has one electron in its outer (third) shell, which can hold eight. With seven vacancies, we can predict that this atom is chemically active.

In fact, this particular sodium atom is not just active, it is extremely so. Why? The shell model shows that a sodium atom has an unpaired electron, but in the real world, electrons really like to be in pairs when they occupy orbitals. Atoms that have unpaired electrons are called **free radicals**. With a few exceptions, free radicals are very unstable, easily forcing electrons upon other atoms or ripping electrons away from them. This property makes free radicals dangerous to life (we return to this topic in Section 5.6). A sodium atom with 11 electrons (a sodium radical) can easily evict the one unpaired electron, so that its second shell—which is full of electrons—becomes its outermost, and no vacancies remain. This is the atom's most stable state. The vast majority of sodium atoms on Earth are like this one, with 11 protons and 10 electrons.

Atoms with an unequal number of protons and electrons are called **ions**. Ions carry a net (or overall) charge. Sodium ions (Na^+) offer an example of how atoms gain a positive charge by losing an electron (FIGURE 2.5A). Other atoms gain a negative charge by accepting an electron. For example, an uncharged chlorine atom has 17 protons and 17 electrons. The outermost shell of this atom can hold eight electrons, but

---

free radical Atom with an unpaired electron.
ion Charged atom.
shell model Model of electron distribution in an atom.

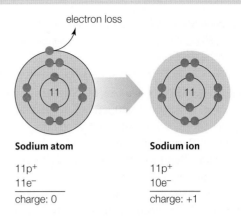

A A sodium atom (Na) becomes a positively charged sodium ion (Na^+) when it loses the single electron in its third shell. The atom's full second shell is now its outermost, so it has no vacancies.

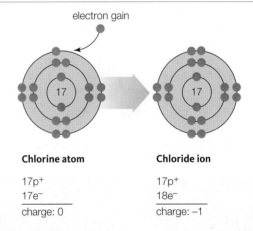

B A chlorine atom (Cl) becomes a negatively charged chloride ion (Cl^-) when it gains an electron and fills the vacancy in its third, outermost shell.

FIGURE 2.5 Ion formation. **FIGURE IT OUT** Does a chloride ion have an unpaired electron?   Answer: No

it has only seven. With one vacancy and one unpaired electron, we can predict—correctly—that this atom is chemically very active. An uncharged chlorine atom (a chlorine radical) easily fills its third shell by accepting an electron. When that happens, the atom becomes a chloride ion (Cl^-) with 17 protons, 18 electrons, and a net negative charge (FIGURE 2.5B).

> ### TAKE-HOME MESSAGE 2.3
> #### Why do atoms interact?
>
> ✔ An atom's electrons are the basis of its chemical behavior.
>
> ✔ Shells represent all electron orbitals at one energy level in an atom. When the outermost shell is not full of electrons, the atom has a vacancy.
>
> ✔ Atoms with vacancies tend to interact with other atoms.

2.4 Chemical Bonds: From Atoms to Molecules

✔ Chemical bonds link atoms into molecules.
✔ The characteristics of a chemical bond arise from the properties of the atoms taking part in it.

An atom can get rid of vacancies by participating in a **chemical bond**, which is an attractive force that arises between two atoms when their electrons interact. When atoms interact, they often form molecules. A molecule consists of atoms held together in a particular number and arrangement by chemical bonds.

Water is an example of a substance made of molecules. Each water molecule consists of three atoms: two hydrogen atoms bonded to the same oxygen atom (**FIGURE 2.6**). Because a water molecule consists of two or more elements, it is called a **compound**. Other molecules, including molecular oxygen (a gas in air), have atoms of one element only.

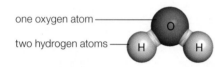

one oxygen atom ———— O
two hydrogen atoms ———— H H

FIGURE 2.6 The water molecule. Each water molecule has two hydrogen atoms bonded to the same oxygen atom.

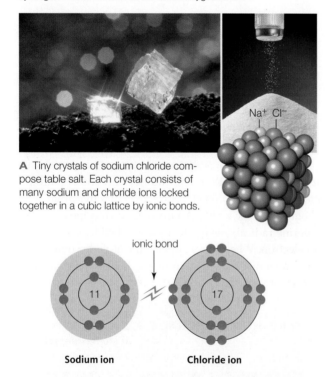

Na⁺ Cl⁻

A Tiny crystals of sodium chloride compose table salt. Each crystal consists of many sodium and chloride ions locked together in a cubic lattice by ionic bonds.

ionic bond

11   17

Sodium ion   **Chloride ion**

B The strong mutual attraction of opposite charges holds a sodium ion and a chloride ion together in an ionic bond.

FIGURE 2.7 ▶**Animated** Ionic bonds in table salt, or NaCl.

The term "bond" applies to a continuous range of atomic interactions. However, we can categorize most bonds into distinct types based on their different properties. Which type forms depends on the atoms taking part in the molecule.

Ionic Bonds

Two ions may be held together by the mutual attraction of their opposite charges, an association called an **ionic bond**. Ionic bonds can be quite strong. Ionically bonded sodium and chloride ions make sodium chloride (NaCl), which we know as table salt; a crystal of this substance consists of a lattice of sodium and chloride ions interacting in ionic bonds (**FIGURE 2.7A**).

Ions retain their respective charges when participating in an ionic bond (**FIGURE 2.7B**). Thus, one "end" of an ionic bond has a positive charge, and the other "end" has a negative charge (right). Any separation of charge into distinct positive and negative regions is called **polarity**.

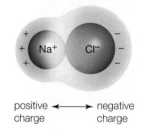

positive ←——→ negative
charge   charge

A sodium chloride molecule is polar because the chloride ion keeps a very strong hold on its extra electron. In other words, it is strongly electronegative. **Electronegativity** is a measure of an atom's ability to pull electrons away from another atom. Electronegativity is not the same as charge. Rather, an atom's electronegativity depends on its size, how many vacancies it has, and its interactions with other atoms.

An ionic bond is completely polar because the atoms participating in it have a very large difference in electronegativity. When atoms with a lower difference in electronegativity interact, they tend to form chemical bonds that are less polar than ionic bonds.

Covalent Bonds

In a **covalent bond**, two atoms share a pair of electrons, so that each atom's vacancy becomes partially filled (**FIGURE 2.8**). Sharing electrons links two atoms just as sharing earphones links two friends (left). Covalent bonds can be stronger than ionic bonds, but they are not always so.

TABLE 2.1 shows different ways of representing covalent bonds. In structural formulas, a line between two atoms represents a single covalent bond. For example, molecular hydrogen (H_2) has one covalent bond

CREDITS: (6, 7A bottom right, 7B, in text top and bottom) © Cengage Learning; (7A) left, Francois Gohier/Science Source; top right, Melica/Shutterstock.

Table 2.1 Ways of Representing Covalent Bonds in Molecules

Common name:	Water	Familiar term.
Chemical name:	Dihydrogen monoxide	Describes elemental composition.
Chemical formula:	H_2O	Indicates unvarying proportions of elements. Subscripts show number of atoms of an element per molecule. The absence of a subscript means one atom.
Structural formula:	H—O—H	Represents each covalent bond as a single line between atoms.
Structural model:		Shows relative sizes and positions of atoms in three dimensions.
Shell model:		Shows how pairs of electrons are shared in covalent bonds.

MOLECULAR HYDROGEN (H—H)

Two hydrogen atoms, each with one proton, share two electrons in a nonpolar covalent bond.

MOLECULAR OXYGEN (O=O)

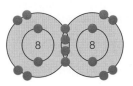

Two oxygen atoms, each with eight protons, share four electrons in a double covalent bond.

WATER (H—O—H)

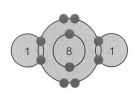

Two hydrogen atoms share electrons with an oxygen atom in two covalent bonds. The bonds are polar because the oxygen exerts a greater pull on the shared electrons than the hydrogens do.

FIGURE 2.8 ▶Animated Covalent bonds, in which atoms fill vacancies by sharing electrons. Two electrons are shared in each covalent bond. When sharing is equal, the bond is nonpolar. When one atom exerts a greater pull on the electrons, the bond is polar.

between hydrogen atoms (H—H). Two, three, or even four covalent bonds may form between atoms when they share multiple pairs of electrons. For example, two atoms sharing two pairs of electrons are connected by two covalent bonds, which are represented by a double line between the atoms. A double bond links the two oxygen atoms in molecular oxygen (O=O). Three lines indicate a triple bond, in which two atoms share three pairs of electrons. A triple covalent bond links the two nitrogen atoms in molecular nitrogen (N≡N). Comparing bonds between the same two atoms: A triple bond is stronger than a double bond, which is stronger than a single bond.

Double and triple bonds are not distinguished from single bonds in structural models, which show the positions and relative sizes of the atoms in three dimensions. The bonds are shown as one stick connecting two balls, which represent atoms. Elements are usually coded by color:

carbon   hydrogen   oxygen   nitrogen   phosphorus

Atoms share electrons unequally in a polar covalent bond. A bond between an oxygen atom and a hydrogen atom in a water molecule is an example. One atom (the oxygen, in this case) is a bit more electronegative. It pulls the electrons a little more toward its side of

the bond, so that atom bears a slight negative charge. The atom at the other end of the bond (the hydrogen) bears a slight positive charge. In most cases, covalent bonds in compounds are polar. By contrast, atoms participating in a nonpolar covalent bond share electrons equally, so there is no difference in charge between the two ends of the bond. The bonds in molecular hydrogen (H_2), oxygen (O_2), and nitrogen (N_2) are examples.

chemical bond An attractive force that arises between two atoms when their electrons interact.
compound Molecule that has atoms of more than one element.
covalent bond Type of chemical bond in which two atoms share a pair of electrons.
electronegativity Measure of the ability of an atom to pull electrons away from other atoms.
ionic bond Type of chemical bond in which a strong mutual attraction links ions of opposite charge.
polarity Separation of charge into positive and negative regions.

TAKE-HOME MESSAGE 2.4
How do atoms interact in chemical bonds?

✔ A chemical bond forms between atoms when their electrons interact. A chemical bond may be ionic or covalent depending on the atoms taking part in it.

✔ An ionic bond is a strong mutual attraction between two ions of opposite charge.

✔ Atoms share a pair of electrons in a covalent bond. When the atoms share electrons unequally, the bond is polar.

2.5 Hydrogen Bonds and Water

✔ The unique properties of liquid water arise because of the water molecule's polarity.

✔ Extensive hydrogen bonds form among water molecules.

Hydrogen Bonding in Water

Water has unique properties that arise from the two polar covalent bonds in each water molecule. Overall, the molecule has no charge, but the oxygen atom carries a slight negative charge; the hydrogen atoms, a slight positive charge. Thus, the molecule itself is polar (**FIGURE 2.9A**).

The polarity of individual water molecules attracts them to one another. The slight positive charge of a hydrogen atom in one water molecule is drawn to the slight negative charge of an oxygen atom in another, an interaction called a hydrogen bond. A **hydrogen bond** is an attraction between a covalently bonded hydrogen atom and another atom taking part in a separate polar covalent bond (**FIGURE 2.9B**). Like ionic bonds, hydrogen bonds form by the mutual attraction of opposite charges. However, unlike ionic bonds, hydrogen bonds do not make molecules out of atoms, so they are not chemical bonds.

Hydrogen bonds are on the weaker end of the spectrum of atomic interactions; they form and break much more easily than covalent or ionic bonds. Even so, many of them form, and collectively they are quite strong. As you will see, hydrogen bonds stabilize the characteristic structures of biological molecules such as DNA and proteins. They also form in tremendous numbers among water molecules (**FIGURE 2.9C**). Extensive hydrogen bonding among water molecules gives liquid water several special properties that make life possible.

Water's Special Properties

Water Is an Excellent Solvent The ability of water molecules to form hydrogen bonds make water an excellent **solvent**, which means that many other substances can dissolve in it. Substances that dissolve easily in water are **hydrophilic** (water-loving). Ionic solids such as sodium chloride (NaCl) dissolve in water because the slight positive charge on each hydrogen atom in a water molecule attracts negatively charged ions (Cl^-), and the slight negative charge on the oxygen atom attracts positively charged

ions (Na^+). Hydrogen bonds among many water molecules are collectively stronger than an ionic bond between two ions, so the solid dissolves as water molecules tug the ions apart and surround each one (left).

Sodium chloride is called a **salt** because it releases ions other than H^+ and OH^- when it dissolves in water (more about this in the next section). When an ionic solid dissolves, its component ions disperse uniformly among molecules of the solvent, and it becomes a **solute**. A uniform mixture such as salt dissolved in water is called a **solution**. Chemical bonds do not form between molecules of solute and solvent, so the proportions of the two substances in a solution can vary. The amount of a solute that is dissolved in a given volume of fluid is its **concentration**.

Many nonionic solids also dissolve easily in water. Sugars are examples. Molecules of these substances have one or more polar covalent bonds, and atoms participating in a polar covalent bond can form hydrogen bonds with water molecules. Hydrogen bonding with water pulls individual molecules of the solid away from one another and keeps them apart. Unlike ionic solids, these substances retain their molecular integrity when they dissolve, which means they do not dissociate into atoms.

Water does not interact with **hydrophobic** (water-dreading) substances such as oils. Oils consist of nonpolar molecules, and hydrogen bonds do not form between nonpolar molecules and water. When you mix oil and water, the water breaks into small droplets, but quickly begins to

---

A Polarity of the water molecule. Each of the hydrogen atoms in a water molecule bears a slight positive charge (represented by a blue overlay). The oxygen atom carries a slight negative charge (red overlay).

slight negative charge

slight positive charge

B A hydrogen bond is an attraction between a hydrogen atom and another atom taking part in a separate polar covalent bond.

a hydrogen bond

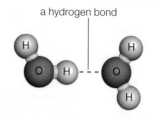

C The many hydrogen bonds that form among water molecules impart special properties to liquid water.

FIGURE 2.9 ▶Animated Hydrogen bonds and water.

cluster into larger drops as new hydrogen bonds form among its molecules. The bonding excludes molecules of oil and pushes them together into drops that rise to the surface of the water. The same interactions occur at the thin, oily membrane that separates the watery fluid inside cells from the watery fluid outside of them. As you will see in Chapter 3, such interactions give rise to the structure of cell membranes.

Water Has Cohesion Molecules of some substances resist separating from one another, and this resistance gives rise to a property called **cohesion**. Water

has cohesion because hydrogen bonds collectively exert a continuous pull on its individual molecules. You can see cohesion in water as surface tension, which means that the surface of liquid water behaves a bit like a sheet of elastic (left).

Cohesion plays a role in many processes that sustain multicelled bodies. Consider how water molecules constantly escape from the surface of liquid water as vapor, a process called **evaporation**. Evaporation is resisted by hydrogen bonding among water molecules. In other words, overcoming water's cohesion takes energy. Thus, evaporation sucks energy (in the form of heat) from liquid water, and this lowers the water's surface temperature. Evaporative water loss helps you and some other mammals cool off when you sweat in hot, dry weather. Sweat, which is about 99 percent water, cools the skin as it evaporates.

Cohesion works inside organisms, too. Consider how plants absorb water from soil as they grow. Water molecules evaporate from leaves, and replacements are pulled upward from roots. Cohesion makes it possible for columns of liquid water to rise from roots to leaves inside narrow pipelines of vascular tissue. In some trees, these pipelines extend hundreds of feet above the soil (Section 28.4 returns to this topic).

---

cohesion Property of a substance that arises from the tendency of its molecules to resist separating from one another.
concentration Amount of solute per unit volume of solution.
evaporation Transition of a liquid to a vapor.
hydrogen bond Attraction between a covalently bonded hydrogen atom and another atom taking part in a separate covalent bond.
hydrophilic Describes a substance that dissolves easily in water.
hydrophobic Describes a substance that resists dissolving in water.
salt Ionic compound that releases ions other than H^+ and OH^- when it dissolves in water.
solute A dissolved substance.
solution Uniform mixture of solute completely dissolved in solvent.
solvent Liquid in which other substances dissolve.
temperature Measure of molecular motion.

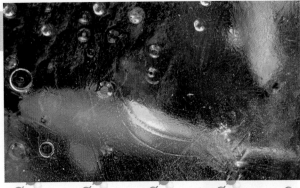

FIGURE 2.10 Hydrogen bonds lock water molecules in a rigid lattice in ice. The molecules in this lattice pack less densely than in liquid water, which is why ice floats on water. A covering of ice can insulate water underneath it, thus keeping aquatic organisms from freezing during cold winters.

Water Stabilizes Temperature All atoms jiggle nonstop, so the molecules they make up jiggle too. We measure the energy of this motion as degrees of **temperature**. Adding energy (in the form of heat, for example) makes the jiggling faster, so the temperature rises. Hydrogen bonding keeps water molecules from moving as much as they would otherwise, so it takes more heat to raise the temperature of water compared with other liquids. Temperature stability is an important part of homeostasis, because most of the molecules of life function properly only within a certain range of temperature.

Below 0°C (32°F), water molecules do not jiggle enough to break hydrogen bonds between them, and they become locked in the rigid, lattice-like bonding pattern of ice (**FIGURE 2.10**). Individual water molecules pack less densely in ice than they do in water, which is why ice floats on water. Sheets of ice that form on the surface of ponds, lakes, and streams can insulate the water under them from subfreezing air temperatures. Such "ice blankets" protect aquatic organisms during long, cold winters.

TAKE-HOME MESSAGE 2.5

What gives water the special properties that make life possible?

✔ Extensive hydrogen bonding among water molecules arises from the polarity of the individual molecules.

✔ Hydrogen bonding among water molecules imparts cohesion to liquid water, and gives it the ability to stabilize temperature and dissolve many substances.

2.6 Acids and Bases

✔ The number of hydrogen ions in a fluid is measured as pH.

✔ Most biological processes occur within a narrow range of pH, typically around pH 7.

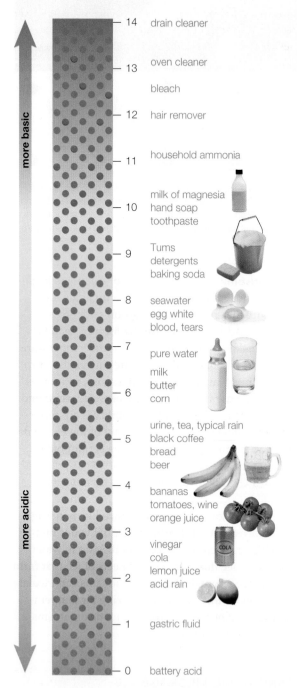

FIGURE 2.11 A pH scale. Here, red dots signify hydrogen ions (H⁺) and gray dots signify hydroxyl ions (OH⁻). Also shown are approximate pH values for some common solutions.

This pH scale ranges from 0 (most acidic) to 14 (most basic). A change of one unit on the scale corresponds to a tenfold change in the amount of H⁺ ions.

FIGURE IT OUT What is the approximate pH of cola?

Answer: 2.5

A hydrogen atom, remember, is just a proton and an electron. When a hydrogen atom participates in a polar covalent bond with a more electronegative atom (such as oxygen), the electron is pulled away from the proton, just a bit. Hydrogen bonding in water tugs on that proton even more, so much that the proton can be pulled right off of the molecule. The electron stays with the rest of the molecule, which becomes negatively charged (ionic), and the proton becomes a hydrogen ion (H^+). For example, a water molecule that loses a proton becomes a hydroxyl ion (OH^-):

$$H_2O \longrightarrow OH^- + H^+$$

water molecule — hydroxide ion — hydrogen ion

The loss is more or less temporary, because these two ions easily get back together to form a water molecule:

$$OH^- + H^+ \longrightarrow H_2O$$

hydroxide ion — hydrogen ion — water molecule

With other molecules, the loss of a hydrogen ion in water is essentially permanent. We use a value called **pH** to measure of the number of hydrogen ions floating around in a water-based fluid. In pure water, the number of H^+ ions is the same as the number of OH^- ions, and the pH is 7, or neutral. The higher the number of hydrogen ions, the lower the pH. A one-unit decrease in pH corresponds to a tenfold increase in the number of H^+ ions, and a one-unit increase corresponds to a tenfold decrease in the number of H^+ ions (**FIGURE 2.11**). One way to get a sense of the pH scale is to taste dissolved baking soda (pH 9), distilled water (pH 7), and lemon juice (pH 2).

An **acid** is a substance that gives up hydrogen ions in water. Acids can lower the pH of a solution and make it acidic (below pH 7). **Bases** accept hydrogen ions from water, so they can raise the pH of a solution and make it basic, or alkaline (above pH 7).

Strong acids ionize completely in water to give up all of their H^+ ions; weak acids give up only some of them. Hydrochloric acid (HCl) is an example of a strong acid. When HCl dissolves in water, all of the molecules give up hydrogen ions, leaving Cl^- ions behind. Hydrogen ions released from HCl makes gastric fluid inside your stomach very acidic (pH 1–2).

---

acid Substance that releases hydrogen ions in water.
base Substance that accepts hydrogen ions in water.
buffer Set of chemicals that can keep the pH of a solution stable by alternately donating and accepting ions that contribute to pH.
pH Measure of the number of hydrogen ions in a fluid.

Mercury Rising (revisited)

All ecosystems now have detectable effects of air pollution, but many of those effects are not as well understood as acid rain. We do know that the concentration of mercury in Earth's waters is rising, and is predicted to double within forty years. We also know that this rise is occurring as a consequence of human activities, which release more than 2,000 tons of mercury into the atmosphere every year.

All human bodies now have detectable amounts of mercury; the average adult living in the U.S. has about 4 micrograms of it circulating in his or her blood. Some comes from dental fillings, imported skin-bleaching cosmetics, and broken fluorescent lamps. However, most comes from dietary seafood: The more fish and shellfish you eat, the more mercury your body has. A diet that consists of a high proportion of seafood can result in a blood mercury content thirty times the average.

Carbonic acid forms when carbon dioxide gas dissolves in plasma, the fluid portion of human blood:

$$CO_2 + H_2O \longrightarrow H_2CO_3$$
carbon dioxide · water molecule · carbonic acid

Carbonic acid is a weak acid, so only some of its molecules give up a hydrogen ion in water. When carbonic acid loses a hydrogen ion, it becomes an ionic molecule called bicarbonate:

$$H_2CO_3 \longrightarrow H^+ + HCO_3^-$$
carbonic acid · hydrogen ion · bicarbonate

Bicarbonate can act like a base by accepting a hydrogen ion. When it does, carbonic acid forms again:

$$H^+ + HCO_3^- \longrightarrow H_2CO_3$$
hydrogen ion · bicarbonate · carbonic acid

Together, carbonic acid and bicarbonate constitute a buffer. A **buffer** is a set of chemicals that can keep the pH of a solution stable by alternately donating and accepting ions that contribute to pH. Consider how the pH of pure water rises when a base is added to it. This is because the base accepts hydrogen ions from the water, thus reducing the number of hydrogen ions floating around in it (and contributing to pH). By contrast, the carbonic acid–bicarbonate buffer system can keep the pH of blood plasma from rising when base is added. The added base causes carbonic acid to give up hydrogen ions (and become bicarbonate). These hydrogen ions replace the ones that the base removed from the solution. The same buffer system can also keep plasma pH from declining when acid is added. Hydrogen ions released by the acid combine with bicarbonate, so they do not contribute to pH. In both cases, the proportion of carbonic acid and bicarbonate molecules in plasma shifts, but the pH stays stable (typically between 7.3 and 7.5).

The addition of too much acid or base can overwhelm a buffer's capacity to stabilize pH. Such buffer failure can be catastrophic in a cell or body because most biological molecules function properly only within a narrow range of pH: Even a slight deviation from that range can halt cellular processes. Consider what happens when breathing is impaired suddenly. Carbon dioxide gas accumulates in tissues, and too much carbonic acid forms in plasma. If the excess acid reduces blood pH below 7.3, a dangerous level of unconsciousness called coma can be the outcome. By contrast, hyperventilation (sustained rapid breathing) causes the body to lose too much CO_2. The loss results in a rise in blood pH. If blood pH rises too much, prolonged muscle spasm (tetany) or coma may occur.

Burning fossil fuels such as coal releases sulfur and nitrogen compounds that affect the pH of rain and other forms of precipitation. These fluids are not buffered, so the addition of acids or bases has a dramatic  effect. In places with a lot of fossil fuel emissions, the rain and fog can be more acidic than vinegar. The corrosive effect of this acid rain is visible in urban areas (left). Acid rain also drastically changes the pH of water in soil, lakes, and streams. Such changes can overwhelm the buffering capacity of fluids inside organisms that live in these environments, with lethal effects. Section 48.5 returns to the topic of acid rain.

TAKE-HOME MESSAGE 2.6

Why are hydrogen ions important in biological systems?

✔ The number of hydrogen ions in a fluid determines its pH. Most biological systems function properly only within a narrow range of pH.

✔ Acids release hydrogen ions in water; bases accept them.

✔ Buffers help keep pH stable. Inside organisms, they play a role in homeostasis.

CREDITS: (in text revisited) Nanisimova/Shutterstock; (in text) left, © Cengage Learning; right, W. K. Fletcher/Science Source.

Section 2.1 Interactions between atoms make the molecules that sustain life, and also some that destroy it. Mercury in air pollution ends up in the bodies of fish, and in turn, in the bodies of humans.

Section 2.2 Atoms consist of **electrons**, which carry a negative **charge**, moving about a **nucleus** of positively charged **protons** and uncharged **neutrons** (**TABLE 2.2**). The **periodic table** lists **elements** in order of **atomic number**. **Isotopes** of an element differ in the number of neutrons. The total number of protons and neutrons is the **mass number**. **Tracers** can be made with **radioisotopes**, which, by a process called **radioactive decay**, emit particles and energy when their nucleus spontaneously breaks up.

Section 2.3 Up to two electrons occupy each orbital (volume of space around a nucleus). Which orbital an electron occupies depends on its energy. A **shell model** represents successive energy levels as concentric

circles. Atoms are in their most stable state when all of their shells are full, so they tend to get rid of vacancies. Many can do so by gaining or losing electrons, thereby becoming charged **ions**. Atoms with unpaired electrons are called **free radicals**. Most free radicals are highly chemically active, easily forcing electrons onto other atoms or pulling electrons away from them.

Section 2.4 A **chemical bond** unites two atoms in a molecule. A **compound** is a molecule that consists of two or more elements. Atoms form different types of bonds depending on their **electronegativity**. An **ionic bond** is a strong association between oppositely charged ions; it arises from the mutual attraction of opposite charges. **Polarity** is a separation of charge into positive and negative regions. Ionic bonds are completely polar. Atoms share a pair of electrons in a **covalent bond**, which is nonpolar if the sharing is equal, and polar if it is not.

Section 2.5 Two polar covalent bonds give each water molecule an overall polarity. **Hydrogen bonds** that form among water molecules in tremendous numbers are the basis of water's unique life-sustaining properties: **cohesion**, resistance to **temperature** change, and the ability to act as a **solvent** that dissolves **salts** and other **solutes**. **Hydrophilic** substances dissolve easily in water to form **solutions**; **hydrophobic** substances do not. The amount of solute in a given volume of fluid is the solute's **concentration**. **Evaporation** is the transition of a liquid to vapor.

Section 2.6 **pH** is a measure of the number of hydrogen ions (H^+) in a liquid. At neutral pH (7), there are an equal number of H^+ and OH^- ions. **Acids** release hydrogen ions in water; **bases** accept them. A **buffer** can keep a solution within a consistent range of pH. Most cell and body fluids are buffered because most molecules of life work only within a narrow range of pH.

Table 2.2 Players in the Chemistry of Life

Atoms	Particles that are basic building blocks of all matter.
Proton (p^+)	Positively charged subatomic particle in the nucleus.
Electron (e^-)	Negatively charged subatomic particle that can occupy a defined volume of space (orbital) around the nucleus.
Neutron	Uncharged subatomic particle of the nucleus.
Element	Pure substance that consists entirely of atoms with the same, characteristic number of protons.
Isotopes	Atoms of an element that differ in the number of neutrons.
Radioisotope	Isotope with an unstable nucleus that emits radiation when it decays (breaks up).
Tracer	Substance with a detectable component (such as a radioisotope) used to track its movement or destination in a biological system.
Ion	Atom that carries a charge after it has gained or lost one or more electrons. A single proton without an electron is a hydrogen ion (H^+).
Molecule	Two or more atoms joined in a chemical bond.
Compound	Molecule of two or more different elements in unvarying proportions (for example, water: H_2O).
Solute	Substance dissolved in a solvent.
Hydrophilic	Refers to a substance that dissolves easily in water.
Hydrophobic	Refers to a substance that resists dissolving in water.
Acid	Compound that releases H^+ when dissolved in water.
Base	Compound that accepts H^+ when dissolved in water.
Salt	Ionic compound that releases ions other than H^+ or OH^- when dissolved in water.
Solvent	Substance that can dissolve other substances.
Buffer	Set of chemicals that can stabilize the pH of a fluid.

self-quiz

Answers in Appendix VII

1. What atom has only one proton?
   a. hydrogen
   b. an isotope
   c. a free radical
   d. a radioisotope

2. A molecule into which a radioisotope has been incorporated can be used as a(n) _____ .
   a. compound
   b. tracer
   c. salt
   d. acid

3. Which of the following statements is incorrect?
   a. Isotopes have the same atomic number and different mass numbers.
   b. Atoms have about the same number of electrons as protons.
   c. All ions are atoms.
   d. Free radicals are dangerous because they emit energy.

data analysis activities

Mercury Emissions by Continent By weight, coal does not contain much mercury, but we burn a lot of it. Several industries besides coal-fired power plants contribute substantially to atmospheric mercury pollution. **FIGURE 2.12** shows mercury emissions by industry from different regions of the world in 2006.

1. About how many tons of mercury were released?

2. Which industry tops the list of mercury emitters? Which industry is next on the list?

3. Which region emitted the most mercury from producing cement?

4. About how many tons of mercury were released from gold production in South America?

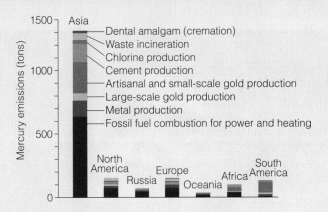

FIGURE 2.12 Global mercury emissions, 2006.

---

4. In the periodic table, symbols for the elements are arranged according to _____ .
   a. size
   b. charge
   c. mass number
   d. atomic number

5. An ion is an atom that has _____ .
   a. the same number of electrons and protons
   b. a different number of electrons and protons
   c. electrons, protons, and neutrons

6. The measure of an atom's ability to pull electrons away from another atom is called _____ .
   a. electronegativity
   b. charge
   c. polarity

7. The mutual attraction of opposite charges holds atoms together as molecules in a(n) _____ bond.
   a. ionic
   b. hydrogen
   c. polar covalent
   d. nonpolar covalent

8. Atoms share electrons unequally in a(n) _____ bond.
   a. ionic
   b. hydrogen
   c. polar covalent
   d. nonpolar covalent

9. A(n) _____ substance repels water.
   a. acidic
   b. basic
   c. hydrophobic
   d. polar

10. A salt does not release _____ in water.
    a. ions
    b. energy
    c. H^+

11. Hydrogen ions (H^+) are _____ .
    a. in blood
    b. protons
    c. indicated by a pH scale
    d. all of the above

12. When dissolved in water, a(n) _____ donates H^+; a(n) _____ accepts H^+.
    a. acid; base
    b. base; acid
    c. buffer; solute
    d. base; buffer

13. A _____ can help keep the pH of a solution stable.
    a. covalent bond
    b. hydrogen bond
    c. buffer
    d. pH

14. A _____ is dissolved in a solvent.
    a. molecule
    b. solute
    c. salt

15. Match the terms with their most suitable description.
    ___ hydrophilic   a. protons > electrons
    ___ atomic number   b. number of protons in nucleus
    ___ hydrogen bonds   c. polar; dissolves easily in water
    ___ positive charge   d. collectively strong
    ___ temperature   e. protons < electrons
    ___ negative charge   f. measure of molecular motion

critical thinking

1. Alchemists were medieval scholars and philosophers who were the forerunners of modern-day chemists. Many spent their lives trying to transform lead (atomic number 82) into gold (atomic number 79). Explain why they never did succeed in that endeavor.

2. Draw a shell model of a lithium atom (Li), which has 3 protons. Predict whether the majority of lithium atoms on Earth are uncharged, positively charged, or negatively charged.

3. Polonium is a rare element with 33 radioisotopes. The most common one, ^{210}Po, has 82 protons and 128 neutrons. When ^{210}Po decays, it emits an alpha particle, which is a helium nucleus (2 protons and 2 neutrons). ^{210}Po decay is tricky to detect because alpha particles do not carry very much energy compared to other forms of radiation. They can be stopped by, for example, a sheet of paper or a few inches of air. This property is one reason why authorities failed to discover toxic amounts of ^{210}Po in the body of former KGB agent Alexander Litvinenko until after he died suddenly and mysteriously in 2006. What element does an atom of ^{210}Po change into after it emits an alpha particle?

4. Some undiluted acids are not as corrosive as when they are diluted with water. That is why lab workers are told to wipe off splashes with a towel before washing. Explain.

CENGAGE To access course materials, please visit
brain.com www.cengagebrain.com

SOURCE: (12) Global Atmospheric Mercury Assessment: Sources, Emissions and Transport, United Nations Environmental Programme, Chemicals Branch. 2008

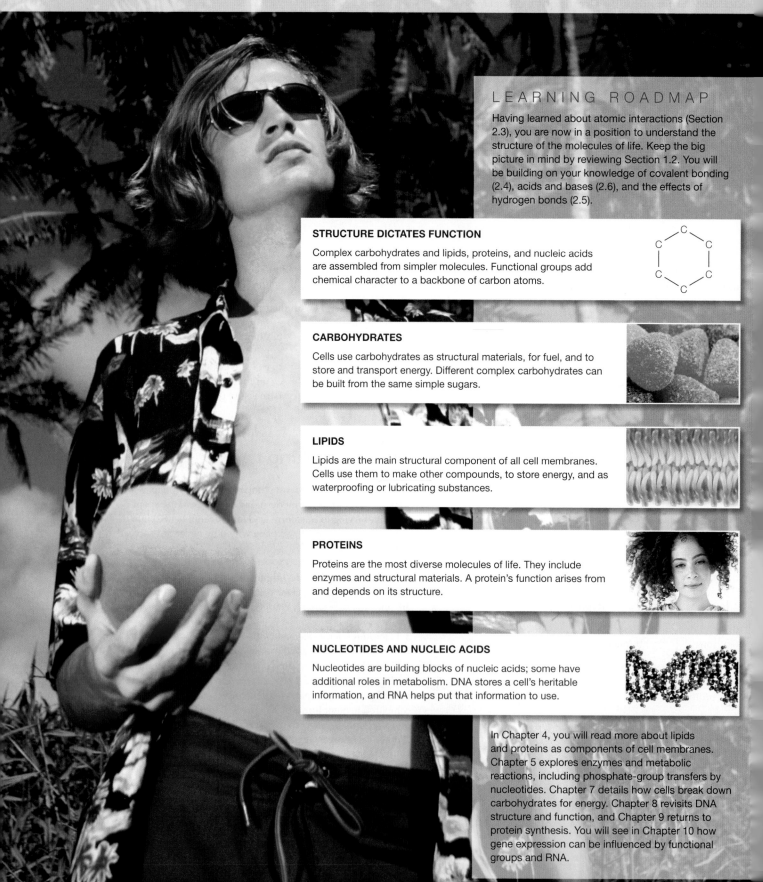

STRUCTURE DICTATES FUNCTION

Complex carbohydrates and lipids, proteins, and nucleic acids are assembled from simpler molecules. Functional groups add chemical character to a backbone of carbon atoms.

CARBOHYDRATES

Cells use carbohydrates as structural materials, for fuel, and to store and transport energy. Different complex carbohydrates can be built from the same simple sugars.

LIPIDS

Lipids are the main structural component of all cell membranes. Cells use them to make other compounds, to store energy, and as waterproofing or lubricating substances.

PROTEINS

Proteins are the most diverse molecules of life. They include enzymes and structural materials. A protein's function arises from and depends on its structure.

NUCLEOTIDES AND NUCLEIC ACIDS

Nucleotides are building blocks of nucleic acids; some have additional roles in metabolism. DNA stores a cell's heritable information, and RNA helps put that information to use.

In Chapter 4, you will read more about lipids and proteins as components of cell membranes. Chapter 5 explores enzymes and metabolic reactions, including phosphate-group transfers by nucleotides. Chapter 7 details how cells break down carbohydrates for energy. Chapter 8 revisits DNA structure and function, and Chapter 9 returns to protein synthesis. You will see in Chapter 10 how gene expression can be influenced by functional groups and RNA.

The human body requires only about a tablespoon of fat each day to stay healthy, but most people in developed countries eat far more than that. The average American eats about 70 pounds of fat per year, which may be part of the reason why the average American is overweight. Being overweight increases one's risk for many chronic illnesses. However, the total quantity of fat in the diet may have less impact on health than the types of fats. Fats are more than inert molecules that accumulate in strategic areas of our bodies. They are the main constituents of cell membranes, and as such they have powerful effects on cell function.

The typical fat molecule has three fatty acid tails, each a long chain of carbon atoms that can vary a bit in structure. Fats with a certain arrangement of hydrogen atoms around those carbon chains are called *trans* fats (**FIGURE 3.1**). Small amounts of *trans* fats occur naturally in red meat and dairy products. However, the main source of these fats in the American diet is an artificial food product called partially hydrogenated vegetable oil.

Hydrogenation is a manufacturing process that adds hydrogen atoms to oils in order to change them into solid fats. In 1908, Procter & Gamble Co. developed partially hydrogenated soybean oil as a substitute for the more expensive solid animal fats they had been using to make candles. However, the demand for candles began to wane as more households in the United States became wired for electricity, and P & G began to look for another way to sell its proprietary fat. Partially hydrogenated vegetable oil looks a lot like lard, so in 1911 the company began aggressively marketing it as a revolutionary new food: a solid cooking fat with a long shelf life, mild flavor, and lower cost than lard or butter.

By the mid-1950s, hydrogenated vegetable oil had become a major part of the American diet. At this writing, it can still be found in many manufactured and fast foods: stick margarines, ready-to-use frostings, french fries, cookies, crackers, cakes and pancakes, peanut butter, pies, doughnuts, muffins, chips, granola bars, breakfast bars, chocolate, microwave popcorn, pizzas, burritos, chicken nuggets, fish sticks, and so on.

For decades, hydrogenated vegetable oil was considered more healthy than animal fats because it was made from plants, but we now know otherwise. The *trans* fats in hydrogenated vegetable oils raise the level of cholesterol in our blood more than any other fat, and they directly alter the function of our arteries and veins. The effects of such changes are quite serious.

Eating as little as 2 grams per day (about 0.4 teaspoon) of hydrogenated vegetable oil measurably increases one's risk of atherosclerosis (hardening of the arteries), heart attack, and diabetes. A small serving of french fries made with hydrogenated vegetable oil contains about 5 grams of *trans* fat.

All organisms consist of the same kinds of molecules, but small differences in the way those molecules are put together can have big effects. With this concept, we introduce you to the chemistry of life. This is your chemistry. It makes you far more than the sum of your body's molecules.

oleic acid has a *cis* bond:

elaidic acid has a *trans* bond:

FIGURE 3.1 *Trans* fats, an unhealthy food. Double bonds in the tail of most naturally occurring fatty acids are *cis*, which means that the two hydrogen atoms flanking the bond are on the same side of the carbon backbone. Hydrogenation creates abundant *trans* bonds, with hydrogen atoms on opposite sides of the tail.

3.2 Organic Molecules

✔ All of the molecules of life are built with carbon atoms.
✔ We use different models to highlight different aspects of the same molecule.

Carbon: The Stuff of Life

The same elements that make up a living body also occur in nonliving things, but their proportions differ.

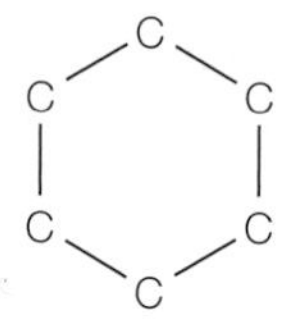

A Carbon's versatile bonding behavior allows it to form a variety of structures, including rings.

B Carbon rings form the framework of many sugars, starches, and fats, such as those found in doughnuts.

FIGURE 3.2 Carbon rings.

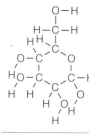

A A structural formula for an organic molecule—even a simple one—can be very complicated. The overall structure is obscured by detail.

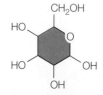

B Structural formulas of organic molecules are typically simplified by using polygons as symbols for rings, omitting some bonds and element labels.

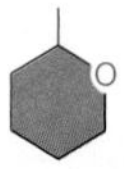

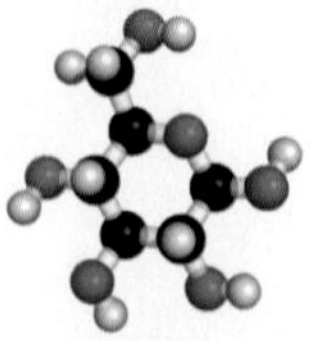

C A ball-and-stick model is often used to show the arrangement of atoms and bonds in three dimensions.

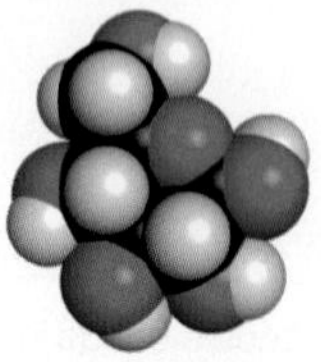

D A space-filling model can be used to show a molecule's overall shape. Individual atoms are visible in this model.

FIGURE 3.3 Modeling an organic molecule. All of these models represent the same molecule: glucose.

For example, compared to sand or seawater, a human body has a much larger proportion of carbon atoms. Why? Unlike sand or seawater, a body contains a lot of the molecules of life—complex carbohydrates and lipids, proteins, and nucleic acids—and these molecules consist of a high proportion of carbon atoms.

Compounds that consist primarily of carbon and hydrogen atoms are said to be **organic**. The term is a holdover from a time when these molecules were thought to be made only by living things, as opposed to the "inorganic" molecules that formed by nonliving processes. We now know that organic compounds were present on Earth long before organisms were, and we can also make them synthetically in laboratories.

Carbon's importance to life arises from its versatile bonding behavior. Carbon has four vacancies in its outer shell (Section 2.3), so it can form four covalent bonds with other atoms, including other carbon atoms. Many organic molecules have a backbone—a chain of carbon atoms—to which other atoms attach. The ends of a backbone may join to form a carbon ring structure (**FIGURE 3.2**). Carbon's ability to form chains and rings, and also to bond with many other elements, means that atoms of this element can be assembled into a wide variety of organic compounds.

Modeling Organic Molecules

As you will see in the next few sections, the function of an organic molecule depends on its structure. Researchers routinely make models of organic molecules such as proteins in order to study (for example) surface properties, structure–function relationships, changes during synthesis or other biochemical processes, and molecular recognition. A molecule's structure can be modeled in various ways. The different models allow us to visualize different characteristics of the same molecule.

Structural formulas of organic molecules can be quite complex, even when the molecules are relatively small (**FIGURE 3.3A**). Thus, formulas of organic molecules are typically simplified. Hydrogen atoms and some of the bonds may not be shown, but are understood to exist where they should. Carbon ring structures such as the ones that occur in glucose and other sugars are often represented as polygons (**FIGURE 3.3B**). If no atom is shown at a corner or at the end of a bond, a carbon atom is implied there.

organic Describes a compound that consists mainly of carbon and hydrogen atoms.

Ball-and-stick models show the positions of individual atoms in three dimensions (**FIGURE 3.3C**). Single, double, and triple covalent bonds are all shown as one stick connecting two balls, which represent atoms. Ball size reflects relative sizes of the atoms, and ball color indicates the element according to a standard code (Section 2.4).

Space-filling models represent atomic volume most accurately (**FIGURE 3.3D**). This type of model shows the overall shape of an organic molecule. Atoms in space-filling models may be color-coded by element using the same scheme as ball-and-stick models.

Many organic molecules are so large that ball-and-stick or space-filling models of them may be incomprehensible. **FIGURE 3.4** shows three different ways to represent hemoglobin, a large molecule that functions as the main oxygen carrier in your blood. Many interesting features of this molecule are not visible in the space-filling model (**FIGURE 3.4A**). Consider that a properly functioning hemoglobin molecule has embedded hemes, which are small carbon-ring structures with an iron atom at their center (Section 5.6 returns to hemes). The hemes are impossible to distinguish in a space-filling model of hemoglobin, but become visible when depicted as in **FIGURE 3.4B**, which shows a surface model. This model reveals the hemes (red sticks) within the molecule's crevices. Surface models are often used to highlight large-scale features such as charge distribution that can be difficult to distinguish in models depicting individual atoms. Other types of models further reduce visual complexity. Proteins and large nucleic acids are typically represented as ribbons that show only the carbon backbone. In a ribbon model of hemoglobin (**FIGURE 3.4C**), you can see that the molecule consists of four coiled protein components, each folded around a heme. Such structural details are clues about function: Oxygen binds at the hemes, so each hemoglobin molecule can carry up to four molecules of oxygen.

A The complexity of a space-filling model of hemoglobin obscures many interesting features of the molecule.

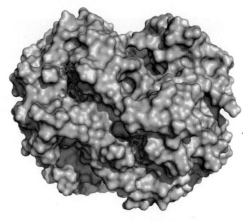

B A surface model of the same molecule reveals crevices and folds that are important for its function. Hemes, in red, are cradled in pockets of the molecule.

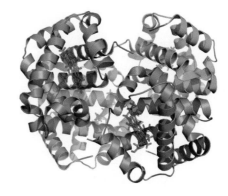

C A ribbon model of hemoglobin reveals all four hemes, also in red. The hemes are held in place by the coiled backbones of the molecule's four protein components.

FIGURE 3.4 Visualizing the structure of hemoglobin, the oxygen-transporting molecule in human blood. Models that show individual atoms usually depict them color-coded by element. Other models may be shown in various colors, depending on which features are being highlighted.

TAKE-HOME MESSAGE 3.2

How are all of the molecules of life alike?

✔ The molecules of life (carbohydrates, lipids, proteins, and nucleic acids) are organic, which means they consist mainly of carbon and hydrogen atoms.

✔ The structure of an organic molecule starts with a chain of carbon atoms (the backbone) that may form a ring.

✔ We use different models to represent different structural characteristics. Considering a molecule's structural features gives us insight into how it functions.

CREDITS: (4A) © National Cancer Institute; (4B) Hemoglobin models: PDF ID: 1GZX; Paoli, M., Liddington, R., Tame, J., Wilkinson, A., Dodson, G., Crystal structure of T state hemoglobin with oxygen bound at all four hems. J.Mol.Bio., v256, pp. 775–792, 1996; (4C) 1BBB, A third quaternary structure of human hemoglobin A at 1.7-A resolution. Silva, M.M., Rogers, P.H., Arnone, A., Journal; (1992) J. Biol. Chem. 267:17248-17256.

3.3 Molecules of Life—From Structure to Function

✔ How an organic molecule functions in a biological system begins with its structure.

Functional Groups

An organic molecule that consists only of hydrogen and carbon atoms is called a **hydrocarbon**. Methane, the simplest hydrocarbon, is one carbon atom bonded to four hydrogen atoms. Other organic molecules, including the molecules of life, have at least one functional group. A **functional group** is an atom (other than hydrogen) or small molecular group covalently bonded to a carbon atom of an organic compound. These groups impart chemical properties such as acidity or polarity.

The chemical behavior of the molecules of life arises mainly from the number, kind, and arrangement of their functional groups. **TABLE 3.1** lists some of the most common functional groups in these molecules. A hydroxyl group adds polar character to an organic compound, thus increasing its solubility in water. A methyl group adds nonpolar character, and may dampen the effect of a polar functional group. Methyl groups added to DNA act like an "off" switch for this molecule; acetyl groups act like an "on" switch (we return to this topic in Chapter 10). Acetyl groups also carry two carbons from one molecule to another in some metabolic reactions.

Table 3.1
Some Functional Groups in Biological Molecules

Group	Structure	Character	Formula	Found In
acetyl	—C(=O)—CH_3	polar, acidic	$-COCH_3$	some proteins, coenzymes
aldehyde	—C—C(=O)—H	polar, reactive	—CHO	simple sugars
amide	—C(=O)—N<	weakly basic, stable, rigid	—C(O)N—	proteins, nucleotide bases
amine	—N(H)—H	very basic	$-NH_2$	nucleotide bases, amino acids
carboxyl	—C—C(=O)—OH	very acidic	—COOH	fatty acids, amino acids
carbonyl	—C(=O)—	polar, reactive	—CO	alcohols, other functional groups
hydroxyl	—O—H	polar	—OH	alcohols, sugars
ketone	—C—C(=O)—C—	polar, acidic	—CO—	simple sugars, nucleotide bases
methyl	—C(H)(H)—H	nonpolar	$-CH_3$	fatty acids, some amino acids
sulfhydryl	—S—H	forms rigid disulfide bonds	—SH	cysteine, many cofactors
phosphate	—O—P(=O)(O⁻)—O⁻	polar, reactive	$-PO_4$	nucleotides, DNA, RNA phospholipids, proteins

Aldehyde and ketone groups are part of simple sugars. Some sugars convert to a ring form when the highly reactive aldehyde group on one carbon of the backbone reacts with a hydroxyl group on another (**FIGURE 3.5**). Carboxyl groups make amino acids and fatty acids acidic; amine and amide groups make nucleotide bases basic.

When a phosphate group is transferred from one molecule to another, energy is transferred along with it. Bonds between sulfhydryl groups stabilize the structure of many proteins, including those that make up human hair. Heat and some kinds of chemicals can temporarily break sulfhydryl bonds, which is why we can curl straight hair and straighten curly hair.

What Cells Do to Organic Compounds

All biological systems are based on the same organic molecules, a similarity that is one of many legacies of life's common origin. However, the details of those molecules differ among organisms. Just as atoms

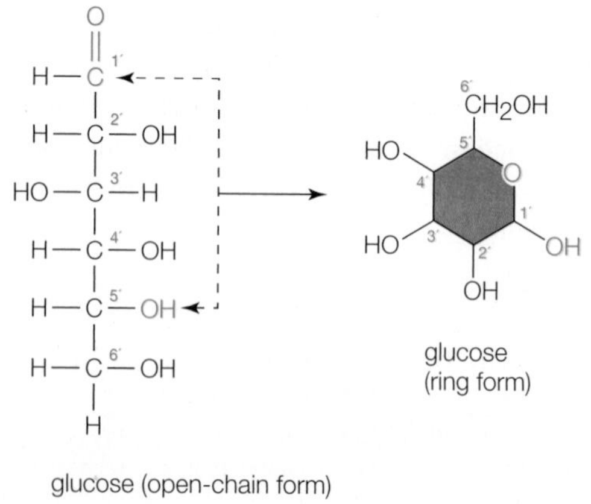

FIGURE 3.5 Glucose. This simple sugar converts from a straight-chain into a ring form when the aldehyde group (on carbon 1) reacts with a hydroxyl group (on carbon 5). In water, the cyclic structure is the more common one.

Note that the carbons in sugars such as glucose are numbered in a standard way: 1′, 2′, 3′, and so on.

A Metabolism refers to processes by which cells acquire and use energy as they make and break down molecules. Humans and other consumers break down the molecules in food. They use energy and raw materials from the breakdown to maintain themselves and to build new components.

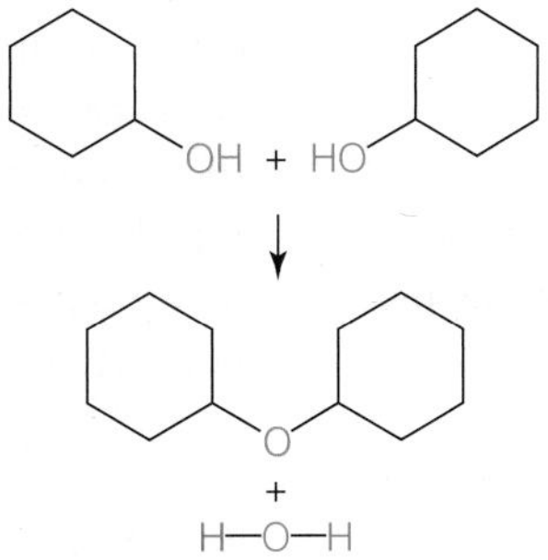

B Condensation. Cells build a large molecule from smaller ones by this reaction. An enzyme removes a hydroxyl group from one molecule and a hydrogen atom from another. A covalent bond forms between the two molecules; water also forms.

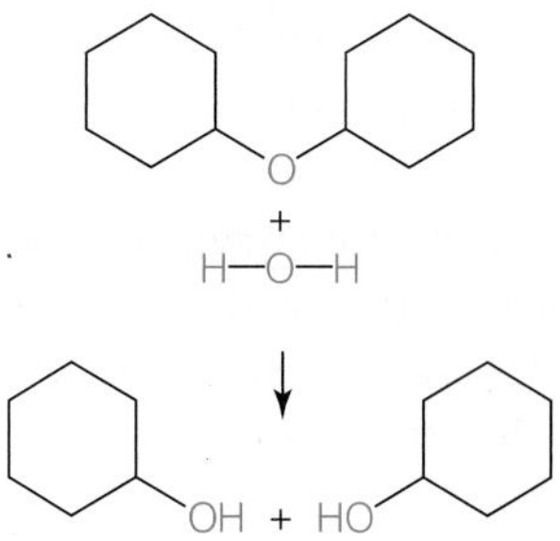

C Hydrolysis. Cells split a large molecule into smaller ones by this water-requiring reaction. An enzyme attaches a hydroxyl group and a hydrogen atom (both from water) at the cleavage site.

FIGURE 3.6 Metabolism. Two common reactions by which cells build and break down organic molecules are shown.

bonded in different numbers and arrangements form different molecules, simple organic building blocks bonded in different numbers and arrangements form different versions of the molecules of life. Cells assemble complex carbohydrates, lipids, proteins, and nucleic acids from small organic molecules. These small organic molecules—simple sugars, fatty acids, amino acids, and nucleotides—are called **monomers** when they are used as subunits of larger molecules. A molecule that consists of multiple monomers is called a **polymer**.

Cells build polymers from monomers, and break down polymers to release monomers. These and other processes of molecular change are called **reactions**. Cells constantly run reactions as they acquire and use energy to stay alive, grow, and reproduce—activities that are collectively called **metabolism** (**FIGURE 3.6A**).

Metabolism requires **enzymes**, which are organic molecules (usually proteins) that speed up reactions without being changed by them. Enzymes drive metabolic reactions in which large organic molecules are assembled from smaller ones. With **condensation**, an enzyme covalently bonds two molecules together. Water (H—O—H) forms as a product of condensation when a hydrogen atom (H—) from one of the molecules combines with a hydroxyl group (—OH) from the other molecule (**FIGURE 3.6B**). With **hydrolysis**, the reverse of condensation, an enzyme breaks apart a large organic molecule into smaller ones. During hydrolysis, a bond between two atoms breaks when a hydroxyl group gets attached to one of the atoms, and a hydrogen atom gets attached to the other (**FIGURE 3.6C**). The hydroxyl group and hydrogen atom come from a water molecule, so this reaction requires water.

We will revisit enzymes and metabolic reactions in Chapter 5. The remainder of this chapter introduces the different types of biological molecules and the monomers from which they are built.

condensation Chemical reaction in which an enzyme builds a large molecule from smaller subunits; water also forms.
enzyme Organic molecule that speeds up a reaction without being changed by it.
functional group An atom (other than hydrogen) or a small molecular group bonded to a carbon of an organic compound; imparts a specific chemical property.
hydrocarbon Compound or region of one that consists only of carbon and hydrogen atoms.
hydrolysis Water-requiring chemical reaction in which an enzyme breaks a molecule into smaller subunits.
metabolism All of the enzyme-mediated chemical reactions by which cells build and break down organic molecules.
monomers Molecules that are subunits of polymers.
polymer Molecule that consists of multiple monomers.
reaction Process of molecular change.

TAKE-HOME MESSAGE 3.3
How do organic molecules work in living systems?

✔ Functional groups of an organic molecule impart chemical characteristics to its hydrocarbon backbone. These groups contribute to the function of a biological molecule.

✔ An organic molecule's structure dictates its function in biological systems.

✔ All life is based on the same types of organic compounds: complex carbohydrates, lipids, proteins, and nucleic acids.

✔ By processes of metabolism, cells assemble the molecules of life from monomers. They also break apart polymers into component monomers.

3.4 Carbohydrates

✔ Carbohydrates are the most plentiful biological molecules.

✔ Cells use some carbohydrates as structural materials; they use others for fuel, or to store or transport energy.

Carbohydrates are organic compounds that consist of carbon, hydrogen, and oxygen in a 1:2:1 ratio. Cells use different kinds as structural materials, for fuel, and for storing and transporting energy.

Carbohydrates in Biological Systems

Simple Sugars "Saccharide" is from *sacchar*, a Greek word that means sugar. **Monosaccharides** (one sugar) are the simplest type of carbohydrate. These molecules have important biological roles. Common monosaccharides have a backbone of five or six carbon atoms, one carbonyl group (—C=O), and two or more hydroxyl groups (—OH). The functional groups, which are polar, make monosaccharides soluble (able to dissolve) in water. Thus, these molecules move easily through the water-based internal environments of all organisms.

Cells break the bonds of sugars to release energy that can be harnessed to power other reactions (we return to this important metabolic process in Chapter 7). Monosaccharides are also used as structural materials to build larger molecules, and as precursors, or parent molecules, that are remodeled into other molecules. For example, cells of plants and many animals make vitamin C from glucose, which is a monosaccharide. Human cells are unable to make vitamin C, so we need to get it from our food.

Short-Chain Carbohydrates Oligosaccharides are short chains of covalently bonded monosaccharides (*oligo*– means a few). **Disaccharides** consist of two sugar monomers. The lactose in milk, with one glucose and one galactose, is a disaccharide. Sucrose, the most plentiful sugar in nature, has a glucose and a fructose unit (**FIGURE 3.7**). Sucrose extracted from sugarcane or sugar beets is our table sugar. Oligosaccharides attached to lipids or proteins function in immunity.

Complex Carbohydrates Foods that we call "complex" carbohydrates consist mainly of **polysaccharides**, which are chains of hundreds or thousands of monosaccharide monomers. The chains may be straight or branched, and can have one or many types of monosaccharides. The most common polysaccharides are cellulose, starch, and glycogen. All consist only of glucose monomers, but as substances their properties are very different. Why? The answer begins with differences in patterns of covalent bonding that link their monomers.

Cellulose, the major structural material of plants, is the most abundant biological molecule on Earth. Hydrogen bonding locks its long, straight chains of covalently bonded glucose monomers into tight, sturdy bundles (**FIGURE 3.8A**). The bundles form tough fibers that act like reinforcing rods inside stems and other plant parts, helping these structures resist wind and other forms of mechanical stress. Cellulose is insoluble (it does not dissolve) in water, and it is not easily broken down. Some bacteria and fungi make enzymes that can break it apart into its component sugars, but humans and other mammals do not. Dietary fiber, or "roughage," usually refers to the cellulose in our vegetable foods. Bacteria that live in the guts of termites and grazers such as cattle and sheep help these animals digest the cellulose in plants.

In **starch**, a different covalent bonding pattern between glucose monomers makes a chain that coils up into a spiral (**FIGURE 3.8B**). Like cellulose, starch does not dissolve readily in water, but it is easier to break down than cellulose. These properties make the molecule ideal for storing sugars in the watery, enzyme-filled interior of plant cells. Most plant leaves make glucose during the day, and their cells store it by building starch. At night, hydrolysis enzymes break the bonds between starch's glucose monomers. The released glucose can be broken down immediately for energy, or converted to sucrose that is transported to other parts of the plant. Humans also have hydrolysis

glucose + fructose ⟶ sucrose + water

FIGURE 3.7 ▶Animated The synthesis of a sucrose molecule is an example of a condensation reaction. You are already familiar with sucrose—it is common table sugar.

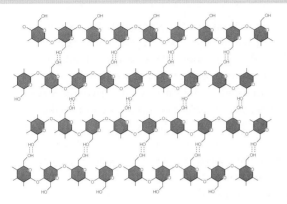

A Cellulose

Cellulose is the main structural component of plants. Above, in cellulose, chains of glucose monomers stretch side by side and hydrogen-bond at many —OH groups. The hydrogen bonds stabilize the chains in tight bundles that form long fibers. Few types of organisms can digest this tough, insoluble material.

B Starch

Starch is the main energy reserve in plants, which store it in their roots, stems, leaves, seeds, and fruits. Below, in starch, a series of glucose monomers form a chain that coils up.

C Glycogen

Glycogen functions as an energy reservoir in animals, including people. It is especially abundant in the liver and muscles. Above, glycogen consists of highly branched chains of glucose monomers.

FIGURE 3.8 ▶Animated Three of the most common complex carbohydrates and their locations in a few organisms. Each polysaccharide consists only of glucose subunits, but different bonding patterns result in substances with very different properties.

enzymes that break down starch, so this carbohydrate is an important component of our food.

Animals store sugars in the form of **glycogen**, a polysaccharide that consists of highly branched chains of glucose monomers (**FIGURE 3.8C**). Muscle and liver cells contain most of the body's glycogen. When the blood sugar level falls, liver cells break down the glycogen, and the released glucose subunits enter the blood.

In chitin, a polysaccharide similar to cellulose, long, unbranching chains of nitrogen-containing monomers are linked by hydrogen bonds (**FIGURE 3.9**). As a structural material, chitin is durable, translucent, and flexible. It strengthens hard parts of many animals, including the outer cuticle of crustaceans, beetles, and ticks, and it reinforces the cell wall of many fungi.

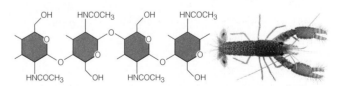

FIGURE 3.9 Chitin. This polysaccharide strengthens the hard parts of many small animals such as lobsters.

carbohydrate Molecule that consists primarily of carbon, hydrogen, and oxygen atoms in a 1:2:1 ratio.
cellulose Tough, insoluble polysaccharide that is the major structural material in plants.
disaccharide Polymer of two sugar subunits.
glycogen Polysaccharide; energy reservoir in animal cells.
monosaccharide Simple sugar; monomer of polysaccharides.
polysaccharide Polymer of many monosaccharides.
starch Polysaccharide; energy reservoir in plant cells.

TAKE-HOME MESSAGE 3.4
What is a carbohydrate?

✔ Cells use simple carbohydrates (sugars) for energy and to build other molecules.

✔ Glucose monomers, bonded different ways, form complex carbohydrates such as cellulose, starch, and glycogen.

CREDITS: (8) art, © Cengage Learning; photo, © JupiterImages Corporation; (9) left, © Cengage Learning; right, David Lettschwager/National Geographic Creative.

✔ Triglycerides, phospholipids, waxes, and steroids are lipids common in biological systems.

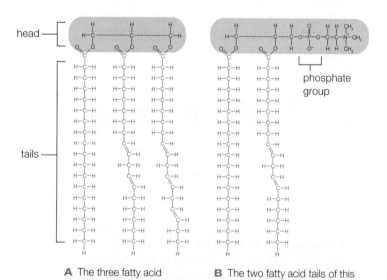

hydrophilic "head" (acidic carboxyl group)

hydrophobic "tail"

A stearic acid (saturated)

B linoleic acid (omega-6)

C linolenic acid (omega-3)

FIGURE 3.10 ▶Animated Fatty acids. **A** The tail of stearic acid is fully saturated with hydrogen atoms. **B** Linoleic acid, with two double bonds, is unsaturated. The first double bond occurs at the sixth carbon from the end, so linoleic acid is called an omega-6 fatty acid. Omega-6 and **C** omega-3 fatty acids are "essential fatty acids." Your body does not make them, so they must come from food.

Lipids are fatty, oily, or waxy organic compounds. They vary in structure, but all are hydrophobic (Section 2.5). Many lipids incorporate **fatty acids**, which are small organic molecules that consist of a long hydrocarbon "tail" with a carboxyl group "head" (**FIGURE 3.10**). The tail is hydrophobic; the carboxyl group makes the head hydrophilic (and acidic). You are already familiar with the properties of fatty acids because these molecules are the main component of soap. The hydrophobic tails of fatty acids in soap attract oily dirt, and the hydrophilic heads dissolve the dirt in water.

Saturated fatty acids have only single bonds linking the carbons in their tails. In other words, their carbon chains are fully saturated with hydrogen atoms (**FIGURE 3.10A**). Saturated fatty acid tails are flexible and they wiggle freely. Double bonds between carbons limit the flexibility of the tails of **unsaturated fatty acids** (**FIGURE 3.10B,C**). **FIGURE 3.1** shows how these bonds are *cis* or *trans*, depending on the way the hydrogens are arranged around them.

Lipids in Biological Systems

Fats The carboxyl group head of a fatty acid can easily form a covalent bond with another molecule. When it bonds to a glycerol, a type of alcohol, it loses its hydrophilic character and becomes part of a fat. **Fats** are lipids with one, two, or three fatty acids bonded to the same glycerol. A fat with three fatty acid tails is called a **triglyceride** (**FIGURE 3.11A**). Triglycerides are entirely hydrophobic, so they do not dissolve in water. Most "neutral" fats, such as butter and vegetable oils, are examples. Triglycerides are the most abundant and richest energy source in vertebrate bodies. Gram for gram, these fats store more energy than carbohydrates.

head

tails

phosphate group

A The three fatty acid tails of a triglyceride are attached to a glycerol head.

B The two fatty acid tails of this phospholipid are attached to a phosphate-containing head.

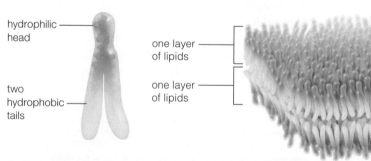

hydrophilic head

two hydrophobic tails

one layer of lipids

one layer of lipids

C A double layer of phospholipids—the lipid bilayer—is the structural foundation of all cell membranes. You will read more about the structure of cell membranes in Chapter 5.

FIGURE 3.11 ▶Animated Lipids with fatty acid tails. **FIGURE IT OUT** Is the triglyceride saturated or unsaturated?

Answer: Unsaturated

Butter, cream, and other high-fat animal products have a high proportion of saturated fats, which means they consist mainly of triglycerides with three saturated fatty acid tails. Saturated fats tend to be solid at room temperature because their floppy saturated tails can pack tightly. Most vegetable oils are unsaturated fats, which means they consist mainly of triglycerides with one or more unsaturated fatty acid tails. Each double bond in a fatty acid tail makes a rigid kink. Kinky tails do not pack tightly, so unsaturated fats are typically liquid at room temperature. The partially hydrogenated vegetable oils that you learned about in Section 3.1 are an exception. These fats are solid at room temperature because the special *trans* double bond keeps fatty acid tails straight, allowing the fat molecules to pack tightly just like saturated fats do.

Phospholipids A **phospholipid** has a phosphate-containing head and two long hydrocarbon tails that are typically derived from fatty acids (**FIGURE 3.11B**). The tails are hydrophobic, but the highly polar phosphate group makes the head hydrophilic. These opposing properties give rise to the basic structure of cell membranes, which consist mainly of phospholipids. In a cell membrane, phospholipids are arranged in two layers—a **lipid bilayer** (**FIGURE 3.11C**). The heads of one layer are dissolved in the cell's watery interior, and the heads of the other layer are dissolved in the cell's fluid surroundings. All of the hydrophobic tails are sandwiched between the hydrophilic heads. You will read more about the structure of cell membranes in Chapters 4 and 5.

Waxes A **wax** is a complex, varying mixture of lipids with long fatty acid tails bonded to long-chain alcohols

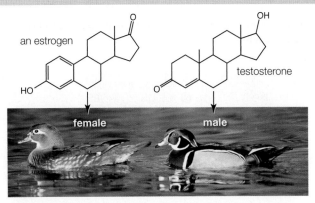

FIGURE 3.12 Steroids. Estrogen and testosterone are steroid hormones that govern reproduction and secondary sexual traits. The two hormones are the source of gender-specific traits in many species, including these wood ducks.

or carbon rings. The molecules pack tightly, so waxes are firm and water-repellent. Plants secrete waxes onto their exposed surfaces to restrict water loss and keep out parasites and other pests. Other types of waxes protect, lubricate, and soften skin and hair. Waxes,  together with fats and fatty acids, make feathers waterproof. Bees store honey and raise new generations of bees inside a honeycomb of secreted beeswax.

Steroids **Steroids** are lipids with no fatty acid tails; they have a rigid backbone that consists of twenty carbon atoms arranged in a characteristic pattern of four rings (**FIGURE 3.12**). These molecules serve varied and important physiological functions in plants, fungi, and animals. Functional groups attached to the rings define the type of steroid. Cholesterol, the most common steroid in animal tissue, is a precursor for many other molecules, including bile salts (which help digest fats), vitamin D (required to keep teeth and bones strong), and steroid hormones.

---

fat Lipid that consists of a glycerol molecule with one, two, or three fatty acid tails. Saturated fats have three saturated fatty acid tails. Unsaturated fats have one or more unsaturated fatty acid tails.
fatty acid Organic compound that consists of an acidic carboxyl group "head" and a long hydrocarbon "tail."
lipid Fatty, oily, or waxy organic compound.
lipid bilayer Double layer of lipids arranged tail-to-tail; structural foundation of cell membranes.
phospholipid A lipid with a phosphate group in its hydrophilic head, and two nonpolar fatty acid tails; main constituent of eukaryotic cell membranes.
saturated fatty acid Fatty acid with only single bonds linking the carbons in its tail.
steroid Type of lipid with four carbon rings and no fatty acid tails.
triglyceride A fat with three fatty acid tails.
unsaturated fatty acid Fatty acid with one or more carbon–carbon double bonds in its tail.
wax Water-repellent mixture of lipids with long fatty acid tails bonded to long-chain alcohols or carbon rings.

> **TAKE-HOME MESSAGE 3.5**
> **What are lipids?**
>
> ✔ Lipids are fatty, waxy, or oily organic compounds.
>
> ✔ Fats have one, two, or three fatty acid tails; triglyceride fats are an important energy reservoir in vertebrate animals.
>
> ✔ Phospholipids arranged in a lipid bilayer are the main component of cell membranes.
>
> ✔ Waxes have complex, varying structures. They are components of water-repelling and lubricating secretions.
>
> ✔ Steroids serve varied and important physiological roles in plants, fungi, and animals.

3.6 Proteins

✔ Of all biological molecules, proteins are the most diverse in both structure and function.

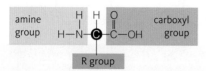

FIGURE 3.13 Generalized structure of an amino acid. See Appendix II for the complete structures of the twenty most common amino acids found in eukaryotic proteins.

Proteins participate in all processes that sustain life. Structural proteins support cell parts and, as part of tissues, multicelled bodies. Most enzymes that drive metabolic reactions are proteins. Proteins move substances, help cells communicate, and defend the body.

With a few exceptions, cells can make all of the thousands of different proteins they need from only twenty kinds of amino acid monomers. An **amino acid** is a small organic compound with an amine group ($-NH_2$), a carboxyl group ($-COOH$, the acid), and a side chain called an "R group" that defines the kind of amino acid. In most amino acids, all three groups are attached to the same carbon atom (**FIGURE 3.13**).

The covalent bond that links amino acids in a protein is called a **peptide bond**. During protein synthesis, a peptide bond forms between the carboxyl group of the first amino acid and the amine group of the second (**FIGURE 3.14 ❶**). Another peptide bond links a third amino acid to the second, and so on (you will learn more about the details of protein synthesis in Chapter 9). A short chain of amino acids is called a **peptide**; as the chain lengthens, it becomes a **polypeptide**. Proteins consist of polypeptides that can be hundreds or even thousands of amino acids long.

The idea that structure dictates function is particularly appropriate as applied to proteins, because the diversity in biological activity among these molecules arises from differences in their three-dimensional shape. Protein structure begins with the linear series of amino acids composing a polypeptide chain ❷. The order of the amino acids in the chain, which is called primary structure, defines the type of protein. The molecule begins to take on shape during protein synthesis, when hydrogen bonds that form between amino acids cause the lengthening polypeptide chain to twist and fold. Hydrogen bonding holds sections of the polypeptide in loops, helices (coils), or flat sheets, and these patterns constitute the protein's secondary structure ❸. The primary structure of each type of protein is unique, but most proteins have similar patterns of secondary structure.

Much as an overly twisted rubber band coils back upon itself, hydrogen bonding also makes the loops, helices, and sheets of a protein fold up into even more compact domains (**FIGURE 3.15A**). These domains are called tertiary structure ❹. Tertiary structure is what makes a protein a working molecule. For example, the helices and loops in a globin chain fold up together to form a pocket that can hold a heme, which is a small compound essential to the finished protein's function. In other proteins, sheets, loops, and helices come together as complex structures that resemble barrels, propellers, sandwiches, and so on. Barrel domains often form tunnels through cell membranes, allowing small molecules to cross. Some proteins have barrel domains that rotate like motors in small molecular machines (**FIGURE 3.15B**). A protein may have several domains, each contributing a particular structural or functional property to the molecule.

FIGURE 3.14 ▶Animated Protein structure.

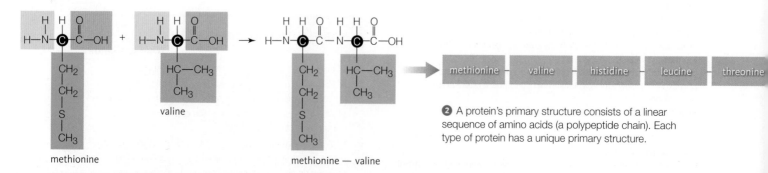

❷ A protein's primary structure consists of a linear sequence of amino acids (a polypeptide chain). Each type of protein has a unique primary structure.

❶ A condensation reaction joins the carboxyl group of one amino acid and the amine group of another to form a peptide bond. In this example, a peptide bond forms between the amino acids methionine and valine.

A In this protein, loops (green), coils (red), and a sheet (yellow) fold up together into a chemically active pocket. The pocket gives this protein the ability to transfer electrons from one molecule to another. Many other proteins have the same pocket structure.

B This barrel domain is part of a rotary mechanism in a larger protein. The protein functions as a molecular motor that pumps hydrogen ions through cell membranes.

FIGURE 3.15 Examples of domains in proteins.

Many proteins also have quaternary structure, which means they consist of two or more polypeptide chains that are closely associated or covalently bonded together. Hemoglobin is like this ❺. So are most enzymes, which have multiple polypeptide chains that collectively form a roughly spherical shape.

Fibrous proteins aggregate by many thousands into much larger structures, with their polypeptide chains organized into strands or sheets. The keratin in your hair is an example ❻. Some fibrous proteins contribute to the structure and organization of cells and tissues. Others, such as the actin and myosin filaments in muscle cells, are part of the mechanisms that help cells, cell parts, and multicelled bodies move.

Carbohydrates, lipids, or both may get attached to a protein after synthesis. A protein with one or more oligosaccharides attached to it is called a glycoprotein. Molecules that allow a tissue or a body to recognize its own cells are glycoproteins, as are other molecules that help cells interact in immunity. A protein with one or more lipids attached to it is called a lipoprotein. Some lipoproteins are aggregate structures that consist of variable amounts and types of proteins and lipids (**FIGURE 3.16**).

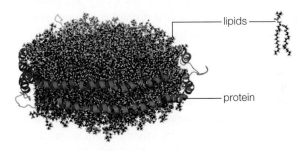

FIGURE 3.16 A lipoprotein particle. The one depicted here (HDL, which is often called "good" cholesterol) consists of thousands of lipids lassoed into a clump by two proteins.

---

amino acid Small organic compound that is a subunit of proteins. Consists of a carboxyl group, an amine group, and a characteristic side group (R), all typically bonded to the same carbon atom.
peptide Short chain of amino acids linked by peptide bonds.
peptide bond A bond between the amine group of one amino acid and the carboxyl group of another. Joins amino acids in proteins.
polypeptide Long chain of amino acids linked by peptide bonds.
protein Organic molecule that consists of one or more polypeptides.

TAKE-HOME MESSAGE 3.6

What is a protein?

✔ A protein is a chain of amino acids. The order of amino acids in a polypeptide chain dictates the type of protein.

✔ Polypeptide chains twist and fold into coils, sheets, and loops, which fold and pack further into functional domains.

✔ A protein's function arises from its shape.

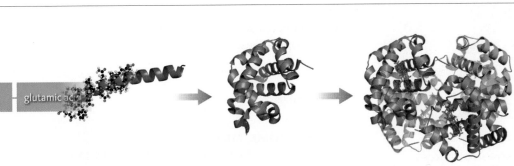

oline — glutamic acid

❸ Secondary structure arises as a polypeptide chain twists into a helix (coil), loop, or sheet held in place by hydrogen bonds.

❹ Tertiary structure arises when loops, helices, and sheets fold up into a domain. In this example, the helices of a globin chain form a pocket.

❺ Many proteins have two or more polypeptide chains (quaternary structure). Hemoglobin, shown here, consists of four globin chains (green and blue). Each globin pocket now holds a heme group (red).

❻ Some types of proteins aggregate into much larger structures. As an example, organized arrays of keratin, a fibrous protein, compose filaments that make up your hair.

CREDITS: (14) #3–5: 1BBB, A third quaternary structure of human hemoglobin A at 1.7-A resolution. Silva, M.M., Rogers, P.H., Arnone, A., Journal: (1992) J.Biol.Chem. 267: 17248–17256; #6: © JupiterImages Corporation. (15A, 16) Castrignanò T, De Meo PD, Cozzetto D, Talamo IG, Tramontano A. (2006). The PMDB Protein Model Database. Nucleic Acids Research, 34: D306-D309. (15B) pdb ID2W5J, Vollmar, M., Shlieper, D., Winn M., Buechner, C., Groth, G. "Structure of the C14 rotor ring of the proton translocating chloroplast ATP synthase." (2009) J. Biol. Chem. 284:18228.

3.7 Why Is Protein Structure So Important?

✔ Changes in a protein's shape may have drastic health consequences.

Protein shape depends on hydrogen bonding, which can be disrupted by heat, some salts, shifts in pH, or detergents. Such disruption can cause proteins to lose their three-dimensional shape, or **denature**. Once a protein's shape unravels, so does its function.

You can see denaturation in action when you cook an egg. A protein called albumin is a major component of egg white. Cooking does not disrupt the covalent bonds of albumin's primary structure, but it does destroy the hydrogen bonds that maintain the protein's shape. When a translucent egg white turns opaque, the albumin has been denatured. For a very few proteins, denaturation is reversible if normal conditions return, but albumin is not one of them. There is no way to uncook an egg.

Prion diseases such as mad cow disease (bovine spongiform encephalitis, or BSE) in cattle, Creutzfeldt–Jakob disease in humans, and scrapie in sheep, are the dire aftermath of a protein that changes shape. These infectious diseases may be inherited, but more often they arise spontaneously. All are characterized by relentless deterioration of mental and physical abilities that eventually causes death (**FIGURE 3.17A**).

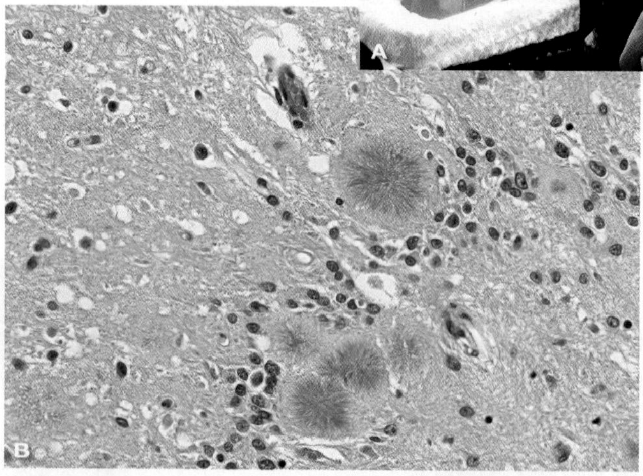

All prion diseases begin with a glycoprotein called PrPC that occurs normally in cell membranes of the mammalian body. This protein is especially abundant in brain cells, but we still know very little about what it does. Sometimes, a PrPC protein misfolds so that part of the molecule forms a sheet instead of a helix. One misfolded molecule should not pose much of a threat, but when this particular protein misfolds it becomes a **prion**, or infectious protein. The shape of a misfolded PrPC protein causes normally folded PrPC proteins to misfold too. Each protein that misfolds becomes infectious, so the number of prions increases exponentially.

The shape of misfolded PrPC proteins allows them to align tightly into long, thin, insoluble fibers, which are called amyloid fibrils. Amyloid fibrils grow in patches from their ends as more PrPC proteins misfold (**FIGURE 3.17B**). These patches disrupt normal brain function, causing symptoms such as confusion, memory loss, and lack of coordination. Holes form in the brain as its cells die. Eventually, the brain becomes so riddled with holes that it looks like a sponge.

In the mid-1980s, an epidemic of mad cow disease in Britain was followed by an outbreak of a new variant of Creutzfeldt–Jakob disease (vCJD) in humans. Researchers isolated a prion similar to the one in scrapie-infected sheep from cows with BSE, and also from humans affected by the new type of Creutzfeldt–Jakob disease. How did the prion get from sheep to cattle to people? Prions resist denaturation, so treatments such as cooking that inactivate other types of infectious agents have little effect on them. The cattle became infected by the prion after eating feed prepared from the remains of scrapie-infected sheep, and people became infected by eating beef from the infected cattle.

Two hundred people have died from vCJD since 1990. The use of animal parts in livestock feed is now banned in many countries, and the number of cases of BSE and vCJD has since declined. Cattle with BSE still turn up, but so rarely that they pose little threat to human populations.

denature To unravel the shape of a protein or other large biological molecule.
prion Infectious protein.

FIGURE 3.17 Variant Creutzfeldt–Jakob disease (vCJD).

A Charlene Singh was one of the three people who developed symptoms of vCJD disease while living in the United States. Like the others, Singh most likely contracted the disease elsewhere; she spent her childhood in Britain. Diagnosed in 2001, she died in 2004.

B Slice of brain tissue from a person with vCJD. Fibers of prion proteins (amyloid fibrils) radiating from several deposits are visible.

TAKE-HOME MESSAGE 3.7
Why is protein structure important?

✔ Protein shape can be unraveled by heat or other conditions that disrupt hydrogen bonding.

✔ A protein's function depends on its shape, so conditions that alter a protein's shape also alter its function.

3.8 Nucleic Acids

✔ DNA and RNA consist of nucleotides.

✔ Some nucleotides also have roles in metabolism.

Nucleotides are small organic molecules that function as energy carriers, enzyme helpers, chemical messengers, and subunits of DNA and RNA. Each consists of a monosaccharide ring bonded to a nitrogen-containing base and one, two, or three phosphate groups (**FIGURE 3.18**). The monosaccharide is a five-carbon sugar, either ribose or deoxyribose; the base, one of five compounds with a flat ring structure (we return to the structure of nucleotide bases in Sections 8.3 and 9.2). When the third phosphate group of a nucleotide is transferred to another molecule, energy is transferred along with it. You will read about such phosphate-group transfers and their important metabolic role in Chapter 5. The nucleotide **ATP** (adenosine triphosphate) serves an especially important role as an energy carrier in cells.

Nucleic acids are polymers, chains of nucleotides in which the sugar of one nucleotide is joined to the phosphate group of the next (**FIGURE 3.19A**). An example is **RNA**, or ribonucleic acid, named after the ribose sugar of its component nucleotides. An RNA molecule is a chain of four kinds of nucleotide monomers, one of which is ATP. RNA molecules carry out protein synthesis, which we discuss in detail in Chapter 9.

DNA, or deoxyribonucleic acid, is a nucleic acid named after the deoxyribose sugar of its component nucleotides. A DNA molecule consists of two chains of nucleotides twisted into a double helix (**FIGURE 3.19B**). Hydrogen bonds between the nucleotides hold the two chains together. Each cell starts life with DNA inherited from a parent cell. That DNA contains all of the information necessary to build a new cell and, in the case of multicelled organisms, an entire individual. The cell uses the order of nucleotide bases in DNA—the DNA sequence—to guide production of RNA and proteins. Parts of the sequence are identical or nearly so in all organisms, and parts are unique to a species or an individual (Chapter 8 returns to DNA structure and function).

ATP Adenosine triphosphate. Nucleotide that consists of an adenine base, a ribose sugar, and three phosphate groups; serves an important role as an energy carrier in cells.

DNA Deoxyribonucleic acid. Nucleic acid that consists of two chains of nucleotides twisted into a double helix; carries hereditary information.

nucleic acid Polymer of nucleotides; DNA or RNA.

nucleotide Monomer of nucleic acids; has a five-carbon sugar, a nitrogen-containing base, and one, two, or three phosphate groups.

RNA Ribonucleic acid. Single-stranded chain of nucleotides. Some types have roles in protein synthesis.

Fear of Frying (revisited)

Trans fatty acids are relatively rare in unprocessed foods, so it makes sense from an evolutionary standpoint that our bodies may not have enzymes to deal with them efficiently. The enzymes that hydrolyze *cis* fatty acids have difficulty breaking down *trans* fatty acids, a problem that may be a factor in the ill effects of *trans* fats. All prepackaged foods in the United States are now required to list *trans* fat content, but may be marked "zero grams of *trans* fats" even when a single serving contains up to half a gram.

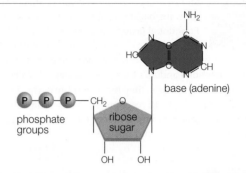

FIGURE 3.18 Example of a nucleotide: ATP. ATP is a monomer of RNA, and also a participant in many metabolic reactions.

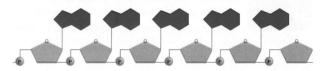

A A chain of nucleotides is a nucleic acid. The sugar of one nucleotide is covalently bonded to the phosphate group of the next, forming a sugar–phosphate backbone.

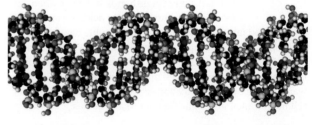

B DNA consists of two chains of nucleotides, twisted into a double helix. Hydrogen bonding maintains the three-dimensional structure of this nucleic acid.

FIGURE 3.19 Nucleic acid structure.

TAKE-HOME MESSAGE 3.8
What are nucleotides and nucleic acids?

✔ Nucleotides are monomers of nucleic acids. ATP has an important metabolic role as an energy carrier.

✔ The nucleic acid DNA holds information necessary to build cells and multicelled individuals.

✔ RNAs are nucleic acids that carry out protein synthesis.

CREDITS: (18, 19) © Cengage Learning; (in text) Kentoh/Shutterstock.com.

summary

Section 3.1 All organisms consist of the same kinds of molecules. Seemingly small differences in the way those molecules are put together can have big effects inside a living organism.

Section 3.2 Molecules of life—complex carbohydrates and lipids, proteins, and nucleic acids—consist mainly of carbon and hydrogen atoms, so they are **organic**. Different models reveal different aspects of their structure.

Section 3.3 **Hydrocarbons** have only carbon and hydrogen atoms. Carbon chains or rings form the backbone of the molecules of life. **Functional groups** attached to the backbone influence the chemical character of these compounds, and thus their function. **Metabolism** includes chemical **reactions** and all other processes by which cells acquire and use energy as they make and break the bonds of organic compounds. In reactions such as **condensation**, **enzymes** build **polymers** from **monomers** of simple sugars, fatty acids, amino acids, and nucleotides. Reactions such as **hydrolysis** release monomers by breaking apart polymers.

Section 3.4 Cells use different kinds of **carbohydrates** for energy, and as structural materials. Enzymes assemble **disaccharides** and **polysaccharides** from **monosaccharide** (simple sugar) monomers. **Cellulose**, **glycogen**, and **starch** are complex carbohydrates that consist of glucose monomers bonded in different patterns.

Section 3.5 **Lipids** are fatty, oily, or waxy compounds. All are nonpolar. A **fatty acid** is a lipid with an acidic head and a long hydrocarbon tail. Only single bonds link the carbons in the tail of a **saturated fatty acid**; the tail of an **unsaturated fatty acid** has one or more double bonds. **Fats** and some other lipids have fatty acid tails; **triglycerides** have three. A **lipid bilayer** (that consists primarily of **phospholipids**) is the basic structure of all cell membranes. **Waxes** are part of water-repellent and lubricating secretions. **Steroids** occur in cell membranes, and some function as hormones.

Section 3.6 Structurally and functionally, **proteins** are the most diverse molecules of life. The shape of a protein is the source of its function. Protein structure begins as a sequence of **amino acids** linked by **peptide bonds** into a **peptide**, then a **polypeptide** (primary structure). Polypeptides twist into loops, helices, and sheets (secondary structure) that can pack further into functional domains (tertiary structure). Many proteins, including most enzymes, consist of two or more polypeptides (quaternary structure). Fibrous proteins aggregate into much larger structures.

Section 3.7 A protein's structure dictates its function, so changes in a protein's structure may also alter its function.

Hydrogen bonds that stabilize a protein's shape may be disrupted by shifts in pH or temperature, or exposure to detergent or some salts. If that happens, the protein unravels, or **denatures**, and so loses its function. **Prion** diseases are a fatal consequence of misfolded proteins.

Section 3.8 **Nucleotides** are small organic molecules that consist of a five-carbon sugar; a nitrogen-containing base; and one, two, or three phosphate groups. Nucleotides are monomers of **DNA** and **RNA**, which are **nucleic acids**. Some, especially **ATP**, have additional functions such as carrying energy. DNA encodes heritable information; RNAs carry out protein synthesis.

self-quiz

Answers in Appendix VII

1. Organic molecules consist mainly of _____ atoms.
   a. carbon
   b. carbon and oxygen
   c. carbon and hydrogen
   d. carbon and nitrogen

2. Each carbon atom can bond with as many as _____ other atom(s).

3. _____ groups are the "acid" part of amino acids and fatty acids.
   a. Hydroxyl (—OH)
   b. Carboxyl (—COOH)
   c. Methyl (—CH_3)
   d. Phosphate (—PO_4)

4. _____ is a simple sugar (a monosaccharide).
   a. Glucose
   b. Sucrose
   c. Ribose
   d. Starch
   e. both a and c
   f. a, b, and c

5. Which three carbohydrates can be built using only glucose monomers?
   a. Starch, cellulose, and glycogen
   b. Glucose, sucrose, and ribose
   c. Cellulose, steroids, and polysaccharides
   d. Starch, chitin, and DNA
   e. Triglycerides, nucleic acids, and polypeptides

6. Unlike saturated fats, the fatty acid tails of unsaturated fats incorporate one or more _____ .
   a. phosphate groups
   b. glycerols
   c. double bonds
   d. single bonds

7. Is this statement true or false? Unlike saturated fats, all unsaturated fats are beneficial to health because their fatty acid tails kink and do not pack together.

8. Steroids are among the lipids with no _____ .
   a. double bonds
   b. fatty acid tails
   c. hydrogens
   d. carbons

9. Which of the following is a class of molecules that encompasses all of the other molecules listed?
   a. triglycerides
   b. fatty acids
   c. waxes
   d. steroids
   e. lipids
   f. phospholipids

Effects of Dietary Fats on Lipoprotein Levels

Cholesterol that is made by the liver or that enters the body from food cannot dissolve in blood, so it is carried through the bloodstream by lipoproteins. Low-density lipoprotein (LDL) carries cholesterol to body tissues such as artery walls, where it can form deposits associated with cardiovascular disease. Thus, LDL is often called "bad" cholesterol. High-density lipoprotein (HDL, **FIGURE 3.16**) carries cholesterol away from tissues to the liver for disposal, so HDL is often called "good" cholesterol.

| | Main Dietary Fats | | | |
	cis fatty acids	*trans* fatty acids	saturated fats	optimal level
LDL	103	117	121	<100
HDL	55	48	55	>40
ratio	1.87	2.44	2.2	<2

In 1990, Ronald Mensink and Martijn Katan published a study that tested the effects of different dietary fats on blood lipoprotein levels. Their results are shown in **FIGURE 3.20**.

1. In which group was the level of LDL ("bad" cholesterol) highest?

2. In which group was the level of HDL ("good" cholesterol) lowest?

3. An elevated risk of heart disease has been correlated with increasing LDL-to-HDL ratios. Which group had the highest LDL-to-HDL ratio?

4. Rank the three diets from best to worst according to their potential effect on heart disease.

FIGURE 3.20 Effect of diet on lipoprotein levels. Researchers placed 59 men and women on a diet in which 10 percent of their daily energy intake consisted of *cis* fatty acids, *trans* fatty acids, or saturated fats.

Blood LDL and HDL levels were measured after three weeks on the diet; averaged results are shown in mg/dL (milligrams per deciliter of blood). All subjects were tested on each of the diets. The ratio of LDL to HDL is also shown.

---

10. _____ are to proteins as _____ are to nucleic acids.
    a. Sugars; lipids
    c. Amino acids; hydrogen bonds
    b. Sugars; proteins
    d. Amino acids; nucleotides

11. A denatured protein has lost its _____ .
    a. hydrogen bonds
    c. function
    b. shape
    d. all of the above

12. _____ consists of nucleotides.
    a. Ribose
    b. RNA
    c. DNA
    d. b and c

13. In the following list, identify the carbohydrate, the fatty acid, the amino acid, and the polypeptide:
    a. NH_2—CHR—COOH
    c. (methionine)$_{20}$
    b. $C_6H_{12}O_6$
    d. $CH_3(CH_2)_{16}COOH$

14. Match the molecules with the best description.
    ____ wax
    a. protein primary structure
    ____ starch
    b. an energy carrier
    ____ triglyceride
    c. water-repellent secretions
    ____ DNA
    d. carries heritable information
    ____ polypeptide
    e. sugar storage in plants
    ____ ATP
    f. richest energy source

15. Match each polymer with the component monomers.
    ____ protein
    a. phosphate, fatty acids
    ____ phospholipid
    b. amino acids, sugars
    ____ glycoprotein
    c. glycerol, fatty acids
    ____ fat
    d. nucleotides
    ____ nucleic acid
    e. glucose only
    ____ wax
    f. sugar, phosphate, base
    ____ nucleotide
    g. amino acids
    ____ lipoprotein
    h. glucose, fructose
    ____ sucrose
    i. lipids, amino acids
    ____ glycogen
    j. fatty acids, carbon rings

critical thinking

1. Lipoproteins are like suitcases that move cholesterol, fatty acid remnants, triglycerides, and phospholipids from one place to another in the body. Given what you know about the insolubility of lipids in water, which of the four kinds of lipids would you predict to be on the outside of a lipoprotein clump, bathed in the water-based fluid portion of blood?

2. In 1976, a team of chemists in the United Kingdom was developing new insecticides by modifying sugars with chlorine (Cl_2), phosgene (Cl_2CO), and other toxic gases. One young member of the team misunderstood his verbal instructions to "test" a new molecule. He thought he had been told to "taste" it. Luckily for him, the molecule was not toxic, but it was very sweet. It became the food additive sucralose.

Sucralose has three chlorine atoms substituted for three hydroxyl groups of sucrose (table sugar). It binds so strongly to the sweet-taste receptors on the tongue that the human brain perceives it as 600 times sweeter than sucrose. Sucralose was originally marketed as an artificial sweetener called Splenda®, but it is now available under several other brand names.

Researchers investigated whether the body recognizes sucralose as a carbohydrate by feeding sucralose labeled with ^{14}C to volunteers. Analysis of the radioactive molecules in the volunteers' urine and feces showed that 92.8 percent of the sucralose passed through the body without being altered. Many people are worried that the chlorine atoms impart toxicity to sucralose. How would you respond to that concern?

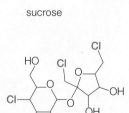

sucrose

sucralose

CREDITS: (20) Source, Mensink RP, Katan MB, "Effect of dietary trans fatty acids on high-density and low-density lipoprotein cholesterol levels in healthy subjects." *NEJM* 323(7):439–45; (in text) © Cengage Learning.

LEARNING ROADMAP

Reflect on life's levels of organization (Section 1.2). In this chapter, you will see how properties of lipids (3.5) give rise to cell membranes, consider the location of DNA (3.8) and the sites of carbohydrate metabolism (3.3, 3.4), and expand your understanding of the roles of proteins (3.6, 3.7). You will also revisit tracers (2.2) and the philosophy of science (1.9).

COMPONENTS OF ALL CELLS

Every cell has a plasma membrane separating its interior from the exterior environment. A cell's interior contains cytoplasm, DNA, and other structures.

PROKARYOTIC CELLS

Archaea and bacteria have no nucleus. In general, they are smaller and structurally more simple than eukaryotic cells. As a group, they are by far the most numerous and diverse organisms.

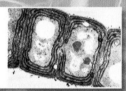

EUKARYOTIC CELLS

Protists, plants, fungi, and animals are eukaryotes. Cells of these organisms differ in internal components and surface specializations, but all start out life with a nucleus.

ORGANELLES OF EUKARYOTES

Membranes around eukaryotic organelles maintain internal environments that allow these structures to carry out specialized functions within a cell.

OTHER CELL COMPONENTS

Cytoskeletal elements organize and move cells and cell components. Cells secrete protective and structural materials, and connect to one another via cell junctions.

Chapter 5 explores cell function, and Chapters 6-8 detail individual metabolic processes. Some cellular structures are required for cell division (Chapter 11). Human genetic disorders return in Chapter 14. Chapter 20 details prokaryotes; Chapter 21, protists. Cell structures introduced in this chapter return in context of the physiology of plants (Chapters 27–30) and animals (Chapters 31–42). You will see some medical consequences of biofilms in Section 37.3.

Cell for cell, microorganisms that live in and on a human body outnumber the person's own cells by about ten to one. Most are bacteria that live in the digestive tract, but these cells are not just stowaways. Gut bacteria help with digestion, make vitamins that mammals cannot, prevent the growth of dangerous germs, and shape the immune system.

One of the most common intestinal bacteria of warm-blooded animals (including humans) is *Escherichia coli*. Most of the hundreds of types, or strains, of *E. coli*, are helpful, but a few make a toxic protein that can severely damage the lining of the intestine. After ingesting as few as ten cells of a toxic strain, a person may become ill with severe cramps and bloody diarrhea that lasts up to ten days. In some people, complications of infection result in kidney failure, blindness, paralysis, and death. Each year, about 265,000 people in the United States become infected with toxin-producing *E. coli*.

Strains of *E. coli* that are toxic to people live in the intestines of other animals—mainly cattle, deer, goats, and sheep—apparently without sickening them. Humans are exposed to the bacteria when they come into contact with feces of animals that harbor them, for example, by eating contaminated ground beef. During slaughter, meat can come into contact with feces. Bacteria in the feces stick to the meat, then get thoroughly mixed into it during the grinding process. Unless contaminated meat is cooked to at least 71°C (160°F), live bacteria will enter the digestive tract of whoever eats it.

People also become infected with toxic *E. coli* by eating fresh fruits and vegetables that have contacted animal feces. Washing produce with water does not remove all of the bacteria because they are sticky (FIGURE 4.1). In 2012, more than 4,000 people in Europe were sickened after eating sprouts, and 49 of them died. The outbreak was traced to a single shipment of contaminated sprout seeds from Egypt.

The impact of such outbreaks, which occur with unfortunate regularity, extends beyond casualties. The contaminated sprouts cost growers in the European Union at least $600 million in lost sales. In 2011 alone, the United States Department of Agriculture (USDA) recalled 36.7 million pounds of ground meat products contaminated with toxic bacteria, at a cost in the billions of dollars. Such costs are eventually passed to taxpayers and consumers.

Food growers and processors are implementing new procedures intended to reduce the number and scope of these outbreaks. Meat and produce are being tested for some bacteria before sale, and improved documentation should allow a source of contamination to be pinpointed more quickly.

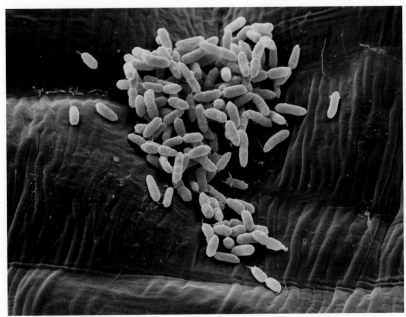

FIGURE 4.1 *Escherichia coli* cells sticking to the surface of a lettuce leaf. Some strains of this bacteria can cause a serious intestinal illness when they contaminate human food (right).

CREDITS: (opposite) Fernan Federici & Jim Haseloff/Wellcome Images; (1) left, © Custom Medical Stock Photo/Getty Images; right, JupiterImages Corporation.

4.2 What Is a Cell?

✔ All cells have a plasma membrane and cytoplasm, and all start out life with DNA.

Cell Theory

Before microscopes were invented, no one knew that cells existed because nearly all are invisible to the naked eye. By 1665, Antoni van Leeuwenhoek had constructed an early microscope that revealed tiny organisms in rainwater, insects, fabric, sperm, feces, and other samples. In scrapings of tartar from his teeth, Leeuwenhoek saw "many very small animalcules, the motions of which were very pleasing to behold." He (incorrectly) assumed that movement defines life, but (correctly) concluded that the moving animalcules he saw were alive. Robert Hooke, a contemporary of Leeuwenhoek, observed cork under a microscope and discovered it to consist of "a great many little Boxes." He called the tiny compartments cells, after the small chambers that monks lived in.

Today we know that a cell carries out metabolism and homeostasis, and reproduces either on its own or as part of a larger organism. By this definition, each cell is alive even if it is part of a multicelled body, and all living organisms consist of one or more cells. We also know that cells reproduce by dividing, so it follows that all existing cells must have arisen by division of other cells (later chapters discuss the processes by which cells divide). As a cell divides, it passes its hereditary material—its DNA—to offspring. Taken together, these generalizations constitute the **cell theory**, which is one of the foundations of modern biology (**TABLE 4.1**).

Components of All Cells

Cells vary in shape and in what they do, but all have a plasma membrane, cytoplasm, and DNA (**FIGURE 4.2**).

Plasma Membrane A cell's **plasma membrane** is its outermost, and this membrane separates the cell's contents from the external environment. Like all other cell membranes, a plasma membrane is selectively permeable, which means that only certain materials can cross it. Thus, a plasma membrane controls exchanges between the cell and its environment. Like all cell membranes, a plasma membrane consists mainly of a phospholipid bilayer (Section 3.5). Many different proteins embedded in this bilayer or attached to one of its surfaces carry out various tasks (**FIGURE 4.3**). For example, some types of membrane proteins form channels through the bilayer; others pump substances across it. You will see in Chapter 5 how a membrane's function arises from its composite structure.

Cytoplasm The plasma membrane encloses a jellylike mixture of water, sugars, ions, and proteins "called **cytoplasm**. A major part of a cell's metabolism

Table 4.1 The Cell Theory
1. Every living organism consists of one or more cells.
2. The cell is the structural and functional unit of all organisms. A cell is the smallest unit of life, individually alive even as part of a multicelled organism.
3. All living cells arise by division of preexisting cells.
4. Cells contain hereditary material, which they pass to their offspring when they divide.

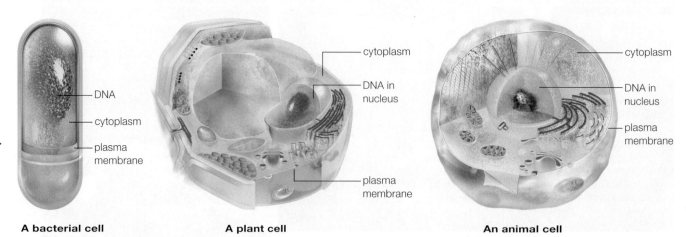

A bacterial cell   **A plant cell**   **An animal cell**

FIGURE 4.2 ▶Animated Overview of the general organization of a cell. All cells start out life with a plasma membrane, cytoplasm, and DNA. Archaea are similar to bacteria in overall structure; both are typically much smaller than eukaryotic cells. If the cells depicted here had been drawn to the same scale, the bacterium would be about this big:

CREDITS: (2, Table 4.1) © Cengage Learning.

occurs in the cytoplasm, and the cell's other internal components, including organelles, are suspended in it. **Organelles** are structures that carry out special functions inside a cell. Membrane-enclosed organelles allow a cell to compartmentalize substances and activities.

DNA Every cell starts out life with DNA. In nearly all bacteria and archaea, that DNA is suspended directly in cytoplasm. By contrast, the DNA of a eukaryotic cell is contained in a **nucleus** (plural, nuclei), an organelle with a double membrane. All protists, fungi, plants, and animals are eukaryotes. Some of these organisms are independent, free-living cells; others consist of many cells working together as a body.

Constraints on Cell Size

A cell exchanges substances with its environment at a rate that keeps pace with its metabolism. These exchanges occur across the plasma membrane, which can handle only so many exchanges at a time. The rate of exchange across a plasma membrane depends on its surface area: The bigger it is, the more substances can cross it during a given interval. Thus, cell size is limited by a physical relationship called the **surface-to-volume ratio**. By this ratio, an object's volume increases with the cube of its diameter, but its surface area increases only with the square.

Apply the surface-to-volume ratio to a round cell. As **FIGURE 4.4** shows, when a cell expands in diameter, its volume increases faster than its surface area does. Imagine that a round cell expands until it is four times its original diameter. The volume of the cell has increased 64 times (4^3), but its surface area has increased only 16 times (4^2). Each unit of plasma membrane must now handle exchanges with four times as much cytoplasm ($64 \div 16 = 4$). If the cell gets too big, the inward flow of nutrients and the outward flow of wastes across that membrane will not be fast enough to keep the cell alive.

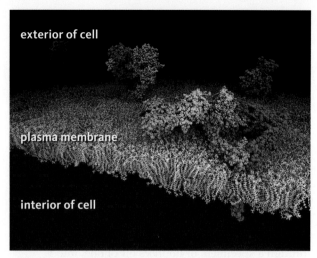

FIGURE 4.3 A plasma membrane separates a cell from its external environment. Proteins (in color) embedded in the lipid bilayer (gray) carry out special membrane functions. Chapter 5 returns to membrane structure and function.

Surface-to-volume limits also affect the form of colonial organisms and multicelled bodies. For example, small cells attach end to end to form strandlike algae, so each can interact directly with the environment. Some muscle cells in your thighs run the length of your upper leg, but each is thin, so it exchanges substances efficiently with fluids in the surrounding tissue.

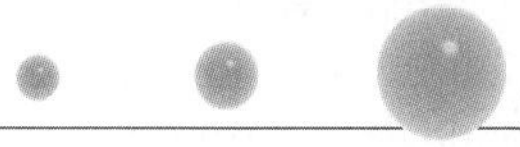

Diameter (cm)	2	3	6
Surface area (cm^2)	12.6	28.2	113
Volume (cm^3)	4.2	14.1	113
Surface-to-volume ratio	3:1	2:1	1:1

FIGURE 4.4 Three examples of the surface-to-volume ratio. This physical relationship between increases in volume and surface area constrains cell size and shape.

TAKE-HOME MESSAGE 4.2

How are all cells alike?

✔ Observations of cells led to the cell theory: All organisms consist of one or more cells; the cell is the smallest unit of life; each new cell arises from another cell; and a cell passes hereditary material to its offspring.

✔ All cells start life with a plasma membrane, cytoplasm, and a region of DNA, which, in eukaryotic cells only, is enclosed within a nucleus.

✔ The surface-to-volume ratio limits cell size and influences cell shape.

cell theory Theory that all organisms consist of one or more cells, which are the basic unit of life; all cells come from division of preexisting cells; and all cells pass hereditary material to offspring.
cytoplasm Jellylike mixture of water and solutes enclosed by a cell's plasma membrane.
nucleus Of a eukaryotic cell, organelle with a double membrane that holds the cell's DNA.
organelle Structure that carries out a specialized metabolic function inside a cell.
plasma membrane A cell's outermost membrane.
surface-to-volume ratio A relationship in which the volume of an object increases with the cube of the diameter, and the surface area increases with the square.

CREDITS: (3) Courtesy of © Johannes Kästner, Universität Stuttgart; (4) © Cengage Learning.

4.3 How Do We See Cells?

✔ We use different types of microscopes to study different aspects of cells and their parts.

Most cells are 10–20 micrometers in diameter, about fifty times smaller than the unaided human eye can perceive (**FIGURE 4.5**). One micrometer (μm) is one-thousandth of a millimeter, which is one-thousandth of a meter (**TABLE 4.2**). We use microscopes to observe cells and other objects in the micrometer range of size.

In a light microscope, visible light illuminates a sample. As you will learn in Chapter 6, all light travels in waves. This property of light causes it to bend when passing through a curved glass lens. Inside a light microscope, such lenses focus light that passes through a specimen, or bounces off of one, into a magnified image (**FIGURE 4.6A**). Microscopes that use polarized light can yield images in which the edges of some structures appear in three-dimensional relief (**FIGURE 4.6B**). Photographs of images enlarged with a microscope are called micrographs; those taken with visible light are called light micrographs (LM).

Most cells are nearly transparent, so their internal details may not be visible unless they are first stained, or exposed to dyes that only some cell parts soak up. Parts that absorb the most dye appear darkest. Staining results in an increase in contrast (the difference between light and dark) that allows us to see a greater range of detail.

Researchers often use light-emitting tracers (Section 2.2) to pinpoint the location of a particular molecule of interest within a cell. When illuminated with laser light, these tracers fluoresce (emit light), and

an image of the emitted light can be captured with a fluorescence microscope (**FIGURE 4.6C**). Such images are called fluorescence micrographs.

Structures smaller than about 200 nanometers across appear blurry under light microscopes because the wavelength of light—the distance from the peak of one wave to the peak behind it—limits the resolving power of even the best light microscope. To observe objects of this size range clearly, we would have to switch to an electron microscope. Electrons travel in wavelengths much shorter than those of visible light, so these microscopes can resolve details thousands of times smaller than light microscopes do.

There are two types of electron microscope; both use magnetic fields as lenses to focus a beam of electrons onto a sample. A transmission electron microscope

Table 4.2 Equivalent Units of Length

Unit		Meter	Inch
			Equivalent
centimeter	cm	1/100	0.4
millimeter	mm	1/1000	0.04
micrometer	μm	1/1,000,000	0.00004
nanometer	nm	1/1,000,000,000	0.00000004
meter	m	100 cm	
		1,000 mm	
		1,000,000 μm	
		1,000,000,000 nm	

FIGURE 4.5 ▶**Animated** Relative sizes. Below, the diameter of most cells is in the range of 1 to 100 micrometers. **TABLE 4.2** shows conversions among units of length; also see Units of Measure, Appendix VI.

Which one is smallest: a protein, a lipid, or a water molecule?

Answer: A water molecule

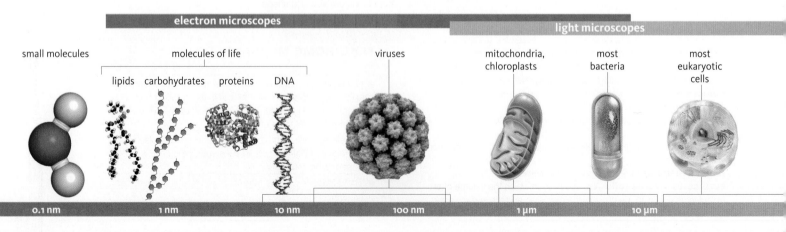

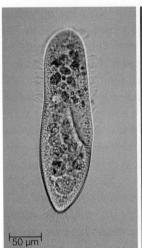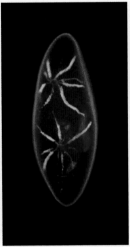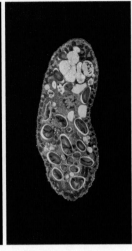

A The green blobs visible in this light micrograph (LM) of a living cell are ingested algal cells. Hairlike structures on the cell's surface are waving cilia that propel this motile organism through its fluid surroundings.

B A light micrograph taken with polarized light shows edges in relief. This technique reveals ingested algal cells, and some internal structures not visible in **A**.

C In this fluorescence micrograph, yellow pinpoints the location of a particular protein in the membrane of organelles called contractile vacuoles. These organelles are also visible in **B**.

D A colorized transmission electron micrograph (TEM) reveals several types of internal structures in a plane (slice). Ingested algal cells are being broken down inside food vacuoles.

E A scanning electron micrograph (SEM) shows details of the cell's surface, including the thick coat of cilia. The cell ingests its food via the indentation (also visible in **A**).

FIGURE 4.6 Different microscopy techniques reveal different characteristics of the same type of organism, a protist (*Paramecium*).

What are the approximate dimensions of a *Paramecium*?

Answer: Approximately 250 μm long and 50 μm wide

directs electrons through a thin specimen, and the specimen's internal details appear as shadows in the resulting image, which is called a transmission electron micrograph, or TEM (**FIGURE 4.6D**). A scanning electron microscope directs a beam of electrons back and forth across the surface of a specimen that has been coated with a thin layer of gold or other metal. The irradiated metal emits electrons and x-rays, which are converted into an image (a scanning electron micrograph, or SEM) of the surface (**FIGURE 4.6E**). SEMs

and TEMs are always black and white; colored versions have been digitally altered to highlight specific details.

TAKE-HOME MESSAGE 4.3

How do we see cells?

✔ Most cells are visible only with the help of microscopes.

✔ We use different microscopes and techniques to reveal different aspects of cell structure.

human eye (no microscope)

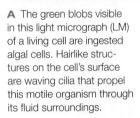

frog eggs

small animals

largest organisms

| 100 μm | 1 mm | 1 cm | 10 cm | 1 m | 10 m | 100 m |

4.4 Introducing Prokaryotes

✔ Prokaryotic cells (bacteria and archaea) have no nucleus.

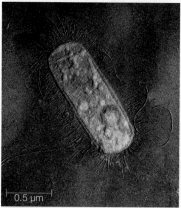

A *Escherichia coli*, a common bacterial inhabitant of human intestines. Short, hair-like structures are pili; longer ones are flagella.

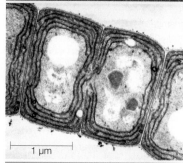

B *Oscillatoria*, a type of cyanobacteria that forms long filaments. Like other members of this ancient lineage, *Oscillatoria* has internal membranes (in green) where photosynthesis occurs. The multi-sided structures (pink) are carboxysomes, protein-enclosed organelles that assist photosynthesis.

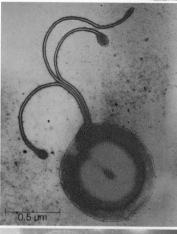

C *Helicobacter pylori*, a bacterium that can cause stomach ulcers when it infects the lining of the stomach. In unfavorable conditions, this species takes on a ball-shaped form (shown) that may offer the cells protection from environmental challenges such as antibiotic treatment.

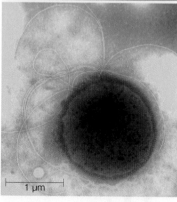

D *Thermococcus gammatolerans*, an archaeon discovered at a deep-sea hydrothermal vent, where it lives under extreme conditions of salt, temperature, and pressure. It is by far the most radiation-resistant organism ever discovered, capable of withstanding thousands of times more radiation than humans can.

FIGURE 4.7 Some representatives of bacteria (**A–C**) and an archaeon (**D**).

All bacteria and archaea are single-celled organisms, although individual cells of many species cluster in filaments or colonies (**FIGURE 4.7**). Outwardly, cells of the two groups appear so similar that archaea were once presumed to be an unusual group of bacteria. Both were classified as prokaryotes, a word that means "before the nucleus." By 1977, it had become clear that archaea are more closely related to eukaryotes than to bacteria, so they were given their own separate domain. The term "prokaryote" is now an informal designation only.

Bacteria and archaea are the smallest and most metabolically diverse forms of life that we know about. Chapter 20 revisits them in more detail; here we present an overview of structures shared by both groups (**FIGURE 4.8**).

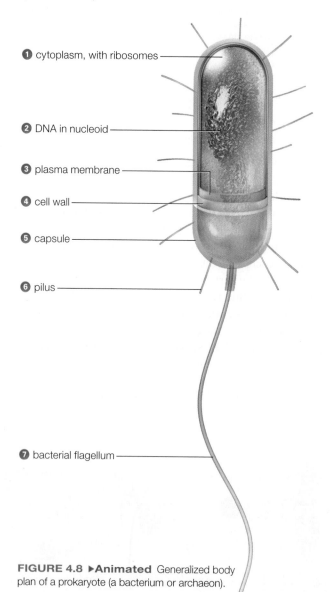

❶ cytoplasm, with ribosomes

❷ DNA in nucleoid

❸ plasma membrane

❹ cell wall

❺ capsule

❻ pilus

❼ bacterial flagellum

FIGURE 4.8 ▶**Animated** Generalized body plan of a prokaryote (a bacterium or archaeon).

CREDITS: (7A) © Biophoto Associates/Science Photo Library; (7B) Biophoto Associates/Science Source; (7C) Biomedical Imaging Unit, Southhampton General Hospital/Science Photo Library; (7D) Archivo Angels Tapias y Fabrice Confalonieri; (8) From Starr/Taggart/Evers/Starr, Biology, 13E. © 2013 Cengage Learning.

Compared with eukaryotic cells, prokaryotes have little in the way of internal framework, but they do have protein filaments under the plasma membrane that reinforce the cell's shape and act as scaffolding for internal structures. The cytoplasm of these cells ❶ contains many **ribosomes** (organelles upon which polypeptides are assembled), and in some species, additional organelles. Cytoplasm also contains **plasmids**, which are small circles of DNA that carry a few genes (units of inheritance). Plasmid genes can provide advantages such as resistance to antibiotics. The cell's remaining genes typically occur on one large circular molecule of DNA located in an irregularly shaped region of cytoplasm called the **nucleoid** ❷. In a few species, the nucleoid is enclosed by a membrane. Other internal membranes carry out special metabolic processes such as photosynthesis in some prokaryotes (**FIGURE 4.7B**).

Like all cells, bacteria and archaea have a plasma membrane ❸. In nearly all prokaryotes, a rigid **cell wall** ❹ surrounding the plasma membrane protects the cell and supports its shape. Most archaeal cell walls consist of proteins; most bacterial cell walls consist of peptides and polysaccharides. Both types are permeable to water, so dissolved substances easily cross.

Polysaccharides form a slime layer or capsule ❺ around the wall of many types of bacteria. These sticky structures help the cells adhere to various surfaces, and they also offer some protection against predators.

Protein filaments called **pili** (singular, pilus) ❻ project from the surface of some prokaryotes. Pili help these cells move across or cling to surfaces. One kind, a "sex" pilus, attaches to another bacterium and then shortens. The attached cell is reeled in, and DNA is transferred from one cell to the other. Many types of bacteria and archaea also have one or more flagella projecting from their surface ❼. **Flagella** (singular, flagellum) are long, slender cellular structures used for motion. A prokaryotic flagellum rotates like a propeller that drives the cell through fluid habitats.

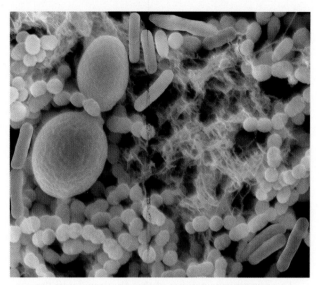

FIGURE 4.9 Oral bacteria in dental plaque, a biofilm. This micrograph shows two species of bacteria (tan, green) and a yeast (red) sticking to one another and to teeth via a gluelike mass of shared, secreted polysaccharides (pink). Other secretions of these organisms cause cavities and periodontal disease.

Biofilms

Bacterial cells often live so close together that an entire community shares a layer of secreted polysaccharides and proteins. A communal living arrangement in which single-celled organisms occupy a shared mass of slime is called a **biofilm**. A biofilm is often attached to a solid surface, and may include bacteria, algae, fungi, protists, and/or archaea. Participating in a biofilm allows the cells to linger in a favorable spot rather than be swept away by fluid currents, and to reap the benefits of living communally. For example, rigid or netlike secretions of some species serve as permanent scaffolding for others; species that break down toxic chemicals allow more sensitive ones to thrive in habitats that they could not withstand on their own; and waste products of some serve as raw materials for others. Later chapters discuss medical implications of biofilms, including the dental plaque that forms on teeth (**FIGURE 4.9**).

biofilm Community of microorganisms living within a shared mass of secreted slime.
cell wall Rigid but permeable structure that surrounds the plasma membrane of some cells.
flagellum Long, slender cellular structure used for motility.
nucleoid Of a bacterium or archaeon, region of cytoplasm where the DNA is concentrated.
pilus A protein filament that projects from the surface of some prokaryotic cells.
plasmid Small circle of DNA in some bacteria and archaea.
ribosome Organelle of protein synthesis.

4.5 Introducing Eukaryotic Cells

✔ Eukaryotic cells carry out much of their metabolism inside organelles enclosed by membranes.

In addition to the nucleus, a typical eukaryotic cell has many other membrane-enclosed organelles, including endoplasmic reticulum, Golgi bodies, and at least one mitochondrion (**TABLE 4.3** and **FIGURES 4.10** and **4.11**). An enclosing membrane allows an organelle to regulate the types and amounts of substances that enter and exit. Through this type of control, an organelle maintains a special internal environment that allows it to carry out a particular function—for example, isolating toxic or sensitive substances from the rest of the cell, moving substances through cytoplasm, maintaining fluid balance, or providing a favorable environment for a special process.

The remaining sections of this chapter detail the structures and functions of organelles typical of eukaryotic cells.

TAKE-HOME MESSAGE 4.5
What do all eukaryotic cells have in common?

✔ All eukaryotic cells start life with a nucleus and other membrane-enclosed organelles.

Table 4.3
Some Organelles in Eukaryotic Cells

Organelles with membranes

Nucleus	Protects, controls access to DNA
Endoplasmic reticulum (ER)	Makes, modifies polypeptides and lipids; other tasks
Golgi body	Modifies and sorts polypeptides and lipids
Vesicle	Transports, stores, or breaks down substances
Mitochondrion	Makes ATP by glucose breakdown
Chloroplast	Makes sugars (in plants, some protists)
Lysosome	Intracellular digestion
Peroxisome	Breaks down fatty acids, amino acids, toxins
Vacuole	Storage, breaks down food or waste

Organelles without membranes

Ribosomes	Assembles polypeptides
Centriole	Anchors cytoskeleton

Other components

Cytoskeleton	Contributes to cell shape, internal organization, movement

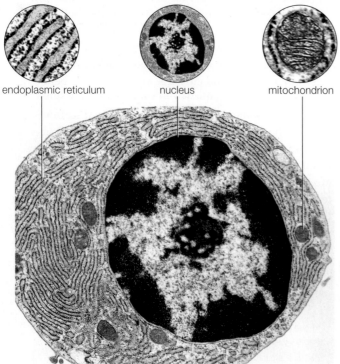

endoplasmic reticulum   nucleus   mitochondrion   cell wall   Golgi body   vacuole

An animal cell (a white blood cell of a guinea pig)

A plant cell (from a root of thale cress)

FIGURE 4.10 Transmission electron micrographs of two eukaryotic cells.

CREDITS: (10) left, Don W. Fawcett/Science Source; right, Biophoto Associates/Science Source; (Table 4.3) © Cengage Learning.

A Typical plant cell components.

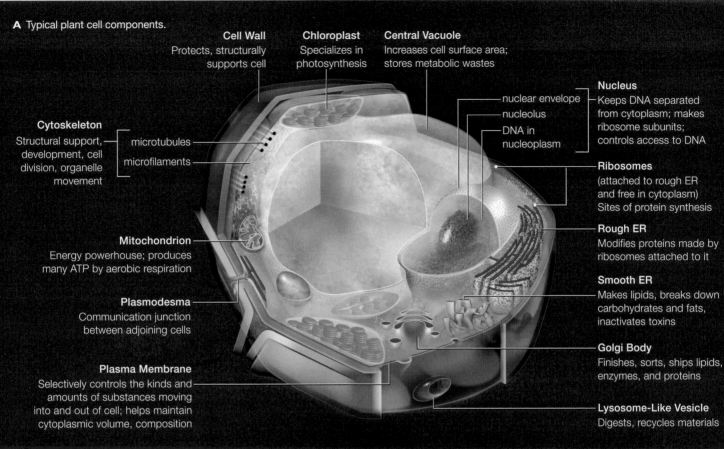

Cell Wall
Protects, structurally supports cell

Chloroplast
Specializes in photosynthesis

Central Vacuole
Increases cell surface area; stores metabolic wastes

nuclear envelope
nucleolus
DNA in nucleoplasm

Nucleus
Keeps DNA separated from cytoplasm; makes ribosome subunits; controls access to DNA

Cytoskeleton
Structural support, development, cell division, organelle movement
microtubules
microfilaments

Ribosomes
(attached to rough ER and free in cytoplasm) Sites of protein synthesis

Rough ER
Modifies proteins made by ribosomes attached to it

Mitochondrion
Energy powerhouse; produces many ATP by aerobic respiration

Smooth ER
Makes lipids, breaks down carbohydrates and fats, inactivates toxins

Plasmodesma
Communication junction between adjoining cells

Golgi Body
Finishes, sorts, ships lipids, enzymes, and proteins

Plasma Membrane
Selectively controls the kinds and amounts of substances moving into and out of cell; helps maintain cytoplasmic volume, composition

Lysosome-Like Vesicle
Digests, recycles materials

B Typical animal cell components.

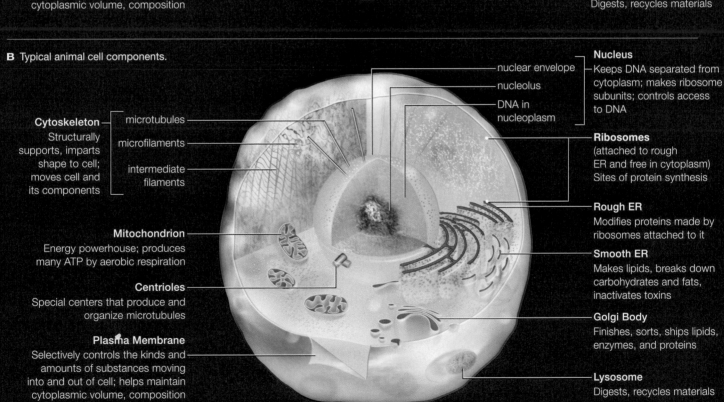

nuclear envelope
nucleolus
DNA in nucleoplasm

Nucleus
Keeps DNA separated from cytoplasm; makes ribosome subunits; controls access to DNA

Cytoskeleton
Structurally supports, imparts shape to cell; moves cell and its components
microtubules
microfilaments
intermediate filaments

Ribosomes
(attached to rough ER and free in cytoplasm) Sites of protein synthesis

Rough ER
Modifies proteins made by ribosomes attached to it

Mitochondrion
Energy powerhouse; produces many ATP by aerobic respiration

Smooth ER
Makes lipids, breaks down carbohydrates and fats, inactivates toxins

Centrioles
Special centers that produce and organize microtubules

Golgi Body
Finishes, sorts, ships lipids, enzymes, and proteins

Plasma Membrane
Selectively controls the kinds and amounts of substances moving into and out of cell; helps maintain cytoplasmic volume, composition

Lysosome
Digests, recycles materials

FIGURE 4.11 ▶Animated Organelles and structures typical of **A** plant cells and **B** animal cells.

4.6 The Nucleus

✔ A nucleus keeps a eukaryotic cell's DNA away from potentially damaging reactions in the cytoplasm.

✔ Pores in the nuclear envelope selectively restrict access to the cell's DNA.

The cell nucleus serves two important functions. First, it keeps the cell's genetic material—DNA—safe from metabolic processes that might damage it. Isolated in its own compartment, DNA stays separated from the bustling activity of the cytoplasm. Second, a nucleus controls the passage of certain molecules across its membrane. **FIGURE 4.12** shows the components of the nucleus. **TABLE 4.4** lists their functions. Let's zoom in on the individual components.

Chromatin

As molecules go, DNA is gigantic. Unraveled and stretched out, the DNA in the nucleus of a single human cell would be about 2 meters (6–1/2 feet) long. All of that DNA fits into a nucleus only six microns in diameter. How? Proteins associate with and organize each DNA molecule so it can pack tightly—and precisely—into the nucleus (Chapter 8 returns to this topic). All of the DNA in a cell's nucleus, together with associated proteins, is collectively called **chromatin**. Chromatin is suspended in **nucleoplasm**, a viscous fluid similar to cytoplasm, that fills the nucleus.

The Nuclear Envelope

The special membrane that encloses the nucleus is called the **nuclear envelope**. It consists of two phospholipid bilayers, folded around an intermembrane space 20–40 nm wide (**FIGURE 4.13**). The outer lipid bilayer is continuous with the lipid bilayer of endoplasmic reticulum, and like rough endoplasmic reticulum it is studded with ribosomes (more about endoplasmic reticulum in the next section). The inner lipid bilayer

Table 4.4	Components of the Nucleus
Chromatin	DNA and associated proteins in a cell nucleus
Nucleoplasm	Semifluid interior portion of the nucleus
Nuclear envelope	Double membrane with nuclear pores that control which substances enter and exit the nucleus
Nucleolus	Dense region of proteins and nucleic acid where ribosomal subunits are being produced

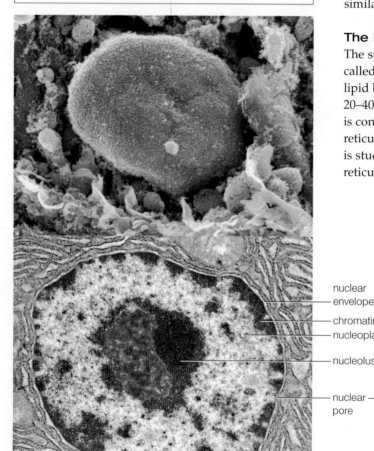

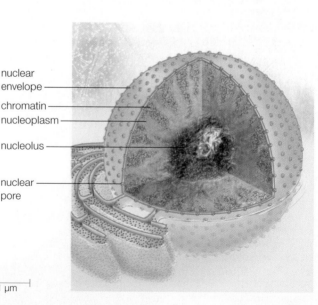

nuclear envelope

chromatin

nucleoplasm

nucleolus

nuclear pore

1 µm

FIGURE 4.12 ▶**Animated** The cell nucleus. The SEM shows the nucleus of a liver cell (in pink); the TEM, the nucleus of a pancreas cell.

CREDITS: (12) top left, Dr. David Furness, Keele University/Science Source; bottom left, © Kenneth Bart (12 right, Table 4.4) © Cengage Learning.

of animal cell nuclei is covered and supported by the nuclear lamina, a dense mesh of fibrous proteins (Section 4.10 returns to these proteins). Membrane proteins connect the lamina to the nuclear envelope.

The two lipid bilayers of a nuclear envelope connect at nuclear pores. Some bacteria have membranes around their DNA, but we do not consider the bacteria to have nuclei because there are no pores in these membranes. By contrast, a eukaryotic cell's nucleus has thousands of nuclear pores, each composed of hundreds of membrane proteins. A nuclear pore forms a hole in the envelope that can be widened or constricted by conformational changes in these proteins.

As you will see in Chapter 5, large molecules, including RNA and proteins, cannot cross lipid bilayers on their own. Nuclear pores function as gateways for these molecules to enter and exit a nucleus. Protein synthesis offers an example of why this movement is important. Protein synthesis occurs in cytoplasm, and it requires the participation of many molecules of RNA. RNA is produced in the nucleus. Thus, RNA molecules must move from nucleus to cytoplasm, and they do so through nuclear pores. Proteins that carry out RNA synthesis must move in the opposite direction, because this process occurs in the nucleus. A cell can regulate the amounts and types of proteins it makes at a given time by selectively restricting the passage of certain molecules through nuclear pores. Chapter 9 returns to the details of protein synthesis; and Chapter 10, to controls over this process.

The Nucleolus

Depending on a cell's metabolic state, its nucleus contains one or more nucleoli. A **nucleolus** (plural, nucleoli) is an irregularly shaped region, dense with proteins and nucleic acids, where subunits of ribosomes are being produced. The subunits pass through nuclear pores into the cytoplasm, where they join and become active in protein synthesis (Section 9.4 returns to ribosome structure). New research is revealing additional roles for nucleoli, for example in cell division, cell death, and responses to cellular stress.

---

chromatin Collective term for all of the DNA and associated proteins in a cell nucleus.
nuclear envelope A double membrane that constitutes the outer boundary of the nucleus. Nuclear pores in the membrane control the entry and exit of large molecules.
nucleolus In a cell nucleus, a dense, irregularly shaped region where ribosomal subunits are being produced.
nucleoplasm Viscous fluid enclosed by the nuclear envelope.

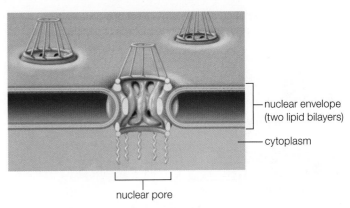

A The nuclear envelope consists of two lipid bilayers that connect at nuclear pores. Each pore is a complex structure formed by hundreds of proteins; each forms a hole in the envelope that can change in diameter. Part of a pore is a basket-shaped, multifunctional scaffold that can, for example, bind to chromatin. Cytoplasmic fibrils guide molecules through the pore.

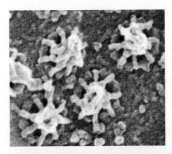

B The proteins that compose a nuclear pore allow only certain molecules to cross the nuclear membrane. The "baskets" on the inside of the nuclear envelope are visible in this SEM.

100 nm

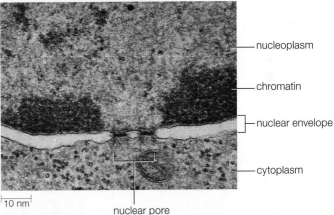

C This TEM shows a pore in the nuclear envelope of a bone marrow cell nucleus. The identity of the "plug" seen in the center of this and many other nuclear pores is unknown; it may be a molecule caught in transit.

FIGURE 4.13 Structure of the nuclear membrane.

TAKE-HOME MESSAGE 4.6
What is the function of the cell nucleus?

✔ A nucleus protects and controls access to a eukaryotic cell's DNA.

✔ The nuclear envelope is a double lipid bilayer. Proteins embedded in the bilayer form pores that control the passage of molecules between the nucleus and cytoplasm.

4.7 The Endomembrane System

✔ The endomembrane system is a set of organelles that makes, modifies, and transports proteins and lipids.

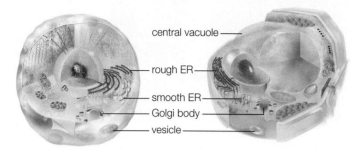

central vacuole

rough ER

smooth ER

Golgi body

vesicle

The **endomembrane system** (above) is a series of interacting organelles between the nucleus and the plasma membrane (**FIGURE 4.14**). Its main function is to make lipids, enzymes, and proteins for insertion into the cell's membranes or secretion to the external environment. The endomembrane system also destroys toxins, recycles wastes, and has other special functions. Components of the system vary among different types of cells, but here we present an overview of the most common ones.

Small, membrane-enclosed sacs called **vesicles** ❶ form by budding from other organelles or when a patch of plasma membrane sinks into the cytoplasm. Many types carry substances from one organelle to another, or to and from the plasma membrane. Some are a bit like trash cans that collect and dispose of waste, debris, or toxins. Enzymes in vesicles called **peroxisomes** break down fatty acids, amino acids, and poisons such as alcohol. They also break down hydrogen peroxide, a toxic by-product of fatty acid breakdown. **Lysosomes**

are vesicles that take part in intracellular digestion. They contain powerful enzymes that break down cellular debris and wastes (carbohydrates, proteins, nucleic acids, and lipids). In cells such as amoebas and white blood cells, ingested bacteria, cell parts, and other particles are delivered to lysosomes for breakdown.

Vacuoles are sacs that form by the fusion of multiple vesicles. Many isolate or break down waste, debris, toxins, or food (**FIGURE 4.15**). Plant cells have a large **central vacuole**, in which amino acids, sugars, ions, wastes, and toxins accumulate. Fluid pressure in a central vacuole keeps plant cells plump, so stems, leaves, and other plant parts stay firm.

Endoplasmic reticulum (**ER**) comprises an interconnected system of tubes and flattened sacs. The ER membrane is an extension of the outer lipid bilayer of the nuclear envelope, so the space it encloses is continuous with the intermembrane space of the nuclear envelope. Two kinds of ER, rough and smooth, are named for their appearance. Thousands of ribosomes that attach to the outer surface of rough ER give this organelle its "rough" appearance. These ribosomes make polypeptides that thread into the interior of the ER as they are assembled ❷. Inside the ER, the polypeptide chains fold and take on their tertiary structure, and many assemble with other polypeptide chains (Section 3.6). Cells that make, store, and secrete proteins have a lot of rough ER. For example, ER-rich cells in the pancreas make digestive enzymes that they secrete into the small intestine.

Some proteins made in rough ER become part of its membrane. Others migrate through the ER compartment to smooth ER. Smooth ER has no ribosomes, so it

FIGURE 4.14 ▸**Animated** Some interactions among components of the endomembrane system.

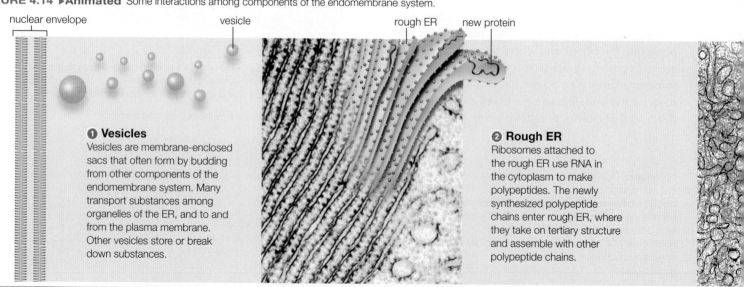

nuclear envelope

vesicle

rough ER

new protein

❶ Vesicles
Vesicles are membrane-enclosed sacs that often form by budding from other components of the endomembrane system. Many transport substances among organelles of the ER, and to and from the plasma membrane. Other vesicles store or break down substances.

❷ Rough ER
Ribosomes attached to the rough ER use RNA in the cytoplasm to make polypeptides. The newly synthesized polypeptide chains enter rough ER, where they take on tertiary structure and assemble with other polypeptide chains.

CREDITS: (in text, 14 art) © Cengage Learning; (14) #2, Don W. Fawcett/Science Source.

does not make its own proteins ❸. Some proteins that arrive in smooth ER are immediately packaged into vesicles for delivery elsewhere. Others are enzymes that stay and become part of the smooth ER. Smooth ER enzymes break down carbohydrates, fatty acids, and some drugs and poisons; and also make lipids for the cell's membranes.

A **Golgi body** has a folded membrane that often looks like a stack of pancakes ❹. Enzymes inside of Golgi bodies put finishing touches on proteins and lipids that have been delivered from ER. These enzymes attach phosphate groups or carbohydrates, and cleave certain proteins. The finished products (such as membrane proteins and proteins for secretion, and other enzymes) are sorted and packaged in new vesicles. Some of the vesicles deliver their cargo to the plasma membrane; others become lysosomes.

---

central vacuole Large fluid-filled organelle in many plant cells.
endomembrane system Series of interacting organelles (endoplasmic reticulum, Golgi bodies, vesicles) between nucleus and plasma membrane; produces lipids, proteins.
endoplasmic reticulum (ER) Membrane-enclosed organelle that is a continuous system of sacs and tubes extending from the nuclear envelope. Smooth ER makes lipids and breaks down carbohydrates and fatty acids; ribosomes on the surface of rough ER make proteins.
Golgi body Membrane-enclosed organelle that modifies proteins and lipids, then packages the finished products into vesicles.
lysosome Enzyme-filled vesicle that breaks down cellular wastes and debris.
peroxisome Enzyme-filled vesicle that breaks down amino acids, fatty acids, and toxic substances.
vacuole A membrane-enclosed organelle filled with fluid; isolates or disposes of waste, debris, or toxic materials.
vesicle Small, membrane-enclosed organelle; different kinds store, transport, or break down their contents.

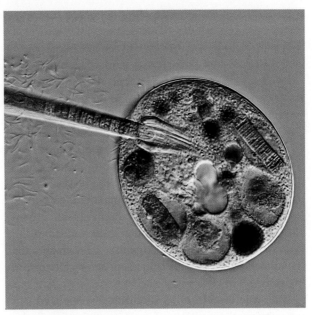

FIGURE 4.15 An example of vacuole function. The protist *Nassula* (round cell) uses a distinctive oral basket to feed on strands of photosynthetic algae. Ingested cells are packaged in food vacuoles that change color (green to purple to brown to gold) as chlorophyll molecules inside them get broken down.

TAKE-HOME MESSAGE 4.7
What is the endomembrane system?

✔ The endomembrane system includes rough and smooth endoplasmic reticulum, vesicles, and Golgi bodies.

✔ Rough ER produces enzymes, membrane proteins, and secreted proteins. Smooth ER produces lipids and breaks down carbohydrates, fatty acids, and toxins.

✔ Golgi bodies modify and ship new proteins and lipids.

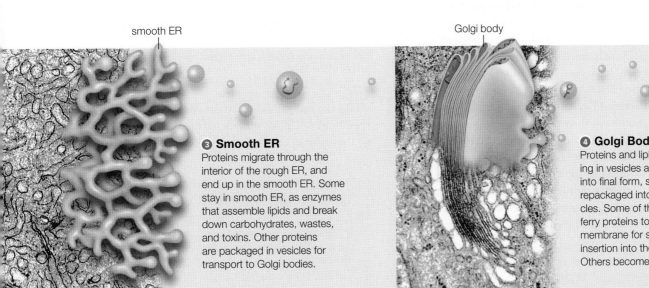

smooth ER

❸ Smooth ER
Proteins migrate through the interior of the rough ER, and end up in the smooth ER. Some stay in smooth ER, as enzymes that assemble lipids and break down carbohydrates, wastes, and toxins. Other proteins are packaged in vesicles for transport to Golgi bodies.

Golgi body

plasma membrane

❹ Golgi Body
Proteins and lipids arriving in vesicles are modified into final form, sorted, and repackaged into new vesicles. Some of these vesicles ferry proteins to the plasma membrane for secretion or insertion into the lipid bilayer. Others become lysosomes.

4.8 Mitochondria

✔ Eukaryotic cells make most of their ATP in mitochondria.

mitochondrion

As you will see in Chapter 5, biologists think of the nucleotide ATP as a type of cellular currency because it carries energy between reactions. Cells require a lot of ATP. The most efficient way they can produce it is by aerobic respiration, a series of oxygen-requiring reactions that harvests the energy in sugars by breaking their bonds. In eukaryotes, aerobic respiration occurs inside organelles called **mitochondria** (singular, mitochondrion). With each breath, you are taking in oxygen mainly for the mitochondria in your trillions of aerobically respiring cells.

The structure of a mitochondrion is specialized for carrying out reactions of aerobic respiration. Each mitochondrion has two membranes, one highly folded inside the other (**FIGURE 4.16**). This arrangement creates two compartments: an outer one (between the two membranes), and an inner one (inside the inner membrane). During aerobic respiration, hydrogen ions accumulate between the two membranes. The buildup causes the ions to flow across the inner mitochondrial membrane, and this flow drives ATP formation (we return to aerobic respiration in Chapter 7).

Nearly all eukaryotic cells (including plant cells) have mitochondria, but the number varies by the type of cell and by the organism. For example, single-celled organisms such as yeast often have only one mitochondrion, but human skeletal muscle cells have a thousand or more. In general, cells that have the highest demand for energy tend to have the most mitochondria.

Typical mitochondria are between 1 and 4 micrometers in length. These organelles can change shape, split in two, branch, or fuse together. They resemble bacteria in size, form, and biochemistry. Mitochondria have their own DNA, which is circular and otherwise similar to bacterial DNA. They divide independently of the cell, and have their own ribosomes. Such clues led to a theory that mitochondria evolved from aerobic bacteria that took up permanent residence inside a host cell (we return to this topic in Section 19.6).

Some eukaryotes that live in oxygen-free environments have hydrogenosomes, which are modified mitochondria that produce hydrogen in addition to

mitochondrion Double-membraned organelle that produces ATP by aerobic respiration in eukaryotes.

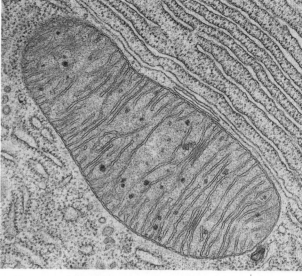

A Mitochondrion in a cell from bat pancreas. 0.5 μm

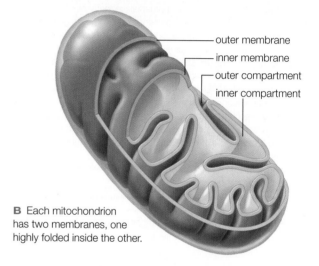

outer membrane
inner membrane
outer compartment
inner compartment

B Each mitochondrion has two membranes, one highly folded inside the other.

FIGURE 4.16 ▶**Animated** The mitochondrion, a eukaryotic organelle that specializes in producing ATP.

What organelle is visible in the upper right-hand corner of the TEM? Answer: Rough ER

ATP. Like mitochondria, hydrogenosomes have two membranes. Unlike mitochondria, they have no DNA, so they cannot divide independently of the cell.

TAKE-HOME MESSAGE 4.8
What do mitochondria do?

✔ Mitochondria are eukaryotic organelles specialized to produce ATP by aerobic respiration, an oxygen-requiring series of reactions that breaks down carbohydrates.

4.9 Chloroplasts and Other Plastids

✔ Plastids function in storage and photosynthesis in plants and some types of algae.

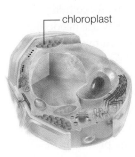
chloroplast

Plastids are double-membraned organelles that function in photosynthesis, storage, or pigmentation in plant and algal cells. Photosynthetic cells of plants and many protists contain **chloroplasts**, which are plastids specialized for photosynthesis (**FIGURE 4.17**). Most chloroplasts are oval or disk-shaped. Each has two outer membranes enclosing a semifluid interior, the stroma, that contains enzymes and the chloroplast's own DNA. In the stroma, a third, highly folded membrane forms a single, continuous compartment. Photosynthesis occurs at this inner membrane.

The innermost membrane of a chloroplast incorporates many pigments, including a green one called chlorophyll (the abundance of chlorophyll in plant cell chloroplasts is the reason most plants are green). During photosynthesis, these pigments capture energy from sunlight, and pass it to other molecules that require energy to make ATP. The resulting ATP is used inside the stroma to build sugars from carbon dioxide and water. (Chapter 6 returns to details of these processes.) In many ways, chloroplasts resemble the photosynthetic bacteria from which they evolved.

Chromoplasts are plastids that make and store pigments other than chlorophylls. They often contain red or orange carotenoids that color flowers, leaves, roots, and fruits (**FIGURE 4.18**). Chromoplasts are related to chloroplasts, and the two types of plastids are interconvertible. For example, as fruits such as tomatoes ripen, green chloroplasts in their cells are converted to red chromoplasts, so the color of the fruit changes.

Amyloplasts are unpigmented plastids that make and store starch grains. They are notably abundant in cells of stems, tubers (underground stems), fruits, and seeds. Like chromoplasts, amyloplasts are related to chloroplasts, and one type can change into the other. Starch-packed amyloplasts are dense and heavy compared to cytoplasm; in some plant cells, they function as gravity-sensing organelles (we return to this topic in Section 30.8).

---

chloroplast Organelle of photosynthesis in the cells of plants and photosynthetic protists.
plastid Double-membraned organelle that functions in photosynthesis, pigmentation, or storage in plants and algal cells; for example, a chloroplast, chromoplast, or amyloplast.

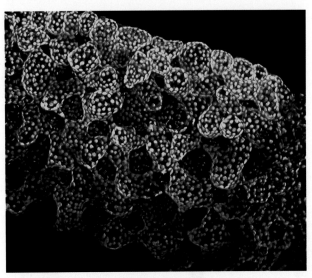

A Chloroplast-packed cells make up a leaf of a flowering plant.

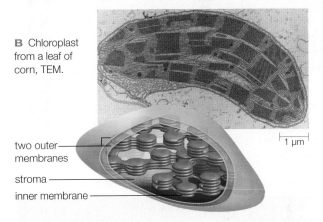

B Chloroplast from a leaf of corn, TEM.

1 µm

two outer membranes

stroma

inner membrane

C Each chloroplast has two outer membranes. Photosynthesis occurs at a much-folded inner membrane.

FIGURE 4.17 ▶**Animated** The chloroplast.

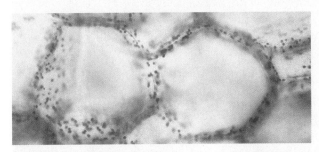

FIGURE 4.18 Chromoplasts in cells of a red bell pepper. The color of the fruit arises from these plastids.

TAKE-HOME MESSAGE 4.9

What are plastids?

✔ Plastids occur in plants and some protists; they function in photosynthesis, storage, and pigmentation.

✔ Chloroplasts are plastids that carry out photosynthesis.

CREDITS: (in text) © Cengage Learning; (17A) Heiti Paves/Science Photo Library; (17B) Science Source; (17C) © Cengage Learning; (18) © David T. Webb.

✔ A cytoskeleton supports a eukaryotic cell and helps it move.

Between the nucleus and plasma membrane of all eukaryotic cells is a system of interconnected protein filaments collectively called the **cytoskeleton**. Elements of the cytoskeleton reinforce, organize, and move cell structures, and often the whole cell. Some are permanent; others form only at certain times.

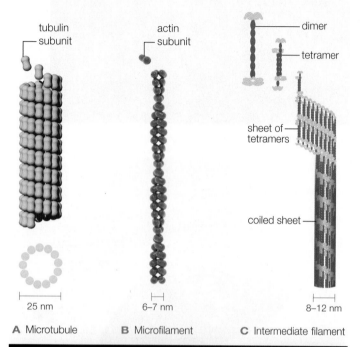

tubulin subunit

actin subunit

dimer

tetramer

sheet of tetramers

coiled sheet

25 nm

6–7 nm

8–12 nm

A Microtubule   **B** Microfilament   **C** Intermediate filament

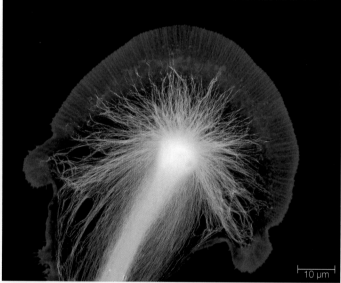

D A fluorescence micrograph shows microtubules (yellow) and microfilaments (blue) in the growing end of a nerve cell. These cytoskeletal elements support and guide the cell's lengthening in a particular direction.

FIGURE 4.19 ▶Animated Cytoskeletal elements.

Microtubules are long, hollow cylinders that consist of subunits of the protein tubulin (**FIGURE 4.19A**). They form a dynamic scaffolding for many cellular processes, rapidly assembling when they are needed, disassembling when they are not. For example, before a eukaryotic cell divides, microtubules assemble, separate the cell's duplicated DNA molecules, then disassemble. As another example, microtubules that form in the growing end of a young nerve cell support its lengthening in a particular direction (**FIGURE 4.19D**).

Microfilaments are fibers that consist primarily of subunits of a protein called actin (**FIGURE 4.19B**). These fine fibers strengthen or change the shape of eukaryotic cells, and have a critical function in cell migration, movement, and contraction. Crosslinked, bundled, or gel-like arrays of them make up the **cell cortex**, a reinforcing mesh under the plasma membrane. Microfilaments also connect plasma membrane proteins to other proteins inside the cell.

Intermediate filaments are the most stable elements of the cytoskeleton, forming a framework that lends structure and resilience to cells and tissues in multicelled organisms. Several types of intermediate filaments are assembled from different proteins (**FIGURE 4.19C**). For example, intermediate filaments that make up your hair consist of keratin, a fibrous protein (Section 3.6). Intermediate filaments that consist of lamins, another type of fibrous protein, form the nuclear lamina of animal cells. In addition to providing structural support, nuclear lamins are part of mechanisms that regulate DNA replication and other processes that take place inside the nucleus.

Motor proteins that associate with cytoskeletal elements move cell parts when energized by a phosphate-group transfer from ATP (Section 3.8). A cell is like a bustling train station, with molecules and structures being moved continuously throughout its interior. Motor proteins are like freight trains, dragging cellular cargo along tracks of microtubules and microfilaments (**FIGURE 4.20**). The motor protein myosin interacts with microfilaments to bring about muscle cell contraction. Another motor protein, dynein, interacts with microtubules to bring about movement of flagella and cilia in eukaryotes. Eukaryotic flagella whip back and forth to propel motile cells such as sperm through fluid (**FIGURE 4.21A**). Cilia (singular, cilium) are short, hair-

FIGURE 4.20 ▶Animated Motor proteins. Here, kinesin (tan) drags a pink vesicle as it inches along a microtubule.

like structures that project from the surface of some cells. The coordinated waving of many cilia propels some cells through fluid, and stirs fluid around other cells that are stationary. The waving movement of eukaryotic flagella and cilia, which differs from the propeller-like rotation of prokaryotic flagella, arises from their internal architecture. Microtubules extend lengthwise through these structures, in what is called a 9+2 array (**FIGURE 4.21B**). The array consists of nine pairs of microtubules ringing another pair in the center. The microtubules grow from a barrel-shaped organelle, the **centriole**, which remains below the finished array as part of a **basal body** (**FIGURE 4.21C**).

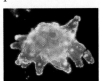

Some eukaryotic cells, including the amoeba at left, form **pseudopods**, or "false feet." As these temporary, irregular lobes bulge outward, they move the cell and engulf a target such as prey. Elongating microfilaments force the lobe to advance in a steady direction. Motor proteins attached to the microfilaments drag the plasma membrane along with them.

basal body Organelle that develops from a centriole.
cell cortex Reinforcing mesh of cytoskeletal elements under a plasma membrane.
centriole Barrel-shaped organelle from which microtubules grow.
cilium Short, movable structure that projects from the plasma membrane of some eukaryotic cells.
cytoskeleton Dynamic framework of protein filaments that support, organize, and move eukaryotic cells and their internal structures.
intermediate filament Stable cytoskeletal element that structurally supports cells and tissues.
microfilament Reinforcing cytoskeletal element that functions in cell movement; a fiber of actin subunits.
microtubule Cytoskeletal element involved in cellular movement; hollow filament of tubulin subunits.
motor protein Type of energy-using protein that interacts with cytoskeletal elements to move the cell's parts or the whole cell.
pseudopod A temporary protrusion that helps some eukaryotic cells move and engulf prey.

TAKE-HOME MESSAGE 4.10
What is a cytoskeleton?

✔ A cytoskeleton of protein filaments is the basis of eukaryotic cell shape, internal structure, and movement.

✔ Microtubules organize eukaryotic cells and help move their parts. Networks of microfilaments reinforce cell shape and function in movement. Intermediate filaments strengthen and maintain the shape of cell membranes and tissues, and form external structures such as hair.

✔ When energized by ATP, motor proteins move along tracks of microtubules and microfilaments. As part of cilia, flagella, and pseudopods, they can move the whole cell.

A Flagella propel motile eukaryotic cells such as sperm through fluid surroundings. The waving motion of eukaryotic flagella and cilia arises from microtubules that run lengthwise through them, in a 9+2 array.

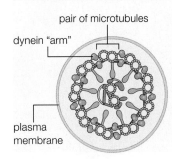

B A 9+2 array consists of a ring of nine pairs of microtubules plus one pair at their core. Stabilizing spokes and linking elements connect the microtubules and keep them aligned in this pattern. Projecting from each pair of microtubules are "arms" of the motor protein dynein.

C Microtubules of a developing 9+2 array grow from a centriole, which remains under the finished array as a basal body. The micrograph below shows basal bodies underlying cilia of the protist pictured in **FIGURE 4.6**.

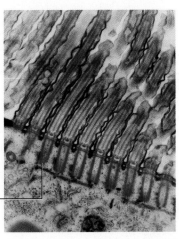

basal body

D Phosphate-group transfers from ATP cause the dynein arms in a 9+2 array to repeatedly bind the adjacent pair of microtubules, bend, and then disengage. The dynein arms "walk" along the microtubules, so adjacent microtubule pairs slide past one another. The short, sliding strokes of the dynein arms occur in a coordinated sequence around the ring, down the length of the microtubules. The movement causes the entire structure to bend.

FIGURE 4.21 ▶**Animated** How eukaryotic flagella and cilia move.

4.11 Cell Surface Specializations

✔ Many cells secrete materials that form a covering or matrix outside their plasma membrane.

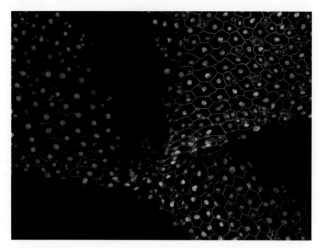

A Nuclei appear orange, and cell walls are green in this fluorescent micrograph of an *Arabidopsis* seedling. The walls, which are rigid and permeable, protect but do not isolate the cells.

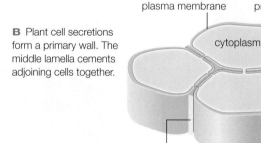

B Plant cell secretions form a primary wall. The middle lamella cements adjoining cells together.

plasma membrane · primary wall · cytoplasm · middle lamella

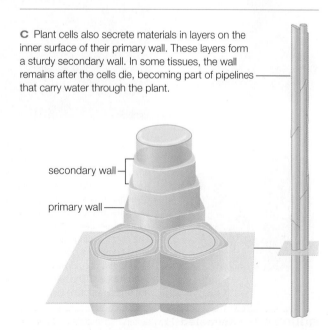

C Plant cells also secrete materials in layers on the inner surface of their primary wall. These layers form a sturdy secondary wall. In some tissues, the wall remains after the cells die, becoming part of pipelines that carry water through the plant.

secondary wall · primary wall

FIGURE 4.22 ▶**Animated** Plant cell walls.

Cell Matrices

Many cells secrete an **extracellular matrix** (**ECM**), a complex mixture of molecules that often includes polysaccharides and fibrous proteins. The composition and function of ECM vary by cell type.

A cell wall is an example of ECM. You learned in Section 4.4 that many prokaryotes have cell walls. Plants have them too (**FIGURE 4.22A**), as do fungi and some protists. The composition of the wall differs among these groups, but in all cases it supports and protects the cell. Like a prokaryotic cell wall, a eukaryotic cell wall is porous: Water and solutes easily cross it on the way to and from the plasma membrane.

In plants, the cell wall forms as a young cell secretes pectin and other polysaccharides onto the outer surface of its plasma membrane. The sticky coating is shared between adjacent cells, and it cements them together. Each cell then forms a **primary wall** by secreting strands of cellulose into the coating. Some pectin remains as the middle lamella, a sticky layer in between the primary walls of abutting plant cells (**FIGURE 4.22B**).

Being thin and pliable, a primary wall allows a growing plant cell to enlarge and change shape. In some plants, mature cells secrete material onto the primary wall's inner surface. These deposits form a firm **secondary wall** (**FIGURE 4.22C**). One of the materials deposited is **lignin**, an organic compound that makes up as much as 25 percent of the secondary wall of cells in older stems and roots. Lignified plant parts are stronger, more waterproof, and less susceptible to plant-attacking organisms than younger tissues.

Animal cells have no walls, but some types secrete an extracellular matrix called basement membrane. Despite the name, basement membrane is not a cell membrane because it does not consist of a lipid bilayer. Rather, it is a sheet of fibrous material that structurally supports and organizes tissues, and it has roles in cell signaling. Bone is an ECM composed mostly of the fibrous protein collagen, and hardened by deposits of calcium and phosphorus.

A **cuticle** is a type of ECM secreted by cells at a body surface. In plants, a cuticle of waxes and proteins helps stems and leaves fend off insects and retain water. Crabs, spiders, and other arthropods have a cuticle that consists mainly of chitin (Section 3.4).

Cell Junctions

In multicelled species, cells can interact with one another and their surroundings by way of cell junctions. **Cell junctions** are structures that connect a cell directly to other cells or to its environment. Cells

CREDITS: (22A) Fernan Federici & Jim Haseloff/Wellcome Images; (22B, C) © Cengage Learning.

send and receive substances and signals through some junctions. Other junctions help cells recognize and stick to each other and to ECM.

Three types of cell junctions are common in animal tissues (**FIGURE 4.23A**). In tissues that line body surfaces and internal cavities, rows of **tight junctions** fasten the plasma membranes of adjacent cells. These junctions prevent body fluids from seeping between the cells. For example, the lining of the stomach is leakproof because tight junctions seal its cells together. These junctions keep gastric fluid, which contains acid and destructive enzymes, safely inside the stomach. If a bacterial infection damages the stomach lining, gastric fluid leaks into and damages the underlying layers. A painful peptic ulcer results.

Adhering junctions fasten cells to one another and to basement membrane. These junctions make a tissue quite strong because they connect to cytoskeletal elements inside the cells. Contractile tissues (such as heart muscle) have a lot of adhering junctions, as do tissues subject to abrasion or stretching (such as skin).

Gap junctions are closable channels that connect the cytoplasm of adjoining animal cells. When open, they permit water, ions, and small molecules to pass directly from the cytoplasm of one cell to another. These channels allow entire regions of cells to respond to a single stimulus. Heart muscle and other tissues in which the cells perform a coordinated action have many gap junctions.

In plants, **plasmodesmata** (singular, plasmodesma) are open channels that connect the cytoplasm of adjoining cells. These cell junctions extend across the cell walls (**FIGURE 4.23B**). Like gap junctions, plasmodesmata allow substances to flow quickly from cell to cell.

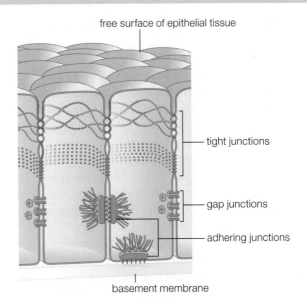

free surface of epithelial tissue

tight junctions

gap junctions

adhering junctions

basement membrane

A Three types of cell junctions in animal tissues.

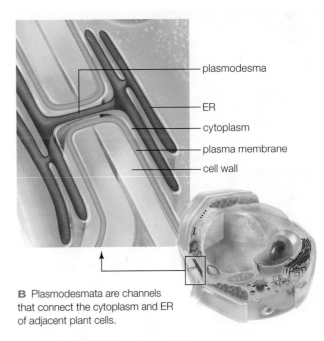

plasmodesma

ER

cytoplasm

plasma membrane

cell wall

B Plasmodesmata are channels that connect the cytoplasm and ER of adjacent plant cells.

FIGURE 4.23 ▶**Animated** Cell junctions.

---

adhering junction Cell junction that fastens an animal cell to another cell, or to basement membrane. Connects to cytoskeletal elements inside the cell.
cell junction Structure that connects a cell to another cell or to extracellular matrix.
cuticle Secreted covering at a body surface.
extracellular matrix (ECM) Complex mixture of cell secretions; its composition and function vary by cell type.
gap junction Cell junction that forms a closable channel across the plasma membranes of adjoining animal cells.
lignin Material that strengthens plant cell walls.
plasmodesma Cell junction that forms an open channel between the cytoplasm of adjacent plant cells.
primary wall The first cell wall of young plant cells.
secondary wall Lignin-reinforced wall that forms inside the primary wall of a plant cell.
tight junction Cell junction that fastens together the plasma membrane of adjacent animal cells; collectively prevent fluids from leaking between the cells.

CREDIT: (23) © Cengage Learning.

4.12 The Nature of Life

✔ We define life by describing the set of properties that is unique to living things.

Carbon, hydrogen, oxygen, and other atoms of organic molecules are the stuff of you, and us, and all of life. Yet it takes more than organic molecules to complete the picture. Life continues only as long as an ongoing flow of energy sustains its organization, because assembling molecules and cells requires energy. Life is no more and no less than a marvelously complex system for prolonging order. With energy and the hereditary codes of DNA, matter becomes organized, generation after generation.

In this chapter, you learned about the structure of cells, which have at their minimum a plasma membrane, cytoplasm, and DNA. Most cells have many other components in addition to these things.

We often use differences in cellular components—the presence or absence of a particular organelle, for example—to categorize life's diversity. What about life's commonality? The cell is the smallest unit with the properties of life, but what is it, exactly, that makes a cell, or an organism that consists of them, alive?

According to evolutionary biologist Gerald Joyce, the simplest definition of life might well be "that which is squishy." He says, "Life, after all, is protoplasmic and cellular. It is made up of cells and organic stuff and is undeniably squishy."

Defining life more unambiguously than "squishy" is challenging, if not impossible. Even deciding what sets the living apart from the nonliving can be tricky. For example, living things have a high proportion of the organic molecules of life, but so do the remains of dead organisms in seams of coal. Living things use energy to reproduce themselves, but computer viruses, which are arguably not alive, can do that too.

So how do biologists, who study life as a profession, define it? The short answer is that their best definition is a long list of descriptions that collectively apply to living things, and not to nonliving things. You already know about two of these properties:

1. They make and use the organic molecules of life.
2. They consist of one or more cells.

The remainder of this book details the others properties of life:

3. Living things engage in self-sustaining biological processes such as metabolism and homeostasis.
4. They change over their lifetime, for example by growing, maturing, and aging.
5. They use DNA as their hereditary material when they reproduce.
6. They have the collective capacity to change over successive generations, for example by adapting to environmental pressures.

TAKE-HOME MESSAGE 4.12
What, exactly, is life?

✔ We describe the characteristic of "life" in terms of properties that are collectively unique to living things.

✔ Organisms make and use the organic molecules of life. DNA is their hereditary material.

✔ In living things, the molecules of life are organized as one or more cells that engage in self-sustaining biological processes.

✔ Living things change over lifetimes, and also over successive generations.

CREDIT: (in text) Kevin Schafer/Getty Images.

Food for Thought (revisited)

One way to ensure food safety involves sterilization, which kills toxic *E. coli* and other bacteria. In the U.S., recalled, contaminated ground beef is often cooked or otherwise sterilized, then processed into ready-to-eat products such as canned chili. Raw beef trimmings, which have a high risk of contact with fecal matter during the butchering process, are effectively sterilized when sprayed with ammonia. Ground to a paste and formed into pellets or blocks, the resulting product is termed "lean finely textured beef" or "boneless lean beef trimmings." Although this product cannot be sold directly to consumers, lean finely textured beef has been and continues to be routinely used as a filler in prepared food products such as hamburger patties, fresh ground beef, hot dogs, lunch meats, sausages, frozen entrees, canned foods, and other items sold to quick service restaurants, hotel and restaurant chains, institutions, and school lunch programs.

A series of news reports in 2012 provoked public outrage at the widespread, unlabeled use of lean finely textured beef, pejoratively nicknamed pink slime. Meat industry organizations and the USDA agree that lean finely textured beef, appetizing or not, is perfectly safe to eat because it has been sterilized—any live bacteria in it have been killed. However, the public controversy prompted some companies to discontinue use of ground beef containing the filler product.

summary

Section 4.1 Bacteria are found in all parts of the biosphere, including the human body. A few types can cause disease. Contamination of food with disease-causing bacteria can result in food poisoning that is sometimes fatal.

Section 4.2 By the **cell theory**, all organisms consist of one or more cells; the cell is the smallest unit of life; each new cell arises from another, preexisting cell; and a cell passes hereditary material to its offspring. All cells start out life with **cytoplasm**, DNA, and a **plasma membrane** that controls the types and kinds of substances that cross it. Most have many additional components. A eukaryotic cell's DNA is contained within a **nucleus**, which is a membrane-enclosed **organelle**. All cell membranes, including the plasma membrane and organelle membranes, consist mainly of phospholipids organized as a lipid bilayer.

A cell's surface area increases with the square of its diameter, while its volume increases with the cube. This **surface-to-volume ratio** limits cell size and influences cell shape.

Section 4.3 Most cells are far too small to see with the naked eye, so we use microscopes to observe them. Different types of microscopes and techniques reveal different internal and external details of cells.

Section 4.4 Bacteria and archaea, informally grouped as prokaryotes, are the most diverse forms of life that we know about. These single-celled organisms have no nucleus, but they do have a **nucleoid** and **ribosomes**. Many also have a protective, rigid **cell wall** and a sticky capsule, and some have motile structures (**flagella**) and other projections (**pili**). There are often **plasmids** in addition to the single circular molecule of DNA. Bacteria and other microbial organisms may live together in a shared mass of slime as a **biofilm**.

Section 4.5 Protists, fungi, plants, and animals are eukaryotic. Cells of these organisms start out life with a nucleus, and typically many other membrane-enclosed organelles. Organelles compartmentalize tasks and substances that are sensitive or dangerous to the rest of the cell.

Section 4.6 A nucleus protects and controls access to a eukaryotic cell's DNA. A double membrane studded with pores constitutes the **nuclear envelope**. The pores serve as gateways for molecules passing into and out of the nucleus. Inside the nuclear envelope, **chromatin** is suspended in viscous **nucleoplasm**. In the nucleus, ribosome subunits are produced in dense, irregularly shaped **nucleoli**.

Section 4.7 The **endomembrane system** is a series of organelles (endoplasmic reticulum, Golgi bodies, vesicles) that interact mainly to make lipids, enzymes, and proteins for insertion into membranes or secretion. **Endoplasmic reticulum (ER)** is a continuous system of sacs and tubes extending from the nuclear envelope. Ribosome-studded rough ER makes proteins; smooth ER makes lipids and breaks down carbohydrates and fatty acids. **Golgi bodies** modify proteins and lipids before sorting them into vesicles.

Different types of **vesicles** store, break down, or transport substances through the cell. Enzymes in **peroxisomes** break down substances such as amino acids, fatty acids, and toxins. **Lysosomes** contain enzymes that break down cellular wastes and debris. Fluid-filled **vacuoles** store or break down waste, food, and toxins. Fluid pressure inside a **central vacuole** keeps plant cells plump, which in turn keeps plant parts firm.

Section 4.8 Double-membraned organelles called **mitochondria** specialize in making many ATP by breaking down organic compounds in the oxygen-requiring pathway of aerobic respiration.

Section 4.9 Different types of **plastids** are specialized for photosynthesis or storage. In eukaryotes, photosynthesis takes place inside **chloroplasts**. Pigment-filled chromoplasts and starch-filled amyloplasts are used for storage; many of these plastids serve additional roles.

Section 4.10 Elements of a **cytoskeleton** reinforce, organize, and move cell structures and often the cell. Cytoskeletal elements include **microtubules**, **microfilaments**, and **intermediate filaments**. Interactions between ATP-driven **motor proteins** and hollow, dynamically assembled microtubules bring about the movement of cell parts. A microfilament mesh called the **cell cortex** reinforces plasma membranes. Elongating microfilaments bring about move-ment of **pseudopods**. Intermediate filaments lend structural support to cells and tissues, and they help support the nuclear membrane. **Centrioles** give rise to a special 9+2 array of microtubules inside **cilia** and eukaryotic flagella, then remain beneath these motile structures as **basal bodies**.

Section 4.11 A secreted mixture of materials forms **extracellular matrix** (**ECM**) that has different functions depending on the cell type. In animals, a secreted basement membrane supports and organizes cells in tissues. Among the eukaryotes, plant cells, fungi, and many protists secrete a cell wall around their plasma membrane. Older plant cells secrete a rigid, **lignin**-containing **secondary wall** inside their pliable **primary wall**. Many eukaryotic cell types also secrete a protective **cuticle**. **Plasmodesmata** are open **cell junctions** that connect the cytoplasm of adjacent plant cells. In animals, **gap junctions** are closable channels between adjacent cells. **Adhering junctions** that connect to cytoskeletal elements fasten cells to one another and to basement membrane. **Tight junctions** form a waterproof seal between cells.

Section 4.12 Differences among cell components (**TABLE 4.5**) allow us to categorize life, but not to define it. We can describe the quality of "life" as a set of properties that are collectively unique to living things. Living things consist of cells that engage in self-sustaining biological processes, pass their hereditary material (DNA) to offspring, and have the capacity to change.

Table 4.5 Comparing Components of Prokaryotic and Eukaryotic Cells

Cell Component	Example(s) of Function	Prokaryotes	Eukaryotes			
			Protists	Fungi	Plants	Animals
Cell wall	Protection, structural support	+	+	+	+	−
Plasma membrane	Control of substances moving into and out of cell	+	+	+	+	+
Nucleus	Physical separation of DNA from cytoplasm	−	+	+	+	+
Nucleolus	Assembly of ribosome subunits	−	+	+	+	+
DNA	Encoding of hereditary information	+	+	+	+	+
RNA	Protein synthesis	+	+	+	+	+
Ribosome	Protein synthesis	+	+	+	+	+
Endoplasmic reticulum	Protein, lipid synthesis; carbohydrate, fatty acid breakdown	−	+	+	+	+
Golgi body	Final modification of proteins, lipids	−	+	+	+	+
Lysosome	Intracellular digestion	−	+	+	+	+
Peroxisome	Breakdown of fatty acids, amino acids, and toxins	−	+	+	+	+
Mitochondrion	Production of ATP by aerobic respiration	−	+	+	+	+
Hydrogenosome	Anaerobic production of ATP	−	+	+	−	+
Chloroplast	Photosynthesis; starch storage	−	+	−	+	−
Central vacuole	Increasing cell surface area; storage	−	−	+	+	−
Flagellum	Locomotion through fluid surroundings	+	+	+	+	+
Cilium	Locomotion through fluid surroundings; movement of surrounding fluid	+	+	−	+	+
Cytoskeleton	Physical reinforcement; internal organization; movement	+	+	+	+	+

+ *found in at least some species;* **−** *not found in any species.*

CREDIT: (Table 4.5) © Cengage Learning.

Abnormal Motor Proteins Cause Kartagener Syndrome An abnormal form of a motor protein called dynein causes Kartagener syndrome, a genetic disorder characterized by chronic sinus and lung infections. Biofilms form in the thick mucus that collects in the airways, and the resulting bacterial activities and inflammation damage tissues.

Affected men can produce sperm but are infertile. Some have become fathers after a doctor injects their sperm cells directly into eggs. Review **FIGURE 4.21** and **FIGURE 4.24**, then explain how abnormal dynein could cause the observed effects.

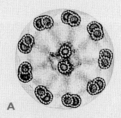 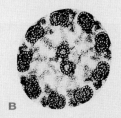

FIGURE 4.24 Cross-section of the flagellum of a sperm cell from **A** a human male affected by Kartagener syndrome and **B** an unaffected male.

self-quiz

1. All cells have these three things in common: _____ .
   a. cytoplasm, DNA, and organelles with membranes
   b. a plasma membrane, DNA, and a nucleus
   c. cytoplasm, DNA, and a plasma membrane
   d. a cell wall, cytoplasm, and DNA

2. Name one major principle of the cell theory.

3. The surface-to-volume ratio _____ .
   a. does not apply to prokaryotic cells
   b. constrains cell size
   c. is part of the cell theory
   d. b and c

4. True or false? Some protists start out life with no nucleus.

5. Cell membranes consist mainly of _____ and _____ .
   a. lipids; carbohydrates
   b. phospholipids; protein
   c. lipids; carbohydrates
   d. phospholipids; ECM

6. True or false? Ribosomes are only found in eukaryotes.

7. Unlike eukaryotic cells, prokaryotic cells _____ .
   a. have no plasma membrane
   b. have RNA but not DNA
   c. have no nucleus
   d. a and c

8. Enzymes contained in _____ break down worn-out organelles, bacteria, and other particles.
   a. lysosomes
   b. amyloplasts
   c. endoplasmic reticulum
   d. peroxisomes

9. Put the following structures in order according to the pathway of a secreted protein:
   a. plasma membrane
   b. Golgi bodies
   c. endoplasmic reticulum
   d. post-Golgi vesicles

10. The main function of the endomembrane system is:
    a. building and modifying proteins and lipids
    b. isolating DNA from toxic substances
    c. secreting extracellular matrix onto the cell surface
    d. producing ATP by aerobic respiration

11. True or false? The plasma membrane is the outermost component of all cells. Explain.

12. Cell junctions called _____ connect the cytoplasm of plant cells.

13. Which of the following organelles contains no DNA?
    a. nucleus
    b. Golgi body
    c. mitochondrion
    d. chloroplast

14. No animal cell has a _____ .
    a. plasma membrane
    b. flagellum
    c. lysosome
    d. cell wall

15. Match each cell component with a function.
    ____ mitochondrion   a. protein synthesis
    ____ chloroplast   b. connection
    ____ cell junction   c. stores starch
    ____ smooth ER   d. breaks down toxins
    ____ Golgi body   e. sorts and ships
    ____ rough ER   f. assembles lipids
    ____ peroxisome   g. photosynthesis
    ____ amyloplast   h. ATP production
    ____ flagellum   i. movement

critical thinking

1. In a classic episode of *Star Trek*, a gigantic amoeba engulfs an entire starship. Spock blows the cell to bits before it can reproduce. Think of at least one inaccuracy that a biologist would identify in this scenario.

2. In plants, the cell wall forms as a young plant cell secretes polysaccharides onto the outer surface of its plasma membrane. Being thin and pliable, this primary wall allows the cell to enlarge and change shape. In mature woody plants, cells in some tissues deposit material onto the primary wall's inner surface. Why doesn't this secondary wall form on the outer surface of the primary wall?

3. Which structures can you identify in the organism below? Is it prokaryotic or eukaryotic? How can you tell?

CREDITS: (24) From "Tissue & Cell", Vol. 27, pp.421–427, Courtesy of Bjorn Afzelius, Stockholm University; (in text CT) P. L. Walne and J. H. Arnott, Planta, 77:325-354, 1967.

LEARNING ROADMAP

In this chapter, you will gain insight into the one-way flow of energy (Section 1.3) through the world of life (1.2). You will learn more about energy, including the laws of nature (1.9) that describe it, and heat (2.5). This chapter also revisits the structure and function of atoms (2.3), molecules (2.4, 3.2, 3.3–3.6), and cells (4.6, 4.7, 4.10, 4.11).

ENERGY FLOW

Each time energy is transferred, some of it disperses. An organism can sustain its life only as long as it continues to harvest energy from the environment.

HOW ENZYMES WORK

Enzymes increase the rate of chemical reactions. They are assisted by cofactors, and affected by temperature, salinity, pH, and other environmental factors.

THE NATURE OF METABOLISM

Sequences of enzyme-mediated reactions build, remodel, and break down organic molecules. Controls that govern steps in these pathways quickly shift cell activities.

MOVEMENT OF FLUIDS

Gradients drive the directional movements of substances across membranes. Water tends to diffuse across a cell membrane to a region of higher solute concentration.

MEMBRANE TRANSPORT

Transport proteins control solute concentrations in cells and organelles by helping substances move across membranes. Substances also move across cell membranes inside vesicles.

The next two chapters detail metabolic pathways of photosynthesis (Chapter 6) and aerobic respiration (Chapter 7). You will reencounter metabolism in the context of cancer (Chapter 10) and inherited disease (Chapter 14); and membrane proteins in processes of immunity (Chapter 37). Later chapters return to membrane transport and calcium ions in cell signaling (Chapters 30, 32, and 35), and enzymes in digestion (Chapter 39).

5.1 A Toast to Alcohol Dehydrogenase

Most college students are under the legal drinking age, but alcohol abuse continues to be the most serious drug problem on college campuses throughout the United States. Before you drink, consider what you are consuming. All alcoholic drinks—beer, wine, hard liquor—contain the same psychoactive ingredient: ethanol. Ethanol molecules move quickly from the stomach and small intestine into the bloodstream. Almost all of the ethanol ends up in the liver, a large organ in the abdomen. Liver cells have impressive numbers of enzymes. One of them, ADH (alcohol dehydrogenase), helps break down ethanol and other toxic compounds (**FIGURE 5.1**).

If you put more ethanol into your body than your enzymes can deal with, then you will damage it. Ethanol and its breakdown products harm liver cells, so the more a person drinks, the fewer liver cells are left to do the breaking down. Ethanol also interferes with normal processes of metabolism. For example, oxygen that would ordinarily take part in breaking down fatty acids is diverted to breaking down ethanol. As a result, fats tend to accumulate as large globules in the tissues of heavy drinkers.

Long-term heavy drinking causes alcoholic hepatitis, a disease characterized by inflammation and destruction of liver tissue. It also causes cirrhosis, a condition in which the liver becomes so scarred, hardened, and filled with fat that it loses its function. (The term cirrhosis is from the Greek *kirros*, meaning "orange-colored," after the abnormal skin color of people with the disease.) The liver is the largest gland in the human body, and it has many important functions. In addition to breaking down fats and toxins, it helps regulate the body's blood sugar level, and it makes proteins that are essential for blood clotting, immune function, and maintaining the solute balance of body fluids. Loss of these functions can be deadly.

Heavy drinking is dangerous in the short term too. Tens of thousands of undergraduate students have been polled about their drinking habits in recent surveys. More than half of them reported that they regularly drink five or more alcoholic beverages within a two-hour period—a self-destructive behavior called binge drinking. Consuming large amounts of alcohol in a brief period of time does far more than damage one's liver. Aside from the related 500,000 injuries from accidents, the 600,000 assaults by intoxicated students, 100,000 cases of date rape, and 400,000 incidences of unprotected sex among students, binge drinking is responsible for killing or causing the death of more than 1,700 college students every year.

With this sobering example, we invite you to learn about how and why your cells break down organic compounds, including toxic molecules such as ethanol.

alcohol
dehydrogenase

FIGURE 5.1 Alcohol dehydrogenase. This enzyme helps the body break down toxic alcohols such as ethanol, making it possible for humans to drink beer, wine, and other alcoholic beverages. The photo shows a tailgate party at a Notre Dame–Alabama football game. During 2012 alone, Indiana State police arrested 138 Notre Dame students for underage drinking at tailgate parties.

5.2 Energy in the World of Life

✔ Sustaining life's organization requires ongoing energy inputs.

Energy Disperses

Energy is formally defined as the capacity to do work, but this definition is not very satisfying. Even brilliant physicists who study energy cannot say exactly what it is. However, we do have an intuitive understanding of energy just by thinking about familiar forms of it, such as light, heat, electricity, and motion. We also understand intuitively that one form of energy can be converted to another. Think about how a lightbulb changes electricity into light, or how an automobile changes the chemical energy of gasoline into the energy of motion, which is also called **kinetic energy**.

The formal study of heat and other forms of energy is thermodynamics (*therm* is a Greek word for heat; *dynam* means energy). By making careful measurements, thermodynamics researchers discovered that the total amount of energy before and after every conversion is always the same. In other words, energy cannot be created or destroyed—a phenomenon that is the **first law of thermodynamics**. Remember, a law of nature describes something that occurs without fail, but our explanation of why it occurs is incomplete (Section 1.9).

Energy also tends to spread out, or disperse, until no part of a system holds more than another part. In a kitchen, for example, heat always flows from a hot pan to cool air until the temperature of both is the same. We never see cool air raising the temperature of a hot pan. **Entropy** is a measure of how much the energy of a particular system has become dispersed. We can use the hot pan in a cool kitchen as an example of a system. As heat flows from the pan into the air, the entropy of the system increases (**FIGURE 5.2**). Entropy continues to increase until the heat is evenly distributed throughout the kitchen, and there is no longer a net (or overall)

FIGURE 5.3 It takes more than 10,000 pounds of soybeans and corn to raise a 1,000-pound steer. Where do the other 9,000 pounds go? About half of the steer's food is indigestible and passes right through it. The animal's body breaks down molecules in the remaining half to access energy stored in chemical bonds. Only about 15 percent of that energy goes toward building body mass. The rest is lost during energy conversions, as metabolic heat.

flow of heat from one area to another. Our system has now reached its maximum entropy with respect to heat. The tendency of entropy to increase is the **second law of thermodynamics**. This is the formal way of saying that energy tends to spread out spontaneously.

Biologists use the concept of entropy as it applies to chemical bonding, because energy flow in living things occurs mainly by the making and breaking of chemical bonds. How is entropy related to chemical bonding? Think about it just in terms of motion. Two unbound atoms can vibrate, spin, and rotate in every direction, so they are at high entropy with respect to motion. A covalent bond between the atoms restricts their movement, so they are able to move in fewer ways than they did before bonding. Thus, the entropy of two atoms decreases when a bond forms between them. Such entropy changes are part of the reason why some reactions occur spontaneously and others require an energy input, as you will see in the next section.

Energy's One-Way Flow

Work occurs as a result of energy transfers. Consider how it takes work to push a box across a floor. In this case, a body (you) transfers energy to another body (the box) to make it move. Similarly, a plant cell works to make sugars. Inside the cell, one set of molecules harvests energy from light, then transfers it to another set of molecules. The second set of molecules uses the energy to build the sugars from carbon dioxide and water. This particular energy transfer involves the conversion of light energy to chemical energy. Most other types of cellular work occur by the transfer of chemical energy from one molecule to another.

---

energy The capacity to do work.
entropy Measure of how much the energy of a system is dispersed.
first law of thermodynamics Energy cannot be created or destroyed.
kinetic energy The energy of motion.
potential energy Stored energy.
second law of thermodynamics Energy tends to disperse spontaneously.

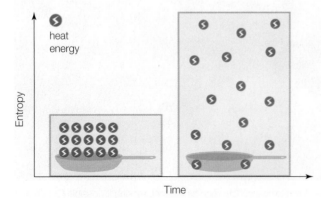

FIGURE 5.2 Entropy. Entropy tends to increase, but the total amount of energy in any system always stays the same.

As you learn about such processes, remember that every time energy is transferred, a bit of it disperses. Energy lost from a transfer is usually in the form of heat. As a simple example, a typical incandescent lightbulb converts only about 5 percent of the energy of electricity into light. The remaining 95 percent of the energy ends up as heat that disperses from the bulb.

Dispersed heat is not useful for doing work, and it is not easily converted to a more useful form of energy (such as electricity). Because some energy in every transfer disperses as heat, and heat is not useful for doing work, we can say that the total amount of energy in the universe available for doing work is always decreasing.

Is life an exception to this inevitable flow? An organized body is hardly dispersed. Energy becomes concentrated in each new organism as the molecules of life organize into cells. Even so, living things constantly use energy—to grow, to move, to acquire nutrients, to reproduce, and so on—and some energy is lost in every one of these processes (**FIGURE 5.3**). Unless those losses are replenished with energy from another source, the complex organization of life will end.

The energy that fuels most life on Earth comes from the sun. That energy flows through producers such as plants, then consumers such as animals (**FIGURE 5.4**). During this journey, the energy is transferred many times. With each transfer, some energy escapes as heat until, eventually, all of it is permanently dispersed. However, the second law of thermodynamics does not say how quickly the dispersal has to happen. Energy's spontaneous dispersal is resisted by chemical bonds. The energy in chemical bonds is a type of **potential energy**, which is energy stored in the position or arrangement of objects in a system (**FIGURE 5.5**). Think of all the bonds in the countless molecules that make up your skin, heart, liver, fluids, and other body parts. Those bonds hold the molecules, and you, together—at least for the time being.

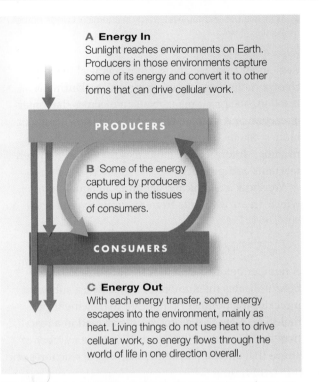

A **Energy In**
Sunlight reaches environments on Earth. Producers in those environments capture some of its energy and convert it to other forms that can drive cellular work.

PRODUCERS

B Some of the energy captured by producers ends up in the tissues of consumers.

CONSUMERS

C **Energy Out**
With each energy transfer, some energy escapes into the environment, mainly as heat. Living things do not use heat to drive cellular work, so energy flows through the world of life in one direction overall.

FIGURE 5.4 ▶**Animated** Energy flows from the environment into living organisms, then back to the environment. The flow drives a cycling of materials among producers and consumers.

FIGURE 5.5 Illustration of potential energy. By opposing gravity's downward pull, the rope attached to the rock keeps the man from falling. Similarly, a chemical bond attaches two atoms and keeps them from flying apart.

TAKE-HOME MESSAGE 5.2

What is energy?

✔ Energy, which is the capacity to do work, cannot be created or destroyed.

✔ Energy disperses spontaneously.

✔ Energy can be transferred between systems or converted from one form to another, but some is lost (as heat, typically) during every such exchange.

✔ Sustaining life's organization requires ongoing energy inputs to counter energy loss. Organisms stay alive by replenishing themselves with energy they harvest from someplace else.

5.3 Energy in the Molecules of Life

✔ All cells store and retrieve energy in chemical bonds of the molecules of life.

Remember from Section 3.3 that chemical reactions change molecules into other molecules. During a reaction, one or more **reactants** (molecules that enter a reaction and become changed by it) become one or more **products** (molecules that are produced by the reaction). Intermediate molecules may form between reactants and products.

We show a chemical reaction as an equation in which an arrow points from reactants to products:

$$2H_2 \ (\text{hydrogen}) \ + \ O_2 \ (\text{oxygen}) \longrightarrow 2H_2O \ (\text{water})$$

A number before a chemical formula in such equations indicates the number of molecules; a subscript indicates the number of atoms of that element per molecule. Note that atoms shuffle around in a reaction, but they never disappear: The same number of atoms that enter a reaction remain at the reaction's end (**FIGURE 5.6**).

Chemical Bond Energy

Every chemical bond holds a certain amount of energy. That is the amount of energy required to break the bond, and it is also the amount of energy released

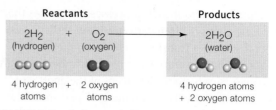

Reactants   **Products**

$$2H_2 \ (\text{hydrogen}) \ + \ O_2 \ (\text{oxygen}) \longrightarrow 2H_2O \ (\text{water})$$

4 hydrogen atoms + 2 oxygen atoms → 4 hydrogen atoms + 2 oxygen atoms

FIGURE 5.6 ▶Animated Chemical bookkeeping. In equations that represent chemical reactions, reactants are written to the left of an arrow that points to the products. A number before a formula indicates the number of molecules. The same number of atoms that enter the reaction remain at its end.

when the bond forms. The particular amount of energy held by a bond depends on which elements are taking part in it. For example, two covalent bonds—one between an oxygen and a hydrogen atom in a water molecule, the other between two oxygen atoms in molecular oxygen (O_2)—both hold energy, but different amounts of it.

Bond energy and entropy both contribute to a molecule's free energy, which is the amount of energy that is available ("free") to do work. In most reactions, the free energy of reactants differs from the free energy of products. If the reactants have less free energy than the products, the reaction will not proceed without a net energy input. Such reactions are **endergonic**, which means "energy in" (**FIGURE 5.7A**). If the reactants have more free energy than the products, the reaction will end with a net release of energy. Such reactions are **exergonic**, which means "energy out" (**FIGURE 5.7B**).

Why Earth Does Not Go Up in Flames

The molecules of life release energy when they combine with oxygen. Think of how a spark ignites tinder-dry wood in a campfire. Wood is mostly cellulose, which consists of long chains of repeating glucose monomers (Section 3.4). A spark starts a reaction that converts cellulose (in wood) and oxygen (in air) to water and carbon dioxide. The reaction is highly exergonic, which means it releases a lot of energy—enough to initiate the same reaction with other cellulose and oxygen molecules. That is why wood keeps burning after it has been lit.

Earth is rich in oxygen—and in potential exergonic reactions. Why doesn't it burst into flames? Luckily, chemical bonds do not break without at least a small input of energy, even in an energy-releasing reaction. We call this input activation energy. **Activation energy**, the minimum amount of energy required to get a chemical reaction started, is a bit like a hill that reactants must climb before they can coast down the other side to become products (**FIGURE 5.8**).

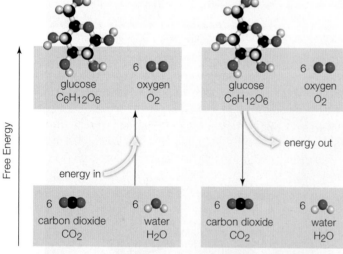

A Endergonic reactions convert molecules with lower free energy to molecules with higher free energy, so they require a net energy input in order to proceed.

B Exergonic reactions convert molecules with higher free energy to molecules with lower free energy, so they end with an energy release.

FIGURE 5.7 ▶Animated The ins and outs of energy in chemical reactions. **FIGURE IT OUT** Which law of thermodynamics explains energy inputs and outputs in chemical reactions?

Answer: The first law

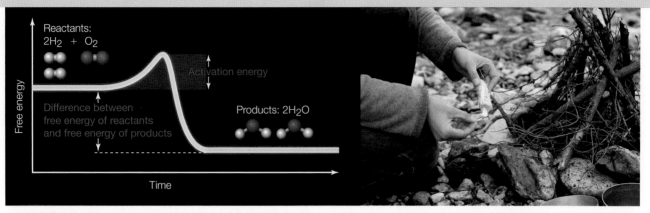

FIGURE 5.8 ▶Animated Activation energy. Most reactions will not begin without an input of activation energy, which is shown in the graph as a bump in a free energy hill. Reactants in this example have more energy than the products. Activation energy keeps this and other exergonic reactions, including the combustion of cellulose in wood, from starting spontaneously.

Both endergonic and exergonic reactions have activation energy, but the amount varies with the reaction. Consider guncotton (nitrocellulose), a highly explosive derivative of cellulose. Christian Schönbein accidentally discovered a way to make it when he used his wife's cotton apron to wipe up a nitric acid spill on his kitchen table, then hung it up to dry next to the oven. The apron exploded. Being a chemist in the 1800s, Schönbein immediately thought of marketing guncotton as a firearm explosive, but it proved to be too unstable to manufacture. So little activation energy is needed to make guncotton react with oxygen that it tends to explode unexpectedly. Several manufacturing plants burned to the ground before guncotton was abandoned for use as a firearm explosive. The substitute? Gunpowder, which has a higher activation energy for a reaction with oxygen.

Energy In, Energy Out

Cells store energy by running endergonic reactions that build organic compounds (**FIGURE 5.9A**). For example, light energy drives the overall reactions of photosynthesis, which produce sugars such as glucose from carbon dioxide and water. Unlike light, glucose can be stored in a cell. Cells harvest energy by running exergonic reactions that break the bonds of organic compounds (**FIGURE 5.9B**). Most cells do this when they carry out the overall reactions of aerobic respiration, which releases the energy of glucose by breaking the bonds between its carbon atoms. You will see in the next few sections how cells use energy released from some reactions to drive others.

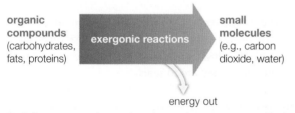

A Cells run endergonic reactions that store energy in the bonds of organic compounds.

B Cells run exergonic reactions that retrieve energy stored in the bonds of organic compounds.

FIGURE 5.9 How cells store and retrieve free energy.

activation energy Minimum amount of energy required to start a chemical reaction.
endergonic Describes a reaction that requires a net input of free energy to proceed.
exergonic Describes a reaction that ends with a net release of free energy.
product A molecule that is produced by a reaction.
reactant A molecule that enters a reaction and is changed by participating in it.

TAKE-HOME MESSAGE 5.3
How do cells use energy?

✔ Endergonic reactions will not run without a net input of energy. Exergonic reactions end with a net release of energy.

✔ Both endergonic and exergonic reactions require an input of activation energy to begin.

✔ Cells store energy in chemical bonds by running endergonic reactions that build organic compounds. To release this stored energy, they run exergonic reactions that break the bonds.

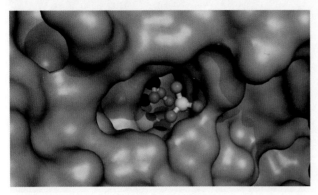

A A glucose molecule meets up with a phosphate in the active site of a hexokinase enzyme.

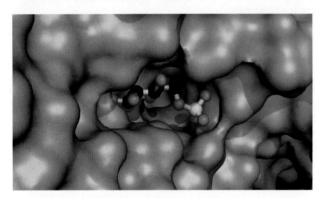

B The enzyme has catalyzed the reaction between glucose and phosphate. The product of this reaction, glucose-6-phosphate, is shown leaving the active site.

FIGURE 5.10 Example of an active site. This one is in hexokinase, an enzyme that adds a phosphate group to glucose and other six-carbon sugars.

enzyme   substrates

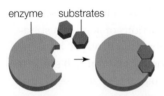

A An active site binds substrates that are complementary in shape, size, polarity, and charge.

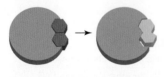

B The binding squeezes substrates together, influences their charge, or causes some change that lowers activation energy, so the reaction proceeds.

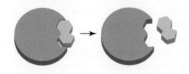

C The product leaves the active site after the reaction is finished. The enzyme is unchanged, so it can work again.

FIGURE 5.11 How an active site works.

✔ Enzymes make specific reactions occur much faster than they would on their own.

The Need for Speed

Metabolism requires enzymes. Why? Consider that a molecule of glucose can break down to carbon dioxide and water on its own, but the process might take decades. That same conversion takes just seconds inside your cells. Enzymes make the difference. In a process called **catalysis**, an enzyme makes a reaction run much faster than it would on its own. The enzyme is unchanged by catalyzing (speeding up) the reaction, so it can work again and again.

Most enzymes are proteins, but some are RNAs. Each kind of enzyme interacts only with specific reactants, or **substrates**, and alters them in a specific way. Such specificity occurs because an enzyme's polypeptide chains fold up into one or more **active sites**, which are pockets where substrates bind and where reactions proceed (**FIGURE 5.10**). An active site is complementary in shape, size, polarity, and charge to the enzyme's substrate (**FIGURE 5.11**). This fit is the reason why each enzyme acts in a specific way on a specific substrate.

The Transition State

When we talk about activation energy, we are really talking about the energy required to bring reactant bonds to their breaking point. At that point, which is called the **transition state**, the reaction can run without any additional energy input. Enzymes bring on the transition state by lowering activation energy (**FIGURE 5.12**). They do so by the following four mechanisms.

Forcing Substrates Together Binding at an active site brings substrates together in close physical proximity. The closer the substrates are to one another, the more likely they are to react.

Orienting Substrates in Positions That Favor Reaction Substrate molecules in a solution collide from random directions. By contrast, binding at an active site positions substrates optimally for reaction.

Inducing a Fit Between Enzyme and Substrate By the **induced-fit model**, an enzyme's active site is not quite complementary to its substrate. Interacting with a substrate molecule causes the enzyme to change shape so that the fit between them improves. The improved fit may result in a stronger bond between enzyme and substrate.

CREDITS: (10) PDB ID: 1GZX; Paoli, M., Liddington, R., Tame, J., Wilkinson, A., Dodson, G., Crystal structure of T state hemoglobin with oxygen bound at all four haems. *J.Mol.Bio.*, v256, pp. 775–792, 1996; (11) © Cengage Learning.

Excluding Water Metabolism occurs in water-based fluids, but water molecules can interfere with certain reactions. The active sites of some enzymes repel water, and keep it away from the reactions.

Enzyme Activity

Environmental factors such as pH, temperature, salt, and pressure influence an enzyme's shape, and so influence its function (Sections 3.6 and 3.7). Each enzyme works best in a particular range of conditions that reflect the environment in which it evolved.

Consider pepsin, a digestive enzyme that works best at low pH (**FIGURE 5.13A**). Pepsin begins the process of protein digestion in the very acidic environment of the stomach (pH 2). During digestion, the stomach's contents pass into the small intestine, where the pH rises to about 7.5. Pepsin denatures (unfolds) above pH 5.5, so this enzyme becomes inactive in the small intestine. Here, protein digestion continues with the assistance of trypsin, an enzyme that functions well at the higher pH.

Adding heat boosts free energy, which is why the jiggling motion of atoms and molecules (Section 2.5) increases with temperature. The greater the free energy of reactants, the closer they are to reaching activation energy. Thus, the rate of an enzymatic reaction typically increases with temperature—but only up to a point. An enzyme denatures above a characteristic temperature. Then, the reaction rate falls sharply as the shape of the enzyme changes and it stops working (**FIGURE 5.13B**). Body temperatures above 42°C (107.6°F) adversely affect the function of many of your enzymes, which is why such severe fevers are dangerous.

The activity of many enzymes is also influenced by the amount of salt in the surrounding fluid. Too little salt, and polar parts of the enzyme attract one another so strongly that the enzyme's shape changes. Too much salt interferes with the hydrogen bonds that hold the enzyme in its characteristic shape, so the enzyme denatures (Section 3.7).

active site Pocket in an enzyme where substrates bind and a chemical reaction occurs.
catalysis The acceleration of a chemical reaction by a molecule that is unchanged by participating in the reaction.
induced-fit model Substrate binding to an active site improves the fit between the two.
substrate Of an enzyme, a reactant that is specifically acted upon by the enzyme.
transition state Point during a reaction at which substrate bonds will break and the reaction will run spontaneously to completion.

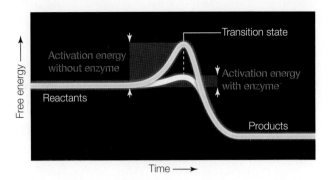

FIGURE 5.12 The transition state. An enzyme enhances the rate of a reaction by lowering activation energy.

FIGURE IT OUT Is this reaction endergonic or exergonic?

Answer: Exergonic

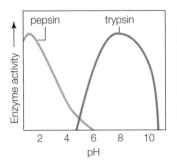

A The pH-dependent activity of two digestive enzymes, pepsin and trypsin. Pepsin acts in the stomach, where the normal pH is 2. Trypsin acts in the small intestine, where the normal pH is around 7.5.

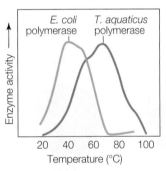

B Comparison of temperature-dependent activity of a DNA synthesis enzyme from two species of bacteria: *E. coli*, which inhabits the human gut (normally 37°C); and *Thermus aquaticus*, which lives in hot springs around 70°C.

FIGURE 5.13 Each enzyme works best within a characteristic range of conditions—generally, the same conditions that occur in the environment in which the enzyme normally functions.

FIGURE IT OUT At what temperature does the *E. coli* DNA polymerase work fastest?

Answer: About 37°C

TAKE-HOME MESSAGE 5.4
How do enzymes work?

✔ Enzymes greatly enhance the rate of specific reactions.

✔ Binding at an enzyme's active site causes a substrate to reach its transition state. In this state, the substrate's bonds are at the breaking point, and the reaction can run spontaneously to completion.

✔ Each enzyme works best within a certain range of environmental conditions that include temperature, pH, pressure, and salt concentration.

✔ ATP, enzymes, and other molecules interact in organized pathways of metabolism.

Metabolism includes all activities by which cells acquire and use energy as they build, break down, or remodel organic molecules (Section 3.3). Such activities often occur stepwise, in a series of enzymatic reactions called a **metabolic pathway**. Some metabolic pathways are linear, meaning that the reactions run straight from reactant to product (**FIGURE 5.14A**). Other reactions are cyclic. In a cyclic pathway, the last step regenerates a reactant required for the first step (**FIGURE 5.14B**). Both linear and cyclic pathways are common in cells; both can involve thousands of molecules and be quite complex. Later chapters detail the steps in some important pathways.

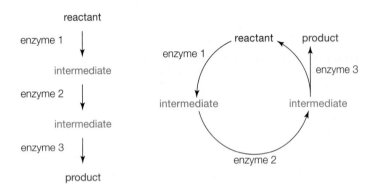

A A linear pathway runs straight from reactant to product.

B The last step of a cyclic pathway regenerates a reactant for the first step.

FIGURE 5.14 ▶**Animated** Linear and cyclic metabolic pathways.

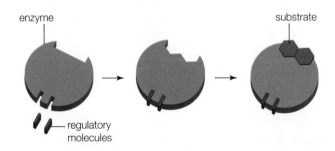

FIGURE 5.15 ▶**Animated** Allosteric regulation, in which regulatory molecules bind to a region of an enzyme that is not the active site. The binding changes the shape of the enzyme, and thus alters its activity.

FIGURE IT OUT Does the binding of regulatory molecules help or hinder this enzyme's function?

Answer: It helps.

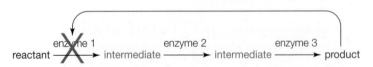

FIGURE 5.16 Feedback inhibition. In this example, three different enzymes act in sequence to convert a substrate to a product. The product inhibits the activity of the first enzyme.

FIGURE IT OUT Is this metabolic pathway cyclic or linear?

Answer: Linear

Controls Over Metabolism

Cells conserve energy and resources by making only what they require at any given moment—no more, no less. Several mechanisms help a cell maintain, raise, or lower its production of thousands of different substances. Consider that reactions do not only run from reactants to products. Many also run in reverse at the same time, with some of the products being converted back to reactants. The rates of the forward and reverse reactions often depend on the concentrations of reactants and products: A high concentration of reactants pushes the reaction in the forward direction, and a high concentration of products pushes it in the reverse direction.

Other mechanisms more actively regulate enzymatic reactions. Certain substances—regulatory molecules or ions—can influence enzyme activity. In some cases, the regulatory substance activates or inhibits an enzyme by binding directly to the active site. In other cases, the regulatory substance binds outside of the active site, a mechanism called **allosteric regulation** (*allo*– means other; –*steric* means structure). Binding of an allosteric regulator alters the shape of the enzyme in a way that enhances or inhibits its function (**FIGURE 5.15**).

Regulation of a single enzyme can affect an entire metabolic pathway. For example, the end product of a series of enzymatic reactions often inhibits the activity of one of the enzymes in the series (**FIGURE 5.16**). This type of regulatory mechanism, in which a change that results from an activity decreases or stops the activity, is called **feedback inhibition**.

Electron Transfers

The bonds of organic molecules hold a lot of energy that can be released in a reaction with oxygen. Burning is one type of reaction with oxygen, and it releases the energy of organic molecules all at once—explosively (**FIGURE 5.17A**). Cells use oxygen to break the bonds of organic molecules, but they have no way to harvest the explosive burst of energy that occurs during burning. Instead, they break the molecules apart in pathways that release the energy in small, manage-

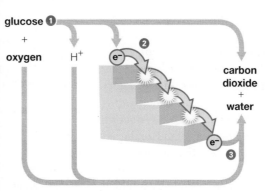

A Left, glucose in a metal spoon reacts (burns) with oxygen inside a glass jar. Energy in the form of light and heat is released all at once as CO_2 and water form.

B In cells, the same overall reaction occurs in a stepwise fashion that involves an electron transfer chain, represented here by a staircase. Energy is released in amounts that cells are able to use.

❶ An input of activation energy splits glucose into carbon dioxide, electrons, and hydrogen ions (H^+).

❷ Electrons lose energy as they move through an electron transfer chain. Energy released by electrons is harnessed for cellular work.

❸ Electrons, hydrogen ions, and oxygen combine to form water.

FIGURE 5.17 ▶**Animated** Comparing uncontrolled (**A**) and controlled (**B**) energy release.

able steps. Most of these steps are oxidation–reduction reactions, or redox reactions for short. A typical **redox reaction** is an electron transfer, in which one molecule accepts electrons (thereby becoming reduced) from another molecule (which becomes oxidized when that happens). To remember what reduced means, think of how the negative charge of an electron "reduces" the charge of a recipient molecule.

Energy is often transferred during redox reactions (**FIGURE 5.18**). In the next two chapters, you will learn about the importance of these energy transfers in electron transfer chains. An **electron transfer chain** is a series of membrane-bound enzymes and other molecules that give up and accept electrons in turn. Electrons are at a higher energy level (Section 2.3) when they enter a chain than when they leave. Energy given off by an electron as it drops to a lower energy level is harvested by molecules of the electron transfer chain to do cellular work (**FIGURE 5.17B**). Electron transfer chains are part of photosynthesis and aerobic respiration. Energy released at certain steps in those chains helps drive the synthesis of ATP.

FIGURE 5.18 Visible evidence of energy transferred during a redox reaction: a glowing protist, *Noctiluca scintillans* (left). The metabolic pathway that produces the blue glow involves an enzyme, luciferase, and its substrate, luciferin. It runs when the cells are mechanically stimulated, as by waves (right) or an attack by a protist-eating predator.

The pathway, summarized below, includes a redox reaction in which luciferin becomes oxidized, and oxygen becomes reduced:

$$\text{luciferin} + 2H^+ + O_2 \xrightarrow{\text{luciferase}} \text{luciferin} = O + H_2O + \textbf{light}$$

Light given off by a living organism is called bioluminescence.

---

allosteric regulation Control of enzyme activity by a regulatory molecule or ion that binds to a region outside the enzyme's active site.
electron transfer chain Array of membrane-bound enzymes and other molecules that accept and give up electrons in sequence, thus releasing the energy of the electrons in steps.
feedback inhibition Regulatory mechanism in which a change that results from some activity decreases or stops the activity.
metabolic pathway Series of enzyme-mediated reactions by which cells build, remodel, or break down an organic molecule.
redox reaction Oxidation–reduction reaction; typically, one molecule accepts electrons (it becomes reduced) from another molecule (which becomes oxidized).

TAKE-HOME MESSAGE 5.5
What is a metabolic pathway?

✔ A metabolic pathway is a stepwise series of enzyme-mediated reactions.

✔ Cells conserve energy and resources by producing only what they need at a given time. Such metabolic control can arise from mechanisms (such as regulatory molecule binding to an enzyme) that start, stop, or alter the rate of a single reaction. Other mechanisms (such as feedback inhibition) influence an entire pathway.

✔ Many metabolic pathways involve electron transfers. Electron transfer chains are sites of energy exchange.

5.6 Cofactors in Metabolic Pathways

✔ Cofactors help enzymes work.

✔ Energy in ATP drives many endergonic reactions.

Most enzymes cannot function properly without assistance from metal ions or small organic molecules. Such enzyme helpers are called **cofactors**. Many dietary vitamins and minerals are essential because they are cofactors or are precursors for them.

Some metal ions that act as cofactors stabilize the structure of an enzyme, in which case the enzyme denatures if the ions are removed. In other cases, metal cofactors play a functional role in a reaction by interacting with electrons in nearby atoms. Atoms of metal elements readily lose or gain electrons, so a metal cofactor can help bring on the transition state by donating electrons, accepting them, or simply tugging on them.

Organic cofactors are called **coenzymes** (**TABLE 5.1** and **FIGURE 5.19**). Coenzymes carry chemical groups, atoms, or electrons from one reaction to another, and

FIGURE 5.19 Example of a coenzyme. Coenzyme Q_{10} (above) is an essential part of the ATP-making machinery in your mitochondria. It carries electrons between enzymes of electron transfer chains during aerobic respiration. Your body makes it, but some foods—particularly red meats, soy oil, and peanuts—are rich dietary sources.

often into or out of organelles. Unlike enzymes, many coenzymes are modified by taking part in a reaction. They are regenerated in separate reactions. Consider NAD^+ (nicotinamide adenine dinucleotide), a coenzyme derived from niacin (vitamin B_3). NAD^+ can accept electrons and hydrogen atoms, thereby becoming reduced to NADH. When electrons and hydrogen atoms are removed from NADH (an oxidation reaction), NAD^+ forms again:

$$NAD^+ + electrons + H^+ \longrightarrow \boxed{NADH} \longrightarrow NAD^+ + electrons + H^+$$

In some reactions, cofactors participate as separate molecules. In others, they stay tightly bound to the enzyme. Catalase, an enzyme of peroxisomes, has four tightly bound cofactors called hemes. A heme is a small organic compound with an iron atom at its center (**FIGURE 5.20**). Catalase's substrate is hydrogen peroxide (H_2O_2), a highly reactive molecule that forms during some normal metabolic reactions. Hydrogen peroxide is dangerous because it can easily oxidize and destroy the organic molecules of life, or form free radicals that do. Catalase neutralizes this threat. Catalase's active site holds a hydrogen peroxide molecule close to a heme. Two H_2O_2 molecules alternately oxidize and then reduce the heme's iron atom, an interaction that causes the molecules to break down and form water.

Substances such as catalase that interfere with the oxidation of other molecules are called **antioxidants**. Antioxidants are essential to health because they limit the amount of damage that cells sustain as a result of oxidation by free radicals or other molecules. Oxidative damage is associated with many diseases, including cancer, diabetes, atherosclerosis, stroke, and neurodegenerative problems such as Alzheimer's disease.

Table 5.1 Some Common Coenzymes

Coenzyme	Example of Function
ATP	Transfers energy with a phosphate group
NAD, NAD^+	Carries electrons during glycolysis
NADP, NADPH	Carries electrons, hydrogen atoms during photosynthesis
FAD, FADH, $FADH_2$	Carries electrons during aerobic respiration
Coenzyme A (CoA)	Carries acetyl group ($COCH_3$) during glycolysis
Coenzyme Q_{10}	Carries electrons in electron transfer chains of aerobic respiration
Heme	Accepts and donates electrons
Ascorbic acid	Carries electrons during peroxide breakdown (in lysosomes)
Biotin (vitamin B_7)	Carries CO_2 during fatty acid synthesis

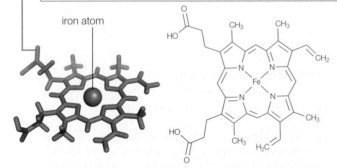

iron atom

FIGURE 5.20 ▶Animated Heme, modeled two ways. This organic molecule is part of the active site in many enzymes (such as catalase). In other contexts, it carries oxygen (e.g., in hemoglobin), or electrons (e.g., in molecules of electron transfer chains).

FIGURE IT OUT Is heme a cofactor or a coenzyme?

Answer: It is both.

ATP—A Special Coenzyme

In cells, the nucleotide ATP (adenosine triphosphate, Section 3.8) functions as a cofactor in many reactions. Bonds between phosphate groups hold a lot of energy compared to other bonds. ATP has two of these

CREDITS: (19) left, © Cengage Learning; right, © Valentyn Volkov/Shutterstock.com; (20, Table 5.1) © Cengage Learning.

bonds holding its three phosphate groups together (**FIGURE 5.21A**). When a phosphate group is transferred to or from a nucleotide, energy is transferred along with it. Thus, the nucleotide can receive energy from an exergonic reaction, and it can contribute energy to an endergonic one. ATP is such an important currency in the energy economy of cells that we use a cartoon coin to symbolize it.

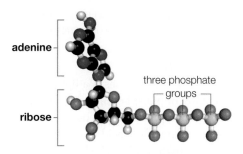

A reaction in which a phosphate group is transferred from one molecule to another is called a **phosphorylation**. ADP (adenosine diphosphate) forms when an enzyme transfers a phosphate group from ATP to another molecule (**FIGURE 5.21B**). Cells constantly run this reaction in order to drive a variety of endergonic reactions. Thus, they must constantly replenish their stockpile of ATP—by running exergonic reactions that phosphorylate ADP. The cycle of using and replenishing ATP is called the **ATP/ADP cycle** (**FIGURE 5.21C**).

The ATP/ADP cycle couples endergonic reactions with exergonic ones (**FIGURE 5.22**). As you will see in Chapter 7, cells harvest energy from organic compounds by running metabolic pathways that break them down. Energy that cells harvest in these pathways is not released to the environment, but rather stored in the high-energy phosphate bonds of ATP molecules and in electrons carried by reduced coenzymes. Both the ATP and the reduced coenzymes that form in these pathways can be used to drive many of the endergonic reactions that a cell runs.

---

antioxidant Substance that prevents oxidation of other molecules.
ATP/ADP cycle Process by which cells regenerate ATP. ADP forms when ATP loses a phosphate group, then ATP forms again as ADP gains a phosphate group.
coenzyme An organic cofactor.
cofactor A coenzyme or metal ion that associates with an enzyme and is necessary for its function.
phosphorylation A phosphate-group transfer.

TAKE-HOME MESSAGE 5.6

How do cofactors work?

✔ Cofactors associate with enzymes and assist their function.

✔ Many coenzymes carry chemical groups, atoms, or electrons from one reaction to another.

✔ The formation of ATP from ADP is an endergonic reaction. ADP forms again when a phosphate group is transferred from ATP to another molecule.

✔ When a phosphate group is transferred from ATP to another molecule, energy is transferred along with it. This energy drives cellular work.

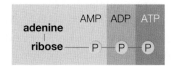

A ATP. Bonds between its phosphate groups hold a lot of energy.

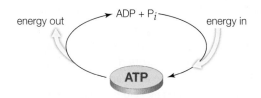

B After ATP loses one phosphate group, the nucleotide is ADP (adenosine diphosphate); after losing two, it is AMP (adenosine monophosphate).

energy out → ADP + P$_i$ → energy in

ATP

C The ATP/ADP cycle. ADP forms in a reaction that removes a phosphate group from ATP (P$_i$ is an abbreviation for phosphate group). Energy released in this reaction drives other reactions that are the stuff of cellular work. ATP forms again in reactions that phosphorylate ADP.

FIGURE 5.21 ATP, an important energy currency in metabolism.

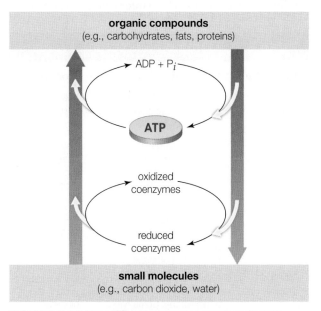

organic compounds
(e.g., carbohydrates, fats, proteins)

ADP + P$_i$

ATP

oxidized coenzymes

reduced coenzymes

small molecules
(e.g., carbon dioxide, water)

FIGURE 5.22 How ATP and coenzymes couple endergonic reactions with exergonic reactions. Yellow arrows indicate energy flow. Compare with **FIGURES 5.9** and **5.21C**.

5.7 A Closer Look at Cell Membranes

✔ A cell membrane is organized as a lipid bilayer with many proteins embedded in it and attached to its surfaces.

The Fluid Mosaic Model

The foundation of cell membranes is a lipid bilayer that consists mainly of phospholipids. Remember from Section 3.5 that a phospholipid has a phosphate-containing head and two fatty acid tails. The head is highly polar and hydrophilic, which means that it interacts with water molecules. The long hydrocarbon tails are very nonpolar and hydrophobic, so they do not interact with water molecules. As a result of these opposing properties, phospholipids swirled into water will spontaneously organize themselves into lipid bilayer sheets or bubbles (left), with hydrophobic tails together, hydrophilic heads facing the watery surroundings (**FIGURE 5.23A**).

fluid

Other molecules, including cholesterol, proteins, glycoproteins, and glycolipids, are embedded in or attached to the lipid bilayer of a cell membrane. Many of these molecules move around the membrane more or less freely. We describe a eukaryotic or bacterial cell membrane as a **fluid mosaic** because it behaves like a two-dimensional liquid of mixed composition. The "mosaic" part of the name comes from the many different types of molecules in the membrane. A cell membrane is fluid because its phospholipids are not chemically bonded to one another; they stay organized in a bilayer as a result of collective hydrophobic and hydrophilic attractions. These interactions are, on an

A In a watery fluid, phospholipids spontaneously line up into two layers: the hydrophobic tails cluster together, and the hydrophilic heads face outward, toward the fluid. This lipid bilayer forms the framework of all cell membranes. Many types of proteins intermingle among the lipids; a few that are typical of plasma membranes are shown opposite.

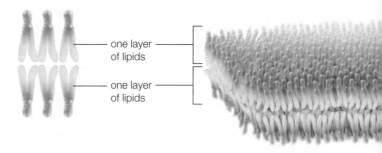

one layer of lipids

one layer of lipids

FIGURE 5.23 Cell membrane structure.

Organization of phospholipids in cell membranes (**A**) and examples of common membrane proteins (**B–E**). For clarity, these proteins are often modeled as blobs or geometric shapes; their structure can be extremely complex.

individual basis, relatively weak. Thus, individual phospholipids in the bilayer drift sideways and spin around their long axis, and their tails wiggle.

A cell membrane's properties vary depending on the types and proportions of molecules composing it. For example, membrane fluidity decreases with increasing cholesterol content. A membrane's fluidity also depends on the length and saturation of its phospholipids' fatty acid tails (Section 3.5).

Archaea do not even use fatty acids to build their phospholipids. Instead, they use molecules with reactive side chains, so the tails of archaeal phospholipids form covalent bonds with one another. As a result of this rigid crosslinking, archaeal phospholipids do not drift, spin, or wiggle in a bilayer. Thus, membranes of archaea are stiffer than those of bacteria or eukaryotes, a characteristic that may help these cells survive in extreme habitats.

Proteins Add Function

Many types of proteins are associated with a cell membrane (**TABLE 5.2**). These proteins can be assigned to one of two categories, depending on the way they are attached to the lipid bilayer. Integral membrane proteins are permanently anchored in the membrane by one or more domains sunk deeply into the lipid bilayer's hydrophobic core. Integral proteins that span the entire bilayer are called transmembrane proteins. By contrast, a peripheral membrane protein temporar-

Table 5.2	Common Membrane Proteins	
Category	**Function**	**Examples**
Adhesion protein	Helps cells stick to one another and to extra-cellular matrix.	Integrins; MHC molecules
Receptor protein	Initiates change in a cell's activity in response to a stimulus (e.g., binding to a hormone or absorbing light energy).	Insulin receptor; B cell receptor
Enzyme	Catalysis. Membranes provide a relatively stable reaction site for many metabolic pathways.	Cytochrome P450
Transport protein	Moves or allows specific ions or molecules across a membrane. Some types require an energy input, as from ATP.	Calcium pump; glucose transporter

CREDITS: (23A, in text, Table 5.2) © Cengage Learning.

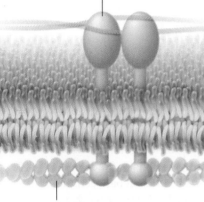

B Adhesion proteins fasten cells together or to extracellular matrix. This one is an integrin that connects microfilaments (inside the cell) to extracellular matrix proteins (outside the cell).

C Receptor proteins initiate a change in cellular activity in response to a stimulus such as binding to a particular substance. This one, a B cell receptor, occurs on cells of the immune system. It consists of an antibody anchored by integral membrane proteins.

D Enzymes catalyze reactions at membranes. This one, a cytochrome P450, is part of electron transfer chains that break down drugs and other organic toxins.

E Transport proteins bind to molecules on one side of the membrane, and release them on the other side. This one transports glucose.

extracellular fluid

lipid bilayer

cytoplasm

microfilament

ily attaches to one of the lipid bilayer's surfaces by interacting with lipid heads or an integral protein.

A cell membrane physically separates an external environment from an internal one, but that is not its only task. Each kind of protein in a membrane imparts a specific function to it. Thus, different cell membranes can have different functions depending on which proteins are associated with them. A plasma membrane incorporates certain proteins that no internal cell membrane has, so it carries out functions that no other membrane does. For example, adhesion proteins occur only on plasma membranes. **Adhesion proteins** fasten cells to one another, or connect extracellular matrix outside the cell to cytoskeletal elements inside of it (**FIGURE 5.23B**). This arrangement strengthens a tissue, and can constrain certain membrane proteins to an upper or lower surface of the cell. Adhesion proteins are the sticky components of adhering and tight junctions in animal tissues (Section 4.11). Many adhesion proteins also have important roles in cell signaling, helping cells sense and respond to external conditions.

Plasma membranes and some internal membranes incorporate **receptor proteins**, which trigger a change in the cell's activities in response to a stimulus (**FIGURE 5.23C**). Each type of receptor protein receives a particular stimulus, for example absorbing light at a certain wavelength, or binding to a certain hormone. Each receptor protein also triggers a specific response inside the cell, which may involve metabolism, movement, division, or even cell death.

All cell membranes incorporate enzymes (**FIGURE 5.23D**). A lipid bilayer provides a scaffold for enzymes that work in series, for example in electron transfer chains. Some membrane enzymes act on other proteins or lipids that are part of the lipid bilayer. All membranes also have **transport proteins**, which move specific substances across the bilayer (**FIGURE 5.23E**). These proteins are important because, as you will see in the next section, lipid bilayers are impermeable to most substances, including ions and polar molecules.

---

adhesion protein Protein that helps cells stick together in animal tissues. Some types form adhering junctions and tight junctions.
fluid mosaic Model of a cell membrane as a two-dimensional fluid of mixed composition.
receptor protein Membrane protein that triggers a change in cell activity in response to a stimulus such as binding a certain substance.
transport protein Protein that passively or actively assists specific ions or molecules across a membrane.

TAKE-HOME MESSAGE 5.7
What is a cell membrane?

✔ The foundation of almost all cell membranes is the lipid bilayer—two layers of lipids (mainly phospholipids), with tails sandwiched between heads.

✔ Proteins embedded in or attached to a lipid bilayer add specific functions to each cell membrane.

5.8 Diffusion and Membranes

✔ Ions and molecules tend to move spontaneously from regions of higher to lower concentration.

✔ Water diffuses across cell membranes by osmosis.

a drop of pink dye diffusing in water

Metabolic pathways require the participation of molecules that must move across membranes and through cells. **Diffusion** (left) is the spontaneous spreading of molecules or ions, and it is an essential way in which substances move into, through, and out of cells. An atom or molecule is always jiggling, and this internal movement causes it to randomly bounce off of nearby objects, including other atoms or molecules. Rebounds from such collisions propel solutes through a liquid or gas, resulting in a gradual and complete mixing. How fast this occurs depends on five factors:

Molecular Size It takes more energy to move a large object than it does to move a small one, so small molecules diffuse more quickly than large ones.

Temperature Atoms and molecules jiggle faster at higher temperature, so they collide more often. Thus, diffusion occurs more quickly at higher temperatures.

Concentration A difference in solute concentration (Section 2.6) between adjacent regions of solution is called a concentration gradient. Solutes tend to diffuse "down" their concentration gradient, from a region of higher concentration to one of lower concentration. Why? Consider that moving objects (such as molecules) collide more often as they get more crowded. Thus, during a given interval, more molecules get bumped out of a region of higher concentration than get bumped into it.

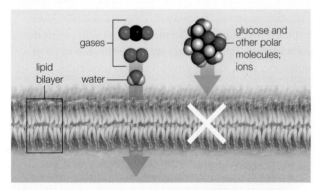

FIGURE 5.24 Selective permeability of lipid bilayers. Hydrophobic molecules, gases, and water molecules can cross a lipid bilayer on their own. Ions in particular and most polar molecules, including glucose, cannot.

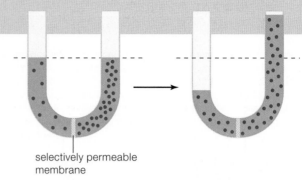

selectively permeable membrane

FIGURE 5.25 ▶**Animated** Osmosis. Water moves across a selectively permeable membrane that separates two fluids of differing solute concentration. The fluid volume changes in the two compartments as water diffuses across the membrane from the hypotonic solution to the hypertonic one.

Charge Each ion or charged molecule in a fluid contributes to the fluid's overall electric charge. A difference in charge between two regions of fluid can affect the rate and direction of diffusion between them. For example, positively charged substances (such as sodium ions) will tend to diffuse toward a region with an overall negative charge.

Pressure The rate of diffusion may be affected by a difference in pressure between two adjoining regions. Pressure squeezes objects—including atoms and molecules—closer together. Atoms and molecules that are more crowded collide and rebound more frequently. Thus, diffusion occurs faster at higher pressures.

Semipermeable Membranes

Lipid bilayers are selectively permeable (Section 4.2); water, gases, and hydrophobic molecules can cross them, but ions and most polar molecules cannot (**FIGURE 5.24**). When two fluids with different solute concentrations are separated by a selectively permeable membrane, water will diffuse across the membrane. The direction of water movement depends on the relative solute concentration of the two fluids, which we describe in terms of tonicity. Fluids that are **isotonic** have the same overall solute concentration. If the overall solute concentrations of the two fluids differ, the fluid with the lower concentration of solutes is said to be **hypotonic** (*hypo–*, under). The other one, with the higher solute concentration, is **hypertonic** (*hyper–*, over).

When a selectively permeable membrane separates two fluids that are not isotonic, water will move across the membrane from the hypotonic fluid into the hypertonic one (**FIGURE 5.25**). The diffusion will continue until the two fluids are isotonic, or until pressure against the hypertonic fluid counters it. The movement of water across membranes is so important in biology that it is given a special name: **osmosis**.

CREDITS: (in text) Andrew Lambert Photography/Science Source; (24, 25) © Cengage Learning.

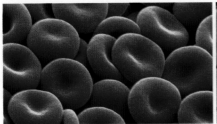 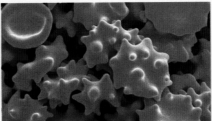 

A Red blood cells in an isotonic solution (such as the fluid portion of blood) have a normal, indented disk shape.

B Water diffuses out of red blood cells immersed in a hypertonic solution, so they shrivel up.

C Water diffuses into red blood cells immersed in a hypotonic solution, so they swell up. Some of these have burst.

FIGURE 5.26 Effects of tonicity in human red blood cells. These cells have no mechanism to compensate for differences in solute concentration between cytoplasm and extracellular fluid.

If a cell's cytoplasm becomes hypertonic with respect to the fluid outside of its plasma membrane, water will diffuse into the cell. If the cytoplasm becomes hypotonic, water will diffuse out. In either case, the solute concentration of the cytoplasm may change. If it changes enough, the cell's enzymes will stop working, with lethal results. Many cells have built-in mechanisms that compensate for differences in solute concentration between cytoplasm and extracellular (external) fluid. In cells with no such mechanism, the volume—and solute concentration—of cytoplasm changes when water diffuses into or out of the cell (**FIGURE 5.26**).

Turgor

The rigid cell walls of plants and many protists, fungi, and bacteria can resist an increase in the volume of cytoplasm even in hypotonic environments. In the case of plant cells, cytoplasm usually contains more solutes than soil water does. Thus, water usually diffuses from soil into a plant—but only up to a point. Stiff walls keep plant cells from expanding very much, so an inflow of water causes pressure to build up inside them. Pressure that a fluid exerts against a structure that contains it is called **turgor**. When enough pressure builds up inside a plant cell, water stops diffusing into

its cytoplasm. The amount of turgor that is enough to stop osmosis is called **osmotic pressure**.

Osmotic pressure keeps walled cells plump, just as high air pressure inside a tire keeps it inflated. A young land plant can resist gravity to stay erect because its cells are plump with cytoplasm (**FIGURE 5.27A**). When soil dries out, it loses water, so the concentration of solutes increases in it. If soil water becomes hypertonic with respect to cytoplasm, water will start diffusing out of the plant's cells, causing their cytoplasm to shrink (**FIGURE 5.27B**). As turgor inside the cells decreases, the plant wilts.

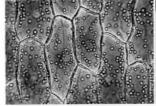 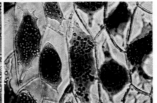

A Osmotic pressure keeps plant parts erect. These cells in an iris petal are plump with cytoplasm.

B Cells from a wilted iris petal. The cytoplasm shrank, and the plasma membrane moved away from the wall.

FIGURE 5.27 Turgor, as illustrated in cells of iris petals.

---

diffusion Spontaneous spreading of molecules or ions.
hypertonic Describes a fluid that has a high solute concentration relative to another fluid separated by a semipermeable membrane.
hypotonic Describes a fluid that has a low solute concentration relative to another fluid separated by a semipermeable membrane.
isotonic Describes two fluids with identical solute concentrations and separated by a semipermeable membrane.
osmosis Diffusion of water across a selectively permeable membrane; occurs in response to a difference in solute concentration between the fluids on either side of the membrane.
osmotic pressure Amount of turgor that prevents osmosis into cytoplasm or other hypertonic fluid.
turgor Pressure that a fluid exerts against a structure that contains it.

TAKE-HOME MESSAGE 5.8
What influences the movement of solutes?

✔ Solutes tend to diffuse into an adjoining region of fluid in which they are not as concentrated. The steepness of a concentration gradient as well as temperature, molecular size, charge, and pressure affect the rate of diffusion.

✔ When two fluids of different solute concentration are separated by a selectively permeable membrane, water diffuses from the hypotonic to the hypertonic fluid. This movement, osmosis, is opposed by turgor.

✔ Many types of molecules and ions can cross a lipid bilayer only with the help of transport proteins.

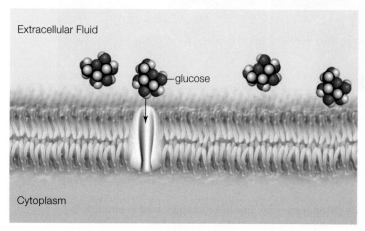

Extracellular Fluid

glucose

Cytoplasm

A A glucose molecule (here, in extracellular fluid) binds to a glucose transporter (gray) in the plasma membrane.

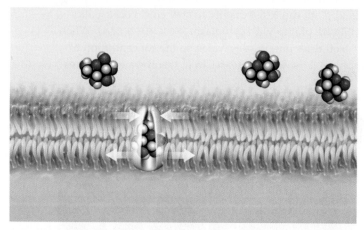

B Binding causes the transport protein to change shape.

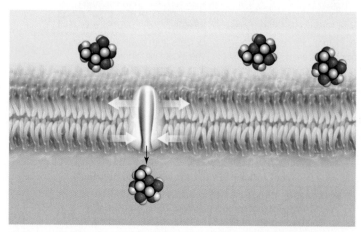

C The transport protein releases the glucose on the other side of the membrane (here, in cytoplasm) and resumes its original shape.

FIGURE 5.28 An example of facilitated diffusion.

FIGURE IT OUT In this example, which fluid is hypotonic: extracellular fluid or the cytoplasm?

Answer: Cytoplasm

Transport Protein Specificity

Substances that cannot diffuse directly through lipid bilayers—ions in particular—cross cell membranes only with the help of transport proteins. Each type of transport protein allows a specific substance to cross: Calcium pumps pump only calcium ions; glucose transporters transport only glucose; and so on. This specificity is an important part of homeostasis. For example, the composition of cytoplasm depends on the movement of particular solutes across the plasma membrane, which in turn depends on the transporters embedded in it. Glucose is an important source of energy for most cells, so they normally take up as much as they can from extracellular fluid. They do so with the help of glucose transporters in the plasma membrane. As soon as a molecule of glucose enters cytoplasm, an enzyme (hexokinase) phosphorylates it. Phosphorylation traps the molecule inside the cell because the transporters are specific for glucose, not phosphorylated glucose. Thus, phosphorylation prevents the molecule from moving back through the transport protein and leaving the cell.

Facilitated Diffusion

Osmosis is an example of **passive transport**, a membrane-crossing mechanism that requires no energy input. The diffusion of solutes through transport proteins is another example. In this case, the movement of the solute (and the direction of its movement) is driven entirely by the solute's concentration gradient. Some transport proteins form pores: permanently open channels through a membrane. Other channels are gated, which means they open and close in response to a stimulus such as a shift in electric charge or binding to a signaling molecule.

With a passive transport mechanism called **facilitated diffusion**, a solute binds to a transport protein, which then changes shape so the solute is released to the other side of the membrane. A glucose transporter is an example of a transport protein that works in facilitated diffusion (**FIGURE 5.28**). This protein changes shape when it binds to a molecule of glucose. The shape change moves the glucose to the opposite side of the membrane, where it detaches from the transport protein. Then, the glucose transporter reverts to its original shape.

active transport Energy-requiring mechanism in which a transport protein pumps a solute across a cell membrane against the solute's concentration gradient.
facilitated diffusion Passive transport mechanism in which a solute follows its concentration gradient across a membrane by moving through a transport protein.
passive transport Membrane-crossing mechanism that requires no energy input.

Active Transport

Maintaining a solute's concentration often means transporting the solute against its gradient, to the side of the membrane where it is more concentrated. This takes energy. In **active transport**, a transport protein uses energy to pump a solute against its gradient across a cell membrane. Typically, an energy input (for example, in the form of a phosphate-group transfer from ATP) changes the shape of an active transport protein. The shape change causes the protein to release a bound solute to the other side of the membrane.

A calcium pump moves calcium ions across cell membranes by active transport (**FIGURE 5.29**). Calcium ions act as potent messengers inside cells, and they also affect the activity of many enzymes. Thus, their concentration in cytoplasm is tightly regulated. Calcium pumps in the plasma membrane of all eukaryotic cells can keep the concentration of calcium ions in cytoplasm thousands of times lower than it is in extracellular fluid.

Another example of active transport involves sodium–potassium pumps (**FIGURE 5.30**). Nearly all cells in your body have these transport proteins. Sodium ions in cytoplasm diffuse into the pump's open channel and bind to its interior. A phosphate-group transfer from ATP causes the pump to change shape so that its channel opens to extracellular fluid, where it releases the sodium ions. Then, potassium ions from extracellular fluid diffuse into the channel and bind to its interior. The transporter releases the phosphate group and reverts to its original shape. The channel opens to the cytoplasm, where it releases the potassium ions.

Bear in mind that the membranes of all cells, not just those of animals, have active transport proteins. In plants, for example, active transport proteins in the plasma membranes of leaf cells pump sucrose into tubes that thread throughout the plant body.

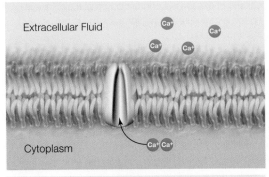

A Two calcium ions (blue) bind to the transport protein (gray).

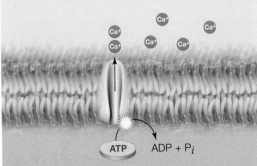

B A phosphate group from ATP causes the protein to change shape so that the calcium ions are ejected to the opposite side of the membrane.

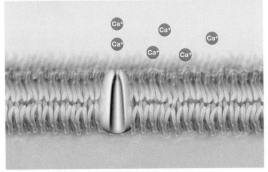

C After it loses the calcium ions, the transport protein resumes its original shape.

FIGURE 5.29 Active transport of calcium ions.

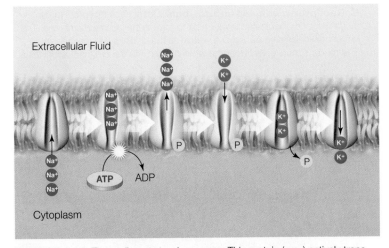

FIGURE 5.30 The sodium–potassium pump. This protein (gray) actively transports sodium ions (Na+) from cytoplasm to extracellular fluid, and potassium ions (K+) in the other direction. The transfer of a phosphate group (P) from ATP provides energy required for transporting the ions against their concentration gradient.

TAKE-HOME MESSAGE 5.9
How do solutes that cannot diffuse through lipid bilayers cross cell membranes?

✔ Transport proteins move specific ions or molecules across a cell membrane. The types and amounts of substances that cross a membrane depend on the transport proteins embedded in it.

✔ In facilitated diffusion (a type of passive transport), a solute binds to a transport protein that releases it on the opposite side of the membrane. The movement is driven by the solute's concentration gradient.

✔ In active transport, a transport protein pumps a solute across a membrane against its concentration gradient. The movement requires energy, as from ATP.

✔ By processes of exocytosis and endocytosis, cells take in and expel particles that are too big for transport proteins, as well as substances in bulk.

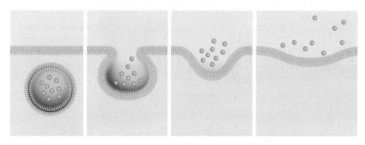

A Exocytosis. A vesicle in cytoplasm fuses with the plasma membrane. Lipids and proteins of the vesicle's membrane become part of the plasma membrane as its contents are expelled to the environment.

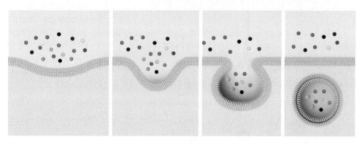

B Pinocytosis. A pit in the plasma membrane traps any fluid, solutes, and particles near the cell's surface in a vesicle as it sinks into the cytoplasm.

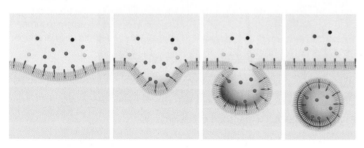

C Receptor-mediated endocytosis. Cell surface receptors (green) bind a target molecule and trigger a pit to form in the plasma membrane. The target molecules are trapped in a vesicle as the pit sinks into the cytoplasm. This mode is more selective about what is taken into the cell than pinocytosis.

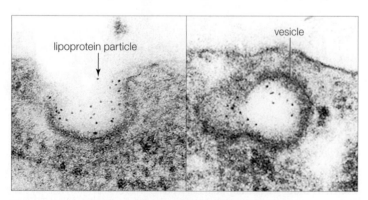

lipoprotein particle

vesicle

D Receptor-mediated endocytosis of lipoprotein particles.

FIGURE 5.31 Exocytosis and endocytosis.

Endocytosis and Exocytosis

Think back on the fluid mosaic structure of a lipid bilayer. When a membrane is disrupted, the fatty acid tails of the phospholipids in the bilayer become exposed to their watery surroundings. Remember, in water, phospholipids spontaneously rearrange themselves so that their nonpolar tails stay together. Thus, a membrane tends to seal itself after a disruption. Vesicles form the same way. When a patch of membrane bulges into the cytoplasm, the hydrophobic tails of the lipids in the bilayer are repelled by the watery fluid on both sides. The fluid "pushes" the phospholipid tails together, which helps round off the bud as a vesicle, and also seals the rupture in the membrane.

Vesicles are constantly carrying materials to and from a cell's plasma membrane. This movement typically requires ATP because it involves motor proteins that drag the vesicles along cytoskeletal elements. We describe the movement based on where and how the vesicle originates, and where it goes.

By **exocytosis**, a vesicle in the cytoplasm moves to the cell's surface and fuses with the plasma membrane (**FIGURE 5.31A**). As the exocytic vesicle loses its identity, its contents are released to the surroundings.

There are several pathways of **endocytosis**, but all take up substances in bulk near the cell's surface (as opposed to one molecule or ion at a time via transport proteins). **Pinocytosis** is an endocytic pathway that brings a drop of extracellular fluid (along with solutes and particles suspended in it) into the cell (**FIGURE 5.31B**). With this pathway, a small patch of plasma membrane balloons inward and then pinches off as it sinks into the cytoplasm. The membrane patch becomes the outer boundary of a vesicle.

Receptor-mediated endocytosis is more selective than pinocytosis about what it brings into the cell (**FIGURE 5.31C**). With this pathway, molecules of a hormone, vitamin, mineral, or another substance bind to receptors on the plasma membrane. The binding triggers a shallow pit to form in the membrane, just under the receptors. The pit sinks into the cytoplasm and traps the target substance in a vesicle as it closes back on itself. LDL and other lipoproteins (Section 3.6) enter cells this way (**FIGURE 5.31D**).

Phagocytosis (which means "cell eating") is a type of receptor-mediated endocytosis in which motile cells engulf microorganisms, cellular debris, or other large particles. Many single-celled protists such as amoebas feed by phagocytosis. Some of your white blood cells use phagocytosis to engulf viruses and bacteria, cancerous body cells, and other threats to health.

CREDITS: (31A–C) © Cengage Learning; (31D) © R.G.W. Anderson, M.S. Brown and J.L. Goldstein. *Cell* 10:351 (1977).

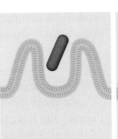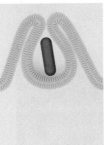

FIGURE 5.32 An example of phagocytosis. The SEM on the left shows a phagocytic white blood cell's pseudopods (extending lobes of cytoplasm) surrounding *Tuberculosis* bacteria (in red). The artwork shows how plasma membrane above the bulging lobes fuses and forms a vesicle. Once inside the cytoplasm, the endocytic vesicle will fuse with a lysosome. Enzymes delivered by the lysosome will break down the contents of the vesicle.

Phagocytosis begins when receptor proteins bind to a particular target. The binding causes microfilaments to assemble in a mesh under the plasma membrane. The microfilaments then contract, forcing a lobe of membrane-enclosed cytoplasm to bulge outward as a pseudopod (Section 4.10). Pseudopods that merge around a target trap it inside a vesicle that sinks into the cytoplasm (**FIGURE 5.32**). Material taken in

by phagocytosis is typically digested by lysosomal enzymes, and the resulting molecular bits may be recycled by the cell, or expelled by exocytosis.

Recycling Membrane

The composition of a plasma membrane begins in the endoplasmic reticulum (ER). There, membrane proteins and lipids are made and modified, and both become part of vesicles that transport them to Golgi bodies for final modification. New plasma membrane forms when the finished proteins and lipids are repackaged as vesicles that travel to the plasma membrane and fuse with it.

FIGURE 5.33 shows what happens when an exocytic vesicle fuses with the plasma membrane. Membrane proteins are oriented toward the interior of a vesicle that buds from a Golgi body, so after the vesicle fuses with the plasma membrane, the proteins face the extracellular environment.

As long as a cell is alive, exocytosis and endocytosis continually replace and withdraw patches of its plasma membrane. If the cell is not enlarging, the total area of the plasma membrane remains more or less constant. Membrane lost as a result of endocytosis is replaced by membrane arriving as exocytic vesicles.

FIGURE 5.33
How membrane proteins become oriented to the inside or the outside of a cell. Proteins of the plasma membrane are assembled in the ER, and finished inside Golgi bodies. The proteins (shown in white) become part of vesicle membranes that bud from the Golgi. The membrane proteins automatically become oriented in the proper direction when the vesicles fuse with the plasma membrane.

FIGURE IT OUT What process does the upper arrow represent?

Answer: Exocytosis

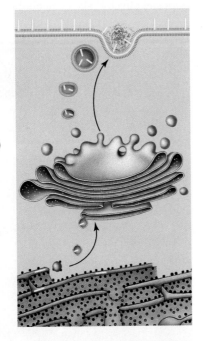

endocytosis Process by which a cell takes in a small amount of extracellular fluid (and its contents) by the ballooning inward of the plasma membrane.
exocytosis Process by which a cell expels a vesicle's contents to extracellular fluid.
phagocytosis "Cell eating"; an endocytic pathway by which a cell engulfs large particles such as microbes or cellular debris.
pinocytosis Endocytic pathway by which fluid and materials in bulk are brought into the cell.

TAKE-HOME MESSAGE 5.10
How do large particles and bulk substances move into and out of cells?

✔ Exocytosis and endocytosis move materials in bulk across plasma membranes.

✔ In exocytosis, a cytoplasmic vesicle fuses with the plasma membrane and releases its contents to the cell's exterior.

✔ In endocytosis, a patch of plasma membrane sinks inward and forms a vesicle in the cytoplasm.

✔ Some cells can engulf large particles by phagocytosis.

A Toast to Alcohol Dehydrogenase (revisited)

In most organisms, the main function of the enzyme alcohol dehydrogenase (ADH) is to detoxify the tiny quantities of alcohols that form in some metabolic pathways. In animals, the enzyme also detoxifies small amounts of alcohols made by gut-inhabiting bacteria, and those in foods such as ripe fruit.

ADH in the human body converts ethanol to acetaldehyde, an organic molecule even more toxic than ethanol and the most likely source of various hangover symptoms:

$$H-\underset{\underset{H}{|}}{\overset{\overset{H}{|}}{C}}-\underset{\underset{H}{|}}{\overset{\overset{H}{|}}{C}}-OH + NAD^+ \xrightarrow{\text{ADH}} H-\underset{\underset{H}{|}}{\overset{\overset{H}{|}}{C}}-\overset{\overset{O}{\|}}{C}-H + NADH$$

ethanol   acetaldehyde

A different enzyme, aldehyde dehydrogenase (ALDH), very quickly converts the toxic acetaldehyde to non-toxic acetate:

$$H-\underset{\underset{H}{|}}{\overset{\overset{H}{|}}{C}}-\overset{\overset{O}{\|}}{C}-H + NAD^+ \xrightarrow{\text{ALDH}} H-\underset{\underset{H}{|}}{\overset{\overset{H}{|}}{C}}-\overset{\overset{O}{\|}}{C}-O^- + NADH + H^+$$

acetaldehyde   acetate

Both ADH and ALDH use the coenzyme NAD$^+$ to accept electrons and hydrogen atoms. Thus, the overall pathway of ethanol metabolism in humans is:

$$\text{ethanol} \xrightarrow[\text{NAD}^+ \quad \text{NADH}]{\text{ADH}} \text{acetaldehyde} \xrightarrow[\text{NAD}^+ \quad \text{NADH}]{\text{ALDH}} \text{acetate}$$

In the average healthy adult human, this metabolic pathway can detoxify between 7 and 14 grams of ethanol per hour. The average alcoholic beverage contains between 10 and 20 grams of ethanol, which is why having more than one drink in any two-hour interval may result in a hangover.

A person's ability to metabolize ethanol in alcoholic drinks depends on the amount and activity of ADH they make. Some people have an overactive form of the enzyme. When they drink, acetaldehyde accumulates in their bodies faster than ALDH can detoxify it:

$$\text{ethanol} \xrightarrow{\text{ADH}} \begin{matrix}\text{acetaldehyde}\\\text{acetaldehyde}\\\text{acetaldehyde}\end{matrix} \xrightarrow{\text{ALDH}} \text{acetate}$$

People who have an overactive form of ADH become flushed and feel ill after drinking even a small amount of alcohol. The unpleasant experience may be part of the reason that these people are statistically unlikely to become alcoholics.

Having an underactive form of ALDH also results in an accumulation of acetaldehyde after drinking:

$$\text{ethanol} \xrightarrow{\text{ADH}} \begin{matrix}\text{acetaldehyde}\\\text{acetaldehyde}\\\text{acetaldehyde}\end{matrix} \xrightarrow{\quad\mathbf{X}\quad} \text{acetate}$$

Underactive ALDH is associated with the same protection from alcoholism as overactive ADH. Both types of variant enzymes are common in people of Asian descent. For this reason, the alcohol flushing reaction is informally called "Asian flush."

Having an underactive ADH enzyme has the opposite effect. It slows alcohol metabolism, so

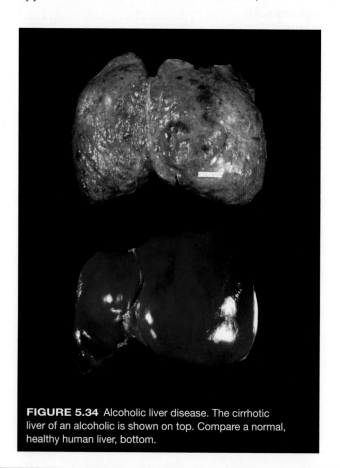

FIGURE 5.34 Alcoholic liver disease. The cirrhotic liver of an alcoholic is shown on top. Compare a normal, healthy human liver, bottom.

people with a low level of ADH activity may not feel the ill effects of drinking alcoholic beverages as much as others do. When these people drink, they have a tendency to become alcoholics. Alcoholism is characterized by compulsive, uncontrolled drinking that damages the individual's health and social relationships. One-quarter of undergraduate students who binge also have other signs of alcoholism.

Alcoholics will continue to drink despite the knowledge that doing so has tremendous negative consequences. In the United States, alcohol abuse is the leading cause of cirrhosis of the liver. A cirrhotic liver is so scarred, hardened, and filled with fat that it no longer functions properly (**FIGURE 5.34**). It no longer produces the protein albumin, so the solute balance of body fluids is disrupted, and the legs and abdomen swell with watery fluid. It can no longer remove drugs and other toxins from the blood, so they accumulate in the brain—which impairs mental functioning and alters personality. Restricted blood flow through the liver causes veins to enlarge and rupture, so internal bleeding is a risk. The damage to the body results in

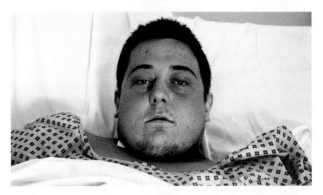

FIGURE 5.35 Gary Reinbach, who died at the age of 22 from alcoholic liver disease shortly after this photograph was taken, in 2009. The odd color of his skin is a symptom of cirrhosis.

Transplantation is a last-resort treatment for a failed liver, but there are not enough liver donors for everyone who needs a transplant. Reinbach was refused a transplant that may have saved his life because he had not abstained from drinking for the prior 6 months.

a heightened susceptibility to diabetes and liver cancer. Once cirrhosis has been diagnosed, a person has about a 50 percent chance of dying within 10 years (**FIGURE 5.35**).

summary

Section 5.1 Currently the most serious drug problem on college campuses is binge drinking, which is often a symptom of alcoholism. Drinking more alcohol than the body's enzymes can detoxify can be lethal in both the short term and the long term.

Section 5.2 **Energy** is the capacity to do work. One form of energy (such as **potential energy**) can be converted to another (such as **kinetic energy**). Energy cannot be created or destroyed (**first law of thermodynamics**), and it tends to disperse spontaneously (**second law of thermodynamics**). **Entropy** is a measure of how much the energy of a system is dispersed. A bit disperses at each energy transfer, usually in the form of heat. Living things maintain their organization only as long as they harvest energy from someplace else. Energy flows in one direction through the biosphere, starting mainly from the sun, then into and out of ecosystems. Producers and then consumers use the captured energy to assemble, rearrange, and break down organic molecules that cycle among organisms in an ecosystem.

Section 5.3 Cells store and retrieve energy by making and breaking chemical bonds in reactions that convert **reactants** to **products**. **Endergonic** reactions require a net input of energy to proceed. **Exergonic** reactions end

with a net release of energy. **Activation energy** is the minimum energy required to start a reaction.

Section 5.4 Enzymes greatly enhance the rate of reactions without being changed by them, a process called **catalysis**. They lower a reaction's activation energy by boosting local concentrations of **substrates**, orienting substrates in positions that favor reaction, inducing the fit between a substrate and the enzyme's **active site** (**induced-fit model**), or excluding water. These mechanisms bring on the substrate's **transition state**. Each type of enzyme works best within a characteristic range of conditions, including temperature, salt concentration, and pH.

Section 5.5 Cells build, convert, and dispose of substances in **metabolic pathways**, which are sequences of enzyme-mediated reactions. Regulating metabolic pathways allows a cell to conserve energy and resources by making only what it needs at a given time. With **allosteric regulation**, a regulatory molecule or ion alters the activity of an enzyme by binding to it in a region other than the active site. The products of some metabolic pathways inhibit their own production, a regulatory mechanism called **feedback inhibition**. **Redox** (oxidation–reduction) **reactions** in **electron transfer chains** allow cells to harvest energy in small, manageable steps.

Section 5.6 Most enzymes require assistance from **cofactors**. Some cofactors are metal ions; organic cofactors are **coenzymes**. Cofactors help some **antioxidant** enzymes prevent dangerous oxidation reactions. ATP is often used as a coenzyme that carries energy between reaction sites in cells. It has two high-energy phosphate bonds. When a phosphate group is transferred from ATP to another molecule, energy is transferred along with it. **Phosphorylations** to and from ATP couple exergonic with endergonic reactions. Cells regenerate ATP in the **ATP/ADP cycle**.

Section 5.7 A cell membrane is a mosaic of proteins and lipids (mainly phospholipids) organized as a lipid bilayer. Membranes of bacteria and eukaryotic cells can be described as a **fluid mosaic**; membranes of archaea are not fluid. Proteins transiently or permanently associated with a membrane carry out most membrane functions. All cell membranes have enzymes, and all have **transport proteins** that help substances move across the membrane. Plasma membranes also incorporate **adhesion proteins** that lock cells together in tissues. Plasma membranes and some internal membranes have **receptor proteins** that trigger a change in cell activities in response to a specific stimulus.

Section 5.8 The rate of **diffusion** is influenced by temperature, solute size, and regional differences in concentration, charge, and pressure. Gases, water, and nonpolar molecules can diffuse across a lipid bilayer. Most other molecules, and ions in particular, cannot.

Osmosis is the diffusion of water across a selectively permeable membrane, from a **hypotonic** fluid toward a **hypertonic** fluid. There is no net movement of water between **isotonic** solutions. **Osmotic pressure** is the amount of **turgor** (fluid pressure against a cell membrane or wall) sufficient to halt osmosis.

Section 5.9 Ions and most polar molecules can cross cell membranes only with the help of a transport protein. With **facilitated diffusion**, a solute follows its concentration gradient across a membrane through a transport protein. Facilitated diffusion is a type of **passive transport** (no energy input is required). With **active transport**, a transport protein uses energy to pump a solute across a membrane against its concentration gradient. A phosphate-group transfer from ATP often supplies the necessary energy for active transport.

Section 5.10 Substances in bulk and large particles are moved across plasma membranes by processes of **exocytosis** and **endocytosis**. With exocytosis, a cytoplasmic vesicle fuses with the plasma membrane, and its contents are released to the outside of the cell. **Pinocytosis** is an endocytic pathway in which a patch of plasma membrane balloons into the cell, and forms a vesicle that sinks into the cytoplasm. Some cells engulf large particles such as prey or cell debris by the endocytic pathway of **phagocytosis**.

self-quiz

Answers in Appendix VII

1. Which of the following statements is *not* correct?
   a. Energy cannot be created or destroyed.
   b. Energy cannot change from one form to another.
   c. Energy tends to disperse spontaneously.

2. _____ is life's primary source of energy.
   a. Food   c. Sunlight
   b. Water   d. ATP

3. Entropy _____ .
   a. disperses   c. always increases, overall
   b. is a measure of disorder   d. b and c

4. If we liken a chemical reaction to an energy hill, then a(n) _____ reaction is, overall, an uphill run.
   a. endergonic   c. catalytic
   b. exergonic   d. both a and c

5. If we liken a chemical reaction to an energy hill, then activation energy is like _____ .
   a. a burst of speed
   b. coasting downhill
   c. a bump at the top of the hill
   d. putting on the brakes

6. _____ are always changed by participating in a reaction. (Choose all that are correct.)
   a. Enzymes   c. Reactants
   b. Cofactors   d. Coenzymes

7. Name one environmental factor that typically influences enzyme function.

8. Which of the following statements is *not* correct?
   a. Metabolic pathways build or break down the organic molecules of life.
   b. All metabolic pathways generate heat.
   c. Electron transfer chains are important sites of energy exchange in many metabolic pathways.
   d. All metabolic pathways require ATP.

9. A molecule that donates electrons becomes _____ , and the one that accepts electrons becomes _____ .
   a. reduced; oxidized   c. oxidized; reduced
   b. ionic; electrified   d. electrified; ionic

10. All antioxidants _____ .
   a. prevent other molecules from being oxidized
   b. are coenzymes
   c. balance charge
   d. deoxidize free radicals

11. Solutes tend to diffuse from a region where they are _____ (more/less) concentrated to another where they are _____ (more/less) concentrated.

12. _____ cannot easily diffuse across a lipid bilayer.
   a. Water   c. Ions
   b. Gases   d. all of the above

13. A transport protein requires ATP to pump sodium ions across a membrane. This is a case of _____ .
   a. passive transport   c. facilitated diffusion
   b. active transport   d. a and c

One Tough Bug The genus *Ferroplasma* consists of a few species of acid-loving archaea. One species, *F. acidarmanus*, was discovered to be the main constituent of slime streamers (a type of biofilm) deep inside an abandoned California copper mine (**FIGURE 5.36**). These cells use an ancient energy-harvesting pathway that combines oxygen with iron-sulfur compounds in minerals such as pyrite. Oxidizing these minerals dissolves them, so groundwater that seeps into the mine ends up with extremely high concentrations of metal ions such as copper, zinc, cadmium, and arsenic. The reaction also produces sulfuric acid, which lowers the pH of the water around the cells to zero.

F. acidarmanus cells maintain their internal pH at a cozy 5.0 despite living in an environment similar to hot battery acid. Thus, researchers investigating *Ferroplasma* were surprised to discover that most of the cells' enzymes function best at very low pH (**FIGURE 5.37**).

1. What does the dashed line signify?

2. Of the four enzymes profiled in the graph, how many function optimally at a pH lower than 5? How many retain significant function at pH 5?

3. What is the optimal pH for *Ferroplasma* carboxylesterase?

FIGURE 5.36 Deep inside one of the most toxic sites in the United States: Iron Mountain Mine, in California. The water in this stream, which is about 1 meter (3 feet) wide in this view, is hot (around 40°C, or 104°F), heavily laden with arsenic and other toxic metals, and has a pH of zero. The slime streamers growing in it are a biofilm dominated by a species of archaea, *Ferroplasma acidarmanus*.

FIGURE 5.37 pH anomaly of *Ferroplasma acidarmanus* enzymes. The graphs (right) show the pH activity profiles of four enzymes isolated from *Ferroplasma*. Researchers had expected these enzymes to function best at the cells' cytoplasmic pH (5.0).

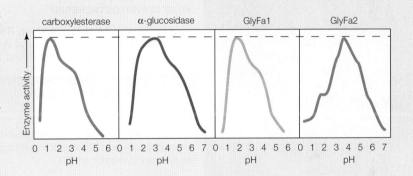

14. Immerse a human red blood cell in a hypotonic solution, and water _____ .
    - a. diffuses into the cell
    - b. diffuses out of the cell
    - c. shows no net movement
    - d. moves in by endocytosis

15. Vesicles are part of _____ .
    - a. endocytosis
    - b. exocytosis
    - c. phagocytosis
    - d. all of the above

16. Match each term with its most suitable description.
    - ____ reactant
    - ____ phagocytosis
    - ____ first law of thermodynamics
    - ____ product
    - ____ cofactor
    - ____ concentration gradient
    - ____ passive transport
    - ____ active transport
    - ____ redox reaction
    - ____ cyclic pathway
    - ____ lipid bilayer

    - a. assists enzymes
    - b. forms at reaction's end
    - c. enters a reaction
    - d. requires energy input
    - e. one cell engulfs another
    - f. energy cannot be created or destroyed
    - g. basis of diffusion
    - h. no energy input required
    - i. phospholipids + water
    - j. goes in circles
    - k. electron exchange

critical thinking

1. Beginning physics students are often taught the basic concepts of thermodynamics with two phrases: First, you can never win. Second, you can never break even. Explain.

2. How is diffusion similar to entropy?

3. How do you think a cell regulates the amount of glucose it brings into its cytoplasm from the extracellular fluid?

4. The enzyme trypsin is sold as a dietary enzyme supplement. Explain what happens to trypsin taken with food.

5. Catalase combines two hydrogen peroxide molecules ($H_2O_2 + H_2O_2$) to make two molecules of water. A gas also forms. What is the gas?

CENGAGE brain.com To access course materials, please visit www.cengagebrain.com.

CREDITS: (36) Katrina J. Edwards; (37) From Golyshina et al., *Environmental Microbiology*, 8(3): 416–425, © 2006 John Wiley and Sons. Used with permission of the publisher.

LEARNING ROADMAP

This chapter explores the main metabolic pathways (Sections 5.5, 5.6) by which organisms harvest energy from the sun (5.2, 5.3). We revisit experimental design (1.6), electrons and energy levels (2.3), bonds (2.4), carbohydrates (3.4), plastids (4.9), plant cell specializations (4.7, 4.11), membrane proteins (5.7), and concentration gradients (5.8).

THE RAINBOW CATCHERS

The main flow of energy through the biosphere starts when photosynthetic pigments absorb light. In plants and other eukaryotes, these pigments occur in chloroplasts.

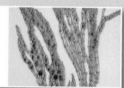

WHAT IS PHOTOSYNTHESIS?

Photosynthesis is a metabolic pathway that occurs in two stages. Light energy harvested in the first stage is used to make molecules that power sugar formation in the second.

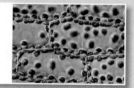

MAKING ATP AND NADPH

The light-dependent reactions produce ATP by either a noncyclic or a cyclic pathway. The noncyclic pathway produces NADPH and oxygen gas in addition to ATP.

MAKING SUGARS

In the second stage of photosynthesis, sugars are assembled from CO_2. The reactions run on ATP and NADPH—molecules that formed in the first stage of photosynthesis.

ALTERNATE PATHWAYS

Metabolic pathways are shaped by evolution. Variations in photosynthetic pathways are evolutionary adaptations that allow plants to thrive in a variety of environments.

You will see in Chapter 7 how molecules originally assembled by photosynthesizers are disassembled to harvest energy stored in their bonds. Chapter 22 returns to evolutionary adaptations of plants; and Chapters 27–30, to plant structure and function. In Chapter 46, you will see how photosynthetic organisms sustain almost all life on Earth and how carbon cycles through the biosphere. Chapter 48 returns to human impacts on the biosphere.

6.1 Biofuels

Today, the expression "food is fuel" is not just about eating. With fossil fuel prices soaring, there is an increasing demand for biofuels, which are oils, gases, or alcohols made from organic matter that is not fossilized. Much of the material currently used for biofuel production in the United States consists of food crops—mainly corn, soybeans, and sugarcane. Growing these crops in large quantities is typically expensive and damaging to the environment, and using them to make biofuel competes with our food supply.

How did we end up competing with our vehicles for food? We both run on the same fuel: energy that plants have stored in chemical bonds. Fossil fuels such as coal and natural gas are the remains of ancient swamp forests that decayed and compacted over millions of years. These fuels consist mainly of molecules originally assembled by ancient plants. By contrast, biofuels—and foods—consist mainly of molecules originally assembled by modern plants.

Autotrophs are organisms that make their own food by harvesting energy directly from the environment (*auto-* means self; *-troph* refers to nourishment). All organisms need carbon; autotrophs obtain it from inorganic molecules such as carbon dioxide (CO_2). Plants and most other autotrophs make their food by the metabolic pathway of photosynthesis (Section 1.3). During this pathway, the energy of sunlight is used to drive the assembly of carbohydrates—sugars—from carbon dioxide and water.

Heterotrophs get their carbon by breaking down organic molecules assembled by other organisms (*hetero-* means other). Heterotrophs are an ecosystem's consumers. We and almost all other heterotrophs

sustain ourselves by extracting energy from organic molecules that photosynthesizers make. Thus, photosynthesis feeds most life on Earth.

A lot of energy is locked up in the chemical bonds of molecules made by plants. That energy can fuel heterotrophs, as when an animal cell powers ATP synthesis by breaking the bonds of sugars (a topic detailed in the next chapter). It can also fuel our cars, which run on energy released by burning biofuels or fossil fuels. Both processes are fundamentally the same: They release energy by breaking the bonds of organic molecules. Both use oxygen to break those bonds, and both produce carbon dioxide.

Corn and other food crops are rich in oils, starches, and sugars that can be easily converted to biofuels. The starch in corn kernels, for example, can be enzymatically broken down to glucose, which is converted to ethanol by heterotrophic bacteria or yeast. Making biofuels from other types of plant matter requires additional steps, because these materials contain a higher proportion of cellulose. Breaking down this tough, insoluble carbohydrate to its glucose monomers adds a lot of cost to the biofuel product. Researchers are currently working on cost-effective ways to break down the abundant cellulose in fast-growing weeds such as switchgrass (**FIGURE 6.1**), and agricultural wastes such as wood chips, wheat straw, cotton stalks, and rice hulls.

---

autotroph Organism that makes its own food using energy from the environment and carbon from inorganic molecules such as CO_2.
heterotroph Organism that obtains carbon from organic compounds assembled by other organisms.

FIGURE 6.1 Making biofuels. Left, Ratna Sharma and Mari Chinn are researching ways to reduce the cost of producing biofuel from renewable sources such as wild grasses and agricultural wastes. Right, switchgrass (*Panicum virgatum*) growing wild in a North American prairie.

6.2 Sunlight as an Energy Source

✔ Photosynthetic organisms use pigments to capture the energy of sunlight.

Properties of Light

Photosynthesizers make their own food by converting light energy to chemical energy. In order to understand how that happens, you have to know a little about the nature of light. Light is electromagnetic radiation, a type of energy that moves through space in waves, a bit like waves move across an ocean. The distance between the crests of two successive waves is called **wavelength**, and it is measured in nanometers (nm).

Light that humans can see is a small part of the spectrum of electromagnetic radiation emitted by the sun (**FIGURE 6.2A**). Visible light travels in wavelengths between 380 and 750 nm, and this is the main form of energy that drives photosynthesis. Our eyes perceive all of these wavelengths combined as white light, and particular wavelengths in this range as different colors. White light separates into its component colors when it passes through a prism, or raindrops that act as tiny prisms (**FIGURE 6.3**). A prism bends longer wavelengths more than it bends shorter ones, so a rainbow of colors forms.

Light travels in waves, but it is also organized in packets of energy called photons. A photon's energy and its wavelength are related, so all photons traveling at the same wavelength carry the same amount of energy. Photons that carry the least amount of energy travel in longer wavelengths; those that carry the most energy travel in shorter wavelengths (**FIGURE 6.2B**).

FIGURE 6.3 A rainbow. Sunlight passing through raindrops separates into its component colors.

Pigments: The Rainbow Catchers

Photosynthesizers use pigments to capture light. A **pigment** is an organic molecule that selectively absorbs light of specific wavelengths. Wavelengths of light that are not absorbed are reflected, and that reflected light gives each pigment its characteristic color.

Chlorophyll a is the most common photosynthetic pigment in plants and photosynthetic protists. It also occurs in some bacteria. Chlorophyll *a* absorbs violet, red, and orange light, and it reflects green light, so it appears green to us. Accessory pigments, including other chlorophylls, collectively harvest a wide range of additional light wavelengths for photosynthesis (**FIGURE 6.4**).

A pigment molecule is a bit like an antenna specialized for receiving light. It has a light-trapping region, in

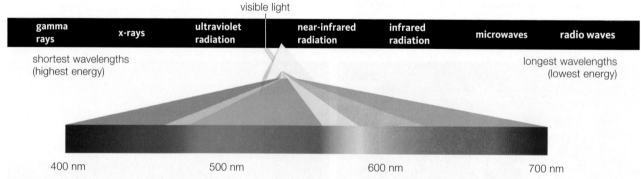

| gamma rays | x-rays | ultraviolet radiation | near-infrared radiation | infrared radiation | microwaves | radio waves |

shortest wavelengths (highest energy)   longest wavelengths (lowest energy)

400 nm   500 nm   600 nm   700 nm

A Electromagnetic radiation moves through space in waves that we measure in nanometers (nm). Visible light makes up a very small part of this energy. Raindrops or a prism can separate visible light's different wavelengths, which we see as different colors. About 25 million nanometers are equal to 1 inch.

B Light is organized as packets of energy called photons. The shorter a photon's wavelength, the greater its energy.

FIGURE 6.2 Properties of light.

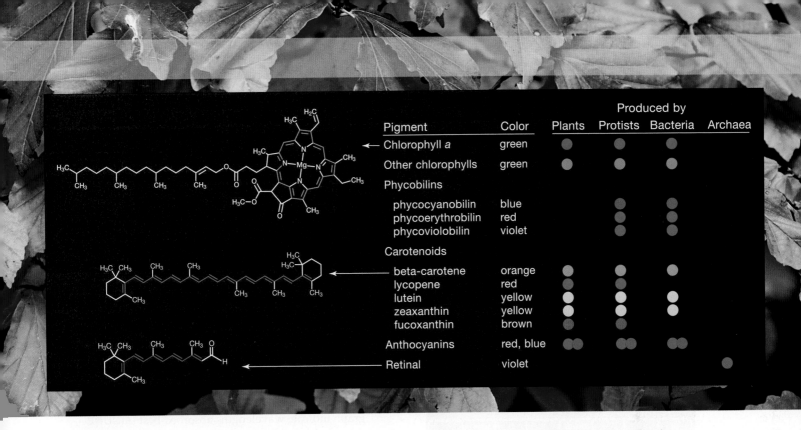

FIGURE 6.4 Examples of photosynthetic pigments. Photosynthetic pigments can collectively absorb almost all visible light wavelengths. Left, the light-catching part of a pigment (shown in color) is the region in which single bonds alternate with double bonds. These and many other pigments (including heme, Section 5.6) are derived from evolutionary remodeling of the same compound. Animals convert dietary beta-carotene into a similar pigment (retinal) that is the basis of vision.

which single bonds alternate with double bonds. Electrons populating the atoms of this region easily absorb a photon—but not just any photon. Only a photon with exactly enough energy to boost an electron to a higher energy level is absorbed (Section 2.3). This is why a pigment absorbs light of only certain wavelengths.

An excited electron (one that has been boosted to a higher energy level) quickly emits its extra energy and returns to a lower energy level. As you will see, photosynthetic cells capture energy emitted from an electron returning to a lower energy level.

Most photosynthetic organisms use a combination of pigments to capture light for photosynthesis—and often for additional purposes. Many accessory pigments are antioxidants that protect cells from the damaging effects of ultraviolet (UV) light in the sun's rays (Section 5.6). Appealing colors attract animals to ripening fruit or pollinators to flowers. You may already be familiar with some of these molecules. Carrots, for example, are orange because they contain beta-carotene (β-carotene); roses are red and violets are blue because of their anthocyanin content.

chlorophyll a Main photosynthetic pigment in plants.
pigment An organic molecule that can absorb light of certain wavelengths.
wavelength Distance between the crests of two successive waves.

In green plants, chlorophylls are usually so abundant that they mask the colors of the other pigments. Plants that change color during autumn are preparing for a period of dormancy; they conserve resources by moving nutrients from tender parts that would be damaged by winter cold (such as leaves) to protected parts (such as roots). Chlorophylls are not needed during dormancy, so they are disassembled and their components recycled. Yellow and orange accessory pigments are also recycled, but not as quickly as chlorophylls. Their colors begin to show as the chlorophyll content declines in leaves. Anthocyanin synthesis also increases in some plants, adding red and purple tones to turning leaf colors. (Chapter 29 returns to the topic of dormancy in plants.)

TAKE-HOME MESSAGE 6.2
How do photosynthesizers absorb light?

✔ The sun emits electromagnetic radiation (light). Visible light is the main form of energy that drives photosynthesis.

✔ Light travels in waves and is organized as photons. We see different wavelengths of visible light as different colors.

✔ Pigments absorb light at specific wavelengths. Photosynthetic species use pigments such as chlorophyll a to harvest the energy of light for photosynthesis.

6.3 Exploring the Rainbow

✔ Photosynthetic pigments work together to harvest light of different wavelengths.

In 1882, botanist Theodor Engelmann designed a series of experiments to test his hypothesis that the color of light affects the rate of photosynthesis. It had

A Each cell in a strand of *Cladophora* is filled with a single chloroplast. Theodor Engelmann used this and other species of green algae in a series of experiments to determine whether some colors of light are better for photosynthesis than others.

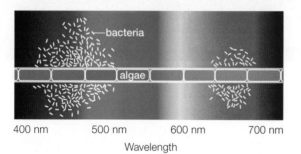

B Engelmann directed light through a prism so that bands of colors crossed a water droplet on a microscope slide. The water held a strand of photosynthetic algae, and also oxygen-requiring bacteria. The bacteria swarmed around the algal cells that were releasing the most oxygen—the ones most actively engaged in photosynthesis. Those cells were under blue and red light.

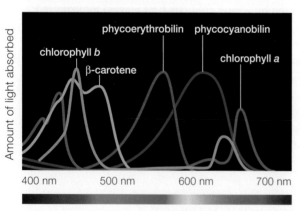

C Absorption spectra of chlorophylls *a* and *b*, β-carotene, and two phycobilins reveal the efficiency with which these pigments absorb different wavelengths of visible light. Line color indicates the characteristic color of each pigment.

FIGURE 6.5 ▶Animated Discovery that photosynthesis is driven by particular wavelengths of light.

Of the five pigments represented in **C**, which three are the main photosynthetic pigments in *Cladophora*?

Answer: Chlorophyll a, chlorophyll b, and β-carotene

long been known that photosynthesis releases oxygen, so Engelmann used oxygen emission as an indirect measure of photosynthetic activity. He directed a spectrum of light across individual strands of green algae suspended in water (**FIGURE 6.5A**). Oxygen-sensing equipment had not yet been invented, so Engelmann used motile, oxygen-requiring bacteria to show him where the oxygen concentration in the water was highest. The bacteria moved through the water and gathered mainly where blue and red light fell across the algal cells (**FIGURE 6.5B**). Engelmann concluded that photosynthetic cells illuminated by light of these colors were releasing the most oxygen—a sign that blue and red light are the best for driving photosynthesis in these algal cells.

Today we can directly measure the efficiency at which a photosynthetic pigment absorbs different wavelengths of light. A graph that shows this efficiency is called an absorption spectrum. Peaks in the graph indicate wavelengths absorbed best (**FIGURE 6.5C**). Engelmann's results (the distribution of bacteria around the algal cells) represent the combined spectra of all the photosynthetic pigments present in this alga.

The combination of pigments used for photosynthesis differs among species. Why? Photosynthesizers are adapted to the environment in which they evolved, and light that reaches different environments varies in its proportions of wavelengths. Consider that seawater absorbs green and blue-green light less efficiently than other colors. Thus, more green and blue-green light penetrates deep ocean water. Algae that can live far below the sea surface (such as *Polysiphonia*, below) tend to be rich in pigments—mainly phycobilins—that absorb green and blue-green light.

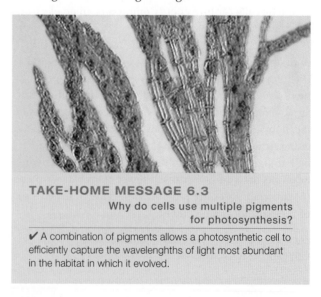

TAKE-HOME MESSAGE 6.3
Why do cells use multiple pigments for photosynthesis?

✔ A combination of pigments allows a photosynthetic cell to efficiently capture the wavelengths of light most abundant in the habitat in which it evolved.

CREDITS: (5A) Jason Sonneman; (5B, C) © Cengage Learning; (in text) © Michael Davidson/The Florida State University.

6.4 Overview of Photosynthesis

✔ In eukaryotes, photosynthesis takes place in chloroplasts.
✔ Photosynthesis occurs in two stages.

All life is sustained by inputs of energy, but not all forms of energy can sustain life. Sunlight, for example, is abundant here on Earth, but it cannot be used to directly power protein synthesis or other energy-requiring reactions that keep organisms alive. Photosynthesis converts the energy of light into the energy of chemical bonds. Unlike light, chemical energy can power the reactions of life, and it can be stored for later use.

In plants and other photosynthetic eukaryotes, photosynthesis takes place in chloroplasts (Section 4.9). Plant chloroplasts have two outer membranes, and they are filled with a thick, cytoplasm-like fluid called **stroma** (**FIGURE 6.6**). Suspended in the stroma are the chloroplast's own DNA, some ribosomes, and an inner, much-folded **thylakoid membrane**. The folds of a thylakoid membrane typically form stacks of interconnected

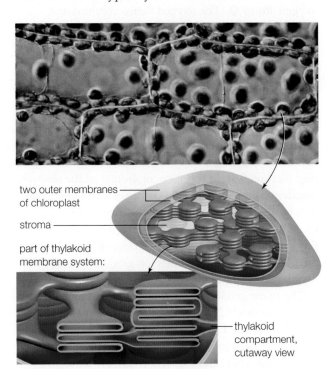

two outer membranes of chloroplast

stroma

part of thylakoid membrane system:

thylakoid compartment, cutaway view

FIGURE 6.6 ▶**Animated** Zooming in on a chloroplast, the site of photosynthesis in a plant cell. The micrograph shows chloroplasts in cells of a moss leaf.

light-dependent reactions First stage of photosynthesis; convert light energy to chemical energy.
light-independent reactions Second stage of photosynthesis; use ATP and NADPH to assemble sugars from water and CO_2.
stroma Cytoplasm-like fluid between the thylakoid membrane and the two outer membranes of a chloroplast.
thylakoid membrane A chloroplast's highly folded inner membrane system; forms a continuous compartment in the stroma.

disks called thylakoids. The space enclosed by the thylakoid membrane is a single, continuous compartment.

Photosynthesis is often summarized by this equation:

$$CO_2 + \text{water} \xrightarrow{\text{light energy}} \text{sugars} + O_2$$

CO_2 (carbon dioxide) and O_2 (oxygen) are gases abundant in air. Keep in mind that photosynthesis is not a single reaction. Rather, it is a metabolic pathway (Section 5.5) comprising many reactions that occur in two stages. Molecules embedded in the thylakoid membrane carry out the reactions of the first stage, which are driven by light and thus called the **light-dependent reactions**. The "photo" in photosynthesis means light, and it refers to the conversion of light energy to the chemical bond energy of ATP during this stage. In addition to making ATP, the main light-dependent pathway in chloroplasts splits water molecules and releases O_2. Hydrogen ions and electrons from the water molecules end up in the coenzyme NADPH:

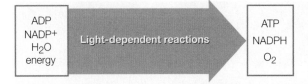

ADP
NADP+
H_2O
energy

Light-dependent reactions

ATP
NADPH
O_2

The "synthesis" part of photosynthesis refers to the reactions of the second stage, which build sugars from CO_2 and water. These sugar-building reactions run in the stroma. They are collectively called the **light-independent reactions** because light energy does not power them. Instead, they run on energy delivered by NADPH and ATP that formed during the first stage:

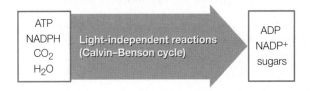

ATP
NADPH
CO_2
H_2O

Light-independent reactions (Calvin–Benson cycle)

ADP
NADP+
sugars

> **TAKE-HOME MESSAGE 6.4**
> **What happens during photosynthesis?**
>
> ✔ In eukaryotic cells, the first stage of photosynthesis occurs at the thylakoid membrane of chloroplasts. During these light-dependent reactions, light energy drives the formation of ATP and NADPH.
>
> ✔ In eukaryotic cells, the second stage of photosynthesis occurs in the stroma of chloroplasts. During these light-independent reactions, ATP and NADPH drive the synthesis of sugars from water and carbon dioxide.

6.5 Light-Dependent Reactions

✔ The reactions of the first stage of photosynthesis convert the energy of light to the energy of chemical bonds.

A chloroplast's thylakoid membrane contains millions of light-harvesting complexes, which are circular arrays of chlorophylls, various accessory pigments, and proteins (**FIGURE 6.7**). When a chlorophyll or accessory pigment in a light-harvesting complex absorbs light, one of its electrons jumps to a higher energy level (shell, Section 2.3). The electron quickly drops back down to a lower shell by emitting its extra energy. Light-harvesting complexes hold onto that emitted energy by passing it back and forth, a bit like volleyball players pass a ball among team members. The reactions of photosynthesis begin when energy being passed around the thylakoid membrane reaches a photosystem. A **photosystem** is a group of hundreds of chlorophylls, accessory pigments, and other molecules that work as a unit to begin the reactions of photosynthesis.

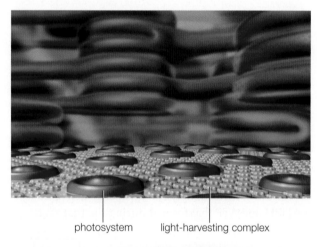

photosystem   light-harvesting complex

FIGURE 6.7 A view of some components of the thylakoid membrane as seen from the stroma. Molecules of electron transfer chains and ATP synthases are also present, but not shown for clarity.

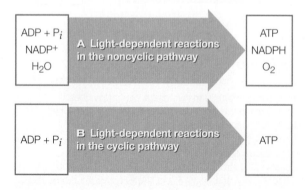

FIGURE 6.8 Summary of the inputs and outputs of the two pathways of light-dependent reactions.

The Noncyclic Pathway

Thylakoid membranes have two kinds of photosystems, type I and type II, that were named in the order of their discovery. In cyanobacteria, plants, and all photosynthetic protists, both photosystem types work together in the noncyclic pathway of photosynthesis (**FIGURE 6.8A**). This pathway begins when energy being passed among light-harvesting complexes reaches a photosystem II (**FIGURE 6.9**). At the center of each photosystem are two very closely associated chlorophyll *a* molecules (a "special pair"). When a photosystem absorbs energy, electrons are ejected from its special pair ❶.

A photosystem can lose only a few electrons before it must be restocked with more. Where do replacements come from? Photosystem II gets more electrons by removing them from water molecules in the thylakoid compartment. This reaction causes the water molecules to dissociate into hydrogen ions and oxygen atoms ❷. The oxygen atoms combine and diffuse out of the cell as oxygen gas (O_2). This and any other process by which a molecule is broken apart by light energy is called **photolysis**.

The actual conversion of light energy to chemical energy occurs when electrons ejected from photosystem II enter an electron transfer chain in the thylakoid membrane ❸. Remember that electron transfer chains can harvest the energy of electrons in a series of redox reactions, releasing a bit of their extra energy with each step (Section 5.5). In this case, molecules of the electron transfer chain use the released energy to actively transport hydrogen ions (H^+) across the membrane, from the stroma to the thylakoid compartment ❹. Thus, the flow of electrons through electron transfer chains sets up and maintains a hydrogen ion gradient across the thylakoid membrane.

The hydrogen ion gradient is a type of potential energy (Section 5.2) that can be tapped to make ATP. The H^+ ions want to follow their concentration gradient by moving back into the stroma, but ions cannot diffuse through the lipid bilayer (Section 5.8). H^+ leaves the thylakoid compartment only by flowing through proteins called ATP synthases embedded in the thylakoid membrane ❼. An ATP synthase is both a transport protein and an enzyme. When hydrogen ions flow through its interior, the protein phosphorylates ADP, so ATP forms in the stroma ❽. The process by which the flow of electrons through electron transfer chains drives ATP formation is called **electron transfer phosphorylation**.

After the electrons have moved through the first electron transfer chain, they are accepted by a photo-

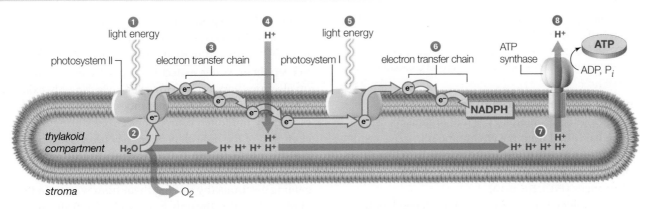

light energy ④ H+ ⑤ light energy ⑧ H+

ATP

⑧ H+ → ATP → ADP, P$_i$

photosystem II ③ electron transfer chain photosystem I ⑥ electron transfer chain ATP synthase

e⁻ e⁻ e⁻ e⁻ e⁻ e⁻

e⁻ e⁻ e⁻ e⁻

NADPH

thylakoid compartment ② H$_2$O H+ H+ H+ H+ H+ H+ H+ H+ ⑦ H+

stroma O$_2$

❶ Light energy ejects electrons from a photosystem II.

❷ The photosystem pulls replacement electrons from water molecules, which then break apart into oxygen and hydrogen ions. The oxygen leaves the cell as O$_2$.

❸ The electrons enter an electron transfer chain in the thylakoid membrane.

❹ Energy lost by the electrons as they move through the chain is used to actively transport hydrogen ions from the stroma into the thylakoid compartment. A hydrogen ion gradient forms across the thylakoid membrane.

❺ Light energy ejects electrons from a photosystem I. Replacement electrons come from an electron transfer chain.

❻ The ejected electrons move through a second electron transfer chain, then combine with NADP⁺ and H⁺, so NADPH forms.

❼ Hydrogen ions in the thylakoid compartment follow their gradient across the thylakoid membrane by flowing through the interior of ATP synthases.

❽ Hydrogen ion flow causes ATP synthases to phosphorylate ADP, so ATP forms in the stroma.

FIGURE 6.9 ▶Animated Light-dependent reactions, noncyclic pathway. ATP and oxygen gas are produced in this pathway. Electrons that travel through two different electron transfer chains end up in NADPH.

system I. When this photosystem absorbs light energy, its special pair of chlorophylls emits electrons ❺. These electrons enter a second, different electron transfer chain. At the end of this chain, the coenzyme NADP⁺ accepts the electrons along with H⁺, so NADPH forms ❻:

$$\text{NADP}^+ + 2e^- + H^+ \longrightarrow \boxed{\textbf{NADPH}}$$

The Cyclic Pathway

As you will see shortly, ATP and NADPH produced in the light-dependent reactions are used to make sugars. On its own, the noncyclic pathway does not yield enough ATP to balance NADPH use in sugar production pathways. The cyclic pathway produces additional ATP for this purpose (**FIGURE 6.8B**).

In the cyclic pathway, electrons that are ejected from photosystem I enter an electron transfer chain, and then return to photosystem I. As in the noncyclic pathway, the electron transfer chain uses electron energy to move hydrogen ions into the thylakoid compartment, and the resulting hydrogen ion gradient drives ATP

formation. However, the cyclic pathway does not produce NADPH or oxygen gas.

The cyclic pathway allows light-dependent reactions to continue when the noncyclic pathway stops, for example under intense illumination. Light energy in excess of what can be used for photosynthesis can result in the formation of dangerous free radicals (Section 2.3). A light-induced structural change in photosystem II prevents this from happening. The photosystem stops initiating the noncyclic pathway, and traps excess energy instead. At such times, the cyclic pathway predominates.

---

electron transfer phosphorylation Process in which electron flow through electron transfer chains sets up a hydrogen ion gradient that drives ATP formation.
photolysis Process by which light energy breaks down a molecule.
photosystem Cluster of pigments and proteins that converts light energy to chemical energy in photosynthesis.

TAKE-HOME MESSAGE 6.5

How do the light-dependent reactions of photosynthesis work?

✔ Photosynthetic pigments in the thylakoid membrane transfer the energy of light to photosystems, which eject electrons that enter electron transfer chains.

✔ In both noncyclic and cyclic pathways, the flow of electrons through the transfer chains sets up hydrogen ion gradients that drive ATP formation.

✔ The noncyclic pathway uses two photosystems. Water molecules are split, oxygen is released, and electrons end up in NADPH.

✔ The cyclic pathway uses only photosystem I. No NADPH forms, and no oxygen is released.

6.6 The Light-Independent Reactions

✔ The chloroplast is a sugar factory operated by enzymes of the Calvin–Benson cycle.

✔ ATP and NADPH from the noncyclic pathway provide energy that powers the light-independent reactions of photosynthesis.

Energy Flow in Photosynthesis

A recurring theme in biology is that organisms use energy harvested from the environment to drive cellular processes. Energy flow in the noncyclic pathway of light-dependent reactions is a classic example of how that happens (**FIGURE 6.10**).

The simpler cyclic pathway of light-dependent reactions was the first one to evolve. Later, the photosynthetic machinery in some organisms evolved so that photosystem II became part of it. Of the two photosystems, photosystem I is the stronger reducing agent, which means it more easily gives up electrons. Those electrons either cycle back to it (in the cyclic pathway), or they end up in NADPH (in the noncyclic pathway). Running the noncyclic pathway requires a continuous input of electrons. Not many electrons float around freely in a thylakoid (or any other biological system); most are associated with atoms or molecules that want to keep them. Water molecules in particular do not give up electrons very easily. However, cells can only exist where there is water, so water molecules provided an essentially unlimited source of electrons during the evolution of the light-dependent reactions. Thus, it is perhaps not suprising that, of the two photosystems, photosystem II is the stronger oxidizer: It is best at gaining electrons (by pulling them from water). In fact, photosystem II is the only biological system strong enough to oxidize water molecules.

Light-Independent Reactions

NADPH is a powerful reducing agent (electron donor), and a lot of it is required to run the second stage of photosynthesis. You learned in Section 6.4 that reactions of this stage are light-independent because light energy does not power them. Energy that drives these

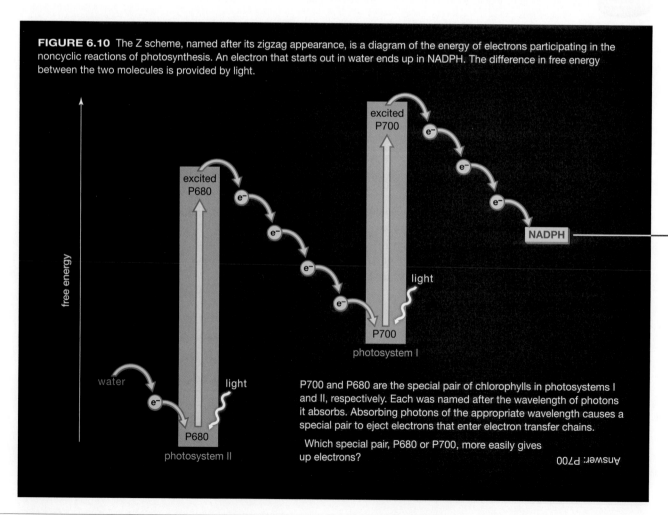

FIGURE 6.10 The Z scheme, named after its zigzag appearance, is a diagram of the energy of electrons participating in the noncyclic reactions of photosynthesis. An electron that starts out in water ends up in NADPH. The difference in free energy between the two molecules is provided by light.

P700 and P680 are the special pair of chlorophylls in photosystems I and II, respectively. Each was named after the wavelength of photons it absorbs. Absorbing photons of the appropriate wavelength causes a special pair to eject electrons that enter electron transfer chains.

Which special pair, P680 or P700, more easily gives up electrons?

Answer: P700

CREDIT: (10) © Cengage Learning.

reactions is provided by phosphate-group transfers from ATP, and electrons from NADPH.

The light-independent reactions, which are collectively called the **Calvin–Benson cycle**, produce sugars in the stroma of chloroplasts (**FIGURE 6.11**). This cyclic pathway uses carbon atoms from CO_2 to build carbon backbones of the sugar molecules. Extracting carbon atoms from an inorganic source (such as CO_2) and incorporating them into an organic molecule is a process called **carbon fixation**.

glucose

In most plants, photosynthetic protists, and some bacteria, the enzyme **rubisco** fixes carbon by attaching CO_2 to RuBP (ribulose bisphosphate), a five-carbon molecule ❶. The six-carbon intermediate that forms by this reaction is unstable, so it splits right away into two three-carbon molecules of PGA (phosphoglycerate). Each PGA receives a phosphate group from ATP, and hydrogen and electrons from NADPH ❷. Thus, ATP energy and the reducing power of NADPH convert

each molecule of PGA into a molecule of PGAL (phosphoglyceraldehyde), a phosphorylated sugar.

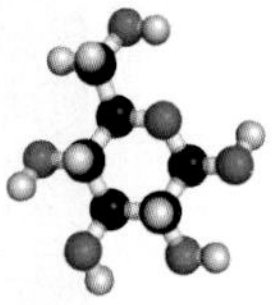

In later reactions, two or more of the three-carbon PGAL molecules can be combined and rearranged to form larger carbohydrates. Glucose has six carbon atoms (left). To make one glucose molecule, six CO_2 must be attached to six RuBP molecules, so twelve PGAL form. Two PGAL combine to form one glucose molecule ❸. The ten remaining PGAL regenerate the starting compound of the cycle, RuBP ❹.

Plants can break down the glucose they make in the Calvin–Benson cycle to access the energy stored in its bonds (we return to this process in Chapter 7). However, most of the glucose is converted at once to sucrose or starch by other pathways that conclude the light-independent reactions. Excess glucose is stored as starch grains in chloroplast stroma. When sugars are needed in other parts of the plant, the starch is broken down and sugar monomers are exported from the cell.

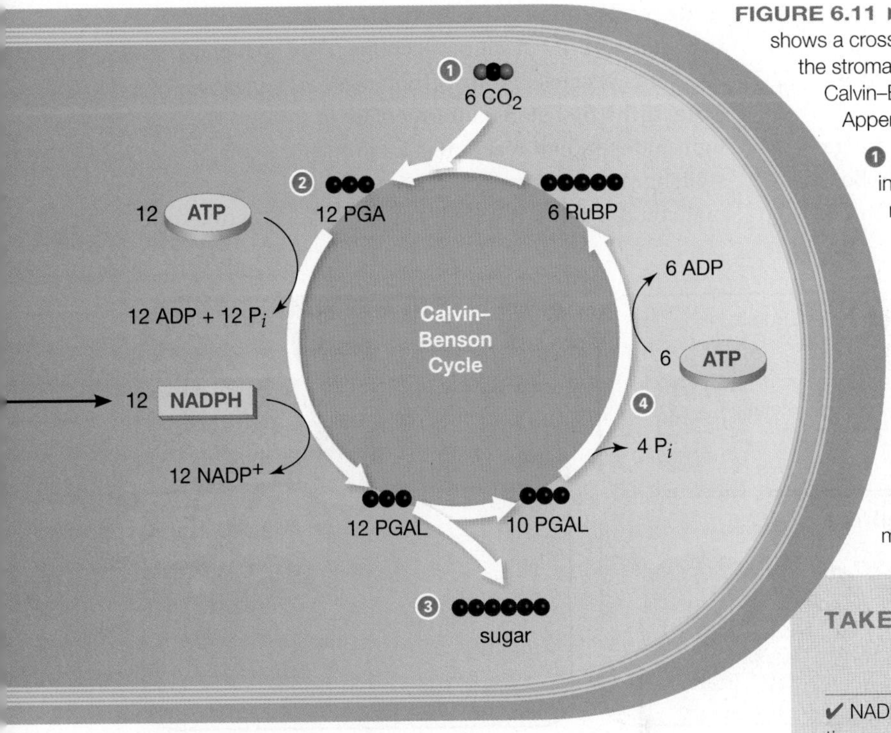

FIGURE 6.11 ▶**Animated** The Calvin–Benson cycle. This sketch shows a cross-section of a chloroplast with the reactions cycling in the stroma. The steps shown are a summary of six cycles of the Calvin–Benson reactions. Black balls signify carbon atoms. Appendix III details the reaction steps.

❶ Six CO_2 diffuse into a photosynthetic cell, and then into a chloroplast. Rubisco attaches each to a RuBP molecule. The resulting intermediates split, so twelve molecules of PGA form.

❷ Each PGA molecule gets a phosphate group from ATP, plus hydrogen and electrons from NADPH, so twelve PGAL form. ❸ Two PGAL may combine to form one six-carbon sugar (such as glucose).

❹ The remaining ten PGAL receive phosphate groups from ATP. The transfer primes them for endergonic reactions that regenerate the 6 RuBP.

How many times does the Calvin–Benson cycle need to run in order to produce one molecule of glucose?

Answer: Six

TAKE-HOME MESSAGE 6.6

How do the light-independent reactions of photosynthesis work?

✔ NADPH and ATP produced by the light-dependent reactions power the light-independent reactions of the Calvin–Benson cycle.

✔ The Calvin–Benson cycle uses atoms of hydrogen (from NADPH), and carbon and oxygen (from CO_2) to build sugars.

✔ Incorporating carbon atoms from an inorganic source (such as CO_2) into an organic molecule is called carbon fixation.

Calvin–Benson cycle Cyclic carbon-fixing pathway that builds sugars from CO_2; light-independent reactions of photosynthesis.
carbon fixation Process by which carbon from an inorganic source such as carbon dioxide gets incorporated into an organic molecule.
rubisco Ribulose bisphosphate carboxylase. Carbon-fixing enzyme of the Calvin–Benson cycle.

6.7 Adaptations: Alternative Carbon-Fixing Pathways

✔ Carbon fixation sustains all life on Earth, but its details differ among species.

Most plants have a thin, waterproof cuticle that limits evaporative water loss from their aboveground parts. Gases cannot diffuse across the cuticle, but oxygen produced by the light-dependent reactions must escape the plant, and carbon dioxide needed for the Calvin–Benson cycle must enter it. Thus, the surfaces of leaves and stems are studded with tiny, closable gaps called **stomata** (**FIGURE 6.12**). When stomata are open, CO_2 diffuses from the air into photosynthetic tissues, and O_2 diffuses out of the tissues into the air. Stomata close to conserve water on hot, dry days. When that happens, gas exchange comes to a halt.

Both stages of photosynthesis run during the day. With stomata closed, the O_2 level in the plant's tissues rises, and the CO_2 level declines. This outcome can reduce the efficiency of sugar production because both gases are substrates of rubisco, and they compete for its active site. Rubisco initiates the Calvin–Benson cycle by attaching CO_2 to RuBP. It also initiates a pathway called **photorespiration**, by attaching O_2 to RuBP. Photorespiration is an inefficient way to produce sugars because it requires more ATP and NADPH than the Calvin–Benson cycle, and it produces CO_2 (**FIGURE 6.13A**).

The detrimental effects of photorespiration are greatest in **C3 plants**, which fix carbon by the Calvin–Benson cycle only (they are called C3 plants because a three-

FIGURE 6.12 Stomata on the surface of a leaf. When these tiny pores are open, gases are exchanged between the plant's internal tissues and air. Stomata close to conserve water on hot, dry days.

carbon molecule, PGA, is the first stable intermediate to form in their light-independent reactions). C3 plants have no way to compensate for a decline in CO_2 level when stomata close during the day, so sugar production becomes less and less efficient with rising daytime temperature. These plants compensate for photorespiration by making a lot of rubisco: It is the most abundant protein on Earth.

C4 plants compensate for rubisco's inefficiency by using an additional set of reactions (they are called C4 plants because a four-carbon molecule, oxaloacetate, is the first stable intermediate to form in their light-independent reactions). Examples are corn, switchgrass, and bamboo. These plants also close stomata on dry days, but their sugar production does not

FIGURE 6.13 ▶Animated Adaptations of C4 and CAM plants minimize photorespiration. Compare **FIGURE 6.14**.

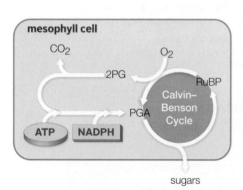

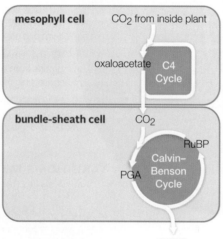

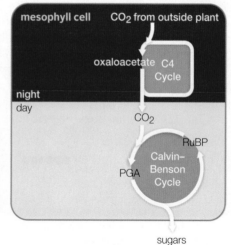

A Photorespiration. When stomata close during photosynthesis, O_2 accumulates in tissues and CO_2 declines. Rubisco initiates more photorespiration, shunting resources away from the sugar-producing reactions of the Calvin–Benson cycle.

B In C4 plants, oxygen also builds up in leaves when stomata close during photosynthesis. An additional pathway in these plants keeps the CO_2 concentration high enough in bundle-sheath cells to prevent photorespiration.

C CAM plants open stomata and fix carbon using a C4 pathway at night. When the plant's stomata are closed during the day, the organic compound made during the night is converted to CO_2 that enters the Calvin–Benson cycle.

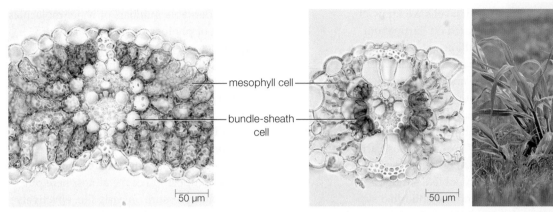

A In a leaf of a C3 plant (here, barley), chloroplasts—the sites of carbon fixation—occur mainly in mesophyll cells. Bundle sheath cells ringing the leaf vein are mostly empty of chloroplasts.

Photorespiration in C3 plants reduces the efficiency of sugar synthesis on hot, dry days.

B In a leaf of a C4 plant (millet, left), carbon is fixed the first time in mesophyll cells, which are near air spaces in the leaf. Carbon fixation occurs for the second time in chloroplast-stuffed bundle-sheath cells.

The photo on the right shows crabgrass "weeds" overgrowing a lawn. Crabgrasses, which are C4 plants, thrive in hot, dry summers, when they easily outcompete Kentucky bluegrass and other fine-leaved C3 grasses commonly planted in residential lawns.

FIGURE 6.14 ▶Animated Comparing the sites of carbon fixation in C3 plants and C4 plants. Micrographs show cross sections. In both types of plants, bundle-sheath cells ring the leaf veins. In C4 plants only, a second carbon fixation occurs in these cells. The adaptation maintains a high CO_2/O_2 ratio near rubisco, minimizing photorespiration when stomata close on hot, dry days.

decline. C4 plants fix carbon twice, in two kinds of cells (**FIGURE 6.13B** and **FIGURE 6.14**). The first set of reactions occurs in mesophyll cells, where carbon is fixed by an enzyme that does not use oxygen even when the carbon dioxide level is low. The product of this C4 pathway, a four-carbon acid, is transported to bundle-sheath cells. There, it is converted to CO_2, and rubisco fixes carbon for the second time as the CO_2 enters the Calvin–Benson cycle.

Bundle-sheath cells of C4 plants have chloroplasts that carry out light-dependent reactions, but only in the cyclic pathway. No oxygen is released, so the O_2 level near rubisco stays low. This, along with the high CO_2 level provided by the C4 reactions, minimizes photorespiration, so sugar production stays efficient in these plants even in hot, dry weather.

Succulents, cacti, and other **CAM plants** use a carbon-fixing pathway that allows them to minimize photorespiration even in desert regions with extremely high daytime temperatures. CAM stands for crassulacean acid metabolism, after the Crassulaceae family of plants in which this pathway was first studied. Like C4 plants, CAM plants fix carbon twice, but the reactions occur at different times rather than in different cells (**FIGURE 6.13C**). Stomata on a CAM plant open only at night, when typically lower temperatures minimize evaporative water loss. The plants use a C4 pathway to fix carbon from CO_2 in the air at this time. The product of this pathway, a four-carbon acid, is stored in the cell's central vacuole. When the stomata close the next day, the acid moves out of the vacuole and becomes broken down to CO_2, which is fixed for the second time when it enters the Calvin–Benson cycle. The high CO_2 level provided by the acid breakdown minimizes photorespiration in these plants.

C3 plant Type of plant that uses only the Calvin–Benson cycle to fix carbon.
C4 plant Type of plant that minimizes photorespiration by fixing carbon twice, in two cell types.
CAM plant Type of C4 plant that minimizes photorespiration by fixing carbon twice, at different times of day.
photorespiration Pathway initiated when rubisco attaches oxygen instead of carbon dioxide to RuBP (ribulose bisphosphate).
stomata Gaps that open on plant surfaces; allow water vapor and gases to diffuse across the epidermis.

TAKE-HOME MESSAGE 6.7
How do carbon-fixing reactions vary?

✔ When stomata close on hot, dry days, they also prevent the exchange of gases between plant tissues and the air.

✔ Rubisco can initiate photorespiration by attaching oxygen (instead of carbon dioxide) to RuBP. Photorespiration reduces the efficiency of sugar production, especially in C3 plants.

✔ Plants adapted to hot, dry conditions limit photorespiration by fixing carbon twice. C4 plants separate the two sets of reactions in space; CAM plants separate them in time.

CREDITS: (14A) Masahiro Yamada, Michio Kawasaki, Tatsuo Sugiyama, Hiroshi Miyake, Mitsutaka Taniguchi; Differential Positioning of C4 Mesophyll and Bundle Sheath Chloroplasts: Aggregative Movement of C4 Mesophyll Chloroplasts in Response to Environmental Stresses: Plant and Cell Physiology; (2009) 50(10): 1736-1749; (14B) left, Eri Maai, Shouu Shimada, Masahiro Yamada, Tatsuo Sugiyama, Hiroshi Miyake, Mitsutaka Taniguchi; The avoidance and aggregative movements of mesophyll chloroplasts in C4 monocots in response to blue light and abscisic acid; Journal of Experimental Botany, doi:10.1093/jxb/err008, by permission of Oxford University Press; right, Image courtesy msuturfweeds.net.

Biofuels (revisited)

The first cells we know of appeared on Earth about 3.4 billion years ago. Like some modern prokaryotes, these ancient organisms did not tap into sunlight: They extracted the energy they needed from simple molecules such as methane and hydrogen sulfide. Both gases were plentiful in the nasty brew that was Earth's early atmosphere (**FIGURE 6.15A**). When the cyclic pathway of photosynthesis first evolved, sunlight offered cells that used it an essentially unlimited supply of energy. Shortly afterward, this pathway became modified. The new noncyclic pathway split water molecules into hydrogen and oxygen. Cells that used the pathway were very successful. Oxygen gas (O_2)

A An artist's view of Earth's early atmosphere, which was abundant in gases such as methane, sulfur, ammonia, and chlorine.

B Today, photosynthesis is now the main pathway by which energy and carbon enter the web of life. The plants in this orchard are producing oxygen and carbon-rich parts (apples) at the Jerzy Boyz farm in Chelan, Washington.

FIGURE 6.15 Then and now—a view of how our atmosphere was irrevocably altered by photosynthesis.

released from uncountable numbers of water molecules began seeping out of photosynthetic prokaryotes.

Oxygen gas reacts easily with metals, so at first, most of it combined with metal atoms in exposed rocks. After the exposed minerals became saturated with oxygen, the gas began to accumulate in the ocean and the atmosphere. From that time on, the world of life would never be the same. Molecular oxygen, which had previously been very rare in the atmosphere, began accumulating. As you will see in Chapter 7, the change in the composition of the atmosphere put tremendous selection pressure on early life, effectively spurring the evolution of aerobic respiration.

Earth's atmosphere is changing again, and this time, the level of carbon dioxide is increasing. To understand why, think about your own body. A human body is about 9.5 percent carbon by weight, which means that you contain an enormous number of carbon atoms. Where did all of those carbon atoms come from?

You and other heterotrophs, remember, ingest tissues of other organisms to get carbon. Thus, your body is built from organic compounds obtained from other organisms. The carbon atoms in those compounds may have passed through other heterotrophs before you ate them, but at some point they were part of photosynthetic organisms (**FIGURE 6.15B**). Plants and almost all other photosynthesizers in the human food chain obtain their carbon from carbon dioxide. Your carbon atoms—and those of most other organisms that live on land—were recently part of Earth's atmosphere, in molecules of CO_2.

Photosynthesis removes carbon dioxide from the atmosphere, and fixes its carbon atoms in organic compounds. When you and other organisms break down organic compounds for energy, carbon atoms are released in the form of CO_2, which then reenters the atmosphere. Since photosynthesis evolved, these two processes have constituted a more or less balanced cycle of the biosphere. You will learn more about the carbon cycle in Section 46.7. For now, know that the amount of carbon dioxide that photosynthesis removes from the atmosphere is roughly the same amount that organisms release back into it. At least it was, until humans came along.

As early as 8,000 years ago, humans began burning forests to clear land for agriculture. When trees and other plants burn, most of the carbon locked in their tissues is released into the atmosphere as carbon dioxide. Fires that occur naturally release carbon dioxide the same way. Today, we burn a lot more than our ancestors ever did. In addition to wood, we are burning fossil fuels—coal, petroleum, and natural gas—to satisfy our greater and greater demands for energy. Fossil fuels are the organic remains of ancient organisms. When we burn fossil fuels, we release the carbon that has been sequestered in their organic molecules for hundreds of millions of years, mainly as CO_2 that reenters the atmosphere.

Our extensive use of fossil fuels has put Earth's atmospheric cycle of carbon dioxide out of balance: We are adding far more CO_2 to the atmosphere than photosynthetic organisms are removing from it. The resulting imbalance is fueling global climate change (we return to this topic in Section 46.8). In 2010 alone, human activities released 10 billion tons of CO_2, an increase of 5.9 percent over 2009, and 49 percent over 1990. Most of this CO_2 comes from burning fossil fuels (**FIGURE 6.16**). How do we know? Researchers can determine how long ago the carbon atoms in a sample of CO_2 were part of a living organism by measuring the ratio of different carbon isotopes in it (you will read more about radioisotope dating techniques in Section 16.6). The results are correlated with global statistics on the extraction, refining, and trade of fossil fuels.

Tiny pockets of Earth's ancient atmosphere remain in Antarctica, preserved in snow and ice that have been accumulating in layers, year after year, for millions of years (**FIGURE 6.17**). Air and dust trapped in each layer reveal the composition of the atmosphere that prevailed when the layer formed. These layers tell us that the atmospheric CO_2 level was relatively stable for about 10,000 years before the industrial revolution began in the mid-1800s. Since then, the CO_2 level has been steadily rising. Today, the atmospheric CO_2 level is higher than it has been for *15 million years*.

Such alarming statistics are why researchers are scrambling to find a way to make cost-effective biofuels. Unlike fossil fuels, biofuels are a renewable source of energy: We can always make more of them simply

FIGURE 6.16 Carbon dioxide is an invisible component of the smog shrouding Lianyungang, China, in December 2013. Most air pollution—here and worldwide—comes from fossil fuel combustion.

FIGURE 6.17 A slice of ancient history. Air bubbles trapped in Antarctic ice core slices such as this one are samples of Earth's atmosphere as it was when the ice formed. The deeper the slice, the older the air in the bubbles.

by growing more plants. Also unlike fossil fuels, biofuels do not contribute to global climate change, because growing plant matter for fuel recycles carbon that is already in the atmosphere.

summary

Section 6.1 Plants and other **autotrophs** make their own food using energy from the environment and carbon from inorganic sources such as CO_2. By metabolic pathways of photosynthesis, plants and most other autotrophs capture the energy of light and use it to build sugars from water and carbon dioxide. **Heterotrophs** get carbon from molecules that other organisms have already assembled; most also get energy from these molecules.

Earth's early atmosphere held very little free oxygen. When the noncyclic pathway of photosynthesis evolved, oxygen released by organisms that used it permanently changed the atmosphere. Photosynthesis removes CO_2 from the atmosphere, and the metabolic activity of most organisms puts it back. This global cycle has been balanced for millions of years. Human activites, especially burning fossil fuels, are disrupting the cycle by adding extra CO_2 to the atmosphere. The resulting imbalance is fueling global climate change.

CREDITS: (16) ChinaFotoPress/Getty Images; (17) www.photo.antarctica.ac.uk.

CHAPTER 6   113
WHERE IT STARTS—PHOTOSYNTHESIS

Sections 6.2, 6.3 Visible light is a very small part of the spectrum of electromagnetic energy radiating from the sun. That energy travels in waves, and it is organized as photons. A photon's wavelength is related to its energy. Wavelengths that we can see—visible light—drive photosynthesis, which begins when photons are absorbed by photosynthetic pigments. A **pigment** absorbs light of particular **wavelengths** only; wavelengths not captured are reflected as its characteristic color. The main photosynthetic pigment, **chlorophyll a**, absorbs violet and red light, so it appears green. Accessory pigments absorb additional wavelengths.

Section 6.4 In chloroplasts, the **light-dependent reactions** of photosynthesis occur at a much-folded **thylakoid membrane**. The membrane forms a compartment in the chloroplast's interior (**stroma**), in which the **light-independent reactions** occur. An overview is shown below:

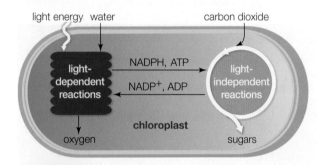

Section 6.5 In the light-dependent reactions, light-harvesting complexes in the thylakoid membrane absorb photons and pass the energy to **photosystems**. Receiving energy causes photosystems to release electrons.

In the noncyclic pathway, electrons released from photosystem II flow through an electron transfer chain, then to photosystem I. Photosystem II replaces lost electrons by pulling them from water, which then splits into H^+ and O_2 (an example of **photolysis**). Electrons released from photosystem I end up in NADPH.

In the cyclic pathway, electrons released from photosystem I enter an electron transfer chain, then cycle back to photosystem I. NADPH does not form and O_2 is not released.

ATP forms by **electron transfer phosphorylation** in both pathways. Electrons flowing through electron transfer chains cause hydrogen ions to accumulate in the thylakoid compartment. The hydrogen ions follow their gradient back across the membrane through ATP synthases, driving ATP synthesis.

Section 6.6 **Carbon fixation** occurs as part of the light-independent reactions of photosynthesis. Inside the stroma of chloroplasts, the enzyme **rubisco** attaches CO_2 to RuBP to start the **Calvin–Benson cycle**. This cyclic pathway builds the carbon backbones of sugars using carbon atoms from CO_2. It is driven by phosphate-group transfers from ATP and electrons from NADPH. Both molecules form during the light-dependent reactions.

Section 6.7 On hot, dry days, a plant conserves water by closing **stomata**, so carbon dioxide for the light-independent reactions cannot enter it, and oxygen produced by the light-dependent reactions cannot leave. The resulting high O_2/CO_2 ratio in the plant's tissues can shift sugar production toward **photorespiration**. This inefficient pathway limits the growth of **C3 plants** in hot, dry climates. Other types of plants minimize photorespiration by fixing carbon twice, thus keeping the CO_2 level high near rubisco. **C4 plants** carry out the two sets of reactions in different cell types; **CAM plants** carry them out at different times.

self-quiz
Answers in Appendix VII

1. A cat eats a bird, which ate a caterpillar that chewed on a weed. Which organisms are autotrophs? Which ones are heterotrophs?

2. Plants use _____ as an energy source to drive photosynthesis.
   a. sunlight   c. O_2
   b. hydrogen ions   d. CO_2

3. Most of the carbon dioxide that plants use for photosynthesis comes from _____ .
   a. glucose   c. rainwater
   b. the atmosphere   d. photolysis

4. Which of the following statements is incorrect?
   a. Pigments absorb light of certain wavelengths only.
   b. Many accessory pigments are multipurpose molecules.
   c. Chlorophyll is green because it absorbs green light.

5. In plants and other photosynthetic eukaryotes, the light-dependent reactions proceed in/at the _____ .
   a. thylakoid membrane   c. stroma
   b. plasma membrane   d. cytoplasm

6. When a photosystem absorbs light, _____ .
   a. sugar phosphates are produced
   b. electrons are transferred to ATP
   c. RuBP accepts electrons
   d. electrons are ejected from its special pair

7. In the light-dependent reactions, _____ .
   a. carbon dioxide is fixed   d. CO_2 accepts electrons
   b. ATP forms   e. b and c
   c. sugars form   f. a and c

8. What accumulates inside the thylakoid compartment of chloroplasts during the light-dependent reactions?
   a. sugars   c. O_2
   b. hydrogen ions   d. CO_2

9. The atoms in the molecular oxygen released during photosynthesis come from _____ molecules.
   a. glucose   c. water
   b. CO_2 d. O_2

10. Light-independent reactions in plants proceed in/at the _____ of chloroplasts.
   a. thylakoid membrane   c. stroma
   b. plasma membrane   d. cytoplasm

Energy Efficiency of Biofuel Production Most of the plant material currently used for biofuel production in the United States consists of food crops—mainly corn, soybeans, and sugarcane. In 2006, David Tilman and his colleagues published the results of a 10-year study comparing the net energy output of various biofuels. The researchers grew a mixture of native perennial grasses without irrigation, fertilizer, pesticides, or herbicides, in sandy soil that was so depleted by intensive agriculture that it had been abandoned. They measured the usable energy in biofuels made from the grasses, and also from corn and soy, then measured the energy it took to grow and produce biofuel from each kind of crop (**FIGURE 6.18**).

1. About how much energy did ethanol produced from one hectare of corn yield? How much energy did it take to grow the corn and make that ethanol?

2. Which of the biofuels tested had the highest ratio of energy output to energy input?

3. Which of the three crops would require the least amount of land to produce a given amount of biofuel energy?

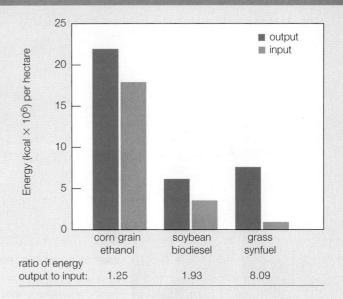

| ratio of energy output to input: | 1.25 | 1.93 | 8.09 |

FIGURE 6.18 Energy inputs and outputs of biofuels made from three different crops. One hectare is about 2.5 acres.

11. The Calvin–Benson cycle starts when _____ .
   a. light is available
   b. carbon dioxide is attached to RuBP
   c. electrons leave a photosystem II

12. Which of the following substances does *not* participate in the Calvin–Benson cycle?
   a. ATP   d. PGAL
   b. NADPH   e. O_2
   c. RuBP   f. CO_2

13. Closed stomata _____ .
   a. limit gas exchange   c. restrict photosynthesis
   b. permit water loss   d. absorb light

14. In C3 plants, _____ makes sugar production inefficient when stomata close during the day.
   a. photosynthesis   c. photorespiration
   b. photolysis   d. carbon fixation

15. Match each with its most suitable description.
   ____ PGAL formation
   ____ CO_2 fixation
   ____ photolysis
   ____ ATP forms; NADPH does not
   ____ photorespiration
   ____ photosynthesis
   ____ pigment
   ____ autotroph

   a. absorbs light
   b. converts light to chemical energy
   c. self-feeder
   d. electrons cycle back to photosystem I
   e. problem in C3 plants
   f. ATP, NADPH required
   g. water molecules split
   h. rubisco function

critical thinking

1. About 200 years ago, Jan Baptista van Helmont wanted to know where growing plants get the materials necessary for increases in size. He planted a tree seedling weighing 5 pounds in a barrel filled with 200 pounds of soil and then watered the tree regularly. After five years, the tree weighed 169 pounds, 3 ounces, and the soil weighed 199 pounds, 14 ounces. Because the tree had gained so much weight and the soil had lost so little, he concluded that the tree had gained all of its additional weight by absorbing the water he had added to the barrel, but of course he was incorrect. What really happened?

2. While gazing into an aquarium, you see bubbles coming from an aquatic plant (left). What are the bubbles?

3. A C3 plant absorbs a carbon radioisotope (as part of $^{14}CO_2$). In which compound does the labeled carbon appear first? Which compound forms first if a C4 plant absorbs the same radioisotope?

4. As you learned in this chapter, cell membranes are required for electron transfer phosphorylation. Thylakoid membranes in chloroplasts serve this purpose in photosynthetic eukaryotes. Prokaryotic cells do not have this organelle, but many are photosynthesizers. How do you think they carry out the light-dependent reactions, given that they have no chloroplasts?

LEARNING ROADMAP

This chapter focuses on metabolic pathways (Section 5.5) that harvest energy (5.2) stored in sugars (3.4). Some reactions (3.3) occur in mitochondria (4.8). You will revisit free radicals (2.3), lipids (3.5), proteins (3.6), electron transfer chains (5.5, 6.5), coenzymes (5.6, 6.6), membrane transport (5.8, 5.9), and photosynthesis (6.4).

ENERGY FROM SUGARS

Most cells can make ATP by breaking down sugars in either aerobic respiration or anaerobic fermentation pathways. Aerobic respiration yields the most ATP.

GLYCOLYSIS

Aerobic respiration and fermentation start in the cytoplasm with glycolysis, a pathway that splits glucose into two pyruvate molecules and yields two ATP.

AEROBIC RESPIRATION

Eukaryotes break down pyruvate to CO_2 in mitochondria. Many coenzymes are reduced; these deliver electrons and hydrogen ions to electron transfer chains that drive ATP formation.

FERMENTATION

Fermentation occurs entirely in cytoplasm, where organic molecules accept electrons from pyruvate. The net yield of ATP is small compared with that from aerobic respiration.

ENERGY FROM OTHER MOLECULES

Cells also make ATP by breaking down molecules other than sugars. Dietary lipids and proteins are converted to substrates of glycolysis or another step in the aerobic respiration pathway.

In Section 11.6, you will see examples of how metabolic pathways are disrupted in cancer cells. Chapter 14 explains why metabolic disorders can be inherited. We return to plant adaptations for photosynthesis in Chapter 27, muscle function in Chapter 35, how the body acquires oxygen for respiration in Chapter 38, and digestion and nutrition in Chapter 39.

7.1 Risky Business

Before photosynthesis evolved, molecular oxygen had been a very small component of Earth's atmosphere. In what may have been the earliest case of catastrophic pollution, the new abundance of this gas exerted tremendous pressure on all life at the time. Why? Then, like now, enzymes that require metal cofactors were a critical part of metabolism. Oxygen reacts with metal cofactors, and free radicals (Section 2.3) form during those reactions. Free radicals damage biological molecules, so they are dangerous to life. Most cells had no way to cope with them, and so were wiped out everywhere except deep water, muddy sediments, and other **anaerobic** (oxygen-free) places.

A very few cells were able to survive in **aerobic** (oxygen-containing) places. By lucky circumstance, these cells made antioxidants that could detoxify or prevent the formation of free radicals. As these organisms evolved, their antioxidant molecules became incorporated into new metabolic pathways. One of the new pathways, aerobic respiration, put the reactive properties of oxygen to use. In modern eukaryotic cells, most of the aerobic respiration pathway takes place inside mitochondria (Section 4.8). An internal folded membrane system allows these organelles to make ATP very efficiently. Electron transfer chains in this membrane set up hydrogen ion gradients that power ATP synthesis. Oxygen molecules accept electrons at the end of these chains.

ATP participates in almost all cellular reactions, so a cell benefits from making a lot of it. However, aerobic respiration is a dangerous occupation. When an oxygen molecule (O_2) accepts electrons from an electron transfer chain, it dissociates into oxygen atoms. Most of the atoms immediately combine with hydrogen ions and end up in water molecules. Occasionally, however, an oxygen atom escapes this final reaction. The atom has an unpaired electron, so it is a free radical.

Mitochondria cannot detoxify free radicals, so they rely on antioxidant enzymes and vitamins in the cell's cytoplasm to do it for them. The system works well, at least most of the time. However, a genetic disorder or an encounter with a toxin or pathogen can tip the normal cellular balance of aerobic respiration and free radical formation. Free radicals accumulate and destroy first the function of mitochondria, then the cell. The resulting tissue damage is called oxidative stress.

At least 83 proteins are directly involved in mitochondrial electron transfer chains. A defect in any one of them—or in any of the thousands of other proteins used by mitochondria—can wreak havoc in the body.

New research is showing that oxidative stress caused by mitochondrial malfunction is also involved in many other illnesses, including cancer, hypertension, Alzheimer's and Parkinson's diseases, and even aging. Hundreds of incurable genetic disorders are associated with mitochondrial defects (**FIGURE 7.1**), and more are being discovered all the time. Nerve cells, which require a lot of ATP, are particularly affected. Symptoms of these disorders range from mild to major progressive neurological deficits, blindness, deafness, diabetes, strokes, seizures, gastrointestinal malfunction, and disabling muscle weakness.

aerobic Involving or occurring in the presence of oxygen.
anaerobic Occurring in the absence of oxygen.

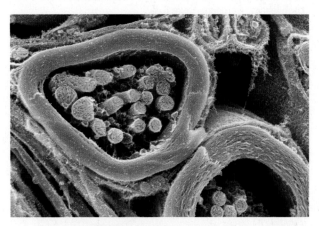

A A cross-section shows mitochondria (yellow) inside nerve cells. These cells are particularly affected by mitochondrial malfunction.

B "Tom does not look sick, but inside his organs are all getting badly damaged," says Martine Martin, pictured here with her eight-year-old son. Tom, who was born with a mitochondrial disease, eats with the help of a machine, suffers intense pain, and will soon be blind. Despite intensive medical intervention, he is not expected to reach his teens.

FIGURE 7.1 Mitochondria, powerhouses of all eukaryotic cells. When they malfunction, the lights go off in cellular businesses.

CREDITS: (opposite) © Maxisport/Shutterstock.com; (1A) © Dr. David Furness/Wellcome Images; (1B) © FairFax Media.

7.2 Overview of Carbohydrate Breakdown Pathways

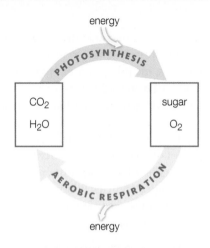

FIGURE 7.2 The global connection between photosynthesis and aerobic respiration. Note the cycling of materials, and the one-way flow of energy (compare Figure 5.4).

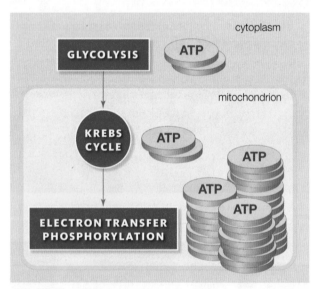

A Aerobic respiration.

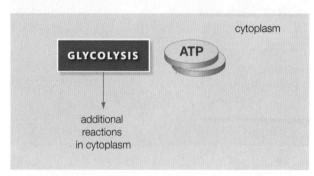

B Fermentation.

FIGURE 7.3 ▶**Animated** Comparison of aerobic respiration and fermentation.

FIGURE IT OUT Which pathway produces more ATP?

Answer: Aerobic respiration

✔ Most organisms, including photosynthetic ones, make ATP by breaking down sugars.

Photosynthetic organisms capture energy from the sun and store it in the form of sugars (Sections 6.5 and 6.6). They and most other organisms use energy stored in sugars to run various endergonic reactions of metabolism that sustain life. However, sugars rarely participate in such reactions, so how do cells harness their energy? In order to use the energy stored in sugars, cells must first transfer it to molecules—ATP in particular—that do participate in energy-requiring reactions. Cells break the bonds between carbon atoms of a sugar molecule, and use energy released as these bonds break to drive ATP synthesis.

In nearly all eukaryotes and some bacteria, the main carbohydrate-breakdown pathway is **aerobic respiration**. Remember, the bonds of organic molecules hold a lot of energy that can be released in a reaction with oxygen (Section 5.5). Aerobic respiration harvests energy from sugars by completely breaking apart their carbon backbones, bond by bond, and it uses oxygen to do this. The pathway's products—carbon dioxide and water—are the raw materials used by the vast majority of photosynthetic organisms to build the sugars in the first place. With this connection, the cycling of carbon, hydrogen, and oxygen through living things comes full circle through the biosphere (**FIGURE 7.2**).

The following equation summarizes the overall pathway of aerobic respiration:

$$C_6H_{12}O_6 \;+\; 6O_2 \longrightarrow 6CO_2 \;+\; 6H_2O$$

glucose oxygen carbon dioxide water

The reactions in the pathway occur in three stages (**FIGURE 7.3A**). The first stage, glycolysis, is a linear pathway that takes place in cytoplasm. Glycolysis begins the breakdown of one sugar molecule for a net yield of 2 ATP. In eukaryotes, the next two stages take place in mitochondria. In the second stage, the Krebs cycle completes the breakdown of the sugar molecule to CO_2. This cyclic pathway produces 2 ATP and reduces many coenzymes. In the third stage, electron transfer phosphorylation (Section 6.5), the coenzymes reduced during glycolysis and the Krebs cycle deliver electrons and hydrogen ions to electron transfer chains. Energy released by electrons as they move through the chains drives the formation of as many as 32 ATP. At the end of the electron transfer chains, water forms when oxygen accepts hydrogen ions and electrons. In exactly the reverse of the photolysis reaction that

FIGURE 7.4 Like you, a whale breathes air to provide its cells with a fresh supply of oxygen for aerobic respiration. Carbon dioxide released from aerobically respiring cells leaves the body in each exhalation.

splits water during the noncyclic, light-dependent reactions of photosynthesis, oxygen combines with electrons and hydrogen ions to form water:

$$O_2 + 4e^- + 4H^+ \longrightarrow 2H_2O$$

Aerobic respiration, which means "taking a breath of air," refers to this pathway's requirement for oxygen as the final acceptor of electrons.

Fermentation refers to sugar breakdown pathways that produce ATP and do not require oxygen (**FIGURE 7.3B**). Like aerobic respiration, fermentation begins with glycolysis in cytoplasm. Unlike aerobic respiration, fermentation includes no electron transfer chains, and, in all organisms that use the pathway, it takes place entirely in cytoplasm. The reactions that conclude it produce no additional ATP, and an organic molecule (not oxygen) accepts electrons.

aerobic respiration Oxygen-requiring metabolic pathway that breaks down sugars to produce ATP.
fermentation A metabolic pathway that breaks down sugars to produce ATP and does not require oxygen.

The breakdown of a sugar molecule by fermentation yields only 2 ATP, but the pathway is efficient enough to sustain many single-celled species. It also helps cells of multicelled species produce ATP under anaerobic conditions, but aerobic respiration is a much more efficient way of harvesting energy from carbohydrates. You and other large, multicelled organisms could not live without its higher yield (**FIGURE 7.4**).

> **TAKE-HOME MESSAGE 7.2**
> **How do cells access the chemical energy in sugars?**
>
> ✔ Most cells can make ATP by breaking down sugars, in aerobic respiration, fermentation, or both.
>
> ✔ Aerobic respiration and fermentation begin in cytoplasm.
>
> ✔ Fermentation does not require oxygen and ends in cytoplasm.
>
> ✔ Aerobic respiration requires oxygen and, in eukaryotes, it ends in mitochondria. This pathway yields much more ATP per sugar molecule than fermentation does.

7.3 Glycolysis—Sugar Breakdown Begins

✔ The reactions of glycolysis convert one molecule of glucose (or other six-carbon sugar) to two molecules of pyruvate.

✔ Glycolysis reactions use 2 ATP and produce 4 ATP, so the net yield is 2 ATP per molecule of sugar.

Glycolysis is a series of reactions that begin the sugar breakdown pathways of both aerobic respiration and fermentation. The word "glycolysis" comes from two Greek words: *glyk-*, sweet, and *-lysis*, loosening; it refers to the release of chemical energy from sugars. The reactions of glycolysis, which occur with some variation in the cytoplasm of almost all cells, convert one molecule of a six-carbon sugar (such as glucose) into two molecules of **pyruvate**, an organic compound with a three-carbon backbone:

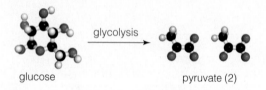

glucose → glycolysis → pyruvate (2)

Other six-carbon sugars (such as galactose and fructose) can enter glycolysis, but we focus here on glucose, for clarity.

Glycolysis begins when a molecule of glucose enters a cell through a glucose transporter, a passive transport protein that you encountered in Section 5.9. The cell invests two ATP in the endergonic reactions that begin the pathway (**FIGURE 7.5**). In the first reaction, a phosphate group is transferred from ATP to the glucose, thus forming glucose-6-phosphate ❶. A model of hexokinase, the enzyme that catalyzes this reaction, is pictured in Section 5.4.

Glycolysis continues as the glucose-6-phosphate accepts a phosphate group from another ATP ❷, then splits to form two PGAL (phosphoglyceraldehyde). Remember from Section 6.6 that this phosphorylated sugar also forms during the Calvin–Benson cycle.

In the next reaction, each PGAL receives a second phosphate group, and each gives up two electrons and a hydrogen ion. Two molecules of PGA (phosphoglycerate) form as products of this reaction ❸. The electrons and hydrogen ions are accepted by two NAD$^+$, which thereby become reduced to NADH. Aerobic respiration's final stage requires this NADH,

GLYCOLYSIS

KREBS CYCLE

ELECTRON TRANSFER PHOSPHORYLATION

FIGURE 7.5 ▶**Animated** Glycolysis. This first stage of sugar breakdown starts and ends in the cytoplasm of all cells. Opposite, for clarity, we track only the six carbon atoms (black balls) that enter the reactions as part of glucose.

Cells invest two ATP to start glycolysis, so the net yield from one glucose molecule is two ATP. Two NADH also form, and two pyruvate molecules are the end products. Appendix III has more details for interested students.

as does fermentation (Sections 7.5 and 7.6 detail the final stages of these pathways).

Next, a phosphate group is transferred from each PGA to ADP, so two ATP form ❹. The direct transfer of a phosphate group from a substrate to ADP is called **substrate-level phosphorylation**. Substrate-level phosphorylation is a completely different process from electron transfer phosphorylation, which is the way ATP forms during the light-dependent reactions of photosynthesis (Section 6.5).

Glycolysis ends with the formation of two more ATP by substrate-level phosphorylation ❺. Remember, two ATP were invested to begin the reactions of glycolysis. A total of four ATP form, so the net yield is two ATP per molecule of glucose ❻. The pathway also produces two three-carbon pyruvate molecules. Pyruvate is a substrate for the second-stage reactions of aerobic respiration, and also for fermentation reactions (Sections 7.4 and 7.6 return to this topic).

---

glycolysis Set of reactions in which a six-carbon sugar (such as glucose) is converted to two pyruvate for a net yield of two ATP.
pyruvate Three-carbon end product of glycolysis.
substrate-level phosphorylation The formation of ATP by the direct transfer of a phosphate group from a substrate to ADP.

TAKE-HOME MESSAGE 7.3
                        What happens during glycolysis?

✔ Glycolysis is the first stage of sugar breakdown in both aerobic respiration and fermentation.

✔ The reactions of glycolysis occur in the cytoplasm.

✔ Glycolysis converts one molecule of glucose to two molecules of pyruvate, with a net energy yield of two ATP. Two NADH also form.

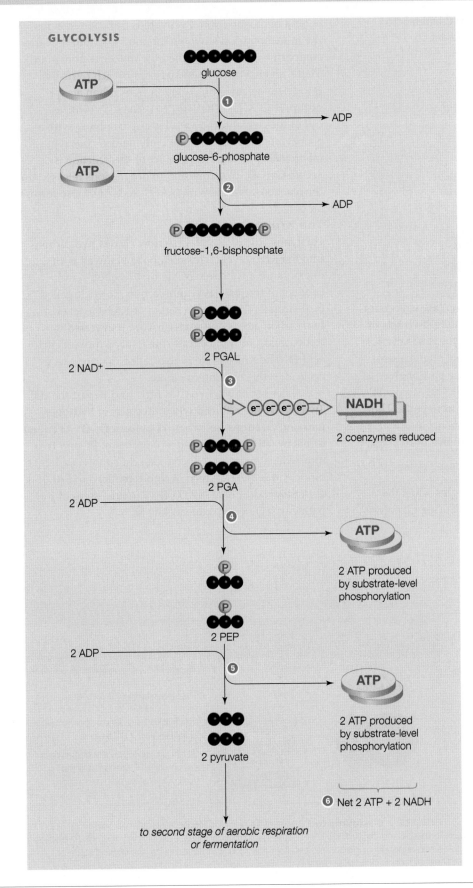

GLYCOLYSIS

glucose

glucose-6-phosphate

fructose-1,6-bisphosphate

2 PGAL

2 NAD⁺

NADH

2 coenzymes reduced

2 PGA

2 ADP

ATP

2 ATP produced
by substrate-level
phosphorylation

2 PEP

2 ADP

ATP

2 ATP produced
by substrate-level
phosphorylation

2 pyruvate

❻ Net 2 ATP + 2 NADH

to second stage of aerobic respiration
or fermentation

ATP-Requiring Steps

❶ An enzyme transfers a phosphate group from ATP to glucose, forming glucose-6-phosphate. (You learned about this enzyme, hexokinase, in **FIGURE 5.10** and Section 5.4.)

❷ A phosphate group from a second ATP is transferred to the glucose-6-phosphate. The resulting molecule is unstable, and it splits into two three-carbon molecules. The molecules are interconvertible, so we will call them both PGAL (phosphoglyceraldehyde).

Two ATP have now been invested in the reactions.

ATP-Generating Steps

❸ An enzyme attaches a phosphate to each PGAL, so two PGA (phosphoglycerate) form. Two electrons and a hydrogen ion (not shown) from each PGAL are accepted by NAD⁺, so two NADH form.

❹ An enzyme transfers a phosphate group from each PGA to ADP, forming two ATP and two intermediate molecules (PEP).

The original energy investment of two ATP has now been recovered.

❺ An enzyme transfers a phosphate group from each PEP to ADP, forming two more ATP and two molecules of pyruvate.

❻ Summing up, glycolysis yields two NADH, two ATP (net), and two pyruvate for each glucose molecule.

Depending on the type of cell and environmental conditions, the pyruvate may enter the second stage of aerobic respiration or it may be used in other ways, such as in fermentation.

7.4 Second Stage of Aerobic Respiration

✔ The second stage of aerobic respiration completes the breakdown of glucose that began in glycolysis.

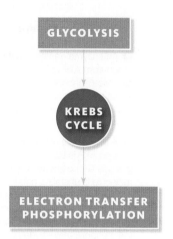

The second stage of aerobic respiration (above) occurs in mitochondria (**FIGURE 7.6**). It includes two sets of reactions, acetyl–CoA formation and the **Krebs cycle**, that break down pyruvate, the product of glycolysis. All of the carbon atoms that were once part of glucose end up in CO_2, which departs the cell. Only two ATP form, but the reactions reduce many coenzymes. The energy of electrons carried by these coenzymes will drive the reactions of aerobic respiration's third stage.

Acetyl–CoA Formation

Aerobic respiration's second stage begins when the two pyruvate molecules that formed during glycolysis enter a mitochondrion. Pyruvate is transported across the mitochondrion's two membranes and into the inner compartment, which is called the mitochondrial matrix. There, an enzyme immediately splits each pyruvate into one molecule of CO_2 and a two-carbon acetyl group (—$COCH_3$, **FIGURE 7.7**). The CO_2 diffuses out of the cell, and the acetyl group combines with a coenzyme rather unimaginatively named coenzyme A (abbreviated CoA). The product of this reaction is acetyl–CoA ❶. Electrons and hydrogen ions released by the reaction combine with NAD^+, so NADH also forms.

The Krebs Cycle

Each molecule of acetyl–CoA now carries two carbons into the Krebs cycle. This metabolic pathway is a cyclic one because a substrate of the first reaction—and a product of the last—is a four-carbon molecule called oxaloacetate. During each round of Krebs reactions, two carbon atoms of acetyl–CoA are transferred to oxaloacetate, forming citrate, the ionized form of citric acid ❷. The Krebs cycle is also called the citric acid cycle after this first intermediate.

In later reactions, two CO_2 form and depart the cell. Two NAD^+ are reduced when they accept hydrogen ions and electrons, so two NADH form ❸, ❹. ATP then forms by substrate-level phosphorylation ❺, and two more coenzymes are reduced: one called flavin adenine dinucleotide (FAD, which becomes reduced to $FADH_2$) ❻, and an additional NAD^+ ❼. The final steps of the pathway regenerate oxaloacetate ❽.

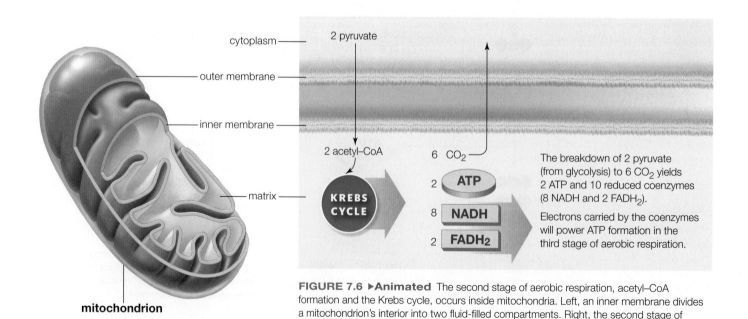

FIGURE 7.6 ▶**Animated** The second stage of aerobic respiration, acetyl–CoA formation and the Krebs cycle, occurs inside mitochondria. Left, an inner membrane divides a mitochondrion's interior into two fluid-filled compartments. Right, the second stage of aerobic respiration takes place in the mitochondrion's innermost compartment, or matrix.

CREDITS: (in text, 6) © Cengage Learning.

Acetyl–CoA Formation and the Krebs Cycle

1 An enzyme splits a pyruvate molecule into a two-carbon acetyl group and CO_2. Coenzyme A binds the acetyl group, forming acetyl–CoA. NAD^+ combines with released electrons and a hydrogen ion, forming NADH.

2 The Krebs cycle starts as two carbon atoms are transferred from acetyl–CoA to oxaloacetate. Citrate forms, and coenzyme A is regenerated.

3 One carbon atom is removed from an intermediate and leaves the cell as CO_2. NAD^+ combines with released electrons and a hydrogen ion, forming NADH.

4 Another carbon atom is removed from another intermediate and leaves the cell as CO_2. NAD^+ combines with released electrons and a hydrogen ion, forming NADH. Pyruvate's three carbon atoms have now exited the cell, in CO_2.

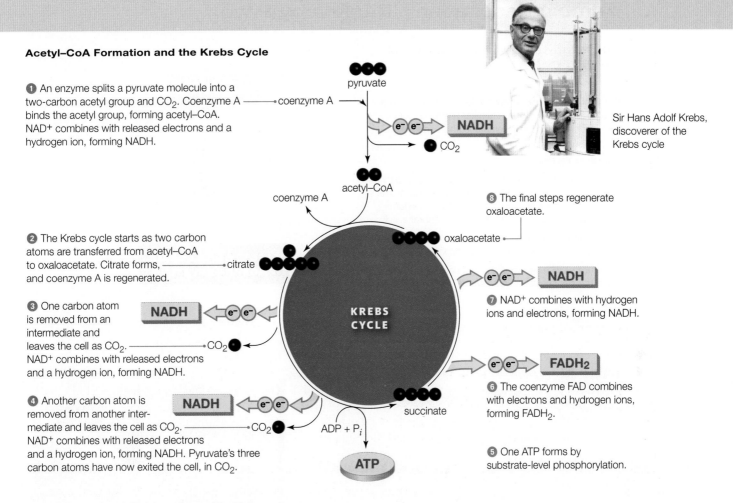

Sir Hans Adolf Krebs, discoverer of the Krebs cycle

8 The final steps regenerate oxaloacetate.

7 NAD^+ combines with hydrogen ions and electrons, forming NADH.

6 The coenzyme FAD combines with electrons and hydrogen ions, forming $FADH_2$.

5 One ATP forms by substrate-level phosphorylation.

FIGURE 7.7 Acetyl–CoA formation and the Krebs cycle.

It takes two cycles of Krebs reactions to break down two pyruvate molecules that formed during glycolysis of one glucose molecule. After two cycles, all six carbons that entered glycolysis in one glucose molecule have left the cell, in six CO_2. Electrons and hydrogen ions are released as each carbon is removed from the backbone of intermediate molecules; ten coenzymes will carry them to the third and final stage of aerobic respiration. Not all reactions are shown; Appendix III has more details.

After two cycles of Krebs reactions, the two carbon atoms carried by each acetyl–CoA end up in CO_2. Thus, the combined second-stage reactions of aerobic respiration break down two pyruvate to six CO_2:

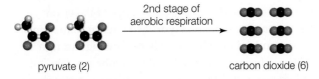

pyruvate (2)   carbon dioxide (6)

Remember, the two pyruvate were a product of glycolysis. So, at this point in aerobic respiration, the carbon backbone of one glucose molecule has been broken down completely, its six carbon atoms having exited the cell in CO_2.

Two ATP that form during the second stage add to the small net yield (two ATP) of glycolysis. However, this stage reduces ten coenzymes—eight NAD^+ and two FAD. Add in the two NAD^+ that were reduced in glycolysis, and the full breakdown of each glucose molecule has a big potential payoff. Twelve reduced coenzymes will deliver electrons—and the energy they carry—to the third stage of aerobic respiration.

Krebs cycle Cyclic pathway that, along with acetyl–CoA formation, breaks down pyruvate to carbon dioxide during aerobic respiration.

TAKE-HOME MESSAGE 7.4
What happens during the second stage of aerobic respiration?

✔ The second stage of aerobic respiration, acetyl–CoA formation and the Krebs cycle, occurs in the inner compartment (matrix) of mitochondria.

✔ The second-stage reactions convert the two pyruvate that formed in glycolysis to six CO_2. Two ATP form, and ten coenzymes (eight NAD^+ and two FAD) are reduced.

7.5 Aerobic Respiration's Big Energy Payoff

✔ Many ATP are formed during the third and final stage of aerobic respiration.

FIGURE 7.8 ▶**Animated** In eukaryotes, the third and final stage of aerobic respiration, electron transfer phosphorylation, occurs at the inner mitochondrial membrane (this page).

❶ NADH and $FADH_2$ deliver electrons to electron transfer chains in the inner mitochondrial membrane.

❷ Electron flow through the chains causes hydrogen ions (H^+) to be pumped from the matrix to the intermembrane space. The activity of the electron transfer chains causes a hydrogen ion gradient to form across the inner mitochondrial membrane.

❸ Hydrogen ion flow back to the matrix through ATP synthases drives the formation of ATP from ADP and phosphate (P_i).

❹ Oxygen combines with electrons and hydrogen ions at the end of the electron transfer chains, so water forms.

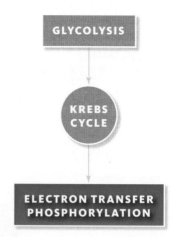

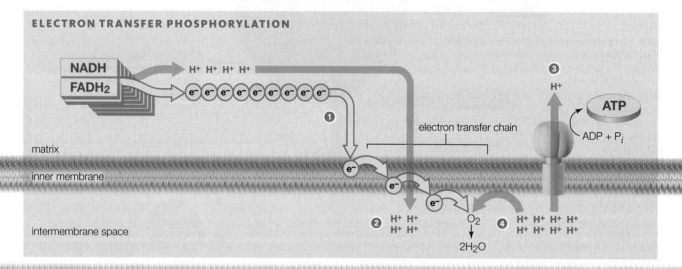

ELECTRON TRANSFER PHOSPHORYLATION

In eukaryotes, the third stage of aerobic respiration, electron transfer phosphorylation, occurs at the inner mitochondrial membrane (**FIGURE 7.8**). The reactions of electron transfer phosphorylation begin with the coenzymes NADH and $FADH_2$, which became reduced in the first two stages of aerobic respiration. These coenzymes donate their cargo of electrons and hydrogen ions to electron transfer chains embedded in the inner mitochondrial membrane ❶.

As the electrons move through the chains, they give up energy little by little (Section 5.5). Some molecules of the electron transfer chains harness that energy to actively transport hydrogen ions across the inner membrane, from the matrix to the intermembrane space ❷. The accumulating hydrogen ions form a gradient across the inner mitochondrial membrane. This gradient attracts the ions back toward the matrix, but ions cannot diffuse through a lipid bilayer on their own (Section 5.8). Hydrogen ions cross the inner mitochondrial membrane only by flowing through ATP synthases embedded in the membrane. The flow of hydrogen ions through ATP synthases causes these proteins to attach phosphate groups to ADP, so ATP forms ❸.

Oxygen accepts electrons at the end of the mitochondrial electron transfer chains ❹. When oxygen accepts electrons, it combines with H^+ to form water, which is a product of the third-stage reactions.

For each glucose molecule that enters aerobic respiration, four ATP form in the first- and second-stage

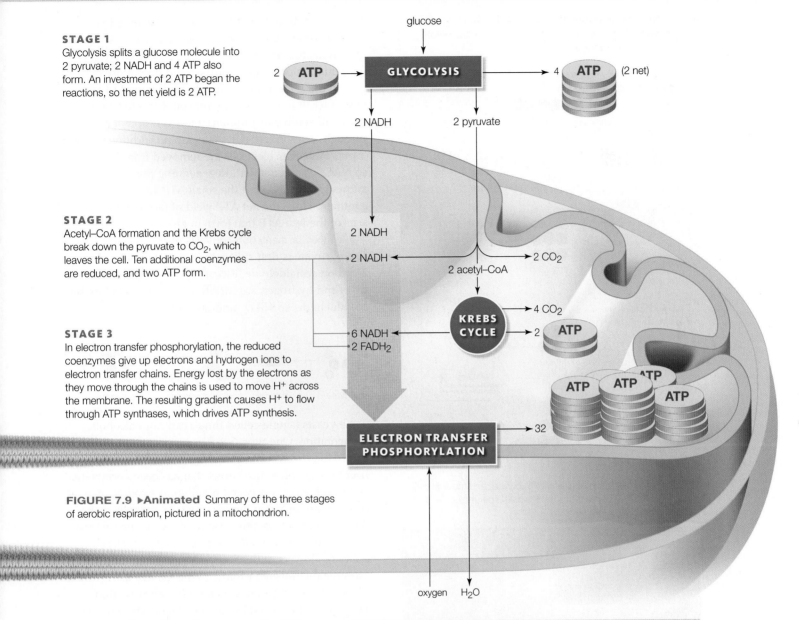

STAGE 1
Glycolysis splits a glucose molecule into 2 pyruvate; 2 NADH and 4 ATP also form. An investment of 2 ATP began the reactions, so the net yield is 2 ATP.

glucose

2 **ATP** → **GLYCOLYSIS** → 4 **ATP** (2 net)

2 NADH   2 pyruvate

STAGE 2
Acetyl–CoA formation and the Krebs cycle break down the pyruvate to CO_2, which leaves the cell. Ten additional coenzymes are reduced, and two ATP form.

2 NADH

2 NADH ← 2 acetyl–CoA → 2 CO_2

KREBS CYCLE → 4 CO_2

STAGE 3
In electron transfer phosphorylation, the reduced coenzymes give up electrons and hydrogen ions to electron transfer chains. Energy lost by the electrons as they move through the chains is used to move H^+ across the membrane. The resulting gradient causes H^+ to flow through ATP synthases, which drives ATP synthesis.

6 NADH ← **KREBS CYCLE** → 2 **ATP**
2 $FADH_2$

ATP   **ATP**   **ATP**   **ATP**

ELECTRON TRANSFER PHOSPHORYLATION → 32

oxygen H_2O

FIGURE 7.9 ▶Animated Summary of the three stages of aerobic respiration, pictured in a mitochondrion.

reactions. The twelve coenzymes reduced in these two stages deliver enough H^+ and electrons to fuel the synthesis of about thirty-two additional ATP in the third stage. Thus, the breakdown of one glucose molecule typically yields thirty-six ATP (**FIGURE 7.9**). The ATP yield varies depending on cell type. For example, aerobic respiration in brain and skeletal muscle cells yields thirty-eight ATP per glucose.

Remember that some energy dissipates with every transfer (Section 5.2). Even though aerobic respiration is a very efficient way of retrieving energy from carbohydrates, about 60 percent of the energy harvested in this pathway disperses as metabolic heat.

TAKE-HOME MESSAGE 7.5
What happens during the third stage of aerobic respiration?

✔ In aerobic respiration's third stage, electron transfer phosphorylation, energy released by electrons moving through electron transfer chains is ultimately captured in the attachment of phosphate to ADP.

✔ Coenzymes that were reduced in the first and second stages deliver electrons and hydrogen ions to electron transfer chains in the inner mitochondrial membrane.

✔ Energy released by electrons as they pass through electron transfer chains is used to pump H^+ from the mitochondrial matrix to the intermembrane space. The H^+ gradient that forms across the inner mitochondrial membrane drives the flow of hydrogen ions through ATP synthases, which results in ATP formation.

✔ About thirty-two ATP form during the third-stage reactions, so a typical net yield of all three stages is thirty-six ATP per glucose.

7.6 Fermentation

✔ Fermentation pathways break down carbohydrates without using oxygen. The final steps in these pathways regenerate NAD⁺ but do not produce ATP.

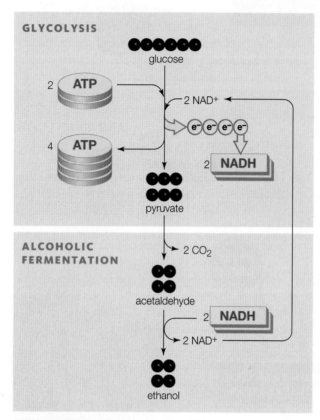

A Alcoholic fermentation begins with glycolysis, and the final steps regenerate NAD⁺. The reactions yield two ATP per molecule of glucose (from glycolysis).

B *Saccharomyces cerevisiae* yeast (top). One product of alcoholic fermentation in these cells (ethanol) makes beer alcoholic; another (CO_2) makes it bubbly. Holes in bread are pockets where CO_2 released by fermenting yeast cells accumulated in the dough.

FIGURE 7.10 Alcoholic fermentation.

Aerobic respiration and fermentation begin with the same set of glycolysis reactions in cytoplasm. After glycolysis, the two pathways differ. The final steps of fermentation occur in cytoplasm and do not use oxygen. In these reactions, pyruvate is converted to other molecules, but it is not fully broken down to CO_2 (as occurs in aerobic respiration). Electrons do not move through electron transfer chains, so no additional ATP forms. However, electrons are removed from NADH, so NAD⁺ is regenerated. Regenerating this coenzyme allows glycolysis—and the small ATP yield it offers—to continue. Thus, the net ATP yield of fermentation consists of the two ATP that form in glycolysis.

Alcoholic fermentation converts pyruvate to ethanol. The pyruvate is first split into carbon dioxide and 2-carbon acetaldehyde (**FIGURE 7.10A**). Then, electrons and hydrogen are transferred from NADH to the acetaldehyde, so NAD⁺ and ethanol form:

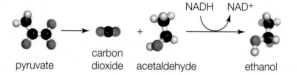

pyruvate → carbon dioxide + acetaldehyde → ethanol

Some yeasts (single-celled fungi) carry out alcoholic fermentation. One type, *Saccharomyces cerevisiae*, helps us produce beer, wine, and bread (**FIGURE 7.10B**). Beer brewers often use barley that has been germinated and dried (a process called malting) as a source of glucose for fermentation by this yeast. As the cells make ATP for themselves, they also produce ethanol (which makes the beer alcoholic) and CO_2 (which makes it bubbly). Flowers of the hop plant add flavor and help preserve the finished product. Winemakers use crushed grapes as a source of sugars for yeast fermentation. The ethanol produced by the cells makes the wine alcoholic, and the CO_2 is allowed to escape to the air.

To make bread, flour is kneaded with water, yeast, and sometimes other ingredients. Flour contains a protein (gluten) and a disaccharide (maltose). Kneading causes the gluten to form polymers in long, interconnected strands that make the resulting dough stretchy and resilient. The yeast cells in the dough first break down the maltose into its two glucose subunits, then use the released sugars for alcoholic fermentation. CO_2 they produce accumulates in bubbles that are trapped by the mesh of gluten strands. As the bubbles expand, they cause the dough to rise. The ethanol product of fermentation evaporates during baking.

In **lactate fermentation**, the electrons and hydrogen ions carried by NADH are transferred directly to pyru-

CREDITS: (in text, 10A) © Cengage Learning; (10B) top, © By London Scientific Films/Oxford Scientific/Getty Images; bottom left, © Elena Boshkovska/Shutterstock.com; bottom right, © Dr. Dennis Kunkel/Visuals Unlimited.

vate (**FIGURE 7.11A**). This reaction converts pyruvate to three-carbon lactate (the ionized form of lactic acid), and also converts NADH to NAD⁺:

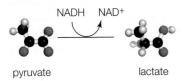

We use lactate fermentation by beneficial bacteria to prepare many foods. Yogurt, for example, is made by allowing species such as *Lactobacillus bulgaricus* and *Streptococcus thermophilus* to grow in milk. Milk contains a disaccharide (lactose) and a protein (casein). The bacterial cells first break down the lactose into its sugar subunits, then use the released glucose for lactate fermentation. Lactate reduces the pH of the milk, which imparts tartness and causes the casein to form a gel.

Cells in animal skeletal muscles are fused as long fibers that carry out aerobic respiration, lactate fermentation, or both. Red fibers have many mitochondria and produce ATP mainly by aerobic respiration. These fibers sustain prolonged activity. They are red because they contain myoglobin, a protein that stores oxygen for aerobic respiration (**FIGURE 7.11B**). White muscle fibers contain few mitochondria and no myoglobin; they make most of their ATP by lactate fermentation. This pathway makes ATP quickly, so it is useful for quick, strenuous bursts of activity (**FIGURE 7.11C**). The low ATP yield does not support prolonged activity.

Most animal muscles are a mixture of white and red fibers, but the proportions vary. For example, great sprinters tend to have more white fibers in their leg muscles; great marathon runners have more red fibers. Chickens cannot fly far because their flight muscles consist mostly of white fibers (thus, the "white" breast meat). A chicken most often walks or runs. Its leg muscles consist mostly of red muscle fibers, the "dark meat." Section 35.7 returns to skeletal muscle fibers.

alcoholic fermentation Anaerobic sugar breakdown pathway that produces ATP, CO_2, and ethanol.
lactate fermentation Anaerobic sugar breakdown pathway that produces ATP and lactate.

TAKE-HOME MESSAGE 7.6
What is fermentation?

✔ Prokaryotes and eukaryotes use fermentation pathways, which are anaerobic, to produce ATP by breaking down carbohydrates. Fermentation's small ATP yield (two per molecule of glucose) occurs by glycolysis.

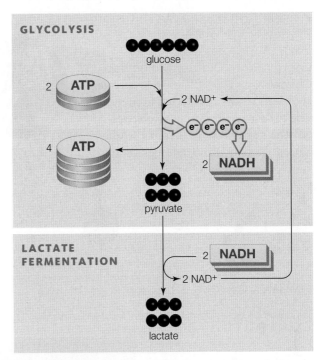

A Lactate fermentation begins with glycolysis, and the final steps regenerate NAD⁺. The reactions yield two ATP per molecule of glucose (from glycolysis).

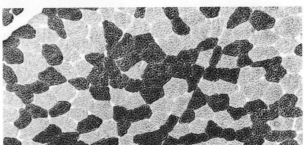

B Lactate fermentation occurs in white muscle fibers, visible in this cross-section of human thigh muscle. The red fibers, which make ATP by aerobic respiration, sustain endurance activities.

C Intense activity such as sprinting quickly depletes oxygen in muscles. Under anaerobic conditions, ATP is produced mainly by lactate fermentation in white muscle fibers. Fermentation does not make enough ATP to sustain this type of activity for long.

FIGURE 7.11 ▶**Animated** Lactate fermentation.

7.7 Alternative Energy Sources in Food

✔ Aerobic respiration can produce ATP from the breakdown of fats and proteins.

Energy From Dietary Molecules

Glycolysis converts glucose to pyruvate, and electrons are transferred from pyruvate to coenzymes during aerobic respiration's second stage. In other words, glucose becomes oxidized (it gives up

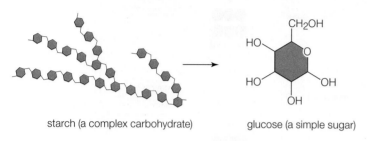

starch (a complex carbohydrate)   glucose (a simple sugar)

A Complex carbohydrates are broken down to their monosaccharide subunits, which can enter glycolysis ➊.

electrons) and coenzymes become reduced (they accept electrons). Oxidizing an organic molecule can break the covalent bonds of its carbon backbone. Aerobic respiration produces a lot of ATP by fully oxidizing glucose, completely dismantling it carbon by carbon.

Cells also dismantle other organic molecules by oxidizing them. Complex carbohydrates, fats, and proteins in food can be converted to molecules that enter glycolysis or the Krebs cycle (**FIGURE 7.12**). As in glucose metabolism, many coenzymes are reduced, and the energy of the electrons they carry ultimately drives the synthesis of ATP in electron transfer phosphorylation.

Complex Carbohydrates In humans and other mammals, the digestive system breaks down starch and other complex carbohydrates to monosaccharides (**FIGURE 7.12A**). The sugars are taken up by cells and

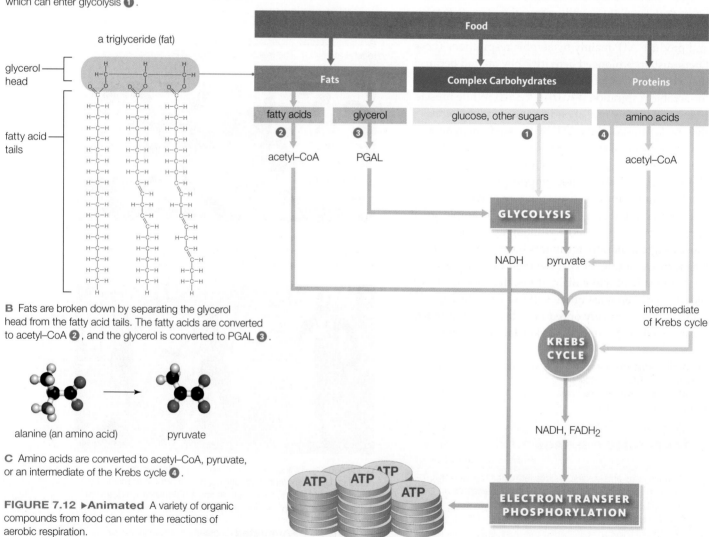

B Fats are broken down by separating the glycerol head from the fatty acid tails. The fatty acids are converted to acetyl–CoA ➋, and the glycerol is converted to PGAL ➌.

alanine (an amino acid)   pyruvate

C Amino acids are converted to acetyl–CoA, pyruvate, or an intermediate of the Krebs cycle ➍.

FIGURE 7.12 ▶**Animated** A variety of organic compounds from food can enter the reactions of aerobic respiration.

Mitochondrial diseases are devastating, incurable, and heritable. Mitochondria, remember, contain their own DNA (Section 4.8). A child inherits them from one parent only: the mother. For a variety of reasons, some women's eggs contain mostly defective mitochondria; children born to these women have a higher than normal risk of mitochondrial disease. Thanks to efforts by Martine Martin and many others, the United Kingdom recently approved a new reproductive technology, three-person IVF, intended to help these women have healthy babies. With this procedure, nuclear DNA from the mother and father is inserted into a donor egg from a woman with healthy mitochondria. The resulting embryo will be implanted into the mother, much as normal IVF works. A baby born from this procedure will have three parents: Its cells will have nuclear DNA inherited from the mother and father, and mitochondria inherited from the donor.

converted to glucose-6-phosphate for glycolysis ❶. When a cell produces more ATP than it uses, ATP accumulates in the cytoplasm. The high concentration of ATP causes glucose-6-phosphate to be diverted away from glycolysis and into a pathway that builds glycogen (Section 3.4). Liver and muscle cells especially favor the conversion of glucose to glycogen, and these cells contain the body's largest stores of it. Between meals, the liver maintains the glucose level in blood by converting the stored glycogen to glucose.

Fats A fat molecule has a glycerol head and one, two, or three fatty acid tails (Section 3.5). Cells dismantle these molecules by first breaking the bonds that connect the fatty acid tails to the glycerol head (**FIGURE 7.12B**). Nearly all cells in the body can oxidize the released fatty acids by splitting their long backbones into two-carbon fragments. These fragments are converted to acetyl–CoA, which can enter the Krebs cycle ❷. Glycerol released by fat breakdown is converted by liver cells to PGAL, an intermediate of glycolysis ❸.

On a per carbon basis, fats are a richer source of energy than carbohydrates. Carbohydrate backbones have many oxygen atoms, so they are partially oxidized. A fat's long fatty acid tails are hydrocarbon chains that typically have no oxygen atoms bonded to them, so they have a longer way to go to become oxidized—more reactions are required to fully break them down. Coenzymes accept electrons in these oxidation reactions. The more reduced coenzymes that form, the more electrons can be delivered to the ATP-forming machinery of electron transfer phosphorylation.

What happens if you eat too many carbohydrates? When the blood level of glucose gets too high, acetyl–CoA is diverted away from the Krebs cycle and into a pathway that makes fatty acids. That is why excess dietary carbohydrate ends up as fat.

Proteins Enzymes in the digestive system split dietary proteins into their amino acid subunits, which are absorbed into the bloodstream. Cells use the amino acids to build proteins or other molecules. When you eat more protein than your body needs for this purpose, the amino acids are broken down. The amino ($—NH_3^+$) group is removed, and it becomes ammonia (NH_3), a waste product that is eliminated in urine. The carbon backbone is split, and acetyl–CoA, pyruvate, or an intermediate of the Krebs cycle forms, depending on the amino acid (**FIGURE 7.12C**). These molecules enter aerobic respiration's second stage ❹.

TAKE-HOME MESSAGE 7.7
Can the body break down any organic molecule for energy?

✔ Oxidizing organic molecules can break their carbon backbones, releasing electrons whose energy can be harnessed to drive ATP formation in aerobic respiration.

✔ First the digestive system and then individual cells convert molecules in food (fats, complex carbohydrates, and proteins) into substrates of glycolysis or aerobic respiration's second-stage reactions.

---

summary

Section 7.1 Photosynthesis changed the composition of Earth's atmosphere, with profound effects on life's evolution. Organisms that could not tolerate the increased atmospheric oxygen persisted only in **anaerobic** habitats. The evolution of antioxidants allowed organisms to thrive under **aerobic** conditions.

Free radicals that form during aerobic respiration are detoxified by antioxidant molecules in the cell's cytoplasm. Missing molecules or heritable defects in mitochondrial electron transfer chain components can cause a buildup of free radicals that damage the cell—and the individual. The resulting oxidative stress plays a role in many illnesses.

 Section 7.2 Most organisms can make ATP by breaking down sugars, either by fermentation or by aerobic respiration. Both pathways begin in cytoplasm. **Aerobic respiration** requires oxygen and, in eukaryotes, ends inside mitochondria. This pathway includes electron transfer chains, and ATP forms by electron transfer phosphorylation. **Fermentation** pathways run entirely in cytoplasm and do not require oxygen. Aerobic respiration yields much more ATP per glucose molecule than fermentation.

 Section 7.3 **Glycolysis**, the first stage of aerobic respiration and fermentation pathways, occurs in cytoplasm. During glycolysis, enzymes use two ATP to convert one molecule of glucose or another six-carbon sugar to two molecules of three-carbon **pyruvate**. Electrons and hydrogen ions are transferred to two NAD^+, which are thereby reduced to NADH. Four ATP also form by **substrate-level phosphorylation**.

 Section 7.4 In eukaryotes, aerobic respiration continues in mitochondria. The second stage of aerobic respiration, acetyl–CoA formation and the **Krebs cycle**, takes place in the inner compartment (matrix) of the mitochondrion. The first steps convert the two pyruvate from glycolysis to two acetyl–CoA and two CO_2. The acetyl–CoA delivers carbon atoms to the Krebs cycle. Then, electrons and hydrogen ions are transferred to NAD^+ and FAD, which are thereby reduced to NADH and $FADH_2$. ATP forms by substrate-level phosphorylation.

Two cycles of Krebs reactions break down the two pyruvate from glycolysis. At this stage of aerobic respiration, the glucose molecule that entered glycolysis has been dismantled completely: All of its carbon atoms have exited the cell in CO_2.

 Section 7.5 In the third stage of aerobic respiration, electron transfer phosphorylation, the many coenzymes reduced in the first two stages now deliver electrons and hydrogen ions to electron transfer chains in the inner mitochondrial membrane. Energy lost by electrons moving through the chains is harnessed to move H^+ from the matrix to the intermembrane space. The resulting hydrogen ion gradient across the inner membrane drives these ions through ATP synthases, which in turn drives ATP synthesis. Oxygen accepts electrons at the end of the chains and combines with hydrogen ions, so water forms. The ATP yield of aerobic respiration varies, but typically it is about thirty-six ATP per glucose.

 Section 7.6 Anaerobic fermentation pathways begin with glycolysis, and they run in the cytoplasm. The electron acceptor at the end of these reactions is an organic molecule such as acetaldehyde (in **alcoholic fermenta-**

tion) or pyruvate (in **lactate fermentation**). The final steps of fermentation regenerate NAD^+, which is required for glycolysis to continue, but they produce no ATP. Thus, the breakdown of one glucose molecule yields only the two ATP from glycolysis. The small yield is enough to sustain single-celled organisms, for example in beneficial microbes that we use in production of foods such as bread and yogurt. Fermentation also supplements metabolism of multicelled eukaryotes, for example in mammalian white skeletal muscle fibers that use lactate fermentation to support intense bursts of activity.

 Section 7.7 Oxidizing an organic molecule can break its carbon backbone. Aerobic respiration fully oxidizes glucose, dismantling its backbone carbon by carbon. Each carbon removed releases electrons that drive ATP formation in electron transfer phosphorylation. Organic molecules other than sugars are also broken down (oxidized) for energy. In humans and other mammals, first the digestive system and then individual cells convert fats, proteins, and complex carbohydrates in food to molecules that are substrates of glycolysis or the second-stage reactions of aerobic respiration.

self-quiz
Answers in Appendix VII

1. True or false? Unlike animals, which make many ATP by aerobic respiration, plants make all of their ATP by photosynthesis.

2. Glycolysis starts and ends in the _____ .
   a. nucleus   c. plasma membrane
   b. mitochondrion   d. cytoplasm

3. Which of the following metabolic pathways require(s) molecular oxygen (O_2)?
   a. aerobic respiration   c. alcoholic fermentation
   b. lactate fermentation   d. all of the above

4. Which molecule does not form during glycolysis?
   a. NADH   b. pyruvate   c. $FADH_2$ d. ATP

5. In eukaryotes, aerobic respiration is completed in the _____ .
   a. nucleus   c. plasma membrane
   b. mitochondrion   d. cytoplasm

6. Which of the following reaction pathways is *not* part of the second stage of aerobic respiration?
   a. electron transfer   c. Krebs cycle
      phosphorylation   d. glycolysis
   b. acetyl–CoA formation   e. a and d

7. After Krebs reactions run through _____ cycle(s), one glucose molecule has been completely broken down to CO_2.
   a. one   b. two   c. three   d. six

8. In the third stage of aerobic respiration, _____ is the final acceptor of electrons.
   a. water   b. hydrogen   c. oxygen   d. NADH

9. _____ accepts electrons in alcoholic fermentation.
   a. Oxygen   c. Acetaldehyde
   b. Pyruvate   d. Sulfate

data analysis activities

Mitochondrial Abnormalities in Tetralogy of Fallot Tetralogy of Fallot (TF) is a genetic disorder in which heart malformations result in abnormal blood circulation, so oxygen does not reach body cells as it should. With insufficient oxygen to accept electrons at the end of miotchondrial electron transfer chains, too many free radicals form. This damages the mitochondria—and the cells. In 2004, Sarah Kuruvilla studied mitochondria in the heart muscle of TF patients. Some of her results are shown in **FIGURE 7.13**.

1. In this study, which abnormality was most strongly associated with TF?

2. What percentage of the TF patients had mitochondria that were abnormal in size?

3. Can you make any correlations between blood oxygen content and mitochondrial abnormalities in these patients?

FIGURE 7.13 Mitochondrial changes in tetralogy of Fallot (TF).

(**A**) Normal heart muscle. Many mitochondria between the fibers provide muscle cells with ATP for contraction. (**B**) Heart muscle from a person with TF has swollen, broken mitochondria.

(**C**) Types of mitochondrial abnormalities in TF patients. SPO_2 is oxygen saturation of the blood. A normal value of SPO_2 is 96%. Abnormalities are marked +.

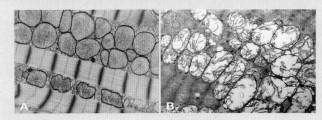

Patient (age)	SPO_2 (%)	Mitochondrial Abnormalities in TF			
		Number	Shape	Size	Broken
1 (5)	55	+	+	−	−
2 (3)	69	+	+	−	−
3 (22)	72	+	+	−	−
4 (2)	74	+	+	−	−
5 (3)	76	+	+	−	+
6 (2.5)	78	+	+	−	+
7 (1)	79	+	+	−	−
8 (12)	80	+	−	+	−
9 (4)	80	+	+	−	−
10 (8)	83	+	−	+	−
11 (20)	85	+	+	−	−
12 (2.5)	89	+	−	+	−

C

---

10. Fermentation pathways make no more ATP beyond the small yield from glycolysis. The remaining reactions serve to regenerate _____ .
   - a. FAD
   - b. NAD⁺
   - c. glucose
   - d. oxygen

11. Most of the energy that is released by the full breakdown of glucose to CO_2 and water ends up in _____ .
   - a. NADH
   - b. ATP
   - c. heat
   - d. electrons

12. Your body cells can break down _____ as a source of energy to fuel ATP production.
   - a. fatty acids
   - b. glycerol
   - c. amino acids
   - d. all of the above

13. Which of the following is *not* produced by an animal muscle cell operating under anaerobic conditions?
   - a. heat
   - b. pyruvate
   - c. NAD⁺
   - d. ATP
   - e. lactate
   - f. all are produced

14. Match the reactions with the events.
   - ___ glycolysis
   - ___ fermentation
   - ___ Krebs cycle
   - ___ electron transfer phosphorylation
   - a. ATP, NADH, FADH₂, and CO_2 form
   - b. glucose to two pyruvate
   - c. NAD⁺ regenerated, little ATP
   - d. H⁺ flow via ATP synthases

15. Match the term with the best description.
   - ___ mitochondrial matrix
   - ___ pyruvate
   - ___ NAD⁺
   - ___ mitochondrion
   - ___ NADH
   - ___ anaerobic
   - a. needed for glycolysis
   - b. inner space
   - c. makes many ATP
   - d. product of glycolysis
   - e. reduced coenzyme
   - f. no oxygen required

critical thinking

1. The higher the altitude, the lower the oxygen level in air. Climbers of very tall mountains risk altitude sickness, a condition characterized by shortness of breath, weakness, dizziness, and confusion. The early symptoms of cyanide poisoning are the same as those for altitude sickness. Cyanide binds tightly to cytochrome *c* oxidase, a protein complex that is the last component of mitochondrial electron transfer chains. Cytochrome *c* oxidase with bound cyanide can no longer transfer electrons. Explain why cyanide poisoning starts with the same symptoms as altitude sickness.

2. As you learned, membranes impermeable to hydrogen ions are required for electron transfer phosphorylation. Membranes in mitochondria serve this purpose in eukaryotes. Bacteria do not have this organelle, but they can make ATP by electron transfer phosphorylation. How do you think they do it, given that they have no mitochondria?

3. The bar-tailed godwit is a type of shorebird that makes an annual migration from Alaska to New Zealand and back. The birds make each 11,500-kilometer (7,145-mile) trip by flying over the Pacific Ocean in about nine days, depending on weather, wind speed, and direction of travel. One bird was observed to make the entire journey uninterrupted, a feat that is comparable to a human running a nonstop seven-day marathon at 70 kilometers per hour (43.5 miles per hour). Would you expect the flight (breast) muscles of bar-tailed godwits to be light or dark colored? Explain your answer.

CENGAGE brain.com To access course materials, please visit www.cengagebrain.com.

LEARNING ROADMAP

Radioisotope tracers (Section 2.2) were used in research that led to the discovery that DNA (3.8), not protein (3.6), is the hereditary material of all organisms (1.3). This chapter revisits free radicals (2.3), the cell nucleus (4.6), and metabolism (5.4–5.6). Your knowledge of carbohydrate ring numbering (3.3) will help you understand DNA replication.

DISCOVERY OF DNA'S FUNCTION

The work of many scientists over nearly a century led to the discovery that DNA, not protein, stores hereditary information in all living things.

STRUCTURE OF DNA MOLECULES

A DNA molecule consists of two long chains of nucleotides coiled into a double helix. The order of the four types of nucleotides in a chain differs among individuals and among species.

CHROMOSOMES

The DNA of eukaryotes is divided among a characteristic number of chromosomes. A living cell's chromosomes contain all of the information necessary to build a new individual.

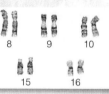

DNA REPLICATION

Before a cell divides, it copies its DNA so both descendant cells will inherit a full complement of chromosomes. Replication of each DNA molecule produces two duplicates.

MUTATIONS

DNA damage by environmental agents can cause replication errors. Newly forming DNA is monitored for errors, most of which are corrected. Uncorrected errors become mutations.

You will revisit DNA structure and function many times, particularly when you learn about how genetic information is converted into parts of a cell (Chapter 9), and how cells control that process (Chapter 10). Chromosome structure will turn up again in the context of cell division in Chapters 11 and 12, and in human inheritance and disease in Chapter 14. Viruses such as bacteriophage will be explained in more detail in Chapter 20, and stem cells in Chapter 31.

8.1 A Hero Dog's Golden Clones

On September 11, 2001, Constable James Symington drove his search dog Trakr from Nova Scotia to Manhattan. Within hours of arriving, the dog led rescuers to the area where the final survivor of the World Trade Center attacks was buried. She had been clinging to life, pinned under rubble from the building where she had worked. Symington and Trakr helped with the search and rescue efforts for three days nonstop, until Trakr collapsed from smoke and chemical inhalation, burns, and exhaustion (**FIGURE 8.1**).

Trakr survived the ordeal, but later lost the use of his limbs from a degenerative neurological disease probably linked to toxic smoke exposure at Ground Zero. The hero dog died in April 2009, but his DNA lives on in his genetic copies—his **clones**. Symington's essay about Trakr's superior nature and abilities as a search and rescue dog won the Golden Clone Giveaway, a contest to find the world's most clone-worthy dog. Trakr's DNA was shipped to Korea, where it was inserted into donor dog eggs, which were then implanted into surrogate mother dogs. Five puppies, all clones of Trakr, were delivered to Symington in July 2009.

Many other adult mammals have been cloned besides Trakr, but the technique is still unpredictable. Depending on the species, few implanted embryos may survive until birth. Of the clones that do survive, many have serious health problems. Why the difficulty? Even though all cells of an individual inherit the same DNA, an adult cell uses only a fraction of it compared with an embryonic cell. To make a clone from an adult cell, researchers must reprogram its DNA to function like the DNA of an egg. They are getting better at it. The research continues because the potential benefits are enormous. Already, cells of cloned human embryos are helping researchers unravel the molecular mechanisms of human genetic diseases. Replacement tissues and organs for people with incurable diseases are being generated from cloned human cells. Endangered animals might be saved from extinction; extinct animals may be brought back.

Perfecting methods for cloning animals brings us closer to the possibility of cloning humans, both technically and ethically. For example, if cloning a lost cat for a grieving pet owner is acceptable, why would it not be acceptable to clone a lost child for a grieving parent? Different people have very different answers to such questions, so controversy over cloning continues even as the techniques improve.

---

clone Genetically identical copy of an organism.

FIGURE 8.1
James Symington and his dog Trakr at Ground Zero, September 2001.

CREDITS: (opposite) Patrick Landmann/Science Source; (1) © James Symington.

✔ Investigations that led to our understanding that DNA is the molecule of inheritance reveal how science advances.

FIGURE 8.2 DNA, the substance, extracted from human cells.

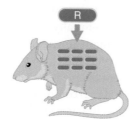

A Griffith's first experiment showed that R cells were harmless. When injected into mice, the bacteria multiplied, but the mice remained healthy.

B The second experiment showed that an injection of S cells caused mice to develop fatal pneumonia. Their blood contained live S cells.

C For a third experiment, Griffith killed S cells with heat before injecting them into mice. The mice remained healthy, indicating that the heat-killed S cells were harmless.

D In his fourth experiment, Griffith injected a mixture of heat-killed S cells and live R cells. To his surprise, the mice became fatally ill, and their blood contained live S cells.

FIGURE 8.3 ▶Animated Fred Griffith's experiments with two strains (R and S) of *Streptococcus pneumoniae* bacteria.

The substance we now call DNA (FIGURE 8.2) was first described in 1869 by Johannes Miescher, a chemist who extracted it from cell nuclei. Miescher determined that DNA is not a protein, and that it is rich in nitrogen and phosphorus, but he never learned its function.

Sixty years after Miescher's work, Frederick Griffith unexpectedly found a clue about DNA's function. Griffith was studying pneumonia-causing bacteria in the hope of creating a vaccine. He isolated two types, or strains, of the bacteria. One was harmless (R); the other lethal (S). Griffith used R and S cells in a series of experiments testing their ability to cause pneumonia in mice (FIGURE 8.3). He discovered that heat destroyed the ability of lethal S bacteria to cause pneumonia, but it did not destroy their hereditary material, including whatever specified "kill mice." That material could be transferred from dead S cells to live R cells, which put it to use. The transformation was permanent and heritable: Even after hundreds of generations, descendants of transformed R cells retained the ability to kill mice.

What substance had caused this transformation? In 1940, Oswald Avery and Maclyn McCarty set out to identify this substance, which they called the "transforming principle." The team extracted lipids, proteins, and nucleic acids from S cells, then used a process of elimination to see which component would cause the transformation. Treating the extract with lipid- and protein-destroying enzymes did not prevent it from transforming R cells, so the transforming principle could not be lipid or protein. Avery and McCarty realized that the substance they were seeking must be nucleic acid—DNA or RNA. DNA-degrading enzymes destroyed the extract's ability to transform cells, but RNA-degrading enzymes did not. Thus, DNA had to be the transforming principle.

The result surprised Avery and McCarty, who, along with most other scientists, had assumed that proteins were the material of heredity. After all, traits are diverse, and proteins are the most diverse of all biological molecules. The two scientists were so skeptical that they published their results only after they had convinced themselves, by years of painstaking experimentation, that DNA was indeed hereditary material. They were also careful to point out that they had not proven DNA was the *only* hereditary material.

Avery and McCarty's tantalizing results prompted a stampede of other scientists into the field of DNA research. The resulting explosion of discovery confirmed the molecule's role as carrier of hereditary information. Key in this advance was the realization that any molecule—DNA or otherwise—had to have certain

CREDITS: (2) Patrick Landmann/Science Source; (3) © Cengage Learning.

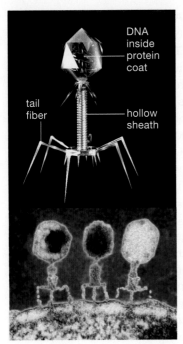

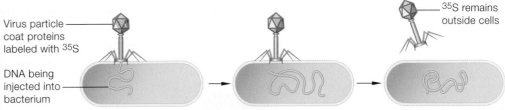

B In one experiment, bacteriophage were labeled with a radioisotope of sulfur (^{35}S), a process that makes their protein components radioactive. The labeled viruses were mixed with bacteria long enough for infection to occur, and then the mixture was whirled in a kitchen blender. Blending dislodged viral parts that remained on the outside of the bacteria. Afterward, most of the radioactive sulfur was detected outside the bacterial cells. The viruses had not injected protein into the bacteria.

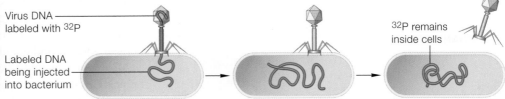

C In another experiment, bacteriophage were labeled with a radioisotope of phosphorus (^{32}P), which makes their DNA radioactive. The labeled viruses were allowed to infect bacteria. After the external viral parts were dislodged from the bacteria, the radioactive phosphorus was detected mainly inside the bacterial cells. The viruses had injected DNA into the cells—evidence that DNA is the genetic material of this virus.

A Top, a model of a bacteriophage. Bottom, micrograph of three viruses injecting DNA into an *E. coli* cell.

FIGURE 8.4 ▶Animated The Hershey–Chase experiments. Alfred Hershey and Martha Chase carried out experiments to determine the composition of the hereditary material that bacteriophage inject into bacteria. The experiments were based on the knowledge that proteins contain more sulfur (S) than phosphorus (P), and DNA contains more phosphorus than sulfur.

properties in order to function as hereditary material. First, a full complement of hereditary information must be transmitted along with the molecule; second, each cell of a given species should contain the same amount of it; third, because the molecule functions as a genetic bridge between generations, it has to be exempt from major change; and fourth, it must be capable of encoding the almost unimaginably huge amount of information required to build a new individual.

In the late 1940s, Alfred Hershey and Martha Chase proved that DNA, and not protein, satisfies the first property of a hereditary molecule: It transmits a full complement of hereditary information. Hershey and Chase specialized in working with **bacteriophage**, a type of virus that infects bacteria (**FIGURE 8.4A**). Like all viruses, these infectious particles carry information about how to make new viruses in their hereditary material. After a virus injects a cell with this material, the cell starts making new virus particles. Hershey and Chase carried out an elegant series of experiments proving that the material a bacteriophage injects into bacteria is DNA, not protein (**FIGURE 8.4B,C**).

bacteriophage Virus that infects bacteria.

The second property expected of a hereditary molecule was pinned on DNA by André Boivin and Roger Vendrely, who meticulously measured the amount of DNA in cell nuclei from a number of species. In 1948, they proved that body cells of any individual of a species contain precisely the same amount of DNA. Daniel Mazia's laboratory discovered that the protein and RNA content of cells varies over time, but not the DNA content, demonstrating that DNA is not involved in metabolism (and proving DNA has the third property expected of a hereditary molecule). The fourth property—that a hereditary molecule must somehow encode a huge amount of information—would be proven along with the elucidation of DNA's structure, a topic we continue in the next section.

TAKE-HOME MESSAGE 8.2

How was the function of DNA discovered?

✔ DNA, the molecule of inheritance, was first discovered in the late 1800s. Its role as the carrier of hereditary information was uncovered over many years, as scientists built upon one another's discoveries.

8.3 The Discovery of DNA's Structure

✔ Watson and Crick's discovery of DNA's structure was based on 150 years of research by other scientists.

ADENINE (A)
deoxyadenosine triphosphate

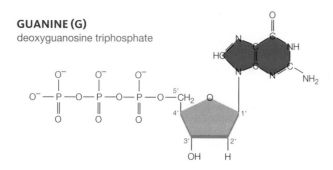

GUANINE (G)
deoxyguanosine triphosphate

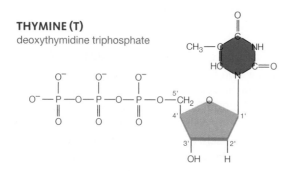

THYMINE (T)
deoxythymidine triphosphate

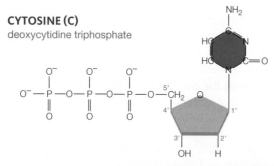

CYTOSINE (C)
deoxycytidine triphosphate

FIGURE 8.5 The four nucleotides in DNA. Each has three phosphate groups, a deoxyribose sugar (orange), and a nitrogen-containing base (blue) after which it is named. Adenine and guanine bases are purines; thymine and cytosine, pyrimidines. Biochemist Phoebus Levene identified the structure of these bases and how they are connected in nucleotides in the early 1900s. Levene worked with DNA for almost 40 years.

Building Blocks of DNA

DNA is a polymer of four types of nucleotides—adenine (A), guanine (G), thymine (T), and cytosine (C). Each has a deoxyribose sugar, three phosphate groups, and a nitrogen-containing base after which it is named (**FIGURE 8.5**). Just how those four nucleotides are arranged in a DNA molecule was a puzzle that took over 50 years to solve.

Clues about DNA's structure started coming together around 1950, when Erwin Chargaff (one of many researchers investigating its function) made two important discoveries about the molecule. First, the amounts of thymine and adenine are identical, as are the amounts of cytosine and guanine (A = T and G = C). We call this discovery Chargaff's first rule. Chargaff's second discovery, or rule, is that the DNA of different species differs in the proportions of adenine and guanine.

Meanwhile, biologist James Watson and biophysicist Francis Crick had been sharing ideas about the structure of DNA. The helical (coiled) pattern of secondary structure that occurs in many proteins (Section 3.6) had just been discovered, and Watson and Crick suspected that the DNA molecule was also a helix. The two spent many hours arguing about the size, shape, and bonding requirements of the four DNA nucleotides. They pestered chemists to help them identify bonds they might have overlooked, fiddled with cardboard cutouts, and made models from scraps of metal connected by suitably angled "bonds" of wire.

Biochemist Rosalind Franklin had also been working on the structure of DNA. Like Crick, Franklin specialized in x-ray crystallography, a technique in which x-rays are directed through a purified and crystallized substance. Atoms in the substance's molecules scatter the x-rays in a pattern that can be captured as an image. Researchers can use the pattern to calculate the size, shape, and spacing between any repeating elements of the molecules—all of which are details of molecular structure. As molecules go, DNA is gigantic, and it was difficult to crystallize given the techniques of the time. Franklin made the first 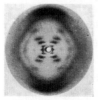 clear x-ray diffraction image of DNA as it occurs in cells (left). From the information in that image, she calculated that DNA is very long compared to its 2-nanometer diameter. She also identified a repeating pattern every 0.34 nanometer along its length, and another every 3.4 nanometers.

Franklin's image and data came to the attention of Watson and Crick, who now had all the information they needed to build the first accurate model of DNA (**FIGURE 8.6**). A DNA molecule has two chains (strands) of nucleotides running in opposite directions and coiled

CREDITS: (5) © Cengage Learning; (in text) NLM.

into a double helix (**FIGURE 8.7**). Bonds between the deoxyribose of one nucleotide and the phosphate of the next form the sugar–phosphate backbone of each chain. Hydrogen bonds between the internally positioned bases hold the two strands together. Only two kinds of base pairings form: A to T, and G to C, and this explains the first of Chargaff's rules. Most scientists had assumed (incorrectly) that the bases had to be on the outside of the helix, because they would be more accessible to DNA-copying enzymes that way. You will see in Section 8.5 how DNA replication enzymes access the bases on the inside of the double helix.

DNA's Base Sequence

A small piece of DNA from a tulip, a human, or any other organism might be:

one base pair

Notice how the two strands of DNA match. They are complementary—the base of each nucleotide on one strand pairs with a suitable partner base on the other. This base-pairing pattern (A to T, G to C) is the same in all molecules of DNA. How can just two kinds of base pairings give rise to the incredible diversity of traits we see among living things? Even though DNA is composed of only four nucleotides, the *order* in which one nucleotide follows the next along a strand—the **DNA sequence**—varies tremendously among species (which explains Chargaff's second rule). DNA molecules can be hundreds of millions of nucleotides long, so their sequence can encode a massive amount of information (we return to the nature of that information in the next chapter). DNA sequence variation is the basis of traits that define species and distinguish individuals. Thus DNA, the molecule of inheritance in every cell, is the basis of life's unity. Variations in its sequence are the foundation of life's diversity.

DNA sequence Order of nucleotides in a strand of DNA.

TAKE-HOME MESSAGE 8.3
What is the structure of DNA?

✔ A DNA molecule consists of two nucleotide chains (strands) running in opposite directions and coiled into a double helix. Internally positioned nucleotide bases hydrogen-bond between the two strands. A pairs with T, and G with C.

✔ The sequence of bases along a DNA strand varies among species and among individuals. This variation is the basis of life's diversity.

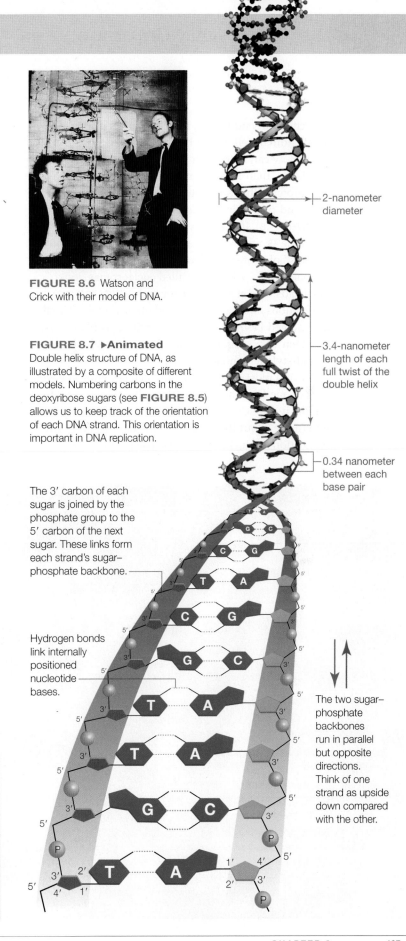

FIGURE 8.6 Watson and Crick with their model of DNA.

FIGURE 8.7 ▶Animated
Double helix structure of DNA, as illustrated by a composite of different models. Numbering carbons in the deoxyribose sugars (see **FIGURE 8.5**) allows us to keep track of the orientation of each DNA strand. This orientation is important in DNA replication.

2-nanometer diameter

3.4-nanometer length of each full twist of the double helix

0.34 nanometer between each base pair

The 3' carbon of each sugar is joined by the phosphate group to the 5' carbon of the next sugar. These links form each strand's sugar–phosphate backbone.

Hydrogen bonds link internally positioned nucleotide bases.

The two sugar–phosphate backbones run in parallel but opposite directions. Think of one strand as upside down compared with the other.

CREDITS: (6) A. C. Barrington Brown © 1968 J. D. Watson; (in text, 7) © Cengage Learning.

8.4 Eukaryotic Chromosomes

✔ In cells, DNA and associated proteins are organized as chromosomes.

Stretched out end to end, the DNA molecules in a single human cell would be about 2 meters (6.5 feet) long. How can that much DNA cram into a nucleus that is less than 10 micrometers in diameter? Inside a cell, proteins that associate with each DNA molecule twist and pack it into a structure called a **chromosome** (**FIGURE 8.8**). In a eukaryotic cell, for example, a DNA molecule ❶ wraps twice at regular intervals around "spools" of proteins called **histones** ❷. These DNA–histone spools, which are called **nucleosomes**, look like beads on a string in micrographs (**FIGURE 8.9**). Interactions among histones and other proteins twist the spooled DNA into a tight fiber ❸. This fiber coils, and then it coils again into a hollow cylinder a bit like an old-style telephone cord ❹.

During most of the cell's life, each chromosome consists of one DNA molecule. When the cell prepares to divide, it duplicates its chromosomes by DNA replication (more about this process in the next section). After replication, each chromosome consists of two DNA

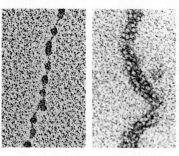

FIGURE 8.9 Chromosome packing. Left, beads-on-a string appearance of DNA–histone spools. Middle, coiled coils of a DNA fiber. Right, a duplicated chromosome just before cell division.

molecules, or **sister chromatids**, attached to one another at a constricted region called the **centromere**:

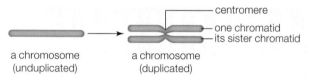

centromere
one chromatid
its sister chromatid

a chromosome (unduplicated)   a chromosome (duplicated)

The chromosomes condense into their familiar "X" shapes ❺ just before the cell divides.

FIGURE 8.8 ▶Animated **Chromosome structure.**

❶ Two strands of DNA twist into a double helix.

❷ At regular intervals, the DNA (blue) wraps around a core of histone proteins (purple).

❸ The DNA and proteins associated with it twist tightly into a fiber.

❹ The fiber coils and then coils again to form a hollow cylinder.

❺ At its most condensed, a duplicated chromosome has an X shape.

❻ The DNA in the nucleus of a eukaryotic cell is typically divided into a number of chromosomes.

CREDITS: (8, in text) © Cengage Learning; (9) left, O. L. Miller, Jr., Steve L. McKnight; middle, B. Hamkalo; right, Andrew Syred/Science Source.

Chromosome Number and Type

The DNA of a eukaryotic cell is divided among some number of chromosomes that differ in length and shape ❻ . That number is called the **chromosome number**, and it is a characteristic of the species. For example, the chromosome number of humans is 46, so our cells have 46 chromosomes.

Actually, human body cells have two sets of 23 chromosomes—two of each type. Having two sets of chromosomes means these cells are **diploid**, or 2*n*. An image of an individual's diploid set of chromosomes is called a **karyotype** (**FIGURE 8.10**). To create a karyotype, cells taken from the individual are treated to make the chromosomes condense, and then stained so the chromosomes can be distinguished under a microscope. A micrograph of a single cell is digitally rearranged so the images of the chromosomes are lined up by centromere location, and arranged according to size, shape, and length.

In a human body cell, all but one pair of chromosomes are **autosomes**, which are the same in both females and males. The two autosomes of a pair have the same length, shape, and centromere location. They also hold information about the same traits. Think of them as two sets of books on how to build a house. Your father gave you one set. Your mother had her own ideas about wiring, plumbing, and so on. She gave you an alternate set that says slightly different things about many of those tasks.

Members of a pair of **sex chromosomes** differ between females and males, and the differences determine an individual's sex. The sex chromosomes of humans are called X and Y. The body cells of typical human females have two X chromosomes (XX, **FIGURE 8.10A**); those of typical human males have

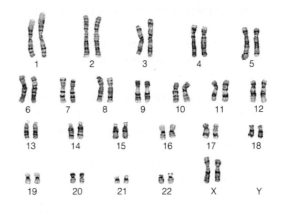

A Karyotype of a female human, with identical sex chromosomes (XX).

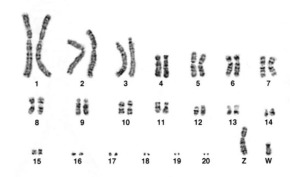

B Karyotype of a female chicken, with nonidentical sex chromosomes (ZW).

FIGURE 8.10 ▶**Animated** Karyotypes.

one X and one Y chromosome (XY). This pattern—XX females and XY males—is the rule among fruit flies, mammals, and many other animals, but there are other patterns (**FIGURE 8.10B**). Female butterflies, moths, birds, and certain fishes have two nonidentical sex chromosomes; the two sex chromosomes of males are identical. Environmental factors (not chromosomes) determine sex in some species of invertebrates, turtles, and frogs. As an example, the temperature of the sand in which sea turtle eggs are buried determines the sex of the hatchlings.

autosome A chromosome that is the same in males and females.
centromere Of a duplicated eukaryotic chromosome, constricted region where sister chromatids attach to each other.
chromosome A molecule of DNA together with associated proteins; carries part or all of a cell's genetic information.
chromosome number The total number of chromosomes in a cell of a given species.
diploid Having two of each type of chromosome characteristic of the species (2*n*).
histone Type of protein that associates with DNA and structurally organizes eukaryotic chromosomes.
karyotype Image of an individual's set of chromosomes arranged by size, length, shape, and centromere location.
nucleosome A length of DNA wound twice around a spool of histone proteins.
sex chromosome Member of a pair of chromosomes that differs between males and females.
sister chromatids The two attached DNA molecules of a duplicated eukaryotic chromosome.

TAKE-HOME MESSAGE 8.4
What is a chromosome?

✔ A chromosome is a molecule of DNA together with associated proteins that organize it and allow it to pack tightly.

✔ A eukaryotic cell's DNA is divided among a characteristic number of chromosomes, which differ in length and shape.

✔ Members of a pair of sex chromosomes differ between males and females. Chromosomes that are the same in both sexes are called autosomes.

CREDITS: (10A) © University of Washington Department of Pathology; (10B) With kind permission from Springer Science+Business Media: *Chromosome Research*, Volume 17, Number 1, 99 133, DOI: 10 1007/s10577-009-9021-6; *Avian comparative genomics reciprocal chromosome painting between domestic chicken (Gallus gallus) and the stone curlew (Burkinus oedicnemus, Charadriiformes)—An atypical species with low diploid number*, Wenhui Nie, Patricia C.M. O'Brien, Bee L. Ng, Biyuan Fu, Vitaly Volobouev, Nigel P. Carter, Malcolm A. Ferguson-Smith and Fengtang Yang; fig 2a.

✔ A cell copies its DNA before it reproduces. Each of the two strands of DNA in the double helix is replicated.

✔ DNA replication is an energy-intensive pathway that requires the participation of many enzymes, including DNA polymerase.

When a cell reproduces, it divides. The two descendant cells must inherit a complete copy of genetic information or they will not function properly. Thus, in preparation for division, the cell copies its chromosomes so that it contains two sets: one for each of its future offspring.

❶ As replication begins, many initiator proteins attach to the DNA at certain sites in the chromosome. Eukaryotic chromosomes have many of these origins of replication; DNA replication proceeds more or less simultaneously at all of them.

❷ Enzymes recruited by the initiator proteins begin to unwind the two strands of DNA from one another.

❸ Primers base-paired with the exposed single DNA strands serve as initiation sites for DNA synthesis.

❹ Starting at primers, DNA polymerases (green boxes) assemble new strands of DNA from nucleotides, using the parent strands as templates.

❺ DNA ligase seals any gaps that remain between bases of the "new" DNA, so a continuous strand forms.

❻ Each parental DNA strand (blue) serves as a template for assembly of a new strand of DNA (magenta). Both strands of the double helix serve as templates, so two double-stranded DNA molecules result. One strand of each is parental (old), and the other is new, so DNA replication is said to be semiconservative.

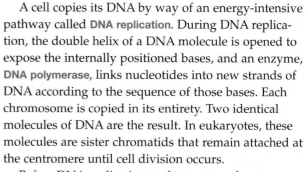

initiator proteins

topoisomerase (untwists the double helix)

helicase (breaks hydrogen bonds between bases)

primer

DNA polymerase

nucleotide

DNA ligase

FIGURE 8.11 ▶**Animated** DNA replication, in which a double-stranded molecule of DNA is copied in its entirety. Green arrows show the direction of synthesis for each strand. The Y-shaped structure of a DNA molecule undergoing replication is called a replication fork.

A cell copies its DNA by way of an energy-intensive pathway called **DNA replication**. During DNA replication, the double helix of a DNA molecule is opened to expose the internally positioned bases, and an enzyme, **DNA polymerase**, links nucleotides into new strands of DNA according to the sequence of those bases. Each chromosome is copied in its entirety. Two identical molecules of DNA are the result. In eukaryotes, these molecules are sister chromatids that remain attached at the centromere until cell division occurs.

Before DNA replication, a chromosome has one molecule of DNA—one double helix (**FIGURE 8.11**). As replication begins, proteins called initiators bind to certain sequences of nucleotides in the DNA ❶. Initiator proteins allow other molecules to bind, including enzymes that pry apart the two DNA strands: one (topoisomerase) that untwists the double helix, and another (helicase) that breaks the hydrogen bonds holding the double helix together. The two DNA strands begin to unwind from one another ❷. Another enzyme (primase) then starts making **primers**, which are short, single strands of nucleotides that serve as attachment points for DNA polymerase. The nucleotide bases of a primer can form hydrogen bonds with the exposed bases of a single strand of DNA ❸. Thus, a primer can base-pair with a complementary strand of DNA:

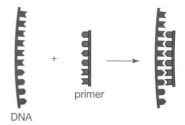

DNA

+ primer

The establishment of base pairing between two strands of DNA (or DNA and RNA) is called **nucleic acid hybridization**. Hybridization is spontaneous and is driven entirely by hydrogen bonding.

DNA polymerases attach to the hybridized primers and begin DNA synthesis. As a DNA polymerase moves along a strand, it uses the sequence of exposed nucleotide bases as a template, or guide, to assemble a new strand of DNA from free nucleotides ❹. Each nucleotide provides energy for its own attachment. The bonds between a nucleotide's phosphate groups hold a lot of energy (Section 5.6). Two of three phosphate groups are removed when a nucleotide is added to a DNA strand. Breaking those bonds releases enough energy to drive the attachment.

A DNA polymerase follows base-pairing rules: It adds a T to the end of the new DNA strand when it reaches an A in the template strand; it adds a G when it reaches a C; and so on. Thus, the DNA sequence of each new strand is complementary to the template (parental) strand. The enzyme **DNA ligase** seals any gaps, so the new DNA strands are continuous ❺. Both of the two strands of the parent molecule are copied at the same time. As each new DNA strand lengthens, it winds up with the template strand into a double helix. So, after replication, two double-stranded molecules of DNA have formed ❻. One strand of each molecule is conserved (parental), and the other is new; hence the name of the process, **semiconservative replication**. Both double-stranded molecules produced by DNA replication are duplicates of the parent molecule.

Numbering the carbons of the deoxyribose sugars in nucleotides allows us to keep track of the orientation of strands in a DNA double helix (see **FIGURES 8.5** and **8.7**). Each strand has two ends. The last carbon atom on one end of the strand is a 5′ (5 prime) carbon of a sugar; the last carbon atom on the other end is a 3′ (three prime) carbon of a sugar:

DNA polymerase can attach a nucleotide only to a 3′ end. Thus, during DNA replication, only one of two new strands of DNA can be constructed in a single piece (**FIGURE 8.12**). Synthesis of the other strand occurs in segments that must be joined by DNA ligase where they meet up. This is why we say that DNA synthesis proceeds only in the 5′ to 3′ direction.

DNA ligase Enzyme that seals gaps in double-stranded DNA.
DNA polymerase DNA replication enzyme. Uses a DNA template to assemble a complementary strand of DNA.
DNA replication Process by which a cell duplicates its DNA before it divides.
nucleic acid hybridization Spontaneous establishment of base-pairing between two nucleic acid strands.
primer Short, single strand of DNA that base-pairs with a specific DNA sequence.
semiconservative replication Describes the process of DNA replication, which produces two copies of a DNA molecule: one strand of each copy is new, and the other is a strand of the original DNA.

TAKE-HOME MESSAGE 8.5

How does a cell copy its DNA?

✔ During replication of a molecule of DNA, each strand of its double helix serves as a template for synthesis of a new, complementary strand of DNA.

✔ Replication of a molecule of DNA produces two double helices that are duplicates of the parent molecule. One strand of each is parental; the other is new.

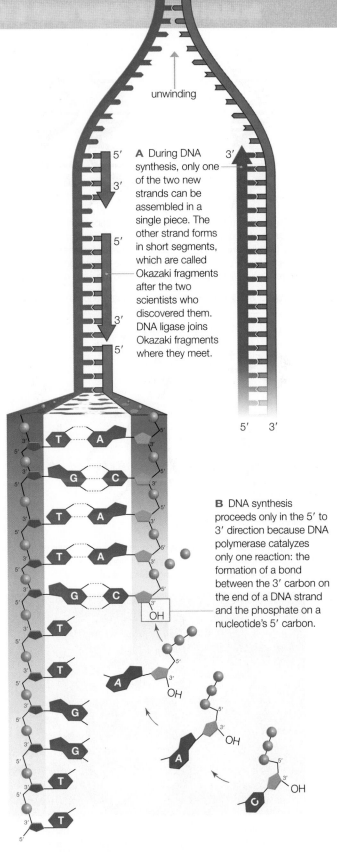

A During DNA synthesis, only one of the two new strands can be assembled in a single piece. The other strand forms in short segments, which are called Okazaki fragments after the two scientists who discovered them. DNA ligase joins Okazaki fragments where they meet.

B DNA synthesis proceeds only in the 5′ to 3′ direction because DNA polymerase catalyzes only one reaction: the formation of a bond between the 3′ carbon on the end of a DNA strand and the phosphate on a nucleotide's 5′ carbon.

FIGURE 8.12 ▶Animated Discontinuous synthesis of DNA. This close-up of a replication fork shows that only one new DNA strand is assembled continuously.

FIGURE IT OUT What do the yellow balls represent? Answer: Phosphate groups

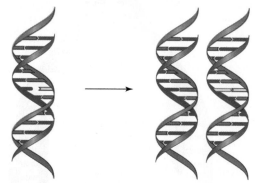

A Repair enzymes can recognize a mismatched base (yellow), but sometimes fail to correct it before DNA replication.

B After replication, both strands base-pair properly. Repair enzymes can no longer recognize the error, which has now become a mutation that will be passed on to the cell's descendants.

FIGURE 8.13 How a replication error can become a mutation.

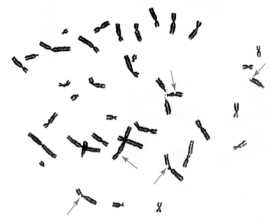

A Major breaks (red arrows) in chromosomes of a human white blood cell after exposure to ionizing radiation. Pieces of broken chromosomes can get lost during DNA replication.

B These *Ranunculus* flowers were grown from plants harvested around Chernobyl, Ukraine, where in 1986 an accident at a nuclear power plant released huge amounts of radiation. A normal flower is shown for comparison, in the inset.

FIGURE 8.14 Exposure to ionizing radiation causes mutations.

✔ Mutations, which are permanent changes in the DNA sequence of a chromosome, can be passed to descendants.

Replication Errors

Sometimes, a new DNA strand is not exactly complementary to its parent strand. A nucleotide may get lost during DNA replication, or an extra one slips in. Occasionally, the wrong nucleotide is added. Most of these replication errors occur simply because DNA polymerases work very fast. Mistakes are inevitable, and some DNA polymerases make a lot of them. Luckily, most DNA polymerases also proofread their work. They can correct a mismatch by reversing the synthesis reaction to remove the mispaired nucleotide, then resuming synthesis in the forward direction.

Replication errors also occur after the cell's DNA gets broken or otherwise damaged, because DNA polymerases do not copy damaged DNA very well. In most cases, repair enzymes and other proteins remove and replace damaged or mismatched bases in DNA before replication begins. When proofreading and repair mechanisms fail, an error becomes a **mutation**, a permanent change in the DNA sequence of a cell's chromosome(s). Repair enzymes cannot fix a mutation after the altered strand has been replicated, because they do not recognize correctly paired bases (**FIGURE 8.13**). Thus, a mutation is passed to one of the cell's offspring and all of its descendants.

Mutations can form in any type of cell. Those that occur during egg or sperm formation can be inherited by offspring, and in fact each human child is born with an average of 36 new ones. Mutations that alter DNA's instructions may have a harmful or lethal outcome; most cancers begin with them (we return to this topic in Section 11.6). However, not all mutations are dangerous. As you will see in Chapter 17, they give rise to the variation in traits that is the raw material of evolution.

Agents of DNA Damage

Electromagnetic energy with a wavelength shorter than 320 nanometers, including x-rays, most ultraviolet (UV) light, and gamma rays, has enough energy to knock electrons out of atoms. Exposure to such ionizing radiation damages DNA, breaking it into pieces that get lost during replication (**FIGURE 8.14A**). Ionizing radiation can also cause covalent bonds to form between bases on opposite strands of the double helix, an outcome that permanently blocks DNA replication. (We consider cancer-causing effects of such cell cycle interruptions in Chapter 11.) The nucleotide bases themselves can be irreparably damaged by ionizing radiation. Repair

CREDITS: (13) © Cengage Learning; (14A) Olga Shovman, Andrew C. Riches, Douglas Adamson, and Peter E. Bryant. *An improved assay for radiation-induced chromatid breaks using a colcemid block and calyculin-induced PCC combination.* Mutagenesis (2008) 23(4): 267-270 first published online March 6, 2008 doi:10.1093/mutage/gen009, by permission of Oxford University Press; (14B) background, Courtesy of Janis Ruksans; inset, Frank Sommariva/image/imagebroker.net/SuperStock.

enzymes remove bases damaged in this way, but they leave an empty space in the double helix or even a strand break. Any of these events can result in mutations (**FIGURE 8.14B**).

UV light in the range of 320 to 380 nanometers does not have enough energy to knock electrons out of atoms. However, it has enough energy to open up the double bond in the ring of a cytosine or thymine base. The open ring can form a covalent bond with

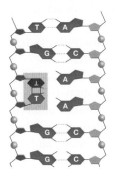

a thymine dimer

the ring of an adjacent pyrimidine (left), forming a dimer that kinks the DNA strand. A specific set of proteins can recognize, remove, and replace pyrimidine dimers before replication begins. Mutations arise if they do not, because DNA polymerase tends to copy kinked DNA incorrectly. Mutations that arise as a result of pyrimidine dimers are the cause of most skin cancers. Exposing unprotected skin to sunlight increases the risk of cancer because UV wavelengths in the light cause dimers to form. For every second a skin cell spends in the sun, 50 to 100 pyrimidine dimers form in its DNA.

Exposure to some chemicals also causes mutations. For instance, several of the fifty-five or more cancer-causing chemicals in tobacco smoke transfer methyl groups ($—CH_3$) to the nucleotide bases in DNA. Nucleotides altered in this way do not base-pair correctly. The body converts other chemicals in the smoke to compounds that bind irreversibly to DNA. In both cases, the resulting replication errors can lead to mutation. Cigarette smoke also contains free radicals, which inflict the same damage on DNA as ionizing radiation.

Rosalind Franklin, X-Rays, and Cancer

Rosalind Franklin arrived at King's College, London, in 1951. The expert x-ray crystallographer had been told she would be the only one in her department working on the structure of DNA, so she did not know that Maurice Wilkins was already doing the same thing just down the hall. No one had told Wilkins about Franklin's assignment; he assumed she was a technician hired to do his x-ray crystallography work. And so a clash began. Wilkins thought Franklin displayed an appalling lack of deference that technicians of the era usually accorded researchers. To Franklin, Wilkins seemed prickly and oddly overinterested in her work.

---

mutation Permanent change in the DNA sequence of a chromosome.

Wilkins and Franklin had been given identical samples of DNA. Franklin's meticulous work with hers yielded the first clear x-ray diffraction image of DNA as it occurs in cells. She gave a presentation on her work in 1952. DNA, she said, had two chains twisted into a double helix, with a backbone of phosphate groups on the outside, and bases arranged in an unknown way on the inside. She had calculated DNA's diameter, the distance between its chains and between its bases, the angle of the helix, and the number of bases in each coil. Crick, with his crystallography background, would have recognized the significance of the work—if he had been there. Watson was in the audience but he was not a crystallographer, and he did not understand the implications of Franklin's data.

Franklin started to write a research paper on her findings. Meanwhile, and perhaps without her knowledge, Watson reviewed Franklin's x-ray diffraction image with Wilkins, and Watson and Crick read a report detailing Franklin's unpublished data. Crick, who had more experience with molecular modeling than Franklin, immediately understood what the image and the data meant. Watson and Crick used that information to build their model of DNA.

On April 25, 1953, Franklin's paper appeared third in a series of articles about the structure of DNA in the journal *Nature*. It supported with solid experimental evidence Watson and Crick's theoretical model, which appeared in the first article of the series.

Rosalind Franklin (left) died in 1958, at the age of 37, of ovarian cancer probably caused by extensive exposure to x-rays during her work. At the time, the link between x-rays, mutations, and cancer was not understood. Because the Nobel Prize is not given posthumously, Franklin did not share in the 1962 honor that went to Watson, Crick, and Wilkins for the discovery of the structure of DNA.

TAKE-HOME MESSAGE 8.6
What are mutations?

✔ Proofreading and repair mechanisms usually maintain the integrity of a cell's genetic information by correcting mispaired bases and fixing damaged DNA before replication.

✔ Mismatched or damaged nucleotides that are not repaired can become mutations—permanent changes in the DNA sequence of a chromosome.

✔ DNA damage by environmental agents such as UV light, chemicals, and free radicals can result in mutations, because damaged DNA is not replicated very well.

✔ Various reproductive interventions produce genetically identical individuals.

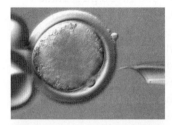

A A cow's egg is held in place by suction through a hollow glass tube called a micropipette. DNA is identified by a purple stain.

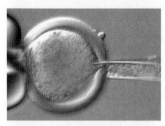

B Another micropipette punctures the egg and sucks out the DNA. All that remains inside the egg's plasma membrane is cytoplasm.

C A new micropipette prepares to enter the egg at the puncture site. The pipette contains a cell grown from the skin of a donor animal.

D The micropipette enters the egg and delivers the skin cell to a region between the cytoplasm and the plasma membrane.

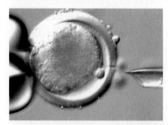

E After the pipette is withdrawn, the donor's skin cell is visible next to the cytoplasm of the egg. The transfer is now complete.

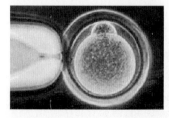

F An electric current causes the foreign cell to fuse with and deposit its nucleus into the cytoplasm of the egg. The egg begins to divide, and an embryo forms.

FIGURE 8.15 ▶Animated Somatic cell nuclear transfer, using cattle cells. This series of micrographs was taken by scientists at Cyagra, a company that specializes in cloning livestock.

The word "cloning" means making an identical copy of something, and it can refer to deliberate interventions in reproduction that produce an exact genetic copy of an organism. Genetically identical organisms occur all the time in nature, arising most often by the process of asexual reproduction (which we discuss in Chapter 11). Embryo splitting, another natural process, results in identical twins. The first few divisions of a fertilized egg form a ball of cells that sometimes splits spontaneously. If both halves of the ball continue to develop independently, identical twins result.

Artificial embryo splitting has been used in research and animal husbandry for decades. With this technique, a tiny ball of cells is grown from a fertilized egg in a laboratory. The ball is teased apart into two halves, each of which goes on to develop as a separate embryo. The embryos are implanted in surrogate mothers, who give birth to identical twins. Artificial twinning and any other technology that yields genetically identical individuals is called **reproductive cloning**.

Twins get their DNA from two parents that typically differ in their DNA sequence. Thus, although twins produced by embryo splitting are identical to one another, they are not identical to either parent. Animal breeders who want an exact copy of a specific individual may turn to a cloning method that starts with a somatic cell taken from an adult organism (a somatic cell is a body cell, as opposed to a reproductive cell; *soma* is a Greek word for body). All cells descended from a fertilized egg inherit the same DNA. Thus, the DNA in each living cell of an individual is like a master blueprint that contains enough information to build an entirely new individual.

Even though a somatic cell contains all the DNA needed to produce a new individual, it will not automatically start dividing and form an embryo. The cell must first be tricked into rewinding its developmental clock. During development, cells in an embryo start using different subsets of their DNA. As they do, the cells become different in form and function, a process called **differentiation**. Differentiation is usually a one-way path in animal cells. Once a cell has become specialized, all of its descendant cells will be specialized the same way.

By the time a liver cell, muscle cell, or other differentiated cell forms, most of its DNA has been turned off, and is no longer used. To clone an adult, scientists transform one of its differentiated cells into an undifferentiated cell by turning the unused DNA back on. One way to do this is **somatic cell nuclear transfer** (**SCNT**), a laboratory procedure in which an unfertil-

CREDIT: (15) Courtesy of Cyagra, Inc., www.cyagra.com.

A Hero Dog's Golden Clones (revisited)

Today, Trakr's clones (right) are search and rescue dogs for Team Trakr Foundation, an international humanitarian organization that Symington established in 2010. The ability to clone dogs is a recent development, but the technique is not. SCNT (the technique used to produce Trakr's clones) first made headlines in 1997, when Scottish geneticist Ian Wilmut and his team produced a clone from the udder cell of an adult sheep. The cloned lamb was named Dolly. At first, Dolly looked and acted like a normal sheep. However, she died early, suffering from health problems that were most likely an outcome of being a clone.

SCNT has also been used to clone mice, rats, rabbits, pigs, cattle, goats, sheep, horses, mules, deer, cats, a camel, a ferret, a monkey, and a wolf. Until recently, many of the clones were unusually overweight or had enlarged organs. Cloned mice developed lung and liver problems, and almost all died prematurely. Cloned pigs tended to limp and have heart problems; some developed without a tail or, even worse, an anus. SCNT technology has improved so much in recent years that such problems are much less common in animals cloned today.

---

ized egg's nucleus is replaced with the nucleus of a donor's somatic cell (**FIGURE 8.15**). If all goes well, the egg's cytoplasm reprograms the transplanted DNA to direct the development of an embryo, which is then implanted into a surrogate mother. The animal born to the surrogate is the donor's clone, genetically identical with the donor of the nucleus.

SCNT is now a common practice among people who breed prized livestock. Among other benefits, many more offspring can be produced in a given time frame by cloning than by traditional breeding methods. Cloned animals have the same championship features as their DNA donors (**FIGURE 8.16**). Offspring can also be produced from a donor animal that is castrated or even dead.

As the techniques become routine, cloning humans is no longer only within the realm of science fiction. SCNT is already being used to produce human embryos for medical purposes, a practice called **therapeutic cloning**. Undifferentiated (stem) cells taken from the cloned human embryos are used to treat human patients and to study human diseases. For example, embryos created using cells from people with genetic heart defects are allowing researchers to study how the defect causes developing heart cells to malfunction. Such research may ultimately lead to treatments

FIGURE 8.16 Champion Holstein dairy cow Nelson's Estimate Liz (right) and her clone, Nelson's Estimate Liz II (left). Liz II was produced by somatic cell nuclear transfer in 2003. She had already begun to win championships by the time she was one year old.

for people who suffer from diseases that are otherwise incurable. (We return to the topic of stem cells and their potential medical benefits in Chapter 31.) Human cloning is not the intent of the research, but if it were, SCNT would indeed be the first step toward that end.

differentiation Process by which cells become specialized during development; occurs as different cells in an embryo begin to use different subsets of their DNA.
reproductive cloning Any of several technologies that produce genetically identical individuals.
somatic cell nuclear transfer (**SCNT**) Reproductive cloning method in which the DNA of an adult donor's body cell is transferred into an unfertilized egg.
therapeutic cloning The use of SCNT to produce human embryos for research purposes.

> ### TAKE-HOME MESSAGE 8.7
> #### How are clones of adult animals produced?
> ✔ Reproductive cloning technologies produce genetically identical individuals.
> ✔ The DNA inside a living cell contains all the information necessary to build a new individual.
> ✔ In somatic cell nuclear transfer (SCNT), the nucleus of a donor's somatic cell is transferred to an egg with no nucleus. The donor DNA directs the cell to develop into an embryo that is implanted into a surrogate. The individual born to the surrogate is a clone of the adult donor.

summary

Section 8.1 Making **clones**, or exact genetic copies, of adult animals is now a common practice. The techniques are improving, but their outcome is still unpredictable. The practice continues to raise ethical questions, particularly about cloning human cells.

Section 8.2 Eighty years of experimentation with cells and **bacteriophage** offered solid evidence that deoxyribonucleic acid (DNA), not protein, is the hereditary material of all life.

Section 8.3 Each DNA nucleotide has a five-carbon sugar (deoxyribose), three phosphate groups, and one of four nitrogen-containing bases after which the nucleotide is named: adenine, thymine, guanine, or cytosine. DNA is a polymer that consists of two strands of these nucleotides coiled into a double helix. Hydrogen bonding between the internally positioned bases holds the strands together. The bases pair in a consistent way: adenine with thymine (A–T), and guanine with cytosine (G–C). The order of bases along a strand of DNA—the **DNA sequence**— varies among species and among individuals, and this variation is the basis of life's diversity.

Section 8.4 The DNA of eukaryotes is typically divided among a number of **chromosomes** that differ in length and shape. In eukaryotic chromosomes, the DNA wraps around **histones** to form **nucleosomes**. When duplicated, a eukaryotic chromosome consists of two **sister chromatids** attached at a **centromere**.

Diploid cells have two of each type of chromosome. **Chromosome number** is the total number of chromosomes in a cell of a given species. A human body cell has twenty-three pairs of chromosomes. Members of a pair of **sex chromosomes** differ among males and females. Chromosomes that are the same in males and females are **autosomes**. Autosomes of a pair have the same length, shape, and centromere location. A **karyotype** is an image of an individual's complete set of chromosomes.

Section 8.5 **DNA replication** is the energy-intensive metabolic pathway in which a cell copies its chromosomes. For each molecule of DNA that is copied, two DNA molecules are produced; each is a duplicate of the parent. One strand of each molecule is new, and the other is parental; thus the name **semiconservative replication**.

During DNA replication, enzymes unwind the double helix. **Primers** base-pair with the exposed single strands of DNA, a process called **nucleic acid hybridization**. Starting at the primers, **DNA polymerase** enzymes use each strand as a template to assemble new, complementary strands of DNA from free nucleotides. Synthesis of one strand necessarily occurs discontinuously. **DNA ligase** seals any gaps to form continuous strands.

Section 8.6 Proofreading by DNA polymerases corrects most DNA replication errors as they occur. DNA damage by environmental agents, including ionizing and nonionizing radiation, free radicals, and some chemicals, can lead to replication errors because DNA polymerase does not copy damaged DNA very well. Most DNA damage is repaired before replication begins. Uncorrected replication errors become **mutations**: permanent changes in the nucleotide sequence of a cell's DNA. Cancer begins with mutations, but not all mutations are harmful.

Section 8.7 **Somatic cell nuclear transfer** (**SCNT**) and other types of **reproductive cloning** technologies produce genetically identical individuals (clones). SCNT using human cells is called **therapeutic cloning**. The DNA in each living cell contains all the information necessary to build a new individual. During development, cells of an embryo become specialized as they begin to use different subsets of their DNA (a process called **differentiation**).

self-quiz   Answers in Appendix VII

1. Which is *not* a nucleotide base in DNA?
   - a. adenine
   - c. glutamine
   - e. cytosine
   - b. guanine
   - d. thymine
   - f. All are in DNA.

2. What are the base-pairing rules for DNA?
   - a. A–G, T–C
   - b. A–C, T–G
   - c. A–T, G–C

3. Variation in _____ is the basis of variation in traits.
   - a. karyotype
   - c. the double helix
   - b. the DNA sequence
   - d. chromosome number

4. One species' DNA differs from others in its _____ .
   - a. nucleotides
   - c. sugar–phosphate backbone
   - b. DNA sequence
   - d. all of the above

5. In eukaryotic chromosomes, DNA wraps around _____ .
   - a. histone proteins
   - c. centromeres
   - b. nucleosomes
   - d. none of the above

6. Chromosome number _____ .
   - a. refers to a particular chromosome in a cell
   - b. is a characteristic feature of a species
   - c. is the number of autosomes in cells of a given type

7. Human body cells are diploid, which means _____ .
   - a. they are complete
   - b. they have two sets of chromosomes
   - c. they contain sex chromosomes

8. When DNA replication begins, _____ .
   - a. the two DNA strands unwind from each other
   - b. the two DNA strands condense for base transfers
   - c. old strands move to find new strands

9. DNA replication requires _____ .
   - a. DNA polymerase
   - c. primers
   - b. nucleotides
   - d. all are required

data analysis activities

Hershey–Chase Experiments The graph shown in **FIGURE 8.17** is reproduced from an original publication by Hershey and Chase. The data are from the experiments described in Section 8.2, in which bacteriophage DNA and protein were labeled with radioactive tracers and allowed to infect bacteria. The virus–bacteria mixtures were then whirled in a blender to dislodge any viral components attached to the exterior of the bacteria. Afterward, radioactivity from the tracers was measured.

1. Before blending, what percentage of each isotope, ^{35}S and ^{32}P, was extracellular (outside the bacteria)?

2. After 4 minutes in the blender, what percentage of each isotope was extracellular?

3. How did the researchers know that the radioisotopes in the fluid came from outside of the bacterial cells and not from bacteria that had been broken apart by whirling in the blender?

4. The extracellular concentration of which isotope increased the most with blending?

5. Do these results imply that viruses inject DNA or protein into bacteria? Why or why not?

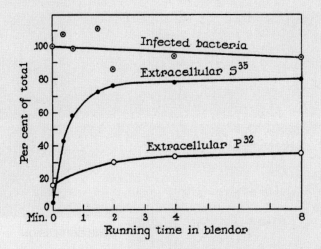

FIGURE 8.17 Detail of Alfred Hershey and Martha Chase's 1952 publication describing their experiments with bacteriophage. "Infected bacteria" refers to the percentage of bacteria that survived the blender.

10. Energy that drives the attachment of a nucleotide to the end of a growing strand of DNA comes from _____ .
   a. ATP   c. the nucleotide
   b. DNA polymerase   d. a and c

11. The phrase "5′ to 3′" refers to the _____ .
   a. timing of DNA replication
   b. directionality of DNA synthesis
   c. number of phosphate groups

12. After DNA replication, a eukaryotic chromosome _____ .
   a. consists of two sister chromatids
   b. has a characteristic X shape
   c. is constricted at the centromere
   d. all of the above

13. All mutations _____ .
   a. cause cancer   c. are caused by radiation
   b. lead to evolution   d. change the DNA sequence

14. _____ is an example of reproductive cloning.
   a. Somatic cell nuclear transfer (SCNT)
   b. Multiple offspring from the same pregnancy
   c. Artificial embryo splitting
   d. a and c

15. Match the terms appropriately.
   ___ bacteriophage   a. nitrogen-containing base,
   ___ clone   sugar, phosphate group(s)
   ___ nucleotide   b. copy of an organism
   ___ diploid   c. does not determine sex
   ___ DNA ligase   d. injects DNA
   ___ DNA polymerase   e. seals breaks in a DNA strand
   ___ autosome   f. can cause cancer
   ___ mutation   g. two chromosomes of each type
                         h. adds nucleotides to a growing
                         DNA strand

critical thinking

1. Show the complementary strand of DNA that forms on this template DNA fragment during replication:

      5′—GGTTTCTTCAAGAGA—3′

2. Woolly mammoths have been extinct for about 10,000 years, but we often find their well-preserved remains in Siberian permafrost. Research groups are now planning to use SCNT to resurrect these huge elephant-like mammals. No mammoth eggs have been recovered so far, so elephant eggs would be used instead. An elephant would also be the surrogate mother for the resulting embryo. The researchers may try a modified SCNT technique used to clone a mouse that had been dead and frozen for sixteen years. Ice crystals that form during freezing break up cell membranes, so cells from the frozen mouse were in bad shape. Their DNA was transferred into donor mouse eggs, and cells from the resulting embryos were fused with mouse stem cells. Four healthy clones were born from the hybrid embryos. What are some of the pros and cons of cloning an extinct animal?

3. Xeroderma pigmentosum is an inherited disorder characterized by rapid formation of many skin sores that develop into cancers. All forms of radiation trigger these symptoms, including fluorescent light, which contains UV light in the range of 320 to 400 nm. In most affected individuals, at least one of nine particular proteins is missing or defective. What is the collective function of these proteins?

LEARNING ROADMAP

Your knowledge of base pairing (Section 8.3) and chromosomes (8.4) will help you understand how cells use nucleic acids (3.8) to build proteins (3.6). This chapter revisits hydrophobicity (2.5), hemoglobin (3.2), pathogenic bacteria (4.1), organelles (4.5, 4.7), free radicals and cofactors (5.6), enzyme function (3.3, 5.4), DNA replication (8.5), and mutations (8.6).

GENE EXPRESSION

The information encoded in DNA occurs in subsets called genes. Converting genetic information to a protein product involves RNA, and it occurs in two stages: transcription and translation.

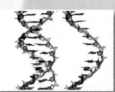

DNA TO RNA: TRANSCRIPTION

During transcription, a gene region in one strand of DNA serves as a template for assembling a strand of RNA. In eukaryotes, a new RNA is modified before leaving the nucleus.

RNA

A messenger RNA carries a gene's protein-building instructions as a string of three-nucleotide codons. Transfer RNA and ribosomal RNA translate those instructions into a protein.

RNA TO PROTEIN: TRANSLATION

During translation, amino acids are assembled into a polypeptide in the order determined by the sequence of codons in an mRNA.

ALTERED PROTEINS

Mutations that change a gene's DNA sequence alter the instructions it encodes. A protein built using altered instructions may function improperly or not at all.

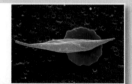

What you learn in this chapter about genes will be the foundation for concepts of gene expression (Chapter 10), inheritance (Chapters 13 and 14), and genetic engineering (Chapter 15). Chapters 16 and 17 will show you how mutations are the raw material of natural selection and other processes of evolution. You will also revisit hemoglobin, sickle-cell anemia, and the circulatory system in Chapter 36, and immunity in Chapter 37.

A dose of ricin as small as a few grains of salt can kill an adult human, and there is no antidote. Ricin is a protein that deters beetles, birds, mammals, and other animals from eating seeds of the castor-oil plant (*Ricinus communis*), which grows wild in tropical regions and is widely cultivated. Castor-oil seeds are the source of castor oil, an ingredient in plastics, cosmetics, paints, soaps, polishes, and many other items. After the oil is extracted from the seeds, the ricin typically is discarded along with the leftover seed pulp.

Ricin's lethal effects were known as long ago as 1888, but using it as a weapon is now banned by most countries under the Geneva Protocol. However, controlling the production of ricin is impossible, because no special skills or equipment are required to manufacture the toxin from easily obtained raw materials. Thus, ricin appears periodically in the news as a tool of criminals. For example, a Texas actress sent ricin-laced letters to President Obama and the mayor of New York City in 2013. Perhaps the most famous example occurred in 1978, at the height of the Cold War when defectors from countries under Russian control were targets for assassination. Bulgarian journalist Georgi Markov had defected to England and was working for the BBC. As he made his way to a bus stop on a London street, an assassin used a modified umbrella to fire a tiny, ricin-laced ball into Markov's leg. Markov died in agony three days later.

Ricin is called a ribosome-inactivating protein (RIP) because it inactivates ribosomes, the organelles that assemble amino acids into proteins. Other RIPs are made by some bacteria, mushrooms, algae, and many plants (including food crops such as tomatoes, barley, and spinach). Most of these proteins are not particularly toxic in humans because they do not cross intact cell membranes very well. Those that do, including ricin, have a domain that binds tightly to plasma membrane glycolipids or glycoproteins (**FIGURE 9.1**). Binding causes the cell to take up the RIP by endocytosis (Section 5.10). Once inside the cell, the second domain of the RIP—an enzyme—begins to inactivate ribosomes. One molecule of ricin can inactivate more than 1,000 ribosomes per minute. If enough ribosomes are affected, protein synthesis grinds to a halt. Proteins are critical to all life processes, so cells that cannot make them die quickly.

Fortunately, few people actually encounter ricin. Other toxic RIPs are more prevalent. Bracelets made from beautiful seeds were recalled from stores in 2011 after a botanist recognized the seeds as jequirity beans.

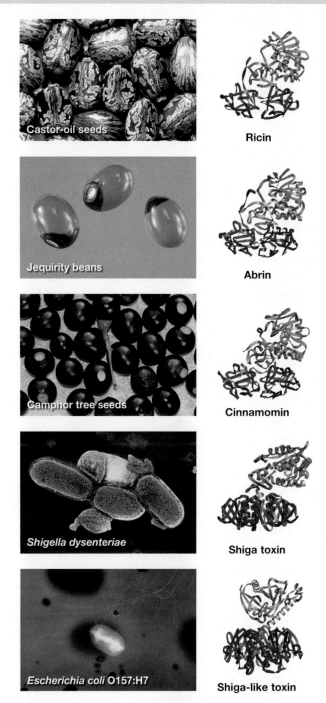

Castor-oil seeds
Ricin

Jequirity beans
Abrin

Camphor tree seeds
Cinnamomin

Shigella dysenteriae
Shiga toxin

Escherichia coli O157:H7
Shiga-like toxin

FIGURE 9.1 Lethal lineup: a few toxic ribosome-inactivating proteins (RIPs) and their sources. One of the two chains of a toxic RIP (brown) helps the molecule cross a cell's plasma membrane; the other (gold) is an enzyme that inactivates ribosomes.

These beans contain abrin, an RIP even more toxic than ricin. Shiga toxin made by *Shigella dysenteriae* bacteria causes dysentery. Some strains of *E. coli* bacteria make Shiga-like toxin, an RIP that is the source of intestinal illness (Section 4.1).

9.2 DNA, RNA, and Gene Expression

✔ Transcription converts information in a gene to RNA; translation converts information in an mRNA to protein.

You learned in Chapter 8 that chromosomes are like a set of books that provide instructions for building and operating an individual. You already know the alphabet used to write those books: the four letters A, T, G, and C, for the four nucleotides in DNA—adenine, thymine, guanine, and cytosine. In this chapter, we investigate the nature of information represented by the sequence of nucleotides in a DNA molecule, and how a cell uses that information.

DNA to RNA

Information encoded within a chromosome's DNA sequence occurs in hundreds or thousands of units called genes. The DNA sequence of a **gene** encodes (contains instructions for building) an RNA or protein

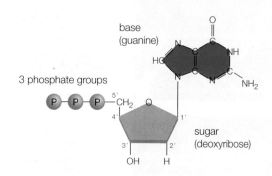

A **The DNA nucleotide guanine (G)**, or deoxyguanosine triphosphate, one of the four nucleotides in DNA. The other nucleotides—adenine, thymine, and cytosine—differ only in their component bases (blue). Three of the four bases in RNA nucleotides are identical to the bases in DNA nucleotides.

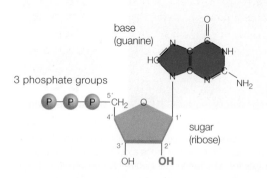

B **The RNA nucleotide guanine (G)**, or guanosine triphosphate. The only difference between the DNA and RNA versions of guanine (or adenine, or cytosine) is that RNA has a hydroxyl group (shown in red) at the 2′ carbon of the sugar.

FIGURE 9.2 ▶**Animated** Comparing nucleotides of DNA and RNA.

product. Converting the information encoded by a gene into a product starts with RNA synthesis, or transcription. During **transcription**, enzymes use the gene's DNA sequence as a template to assemble a strand of RNA:

$$\text{DNA} \xrightarrow{\textit{transcription}} \text{RNA}$$

Most of the RNA inside cells occurs as a single strand that is similar in structure to a single strand of DNA. For example, both are chains of four kinds of nucleotides. Like a DNA nucleotide, an RNA nucleotide has three phosphate groups, a sugar, and one of four bases. However, the sugar in an RNA nucleotide is a ribose, which differs just a bit from deoxyribose, the sugar in a DNA nucleotide (**FIGURE 9.2**). Three bases (adenine, cytosine, and guanine) occur in both RNA and DNA nucleotides, but the fourth base differs between the two molecules (**FIGURE 9.3**). In DNA, the fourth base is thymine (T); in RNA, it is uracil (U).

Despite these small differences in structure, DNA and RNA have very different functions. DNA's important but only role is to store a cell's genetic information. By contrast, a cell makes several kinds of RNAs, each with a different function. MicroRNAs are important in gene control, which is the subject of the next chapter. Three other types of RNA have roles in protein synthesis. **Ribosomal RNA (rRNA)** is the main component of ribosomes (Section 4.4), which assemble amino acids into polypeptide chains (Section 3.6). **Transfer RNA (tRNA)** delivers amino acids to ribosomes, one by one, in the order specified by a **messenger RNA (mRNA)**.

RNA to Protein

Messenger RNA was named for its function as the "messenger" between DNA and protein. An mRNA carries a protein-building message that is encoded by sets of three nucleotides, "genetic words" that occur one after another along its length. Like the words of a sentence, a series of these genetic words can form a meaningful parcel of information—in this case, the sequence of amino acids of a protein.

By the process of **translation**, the protein-building information in an mRNA is decoded (translated) into a sequence of amino acids. The result is a polypeptide chain that twists and folds into a protein:

$$\text{mRNA} \xrightarrow{\textit{translation}} \text{protein}$$

Transcription and translation are both part of **gene expression**, the multistep process by which information

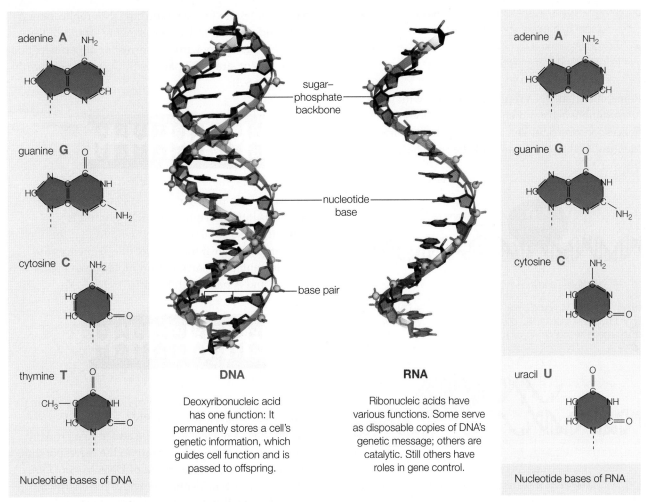

adenine **A**

guanine **G**

cytosine **C**

thymine **T**

Nucleotide bases of DNA

DNA

Deoxyribonucleic acid has one function: It permanently stores a cell's genetic information, which guides cell function and is passed to offspring.

sugar–
phosphate
backbone

nucleotide
base

base pair

RNA

Ribonucleic acids have various functions. Some serve as disposable copies of DNA's genetic message; others are catalytic. Still others have roles in gene control.

adenine **A**

guanine **G**

cytosine **C**

uracil **U**

Nucleotide bases of RNA

FIGURE 9.3 Comparing the structure and function of DNA and RNA.

in a gene guides the assembly of an RNA or protein product. Expression of genes that encode RNA products (such as tRNA and rRNA) involves transcription. Expression of genes that encode protein products involves both transcription and translation:

$$\text{DNA} \xrightarrow{\textit{transcription}} \textbf{mRNA} \xrightarrow{\textit{translation}} \textbf{protein}$$

gene A part of a chromosome that encodes an RNA or protein product in its DNA sequence.
gene expression Process by which the information in a gene guides assembly of an RNA or protein product.
messenger RNA (mRNA) RNA that has a protein-building message.
ribosomal RNA (rRNA) RNA that becomes part of ribosomes.
transcription Process by which enzymes assemble an RNA using the nucleotide sequence of a gene as a template.
transfer RNA (tRNA) RNA that delivers amino acids to a ribosome during translation.
translation Process by which a polypeptide chain is assembled from amino acids in the order specified by an mRNA.

The DNA sequence of a cell's chromosome(s) contains all the information it needs to make the molecules of life. Each gene encodes an RNA, and RNAs interact to assemble proteins from amino acids (Section 3.6). Proteins (enzymes, in particular) assemble lipids and carbohydrates, replicate DNA, make RNA, and perform many other functions that keep the cell alive.

TAKE-HOME MESSAGE 9.2
What is the nature of the information carried by a DNA sequence?

✔ Information in a DNA sequence occurs in units that are called genes.

✔ A cell uses the information encoded in a gene to make an RNA or protein product, a process called gene expression.

✔ The DNA sequence of a gene is transcribed into RNA.

✔ Information carried by a messenger RNA (mRNA) is translated into a protein.

9.3 Transcription: DNA to RNA

✔ RNA polymerase links RNA nucleotides into a chain, in the order dictated by the sequence of a gene.

✔ A new RNA strand is complementary in sequence to the DNA strand from which it was transcribed.

Remember that DNA replication begins with one DNA double helix and ends with two DNA double helices (Section 8.5). The two double helices are identical to the parent molecule because base-

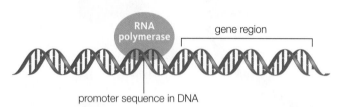

promoter sequence in DNA

❶ The enzyme RNA polymerase binds to a promoter in the DNA. The binding positions the polymerase near a gene. Only the DNA strand complementary to the gene sequence will be translated into RNA.

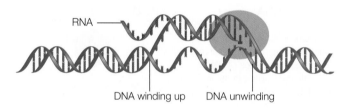

DNA winding up   DNA unwinding

❷ RNA polymerase begins to move along the gene and unwind the DNA. As it does, it links RNA nucleotides in the order specified by the sequence of the complementary (noncoding) DNA strand. The DNA winds up again after the polymerase passes.

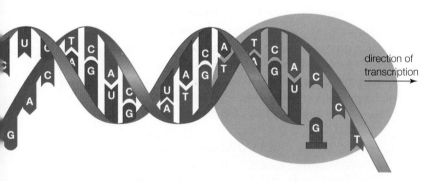

direction of transcription

❸ Zooming in on the site of transcription, we see that RNA polymerase covalently bonds successive nucleotides into a new strand of RNA. The new RNA is complementary in sequence to the template DNA strand, so it is an RNA copy of the gene.

FIGURE 9.4 ▶Animated Transcription. By this process, a strand of RNA is assembled from nucleotides according to a template: a gene region in DNA. **FIGURE IT OUT** After the guanine, what is the next nucleotide that will be added to this growing strand of RNA?

Answer: Another guanine (G)

pairing rules are followed during DNA replication. A nucleotide can be added to a growing strand of DNA only if it base-pairs with the corresponding nucleotide of the parent strand: G pairs with C, and A pairs with T (Section 8.3):

The same base-pairing rules also govern RNA synthesis in transcription. An RNA strand is structurally so similar to a DNA strand that the two can base-pair if their nucleotide sequences are complementary. In such hybrid molecules, G pairs with C, and A pairs with U (uracil):

During transcription, a strand of DNA acts as a template upon which a strand of RNA is assembled from nucleotides. A nucleotide can be added to a growing RNA only if it is complementary to the corresponding nucleotide of the parent strand of DNA. As in DNA replication, each nucleotide provides the energy for its own attachment to the end of a growing strand.

Transcription is also similar to DNA replication in that one strand of a nucleic acid serves as a template for synthesis of another. However, in contrast with DNA replication, only part of one DNA strand, not the whole molecule, is used as a template for transcription. The enzyme **RNA polymerase**, not DNA polymerase, adds nucleotides to the end of a growing RNA. Also, transcription produces a single strand of RNA, not two DNA double helices.

In eukaryotic cells, transcription occurs in the nucleus; in prokaryotes, it occurs in cytoplasm. In both cases, the process begins at a regulatory site called a **promoter** (**FIGURE 9.4 ❶**), a short DNA sequence close to a gene and upstream from it (in the 5′ direction). A promoter is recognized by DNA-binding proteins that in turn bind RNA polymerase. After the polymerase binds, it moves along the DNA toward the gene and over it, unwinding the double helix just a bit so it can "read" the base sequence of the noncoding strand (the strand complementary to the gene) ❷. As it does, the polymerase joins free RNA nucleotides into a chain,

CREDITS: (4, in text) © Cengage Learning.

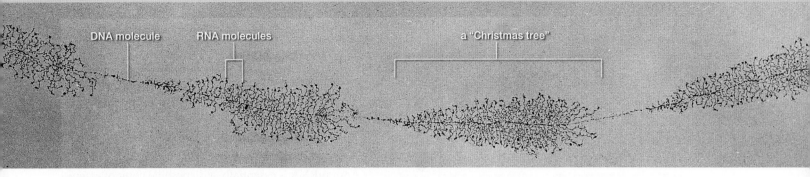

FIGURE 9.5 Typically, many RNA polymerases simultaneously transcribe the same gene, producing a structure often called a "Christmas tree" after its shape. Here, four genes next to one another on the same chromosome are being transcribed.

FIGURE IT OUT Are the polymerases transcribing this DNA molecule moving from left to right or from right to left?

Answer: Left to right (the RNAs get longer as the polymerases move along the DNA)

in the order dictated by that base sequence. As in DNA replication, the synthesis is directional: An RNA polymerase adds nucleotides only to the 3′ end of the growing strand of RNA.

When the polymerase reaches the end of the gene region, the DNA and the new RNA are released. RNA polymerase follows base-pairing rules, so the new RNA is complementary in base sequence to the DNA strand that served as its template ❸. It is an RNA copy of a gene, the same way that a paper transcript of a conversation carries the same information in a different format. Typically, many polymerases transcribe a particular gene region at the same time, so many new RNA strands can be produced quickly (FIGURE 9.5).

Post-Transcriptional Modifications

Just as a dressmaker may snip off loose threads or add bows to a dress before it leaves the shop, so do eukaryotic cells tailor their RNA before it leaves the nucleus. Consider that most eukaryotic genes contain intervening sequences called **introns**. Intron sequences are removed in chunks from a newly transcribed RNA before it leaves the nucleus. Sequences that remain in the RNA after this process are called **exons** (FIGURE 9.6). Exons can be rearranged and spliced together in different combinations—a process called **alternative splicing**—so one gene may encode two or more versions of the same product.

---

alternative splicing Post-transcriptional RNA modification process in which some exons are removed or joined in different combinations.
exon Nucleotide sequence that remains in an RNA after post-transcriptional modification.
intron Nucleotide sequence that intervenes between exons and is removed during post-transcriptional modification.
promoter DNA sequence that is a site where transcription begins.
RNA polymerase Enzyme that carries out transcription.

A newly transcribed RNA that will become an mRNA is further tailored after splicing. Enzymes attach a modified guanine "cap" to the 5′ end; later, this cap will help the finished mRNA bind to a ribosome. Between 50 and 300 adenines are also added to the 3′ end of a new mRNA. This poly-A tail is a signal that allows an mRNA to be exported from the nucleus, and as you will see in Chapter 10, it helps regulate the timing and duration of the mRNA's translation.

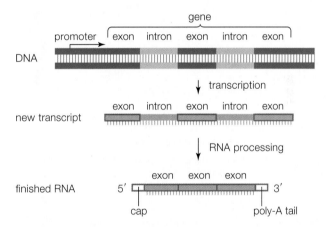

FIGURE 9.6 ▶Animated Post-transcriptional modification of RNA. Introns are removed and exons spliced together. Messenger RNAs also get a poly-A tail and modified guanine "cap."

TAKE-HOME MESSAGE 9.3
How is RNA assembled?

✔ Transcription produces an RNA from a gene.

✔ RNA polymerase uses a gene region in a chromosome as a template to assemble a strand of RNA. The new strand is an RNA copy of the gene from which it was transcribed.

✔ Post-transcriptional modification of RNA occurs in the nucleus of eukaryotes.

9.4 RNA and the Genetic Code

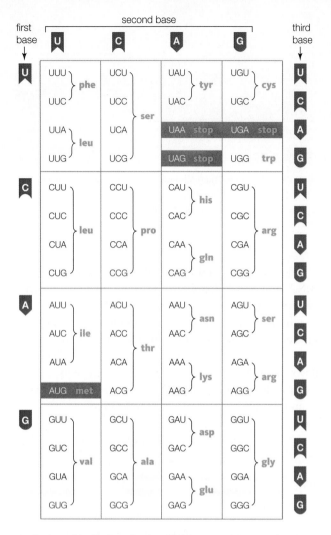

A Codon table. Each codon in mRNA is a set of three nucleotide bases. The left column lists a codon's first base, the top row lists the second, and the right column lists the third.

Sixty-one of the triplets encode amino acids; one of those, AUG, both codes for methionine and serves as a signal to start translation. Three codons are signals that stop translation.

ala alanine (A)	**gly** glycine (G)	**pro** proline (P)
arg arginine (R)	**his** histidine (H)	**ser** serine (S)
asn asparagine (N)	**ile** isoleucine (I)	**thr** threonine (T)
asp aspartic acid (D)	**leu** leucine (L)	**trp** tryptophan (W)
cys cysteine (C)	**lys** lysine (K)	**tyr** tyrosine (Y)
glu glutamic acid (E)	**met** methionine (M)	**val** valine (V)
gln glutamine (Q)	**phe** phenylalanine (F)	

B Names and abbreviations of the 20 naturally occurring amino acids specified by the genetic code (**A**).

FIGURE 9.7 The genetic code.

FIGURE IT OUT Which codons specify the amino acid lysine?

Answer: AAA and AAG

✔ Nucleotide base triplets in an mRNA encode a protein-building message. Ribosomal RNA and transfer RNA translate that message into a polypeptide chain.

DNA stores heritable information about proteins, but making those proteins requires messenger RNA (mRNA), transfer RNA (tRNA), and ribosomal RNA (rRNA). The three types of RNA interact to translate DNA's information into a protein.

An mRNA is essentially a disposable copy of a gene. Its job is to carry the gene's protein-building information to the other two types of RNA during translation. That protein-building information consists of a linear sequence of genetic "words" spelled with an alphabet of the four nucleotide bases A, C, G, and U. Each of the genetic "words" carried by an mRNA is three bases long, and each is a code—a **codon**—for a particular amino acid. With four possible bases in each of the three positions of a codon, there are a total of sixty-four (or 4^3) mRNA codons. Collectively, the sixty-four codons constitute the **genetic code** (**FIGURE 9.7**). The sequence of bases in a triplet determines which amino acid the codon specifies. For instance, the codon UUU codes for the amino acid phenylalanine (phe), and UUA codes for leucine (leu).

Codons occur one after another along the length of an mRNA. When an mRNA is translated, the order of its codons determines the order of amino acids in the resulting polypeptide. Thus, the DNA sequence of a gene is transcribed into the nucleotide sequence of an mRNA, which is in turn translated into an amino acid sequence (**FIGURE 9.8**).

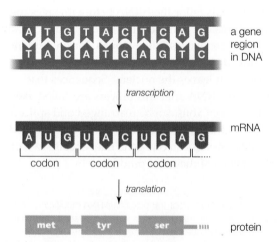

FIGURE 9.8 Example of the correspondence between DNA, RNA, and protein. A gene region in a strand of chromosomal DNA is transcribed into an mRNA, and the codons of the mRNA specify a chain of amino acids—a protein.

CREDITS: (7, 8) © Cengage Learning.

With a few exceptions, twenty naturally occurring amino acids are encoded by the sixty-four codons in the genetic code. Some amino acids are specified by more than one codon. For instance, the amino acid tyrosine (tyr) is specified by two codons: UAU and UAC. Other codons signal the beginning and end of a protein-coding sequence. The first AUG in an mRNA usually serves as the signal to start translation. AUG is the codon for methionine, so methionine is always the first amino acid in new polypeptides of such organisms. The codons UAA, UAG, and UGA do not specify an amino acid. These are signals that stop translation, so they are called stop codons. A stop codon marks the end of the protein-coding sequence in an mRNA.

The genetic code is highly conserved, which means that most organisms use the same code and probably always have. Bacteria, archaea, and some protists have a few codons that differ from the eukaryotic code, as do mitochondria and chloroplasts—a clue that led to a theory of how these two organelles evolved (we return to this topic in Section 19.6).

Ribosomes interact with transfer RNAs (tRNAs) to translate the sequence of codons in an mRNA into a polypeptide. A ribosome has two subunits, one large and one small (**FIGURE 9.9**). Both subunits consist mainly of rRNA, with some associated structural proteins. During translation, a large and a small ribosomal subunit converge as an intact ribosome on an mRNA. Ribosomal RNA is one example of RNA with enzymatic activity: During translation, rRNA catalyzes formation of peptide bonds between amino acids.

Amino acids are delivered to ribosomes by tRNAs. Each tRNA has two attachment sites. The first is an **anticodon**, which is a triplet of nucleotides that base-pairs with an mRNA codon (**FIGURE 9.10A**). The other attachment site binds to an amino acid—the one specified by the codon. Transfer RNAs with different anticodons carry different amino acids.

During translation, tRNAs deliver amino acids to a ribosome, one after the next in the order specified by the codons in an mRNA (**FIGURE 9.10B**). As the amino acids are delivered, the ribosome joins them via peptide bonds into a new polypeptide (Section 3.6). Thus, the order of codons in an mRNA—DNA's protein-building message—becomes translated into a new protein.

---

anticodon In a tRNA, set of three nucleotides that base-pairs with an mRNA codon.
codon In an mRNA, a nucleotide base triplet that codes for an amino acid or stop signal during translation.
genetic code Complete set of sixty-four mRNA codons.

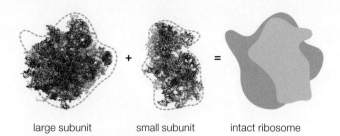

large subunit   small subunit   intact ribosome

FIGURE 9.9 ▶**Animated** Ribosome structure. Each intact ribosome consists of a large and a small subunit. The structural protein components of the two subunits are shown in green; the catalytic rRNA components, in brown.

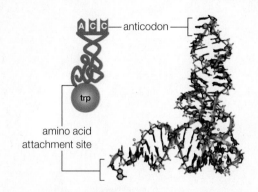

A Icon and model of the tRNA that carries the amino acid tryptophan. Each tRNA's anticodon is complementary to an mRNA codon. Each also carries the amino acid specified by that codon.

B During translation, tRNAs dock at an intact ribosome (for clarity, only the small subunit is shown, in tan). Here, the anticodons of two tRNAs have base-paired with complementary codons on an mRNA (red). The amino acids they carry are not shown, for clarity.

FIGURE 9.10 tRNA structure.

TAKE-HOME MESSAGE 9.4

What roles do mRNA, tRNA, and rRNA play during translation?

✔ The sequence of nucleotide triplets (codons) in an mRNA encodes a gene's protein-building message.

✔ The genetic code consists of sixty-four codons. Three are signals that stop translation; the remaining codons specify an amino acid. In most mRNAs, the first occurrence of the codon that specifies methionine is a signal to begin translation.

✔ Ribosomes, which consist of two subunits of rRNA and proteins, assemble amino acids into polypeptide chains.

✔ A tRNA has an anticodon complementary to an mRNA codon, and a binding site for the amino acid specified by that codon. During translation, tRNAs deliver amino acids to ribosomes.

9.5 Translation: RNA to Protein

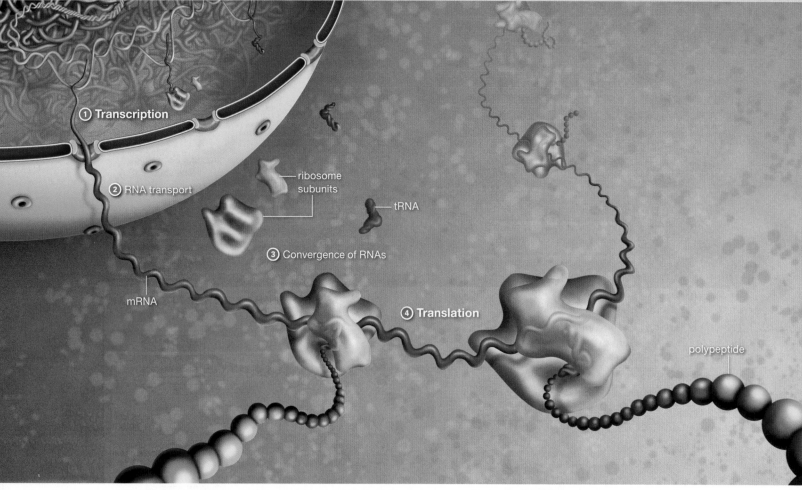

① Transcription

② RNA transport

ribosome
subunits

tRNA

③ Convergence of RNAs

④ Translation

mRNA

polypeptide

FIGURE 9.11 Overview of translation in a eukaryotic cell. RNAs are transcribed in the nucleus, then transported into the cytoplasm through nuclear pores. Translation begins when ribosomal subunits and tRNA converge on an mRNA in cytoplasm.

✔ Translation converts the information carried by an mRNA into a new polypeptide chain.

✔ The order of codons in an mRNA determines the order of amino acids in the polypeptide chain translated from it.

Translation, the second part of protein synthesis, proceeds in three stages: initiation, elongation, and termination. In all cells, translation occurs in cytoplasm. Cytoplasm contains many free amino acids, tRNAs, and ribosomal subunits available to participate in protein synthesis. In a eukaryotic cell, RNAs that carry out transcription are produced in the nucleus (**FIGURE 9.11 ❶**), then transported through nuclear pores into the cytoplasm ❷. Translation is initiated when ribosomal subunits and tRNAs converge on an mRNA. First, a small ribosomal subunit binds to an mRNA, and the anticodon of a special tRNA called an initiator base-pairs with the mRNA's first AUG codon. Then, a large ribosomal subunit joins the small subunit ❸.

The complex of molecules is now ready to carry out protein synthesis. In the elongation stage, the intact ribosome moves along the mRNA and assembles a polypeptide chain ❹. **FIGURE 9.12** shows how this works. Initiator tRNAs carry methionine, so the first amino acid of the new polypeptide chain is a methionine. Another tRNA joins the complex when its anticodon base-pairs with the second codon in the mRNA ❺. This tRNA brings with it the second amino acid. The ribosome then catalyzes formation of a peptide bond between the first two amino acids ❻.

As the ribosome moves to the next codon, it releases the first tRNA. Another tRNA brings the third amino acid to the complex as its anticodon base-pairs with the third codon of the mRNA ❼. The ribosome catalyzes the formation of a peptide bond between the second and third amino acids ❽.

The second tRNA is released and the ribosome moves to the next codon. Another tRNA brings the

CREDIT: (11) © Cengage Learning.

fourth amino acid to the complex as its anticodon base-pairs with the fourth codon of the mRNA ❾. The ribosome catalyzes the formation of a peptide bond between the third and fourth amino acids.

The new polypeptide chain continues to elongate as amino acids are delivered by successive tRNAs. Translation terminates when the ribosome reaches a stop codon in the mRNA ❿. The mRNA and the polypeptide detach from the ribosome, and the ribosomal subunits separate from each other. Translation is now complete. The new polypeptide joins the pool of proteins in cytoplasm, or it enters rough ER (Section 4.7).

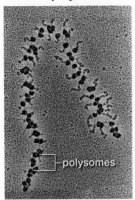
polysomes

Many ribosomes may simultaneously translate the same mRNA, in which case they are called polysomes (left). In bacteria and archaea, transcription and translation both occur in the cytoplasm, and these processes are closely linked in time and space. Translation begins before transcription ends, so a transcription "Christmas tree" is often decorated with polysome "balls."

Given that many polypeptides can be translated from one mRNA, why would any cell also make many copies of an mRNA? Compared with DNA, RNA is not very stable. An mRNA may last only a few minutes in cytoplasm before enzymes disassemble it. The fast turnover allows cells to adjust their protein synthesis quickly in response to changing needs.

Translation is energy intensive. Most of that energy is provided in the form of phosphate-group transfers from the RNA nucleotide GTP (shown in **FIGURE 9.2B**) to molecules that help the ribosome move from one codon to the next along the mRNA.

TAKE-HOME MESSAGE 9.5

How is mRNA translated into protein?

✔ Translation is an energy-requiring process that converts a sequence of codons in an mRNA into a polypeptide.

✔ During initiation, an mRNA joins with an initiator tRNA and two ribosomal subunits.

✔ During elongation, amino acids are delivered to the ribosome by tRNAs in the order dictated by successive mRNA codons. As amino acids arrive, the ribosome joins each to the end of the polypeptide.

✔ Termination occurs when the ribosome encounters a stop codon in the mRNA. The mRNA and the polypeptide are released, and the ribosome disassembles.

---

❺ Ribosomal subunits and an initiator tRNA converge on an mRNA. A second tRNA binds to the second codon.

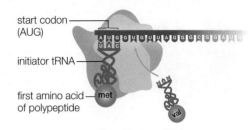

start codon (AUG)
initiator tRNA
first amino acid of polypeptide

❻ A peptide bond forms between the first two amino acids.

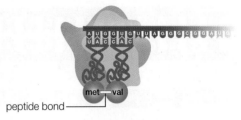

peptide bond

❼ The first tRNA is released and the ribosome moves to the next codon. A third tRNA binds to the third codon.

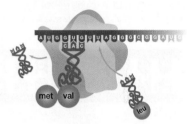

❽ A peptide bond forms between the second and third amino acids.

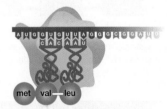

❾ The second tRNA is released and the ribosome moves to the next codon. A fourth tRNA brings the next amino acid to be added to the polypeptide chain.

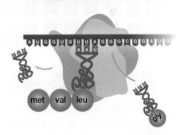

❿ The process repeats until the ribosome encounters a stop codon. Then, the new polypeptide is released and the ribosomal subunits separate.

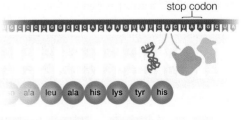

stop codon

ala leu ala his lys tyr his

FIGURE 9.12 ▶Animated Translation. Translation begins when ribosomal subunits and an initiator tRNA converge on an mRNA. Then, tRNAs deliver amino acids in the order dictated by successive codons in the mRNA. The ribosome links the amino acids together as it moves along the mRNA, so a polypeptide forms and elongates. Translation ends when the ribosome reaches a stop codon.

9.6 Mutated Genes and Their Protein Products

A Hemoglobin, an oxygen-binding protein in red blood cells. This protein consists of four polypeptides: two alpha globins (blue) and two beta globins (green). Each globin has a pocket that cradles a heme (red). Oxygen molecules bind to the iron atom at the center of each heme.

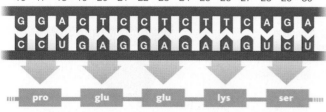

B Part of the DNA (blue), mRNA (brown), and amino acid sequence of human beta globin. Numbers indicate nucleotide position in the mRNA.

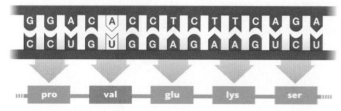

C A base-pair substitution replaces a thymine with an adenine. When the altered mRNA is translated, valine replaces glutamic acid as the sixth amino acid. Hemoglobin with this form of beta globin is called HbS, or sickle hemoglobin.

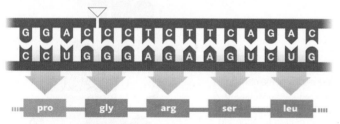

D A base-pair deletion shifts the reading frame for the rest of the mRNA, so a completely different protein product forms. The mutation shown results in a defective beta globin. The outcome is beta thalassemia, a genetic disorder in which a person has an abnormally low amount of hemoglobin.

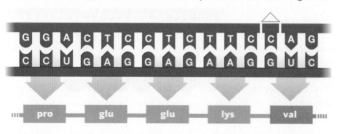

E An insertion of one nucleotide causes the reading frame for the rest of the mRNA to shift. The protein translated from this mRNA is too short and does not assemble correctly into hemoglobin molecules. As in **D**, the outcome is beta thalassemia.

FIGURE 9.13 ▶**Animated** Examples of mutations.

✔ If the nucleotide sequence of a gene changes, it may result in an altered gene product, with harmful effects.

Mutations, remember, are permanent changes in the DNA sequence of a chromosome (Section 8.6). A mutation in which one nucleotide and its partner are replaced by a different base pair is called a **base-pair substitution**. Other mutations involve the loss of one or more base pairs (a **deletion**) or the addition of extra base pairs (an **insertion**).

Mutations are relatively uncommon events in a normal cell. Consider that the chromosomes in a diploid human cell collectively consist of about 6.5 billion nucleotides, any of which may become mutated each time that cell divides. On average, about 175 nucleotides do change during DNA replication. However, only about 3 percent of the cell's DNA encodes protein products, so there is a low probability that any of those mutations will be in a protein-coding region.

When a mutation does occur in a protein-coding region, the redundancy of the genetic code offers the cell a margin of safety. For example, a mutation that changes a CCU codon to CCC may have no further effect, because both of these codons specify proline. Other mutations may change an amino acid in a protein, or result in a premature stop codon that shortens it.

Mutations that alter a protein can have drastic effects on an organism. Consider the effects of such mutations on hemoglobin, an oxygen-transporting protein in your red blood cells (Section 3.2). Hemoglobin's structure allows it to bind and release oxygen. In adult humans, a hemoglobin molecule consists of four polypeptides called globins: two alpha globins and two beta globins (**FIGURE 9.13A**). Each globin folds around a heme, a cofactor with an iron atom at its center (Section 5.6). Oxygen molecules bind to hemoglobin at those iron atoms.

Mutations in the genes for alpha or beta globin cause a condition called anemia, in which a person's blood is deficient in red blood cells or in hemoglobin. Both outcomes limit the blood's ability to carry oxygen, and the resulting symptoms can range from mild to life-threatening.

Sickle-cell anemia, a type of anemia that is most common in people of African ancestry, arises because of a base-pair substitution in the beta globin gene. The substitution causes the body to produce a version of beta globin in which the sixth amino acid is valine instead of glutamic acid (**FIGURE 9.13B,C**). Hemoglobin assembled with this altered beta globin chain is called sickle hemoglobin, or HbS.

Unlike glutamic acid, which carries a negative charge, valine carries no charge. As a result of that one base-pair substitution, a tiny patch of the beta globin polypeptide that is normally hydrophilic becomes hydrophobic. This change slightly alters hemoglobin behavior. Under certain conditions, HbS molecules stick together and form large, rodlike clumps. Red blood cells that contain the clumps become distorted into a crescent (sickle) shape (**FIGURE 9.14**). Sickled cells clog tiny blood vessels, thus disrupting blood circulation throughout the body. Over time, repeated episodes of sickling can damage organs and eventually cause death.

A different type of anemia, beta thalassemia, is caused by the deletion of the twentieth nucleotide in the coding region of the beta globin gene (**FIGURE 9.13D**). Like many other deletions, this one causes the reading frame of the mRNA codons to shift. A frameshift usually has drastic consequences because it garbles the genetic message, just as incorrectly grouping a series of letters garbles the meaning of a sentence:

The fat cat ate the sad rat
T hef atc ata tet hes adr at

The frameshift caused by the beta globin deletion results in a polypeptide that differs drastically from normal beta globin in amino acid sequence and in length. This outcome is the source of the anemia. Beta thalassemia can also be caused by insertion mutations, which, like deletions, often result in frameshifts (**FIGURE 9.13E**).

Not all mutations that affect protein structure disrupt codons for amino acids. DNA also contains special nucleotide sequences that influence the expression of nearby genes (we return to this topic in the next chapter). A promoter is one example; an intron–exon splice site is another. Consider a mutation that causes the hairless appearance of sphynx cats (**FIGURE 9.15**). In this case, a base-pair substitution disrupts an intron–exon splice site in a gene for keratin, a fibrous protein (Section 3.6). The intron is not correctly removed during post-transcriptional processing. The altered protein translated from the resulting mRNA cannot properly assemble into filaments that make up hair. Cats with this mutation still make hair, but it falls out before it gets very long.

base-pair substitution Type of mutation in which a single base pair changes.
deletion Mutation in which one or more nucleotides are lost.
insertion Mutation in which one or more nucleotides become inserted into DNA.

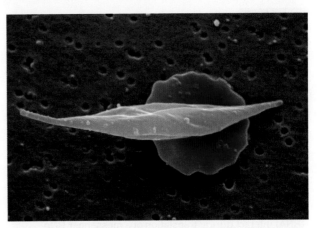

FIGURE 9.14 ▶Animated A sickled red blood cell compared with a normal one. A single base-pair substitution gives rise to an abnormal beta globin chain that, when assembled in hemoglobin molecules, forms HbS. The sixth amino acid in these abnormal beta globin chains is valine, not glutamic acid. In the body, the difference causes HbS molecules to form rod-shaped clumps that distort normally round blood cells (red) into the sickle shape (tan) characteristic of sickle-cell anemia. Sickled cells clog small blood vessels.

FIGURE 9.15 The hairless appearance of a sphynx cat arises from a single base-pair mutation in a gene for keratin, a fibrous protein that makes up hair. The altered keratin that results from the mutation does not assemble correctly into filaments. Sphynx cats are not truly hairless; they produce hair, but it is easily dislodged.

TAKE-HOME MESSAGE 9.6
What happens after a gene becomes mutated?

✔ Mutations that result in an altered protein can have drastic consequences.

✔ A base-pair substitution may change an amino acid in a protein, or it may introduce a premature stop codon.

✔ Frameshifts that occur after an insertion or deletion can change an mRNA's codon reading frame, thus garbling its protein-building instructions.

summary

 Section 9.1 The ability to make proteins is critical to all life processes. Ribosome-inactivating proteins (RIPs) have an enzyme domain that alters ribosomes. Ricin and other toxic RIPs have an additional protein domain that triggers endocytosis. Once in cytoplasm, the enzyme domain destroys the cell's ability to make proteins.

 Section 9.2 Information encoded by the nucleotide sequence of DNA occurs in subsets called **genes**. **Gene expression** is the conversion of information in a gene to an RNA or protein product. RNA is produced during **transcription**. **Ribosomal RNA (rRNA)** and **transfer RNA (tRNA)** interact during **translation** of a **messenger RNA (mRNA)** into a protein product.

 Section 9.3 During transcription, **RNA polymerase** binds to a **promoter** near a gene region, then links RNA nucleotides in the order dictated by the nucleotide base sequence of the noncoding DNA strand. The resulting RNA strand is an RNA copy of the gene.

In eukaryotes, mRNA is modified before it leaves the nucleus. **Intron** sequences are removed, and the remaining **exon** sequences may be rearranged and spliced in different combinations, a process called **alternative splicing**. New mRNAs also receive a cap and poly-A tail.

 Section 9.4 An mRNA's protein-building information consists of a series of **codons**. Sixty-four codons constitute the **genetic code**. Three codons are signals that terminate translation. The remaining codons specify a particular amino acid; one of those also serves as a signal to start translation. Some amino acids are specified by multiple codons. Each tRNA has an **anticodon** that can base-pair with a codon, and it binds to the amino acid specified by that codon. Proteins and enzymatic rRNA make up the two subunits of ribosomes.

 Section 9.5 During translation, codons in an mRNA direct synthesis of a polypeptide. First, the mRNA, an initiator tRNA, and two ribosomal subunits converge. Next, amino acids are delivered by tRNAs in the order specified by the codons in the mRNA. The intact ribosome catalyzes formation of a peptide bond between the successive amino acids, so a polypeptide forms. Translation ends when the ribosome encounters a stop codon in the mRNA.

 Section 9.6 **Insertions**, **deletions**, and **base-pair substitutions** are mutations. A mutation that changes a gene's product may have harmful effects. In humans, an example is sickle-cell anemia, a disorder caused by a single base-pair substitution in the gene for the beta globin chain of hemoglobin.

Ricin, RIP (revisited)

 RIPs remove a particular adenine base from one of the rRNAs in the ribosome's large subunit. The adenine is part of a binding site for proteins involved in GTP-requiring steps of elongation. After the base has been removed, the ribosome can no longer bind to these proteins, and elongation stops.

Despite their toxicity, the main function of RIPs may not be destroying ribosomes. Many are part of plant immune systems, but it is their antiviral and anticancer activity that has researchers abuzz. Plants that make RIPs have been used as traditional medicines for many centuries; now, Western scientists are investigating RIPs as drugs to combat HIV and cancer. For example, researchers who design drugs for cancer therapy have modified ricin's glycolipid-binding domain to recognize plasma membrane proteins (Section 5.7) especially abundant in cancer cells. The modified ricin preferentially enters—and kills—cancer cells. Ricin's toxic enzyme has also been attached to an antibody that can find cancer cells in a person's body. The intent of both strategies: to assassinate the cancer cells without harming normal ones.

self-quiz   Answers in Appendix VII

1. A chromosome contains many different gene regions that are transcribed into different _____ .
   a. proteins   c. RNAs
   b. polypeptides   d. a and b

2. A binding site for RNA polymerase is called a _____ .
   a. gene   c. codon
   b. promoter   d. protein

3. An RNA molecule is typically _____ ; a DNA molecule is typically _____ .
   a. single-stranded; double-stranded
   b. double-stranded; single-stranded

4. RNAs form by _____ ; proteins form by _____ .
   a. replication; translation
   b. translation; transcription
   c. transcription; translation
   d. replication; transcription

5. The main function of a DNA molecule is to _____ .
   a. store heritable information
   b. carry a translatable message
   c. form peptide bonds between amino acids

6. The main function of an mRNA molecule is to _____ .
   a. store heritable information
   b. carry a translatable message
   c. form peptide bonds between amino acids

7. Most codons specify a(n) _____ .
   a. protein   b. mRNA   c. amino acid

CREDITS: (in text revisited) Vaughan Fleming/Science Source.

RIPs as Cancer Drugs Researchers are taking a page from the structure–function relationship of RIPs in their quest for cancer treatments. The most toxic RIPs, remember, have one domain that interferes with ribosomes, and another that carries them into cells. Melissa Cheung and her colleagues incorporated a peptide that binds to skin cancer cells into the enzymatic part of an RIP, the *E. coli* Shiga-like toxin. The researchers created a new RIP that specifically kills skin cancer cells, which are notoriously resistant to established therapies. Some of their results are shown in **FIGURE 9.16**.

1. Which cells had the greatest response to an increase in concentration of the engineered RIP?

2. At what concentration of RIP did all of the different kinds of cells survive?

3. Which cells survived best at 10^{-6} grams per liter RIP?

4. Why are some of the data points linked by curved lines?

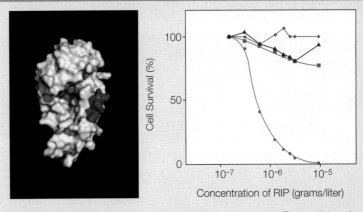

FIGURE 9.16 Effect of an engineered RIP on cancer cells. The model on the left shows the enzyme portion of *E. coli* Shiga-like toxin engineered to carry a small sequence of amino acids (in blue) that targets skin cancer cells. (Red indicates the active site.) The graph (right) shows the effect of this engineered RIP on human cancer cells of the skin (●); breast (◆); liver (▲); and prostate (■).

8. Energy that drives transcription is provided mainly by _____ .
    a. ATP
    b. RNA nucleotides
    c. GTP
    d. RNA polymerase

9. Anticodons pair with _____ .
    a. mRNA codons
    b. DNA codons
    c. RNA anticodons
    d. amino acids

10. Up to _____ amino acids can be encoded by an mRNA that consists of 45 nucleotides plus a stop codon.
    a. 15   b. 45   c. 90   d. 135

11. _____ are removed from new mRNAs.
    a. Introns
    b. Exons
    c. Telomeres
    d. Amino acids

12. Where does transcription take place in a typical eukaryotic cell?
    a. the nucleus
    b. ribosomes
    c. the cytoplasm
    d. b and c are correct

13. Where does translation take place in a typical eukaryotic cell?
    a. the nucleus
    b. the cytoplasm
    c. a and b
    d. neither a nor b

14. Energy that drives translation is provided mainly by _____ .
    a. ATP
    b. amino acids
    c. GTP
    d. all are correct

15. Match the terms with the best description.
    ____ genetic message
    ____ promoter
    ____ polysome
    ____ exon
    ____ genetic code
    ____ intron
    a. protein-coding segment
    b. RNA polymerase binding site
    c. read as base triplets
    d. removed before translation
    e. occurs only in groups
    f. complete set of 64 codons

critical thinking

1. Researchers are designing and testing antisense drugs as therapies for a variety of diseases, including cancer, AIDS, diabetes, and muscular dystrophy. The drugs are also being tested to fight infection by deadly viruses such as Ebola. Antisense drugs consist of short mRNA strands that are complementary in base sequence to mRNAs that form during the progression of disease. Speculate on how these drugs work.

2. An anticodon has the sequence GCG. What amino acid does this tRNA carry? What would be the effect of a mutation that changed the C of the anticodon to a G?

3. Each position of a codon can be occupied by one of four (4) nucleotides. What is the minimum number of nucleotides per codon necessary to specify all 20 of the amino acids that are found in proteins?

4. Refer to **FIGURE 9.7**, then translate the following mRNA nucleotide sequence into an amino acid sequence, starting at the first base:

5′—UGUCAUGCUCGUCUUGAAUCUUGUGAUGC
UCGUUGGAUUAAUUGU—3′

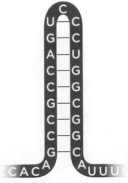

5. Translate the sequence of bases in the previous question, starting at the second base.

6. Bacteria use the same stop codons as eukaryotes. However, bacterial transcription is also terminated in places where the mRNA folds back on itself to form a hairpin-looped structure like the one shown on the left. How do you think that this structure stops transcription?

LEARNING ROADMAP

This chapter explores metabolism (Sections 5.5, 5.6) in the context of gene expression (9.2). You will be applying what you know about DNA (8.2, 8.4, 8.5), mutations (8.6, 9.6), and differentiation (8.7), as well as transcription (9.3) and translation (9.5). You will also revisit functional groups (3.3), carbohydrates (3.4), glycolysis (7.3), and fermentation (7.6).

MECHANISMS OF GENE CONTROL

Every step of gene expression is regulated. This control is critical for development, and it allows individual cells to respond to changes in external and internal conditions.

MASTER GENES

During development, the orderly, localized expression of master genes gives rise to the body plan of complex multicelled animals.

GENE CONTROL IN EUKARYOTES

Examples of gene control in eukaryotes include X chromosome inactivation and male sex determination in mammals, and flower formation in plants.

GENE CONTROL IN PROKARYOTES

Most gene control in prokaryotes occurs at the level of transcription. Fast adjustment of transcription allows these cells to respond quickly to changes in external conditions.

EPIGENETICS

New research is revealing how gene expression patterns that arise during an individual's lifetime in response to environmental pressures can be passed to descendants.

Chapter 11 discusses the cell cycle and how it goes awry in cancer; Chapter 12, the genetic basis of reproduction; Chapter 13, inheritance; Chapter 14, human inheritance patterns; and Chapter 30, plant gene controls. The study of genomes is explained in Section 15.5, and evolutionary processes involving mutation will be discussed more fully in Chapter 17. Section 19.4 discusses the hypothesis that RNA, not DNA, was genetic material in the distant past.

You are in college, your whole life ahead of you. Your risk of developing cancer is as remote as old age, an abstract statistic that is easy to forget. "There is a moment when everything changes—when the width of two fingers can suddenly be the total distance between you and eternity." Robin Shoulla wrote those words after being diagnosed with breast cancer. She was seventeen years old.

At an age when most young women are thinking about school, friends, parties, and potential careers, Robin was dealing with radical mastectomy: the removal of a breast, all lymph nodes under the arm, and skeletal muscles in the chest wall under the breast. She was pleading with her oncologist not to use her jugular vein for chemotherapy and wondering if she would survive to see the next year (FIGURE 10.1).

Robin's ordeal became part of a statistic, one of more than 200,000 new cases of breast cancer diagnosed in the United States each year. About 5,700 of those cases occur in women and men under thirty-four years of age.

Every second, millions of cells in your skin, bone marrow, gut lining, liver, and elsewhere are dividing and replacing their worn-out, dead, and dying predecessors. They do not divide at random; in normal cells, growth and division is tightly regulated. When this control fails, cancer is the outcome.

Cancer is a process in which abnormally growing and dividing cells disrupt body tissues. Mechanisms that normally keep cells from getting overcrowded in tissues are lost, so cancer cell populations may reach extremely high densities. Unless chemotherapy, surgery, or another procedure eradicates them, cancer cells can put an individual on a painful road to death. In developed countries, cancers cause 15 to 20 percent of all human deaths.

Cancer typically begins with a mutation in a gene whose product is part of a system that controls cell growth and division. Such controls govern when and how fast specific genes are transcribed and translated. A cancer-causing mutation may be inherited, or it may arise after birth, as when DNA becomes damaged by environmental agents. If the mutation alters the gene's protein product so that it no longer works properly, one level of control over the cell's growth and division has been lost. You will be considering the impact of gene controls in chapters throughout this book, and also in some chapters of your life.

Robin Shoulla survived. Radical mastectomy is rarely performed today, but it was her only option. Now, seventeen years later, she has what she calls a normal life: career, husband, children. Her goal as a cancer survivor: "To grow very old with gray hair and spreading hips, smiling."

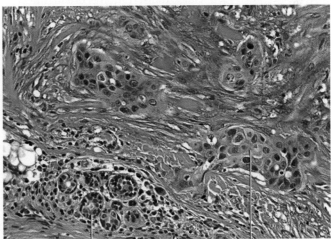

normal cells in organized clusters   disorganized clusters of malignant cells

FIGURE 10.1 A case of breast cancer. Left, Robin Shoulla. Right, this light micrograph shows irregular clusters of cancer cells that have infiltrated milk ducts in human breast tissue. Diagnostic tests revealed abnormal cells such as these in Robin's body when she was just seventeen years old.

✔ Controls over gene expression govern the kinds and amounts of substances that are present in a cell at any given time.

A typical cell in your body uses only about 10 percent of its genes at a time. Some of the active genes affect structures and metabolic pathways common to all cells; others are expressed only by certain subsets of cells. For example, most body cells express genes that encode the enzymes of glycolysis, but only immature red blood cells express globin genes. Differentiation (Section 8.7) occurs as different cell lineages begin to express different subsets of their genes during development. Which genes a cell uses determines the molecules it will produce, which in turn determines what kind of cell it will be. Thus, control over gene expression is necessary for proper development of complex, multicelled bodies. It also allows individual cells to respond appropriately to changes in their internal and external environments.

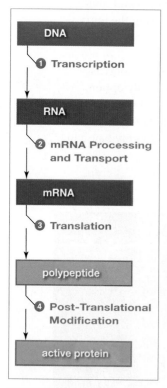

FIGURE 10.2 ▶Animated Points of control over gene expression.

Gene Expression Control

The "switches" that turn a gene on or off are molecules or processes that affect individual steps of its expression (**FIGURE 10.2**).

❶ **Transcription** Proteins called **transcription factors** affect whether and how fast a gene is transcribed by binding directly to the DNA. In eukaryotes especially, transcription is often governed by multiple transcription factors; overlapping and opposing effects of these proteins give the cell a nuanced level of control over RNA production.

Transcription factors called **repressors** shut off transcription or slow it down, either by preventing RNA polymerase from accessing a promoter or by impeding its progress along the DNA strand. Some repressors work by binding directly to the promoter. Others bind to a silencer—a region in the DNA that may be thousands of base pairs away from the gene. **Activators** are transcription factors that help RNA polymerase bind to a promoter, so they speed up transcription. Some eukaryotic activators work by binding to DNA sequences called **enhancers**, which, like silencers, may be far away from the gene

FIGURE 10.3 Hypothetical part of a chromosome that contains a gene. Molecules that affect the rate of transcription of the gene bind at promoter (yellow) or enhancer (green) sequences.

they affect (**FIGURE 10.3**). An enhancer operating on a distant gene can inappropriately affect a nearby gene; insulators prevent this from occurring. An insulator is a region of DNA that, upon binding a transcription factor, can block the effect of an enhancer on a neighboring gene (**FIGURE 10.4**).

Chromatin structure also affects transcription. The DNA of eukaryotic cells and some archaea wraps around histones (Section 8.4). In these cells, only the regions of DNA that have been unwound from histones are accessible to RNA polymerase. Modifications to histone proteins change the way they interact with DNA wrapped around them, thus affecting transcription. Some modifications make histones release their grip on DNA; others make them tighten it. For example, adding acetyl groups ($-COCH_3$) to a histone loosens the DNA, so enzymes that acetylate histones allow transcription to proceed. Conversely, adding methyl groups ($-CH_3$) to a histone tightens DNA, so enzymes that methylate histones shut down transcription.

In some specialized eukaryotic cells, a high level of gene expression can be achieved by copying DNA repeatedly without cell division. The result is a polytene chromosome consisting of hundreds or thousands of side-by-side copies of the same DNA molecule (the chromosome shown in **FIGURE 10.4** is polytene). Transcription of one gene occurs simultaneously on all of the DNA strands, quickly producing a lot of mRNA.

❷ **mRNA Processing and Transport** As you know, transcription in eukaryotes occurs in the nucleus, and translation occurs in cytoplasm (Section 9.5). A newly transcribed mRNA can pass through pores of a nuclear envelope only after it has been processed appropriately—spliced, capped, and finished with a poly-A tail. Mechanisms that delay these post-transcriptional modifications also delay the mRNA's appearance in cytoplasm for translation.

Control over post-transcriptional modification can also affect the form of a protein. Consider RNA transcribed from the gene for fibronectin, a protein produced by cells of vertebrate animals. Two cell types splice this RNA alternatively, so they produce different mRNAs—and different forms of fibronectin. Liver cells produce a soluble form that circulates in blood plasma.

transcription start site   transcription end

Fibroblasts produce an insoluble form that is a major component of extracellular matrix (Section 4.11).

In eukaryotic cells, most mRNAs are delivered to organelles or specific regions of cytoplasm, thus allowing proteins to be produced close to where they are being used. In an egg, cytoplasmic localization of mRNA is crucial for proper development of the future embryo. How does localization occur? A short nucleotide sequence near an mRNA's poly-A tail is like a zip code that specifies a particular destination. Proteins that attach to the zip code drag an mRNA along cytoskeletal elements to its destination. Other proteins influence localization by interacting with mRNA-binding proteins. mRNA localization also occurs in prokaryotes, but the mechanism is not yet understood.

❸ **Translation** Production of the many molecules that participate in translation is a major point of control in eukaryotic cells. An mRNA's sequence also affects translation. For example, proteins that bind to an mRNA's zip code region prevent transcription from occurring before the mRNA is delivered to its final destination. As another example, consider that the longer an mRNA lasts, the more protein can be made from it. Enzymes begin disassembling an mRNA as soon as it arrives in cytoplasm. The fast turnover allows a cell to

adjust protein synthesis quickly. How long an mRNA persists depends on its base sequence, the length of its poly-A tail, and which proteins are attached to it.

In eukaryotes, translation of a particular mRNA can be shut down by microRNAs, which are tiny bits of noncoding RNA. A microRNA is complementary in sequence to part of an mRNA, and when the two show up together in cytoplasm they base-pair to form a small double-stranded region of RNA. By a process called RNA interference, any double-stranded RNA is cut up into small bits that are taken up by special enzyme complexes. These complexes then destroy every RNA in a cell that can base-pair with the bits. Thus, expression of a microRNA results in the destruction of all mRNA complementary to it.

Double-stranded RNA is also a factor in control over translation in prokaryotes. For example, bacteria can shut off translation of a particular mRNA by expressing an antisense RNA (one that is complementary in sequence to the mRNA). The two molecules hybridize (Section 8.5) to form a double-stranded RNA that ribosomes cannot translate. As another example, some bacterial mRNAs can loop back on themselves to form a small double-stranded region. Translation only occurs when this structure is unraveled, for example by exposure to heat.

❹ **Post-Translational Modification** Many newly synthesized polypeptide chains must be modified before they become functional. For example, some enzymes become active only after they have been phosphorylated (Section 5.6). Such post-translational modifications inhibit, activate, or stabilize many molecules, including enzymes that participate in transcription and translation.

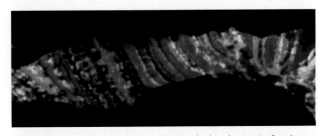

FIGURE 10.4 Fluorescence micrograph showing part of a chromosome in a fruit fly salivary gland cell. DNA appears blue. Red and green show the locations of two transcription factors bound to insulator sequences (yellow shows where these two colors overlap). The transcription factors are inhibiting the effect of nearby enhancers.

activator Regulatory protein that increases the rate of transcription when it binds to a promoter or enhancer.
enhancer Binding site in DNA for proteins that enhance the rate of transcription.
repressor Regulatory protein that blocks transcription.
transcription factor Regulatory protein that influences transcription; e.g., an activator or repressor.

TAKE-HOME MESSAGE 10.2
What is gene expression control?

✔ Gene expression can be switched on or off, or speeded or slowed, by molecules and processes that operate at each step.

✔ These gene expression controls allow cells to respond appropriately to environmental change. They are also critical for differentiation and development in multicelled eukaryotes.

CREDIT: (4) *Journal of Bioscience*, Volume 36, Number 3, August 2011, Indian Academy of Sciences, Springer.

CHAPTER 10
CONTROL OF GENE EXPRESSION
165

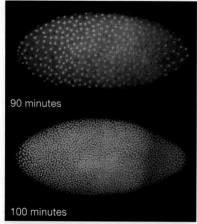

A 100 minutes after fertilization, the master gene *even-skipped* is expressed (in red) only where two transcription factors (green and blue) overlap. The transcription factors are the protein products of maternal mRNAs.

90 minutes

100 minutes

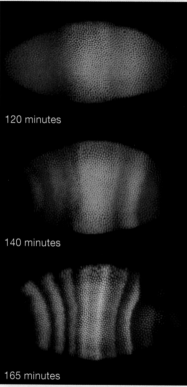

B By 165 minutes after fertilization, the products of several master genes, including the two shown here in green and blue, have confined the expression of *even-skipped* (red) to seven stripes. (Pink and yellow areas are regions in which red fluorescence has overlapped with blue or green.)

120 minutes

140 minutes

165 minutes

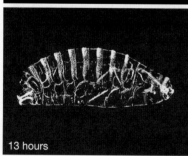

C One day later, seven segments have developed. The position of the segments corresponds to the position of the *even-skipped* stripes.

13 hours

FIGURE 10.5 How gene expression control makes a fly, as illuminated by the formation of segments in a *Drosophila* embryo.

Expression of different master genes is shown by different colors in fluorescence microscopy images of whole embryos at successive stages of development (time after fertilization is indicated). Bright dots are individual nuclei.

✔ Cascades of gene expression govern the development of a complex, multicelled body.

As an animal embryo develops, its cells differentiate and form tissues, organs, and body parts. The entire process of development is driven by cascades of master gene expression. The products of **master genes** affect the expression of many other genes. Expression of a master gene causes other genes to be expressed, which in turn cause other genes to be expressed, and so on.

The orchestration of gene expression during animal development begins when different maternal mRNAs localize to different regions of cytoplasm in an egg as it forms. These mRNAs are translated only after the egg is fertilized. Then, their protein products diffuse away, forming gradients that span the developing embryo. The position of a cell within the embryo determines how much of these proteins it is exposed to. This in turn determines which master genes it turns on. The products of those master genes also form in gradients that diffuse away from cells expressing them. Still other master genes are transcribed depending on where a cell falls within these gradients, and so on. Eventually, the products of master genes cause undifferentiated cells to differentiate, and specialized structures form in specific regions of the embryo (**FIGURE 10.5**).

Homeotic Genes

A **homeotic gene** is a type of master gene whose expression directs the formation of a specific body part such as an eye, leg, or wing. Animal homeotic genes encode transcription factors with a homeodomain, which is a region of about sixty amino acids that can bind directly to a promoter or some other sequence of nucleotides in a chromosome (**FIGURE 10.6A**).

The function of many homeotic genes has been discovered by deliberately manipulating their expression. In an experiment called a **knockout**, researchers inactivate a gene by introducing a mutation that prevents its expression, or by deleting it entirely. A knockout organism (one that has had a gene knocked out) may differ from normal individuals, and the differences are clues to the function of the missing gene product.

Homeotic genes are often named for what happens when a mutation alters their function. For example, fruit flies with a mutation that affects their *antennapedia* gene (*ped* means foot) have legs in place of antennae (**FIGURE 10.6B**). The *dunce* gene is required for learning and memory. *Wingless*, *wrinkled*, and *minibrain* are self-explanatory. *Tinman* is necessary for development of a heart. Flies with a mutated *groucho* gene have extra

CREDITS: (5A, B) © Maria Samsonova and John Reinitz; (5C) © Jim Langeland, Jim Williams, Julie Gates, Kathy Vorwerk, Steve Paddock and Sean Carroll, HHMI, University of Wisconsin-Madison.

groucho mutation

normal fly head

bristles that resemble bushy eye-brows (left). Flies with a mutated *eyeless* gene lack eyes (**FIGURE 10.6C**). One gene was named *toll*, after what its German discoverer exclaimed upon seeing the disastrous effects of the mutation (*toll* is German slang that means "cool!").

Homeotic genes control development by the same mechanisms in all multicelled eukaryotes, and many are interchangeable among different species. Thus, we can infer that they evolved in the most ancient eukaryotic cells. Homeodomains often differ among species only in conservative substitutions (one amino acid has replaced another with similar chemical properties). Consider the *eyeless* gene. Eyes form in embryonic fruit flies wherever this gene is expressed, which is normally in tissues of the head only. If the *eyeless* gene is expressed in another part of the developing embryo, eyes form there too (**FIGURE 10.6D**).

Humans, squids, mice, fishes, and many other animals have a gene called *PAX6*, which is very similar in DNA sequence to the *eyeless* gene of flies. In humans, mutations in *PAX6* cause eye disorders such as aniridia, in which the irises are underdeveloped or missing (**FIGURE 10.6E**). If a functional *PAX6* gene from a human or a mouse is inserted into a fly, it has the same effect as the *eyeless* gene: An eye forms wherever it is expressed. (Because *PAX6* is just a switch, the eye that forms is a fly eye, not a human or mouse eye.) The same principle applies in reverse: The *eyeless* gene from flies switches on eye formation in frogs. Such studies are evidence of shared ancestry among these evolutionarily distant animals.

homeotic gene Type of master gene with a homeodomain; its expression directs formation of a specific body part in development.
knockout An experiment in which a gene is deliberately inactivated in a living organism; also, an organism that has a knocked-out gene.
master gene Gene encoding a product that affects the expression of many other genes.

TAKE-HOME MESSAGE 10.3

How do genes control development in animals?

✔ Animal development is orchestrated by cascades of master gene expression in embryos.

✔ The expression of homeotic genes during development directs the formation of specific body parts.

✔ Homeotic genes that function in similar ways in evolutionarily distant animals are evidence of shared ancestry.

A A model of the protein product (in gold) of the homeotic gene *antennapedia* attached to a promoter. The homeodomain is the region that binds to the DNA. Expression of *antennapedia* in embryonic tissues of the insect thorax causes legs to form.

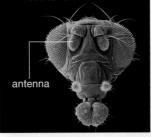

B The head of a normal fruit fly (left) has two antennae. Right, a mutation that triggers expression of the *antennapedia* gene in embryonic tissues of the head causes legs to form instead of antennae.

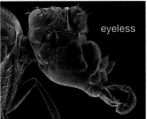

C A normal fruit fly (left) has large, round eyes. A fruit fly with a mutation in its *eyeless* gene (right) develops without eyes.

D Eyes form wherever the *eyeless* gene is expressed in fly embryos—here, on the head and also on the wing.

The *PAX6* gene of humans, mice, squids, and some other animals is so similar to *eyeless* that it similarly triggers eye development in fruit flies.

E A normal human eye has a colored iris surrounding the pupil (dark area where light enters). Mutations in *PAX6* cause eyes to develop without an iris, a condition called aniridia.

FIGURE 10.6 Some effects of homeotic gene mutations.

CREDITS: (in text top and bottom) Courtesy of Dr. Barbara Jennings, UCL Cancer Institute, www.ucl.ac.uk; (6A) © Cengage Learning; (6B) left, © Jürgen Berger, Max-Planck-Institute for Developmental Biology, Tübingen; right, © Visuals Unlimited; (6C) David Scharf/Science Source; (6D) Eye of Science/Science Source; (6E) right, Courtesy of the Aniridia Foundation International, www.aniridia.net; left, M. Bloch.

✔ Selective gene expression gives rise to many traits.

Gene control influences many traits that are characteristic of humans and other eukaryotic organisms, as the following examples illustrate.

X Marks the Spot

In humans and other mammals, a female's cells have two X chromosomes, one inherited from her mother, the other one from her father. In each cell, one X chromosome is always tightly condensed (**FIGURE 10.7A**). We call the condensed X chromosomes **Barr bodies**, after Murray Barr, who discovered them. Condensation inhibits transcription, so most of the genes on a Barr body are not expressed. This **X chromosome inactivation** ensures that only one of the two X chro-

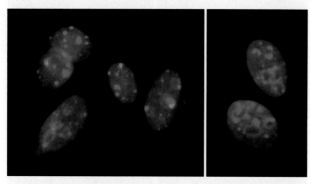

A Barr bodies are visible as red spots in the nucleus of the four XX cells on the left. Compare the nucleus of two XY cells to the right.

B When this calico cat was an embryo, one of the two X chromosomes was inactivated in each of her cells. The descendants of the cells formed her adult body, which is a mosaic for expression of X chromosome genes. Black fur arises in patches where genes on the X chromosome inherited from one parent are expressed; orange fur arises in patches where genes on the X chromosome inherited from the other parent are expressed.

FIGURE 10.7 ▶**Animated** X chromosome inactivation.

mosomes in a female's cells is active, thus equalizing expression of X chromosome genes between the sexes—a mechanism called **dosage compensation**. The body cells of male mammals (XY) have one set of X chromosome genes. Body cells of female mammals (XX) have two sets, but female embryos do not develop properly when both sets are expressed.

X chromosome inactivation occurs when an embryo is a ball of about 200 cells. In humans and most other mammals, it occurs independently in every cell of a female embryo. Which X chromosome condenses is random: The paternal or maternal X chromosome may get inactivated in any cell. Once the selection is made in a cell, it is permanent. All of that cell's descendants make the same selection as they continue dividing and forming tissues. As a result of random inactivation of maternal and paternal X chromosomes, an adult female mammal is a "mosaic" for the expression of X chromosome genes. She has patches of tissue in which genes of the maternal X chromosome are expressed, and patches in which genes of the paternal X chromosome are expressed (**FIGURE 10.7B**).

How does just one of two X chromosomes get inactivated? An X chromosome gene called *XIST* is transcribed on only one of the two X chromosomes. The gene's product, a long noncoding RNA, sticks to the chromosome that expresses the gene. The RNA coats the chromosome, and by an unknown mechanism causes it to condense into a Barr body. Thus, transcription of the *XIST* gene keeps the chromosome from transcribing other genes. The other chromosome does not express *XIST*, so it does not get coated with RNA; its genes remain available for transcription. How a cell "chooses" which X chromosome will express *XIST* is still unknown.

Male Sex Determination in Humans

Only a few of the 1,113 genes on the human X chromosome are associated with traits such as body fat distribution that differ between males and females. Most X chromosome genes govern nonsexual traits such as blood clotting and color perception. Such genes are expressed in both males and females. Males, remember, also inherit one X chromosome.

The human Y chromosome carries only 128 genes, but one of them is *SRY*—the master gene for male sex determination in mammals. Its expression in XY embryos triggers the formation of testes, which are male reproductive organs. Some of the cells in testes make testosterone, a sex hormone that controls the emergence of male secondary sexual traits such as

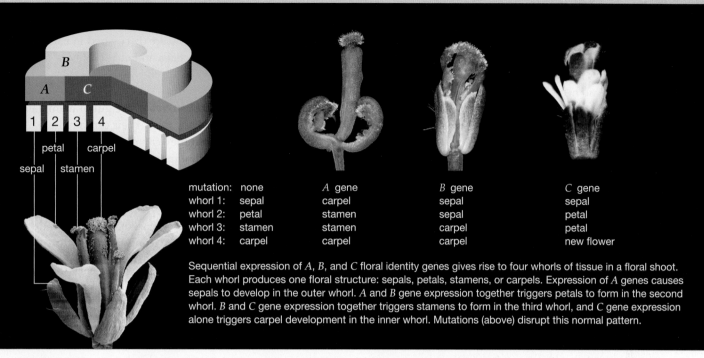

mutation:	none	A gene	B gene	C gene
whorl 1:	sepal	carpel	sepal	sepal
whorl 2:	petal	stamen	sepal	petal
whorl 3:	stamen	stamen	carpel	petal
whorl 4:	carpel	carpel	carpel	new flower

Sequential expression of *A*, *B*, and *C* floral identity genes gives rise to four whorls of tissue in a floral shoot. Each whorl produces one floral structure: sepals, petals, stamens, or carpels. Expression of *A* genes causes sepals to develop in the outer whorl. *A* and *B* gene expression together triggers petals to form in the second whorl. *B* and *C* gene expression together triggers stamens to form in the third whorl, and *C* gene expression alone triggers carpel development in the inner whorl. Mutations (above) disrupt this normal pattern.

FIGURE 10.8 ▶Animated Control of flower formation, as revealed by mutations in *Arabidopsis thaliana*.

facial hair, increased musculature, and deepened voice. We know that *SRY* is the master gene that controls emergence of male sexual traits because mutations in this gene cause XY individuals to develop external genitalia that appear female. An XX embryo has no Y chromosome, no *SRY* gene, and much less testosterone, so primary female reproductive organs (ovaries) form instead of testes. Ovaries make estrogens and other sex hormones that will govern the development of female secondary sexual traits, such as enlarged, functional breasts, and fat deposits around the hips and thighs.

Flower Formation

In flowering plants, populations of cells in a shoot tip may give rise to a flower instead of leaves. Studies of mutations in thale cress, *Arabidopsis thaliana*, revealed the gene control behind this switch in development. Transcription factors produced by three sets of floral identity genes (called *A*, *B*, and *C*) guide the process. These genes are switched on by environmental cues such as seasonal changes in the length of night, as you will see in Section 30.9.

Barr body Inactivated X chromosome in a cell of a female mammal. The other X chromosome is active.
dosage compensation Mechanism in which X chromosome inactivation equalizes gene expression between males and females.
X chromosome inactivation Developmental shutdown of one of the two X chromosomes in the cells of female mammals.

When a flower forms at the tip of a shoot, differentiating cells form whorls of tissue, one over the other like layers of an onion. Each whorl produces one type of floral structure—sepals, petals, stamens, or carpels. This pattern is dictated by sequential, overlapping expression of the *ABC* genes (**FIGURE 10.8**). The *A* genes are switched on first in a shoot tip, and their products trigger events that cause the outer whorl to form and give rise to sepals. *B* genes switch on before *A* genes turn off. Together, *A* and *B* gene products cause the second whorl to form and produce petals. Next, *A* genes turn off and the *C* gene switches on (there is only one *C* gene). Together, the products of the *B* and *C* genes trigger formation of the third whorl, which makes stamens. Finally, the *B* gene turns off, and the *C* gene product on its own gives rise to the fourth, inner whorl, which produces the carpel.

TAKE-HOME MESSAGE 10.4
   How do genes control development in animals?

✔ X chromosome inactivation balances expression of X chromosome genes between female (XX) and male (XY) mammals.

✔ *SRY* gene expression triggers the development of male traits in mammals.

✔ In plants, expression of *ABC* floral identity genes governs development of the specialized parts of a flower.

CREDITS: (8) top left, © Cengage Learning; bottom left, Juergen Berger, Max Planck Institute for Developmental Biology, Tuebingen, Germany; (8A, B gene) © Jose Luis Riechmann; (8C gene) Image by Marty Yanofsky.

✔ Bacteria control gene expression mainly by adjusting the rate of transcription.

Prokaryotes do not undergo development, so these cells have no need for master genes. However, they do respond to environmental fluctuations by adjusting gene expression, mainly at the level of transcription. For example, when a preferred nutrient becomes available, a bacterium begins transcribing genes whose products allow the cell to use the nutrient. When the nutrient is no longer available, transcription of those genes stops. Thus, the cell does not waste energy and resources producing gene products that are not needed.

In bacteria, genes that are used together often occur together on the chromosome, one after the other. One promoter precedes the genes, so all are transcribed together into a single RNA strand. Thus, their transcription can be controlled in a single step that typically involves repressor binding to an **operator**, which is a type of silencer sequence in DNA. A group of genes together with a promoter and one or more opera-

tors that control their transcription are collectively called an **operon**. Operons were discovered in bacteria, but they also occur in archaea and eukaryotes.

The *lac* Operon

Escherichia coli bacteria that live in the gut of mammals dine on nutrients traveling past. Their carbohydrate of choice is glucose, but they can make use of other sugars such as the lactose in milk. An operon called *lac* allows *E. coli* cells to metabolize lactose (**FIGURE 10.9**). The *lac* operon includes three genes and a promoter flanked by two operators ❶.

One gene in the *lac* operon encodes an active transport protein (Section 5.7) that brings lactose across the plasma membrane into the cell. Another encodes an enzyme (β-galactosidase) that breaks the bond between lactose's two monosaccharide monomers, glucose and galactose. The third gene encodes an enzyme whose function is still being investigated. Bacteria make these three proteins only when lactose is present. When lactose is not present, a repressor binds to the two opera-

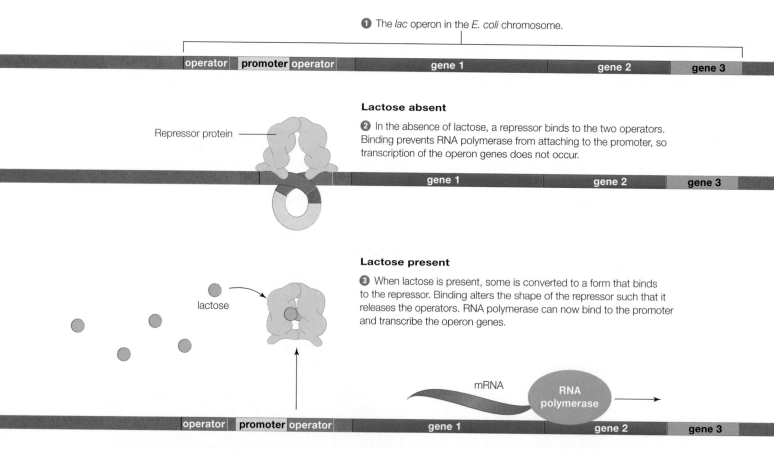

❶ The *lac* operon in the *E. coli* chromosome.

operator | promoter | operator | gene 1 | gene 2 | gene 3

Lactose absent

❷ In the absence of lactose, a repressor binds to the two operators. Binding prevents RNA polymerase from attaching to the promoter, so transcription of the operon genes does not occur.

Repressor protein

gene 1 | gene 2 | gene 3

Lactose present

❸ When lactose is present, some is converted to a form that binds to the repressor. Binding alters the shape of the repressor such that it releases the operators. RNA polymerase can now bind to the promoter and transcribe the operon genes.

lactose

mRNA

RNA polymerase

operator | promoter | operator | gene 1 | gene 2 | gene 3

FIGURE 10.9 ▶Animated Example of gene control in bacteria: the lactose operon on a bacterial chromosome. The operon consists of a promoter flanked by two operators, and three genes for lactose-metabolizing enzymes. **FIGURE IT OUT** What portion of the operon binds RNA polymerase when lactose is present?

Answer: The promoter

tors and twists the region of DNA with the promoter into a loop ❷. RNA polymerase cannot bind to the twisted-up promoter, so the *lac* operon's genes cannot be transcribed. When lactose is present, some of it is converted to another sugar that binds to the repressor and changes its shape. The altered repressor releases the operators and the looped DNA unwinds. The promoter is now accessible to RNA polymerase, and transcription of lactose-metabolizing genes begins ❸.

In bacteria, glucose metabolism requires fewer enzymes than lactose metabolism does, so it requires less energy and fewer resources. Accordingly, when both lactose and glucose are present, the cells will use up all of the available glucose before switching to lactose metabolism. How does a cell ignore the presence of one sugar while it uses another? Another level of gene control over the *lac* operon regulates this metabolic switch, but how it works is still being debated.

Lactose Intolerance

Like *E. coli*, humans and other mammals break down lactose, but most do so only when young. An individual's ability to digest lactose ends at a species-specific age. In the majority of humans worldwide, this switch occurs at about age five, when transcription of the gene for lactase (a type of β-galactosidase) slows. The resulting decline in production of this enzyme causes a common condition known as lactose intolerance.

Cells in the intestinal lining secrete lactase into the small intestine, where the enzyme cleaves lactose into its glucose and galactose monomers. These sugars are absorbed directly by the small intestine, but lactose and other disaccharides are not. Thus, when lactase production slows, lactose passes undigested through the small intestine. The lactose ends up in the large intestine, which hosts huge numbers of *E. coli* and a variety of other bacteria. These resident organisms respond to the presence of lactose by switching on their *lac* operons. Carbon dioxide, methane, hydrogen, and other gaseous products of their various fermentation reactions accumulate quickly in the large intestine, distending its wall and causing pain. Other products of their metabolism disrupt the solute–water balance inside the large intestine, and diarrhea results.

Not everybody is lactose intolerant. About one-third of human adults carry a mutation that allows them to digest milk; this mutation is more common in some

got lactase?

populations than in others. Recent analyses of DNA from well-preserved skeletons shows that the vast majority of adult humans living about 8,000 years ago in Europe were lactose intolerant. Around that time, a mutation appeared in the DNA of prehistoric people inhabiting a region between what is now central Europe and the Balkans. This mutation allowed its bearers to continue digesting milk as adults, and it spread rapidly to the rest of the continent along with the practice of dairy farming. Today, most adults of northern and central European ancestry are able to digest milk because they carry this mutation, a single base-pair substitution in an enhancer that controls the lactase gene promoter. Other mutations in the same enhancer arose independently in North Africa, southern Asia, and the Middle East. Some people descended from these populations can continue to digest milk as adults.

Riboswitches

mRNAs that regulate their own translation are common in bacteria. These mRNAs have small sequences called riboswitches that bind to a metal ion, cofactor, or metabolic product. The binding causes a conformational change in the mRNA that affects its translation. Consider what happens when bacteria make vitamin B_{12}. The enzymes involved in synthesis of this vitamin are produced from mRNAs that have riboswitches. In this case, the riboswitches bind to vitamin B_{12}. Binding changes the shape of the mRNA so that ribosomes can no longer attach to it, so production of B_{12}-making enzymes stops. This example also illustrates feedback inhibition (Section 5.5).

TAKE-HOME MESSAGE 10.5
What are some outcomes of gene control in prokaryotes?

✔ The bacterial *lac* operon allows expression of lactose-metabolizing proteins only when lactose is present.

✔ Most adult humans do not produce the enzyme that breaks down lactose. When undigested lactose enters the large intestine, resident bacteria give rise to symptoms of lactose intolerance.

---

operator Part of an operon; a DNA binding site for a repressor.
operon Group of genes together with a promoter–operator DNA sequence that controls their transcription.

10.6 Epigenetics

✔ Methylations and other modifications that accumulate in DNA during an individual's lifetime can be passed to offspring.

You learned in Section 10.2 that the addition of methyl groups to histone proteins suppresses transcription. Direct methylation of DNA nucleotides also suppresses transcription, often more permanently than histone modifications. Once a particular nucleotide has become methylated in a cell's DNA, it will usually stay methylated in the DNA of the cell's descendants. Methylation and other heritable modifications to DNA that affect its function but do not alter the nucleotide sequence are said to be **epigenetic**.

DNA methylation is necessarily a part of differentiation, so it begins very early in embryonic development: Genes actively expressed in a zygote (the first cell of a new individual) become silenced as their promoters get methylated. This silencing is the basis of selective gene expression that drives differentiation. During development, and also during the remainder of the individual's life, each cell's DNA continues to acquire methylations. Between 3 and 6 percent of the DNA has been methylated in a normal, differentiated body cell.

In eukaryotes, methyl groups are usually added to a cytosine that is followed by a guanine (**FIGURE 10.10**), but which of these cytosines are methylated varies by the individual. This is because methylation is influenced by environmental factors. For instance, humans conceived during a famine end up with an unusually low number of methyl groups attached to the nucleotides of certain genes. The product of one of those genes is a hormone that fosters prenatal growth and development. The resulting increase in expression of this gene may offer a survival advantage in a poor nutritional environment.

Methyl groups are also added to nucleotides by chance during DNA replication, so cells that divide a lot tend to have more methyl groups in their DNA than inactive cells. Free radicals and toxic chemicals add more methyl groups. These and other factors that influence DNA methylation can have multigenerational effects. When an organism reproduces, it passes its DNA to offspring. Methylation of parental DNA is normally "reset" in the zygote, with new methyl groups being added and old ones being removed. This reprogramming does not remove all of the parental methyl groups, however, so methylations acquired during an individual's lifetime can be passed to future offspring.

epigenetic Refers to heritable changes in gene expression that are not the result of changes in DNA sequence.

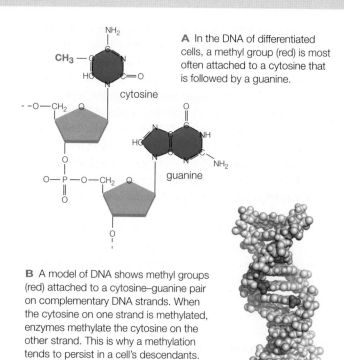

A In the DNA of differentiated cells, a methyl group (red) is most often attached to a cytosine that is followed by a guanine.

B A model of DNA shows methyl groups (red) attached to a cytosine–guanine pair on complementary DNA strands. When the cytosine on one strand is methylated, enzymes methylate the cytosine on the other strand. This is why a methylation tends to persist in a cell's descendants.

FIGURE 10.10 DNA methylation.

FIGURE 10.11 An epigenetic effect. Grandsons of men who endured a famine when they were boys tend to live about 32 years longer than grandsons of men who ate well during the same winter.

Between You and Eternity (revisited)

An effective cancer treatment must eliminate cancerous cells from a person's body. However, cancer cells are body cells, so drugs and other therapies that kill them also kill normal body cells. There is often a fine line between eliminating cancer from a patient's body, and killing the patient.

As you will see in Chapter 11, a normal cell has layers upon layers of gene expression controls and fail-safe mechanisms that determine when division occurs and when it does not. This finely tuned system becomes unbalanced in a cancer cell, so that some genes are expressed at a higher level than they should be, and some are expressed at a lower level.

For decades, we have been studying controls over cell division and how they go awry in cancer, because understanding how a cancer cell differs from a normal cell at a molecular level offers our best chance for developing a cure. In the early 1980s, researchers discovered that mutations in some genes predispose individuals to develop certain kinds of cancer. These genes are tumor suppressors, so named because tumors are more likely to occur when these genes mutate. Two examples are *BRCA1* and *BRCA2*: A mutated version of one or both of these genes is often found in breast and ovarian cancer cells. Because mutations in genes such as *BRCA* can be inherited, cancer is not only a disease of the elderly, as Robin Shoulla's story illustrates. Robin is one of the unlucky people who carry mutations in both of her *BRCA1* and *BRCA2* genes. A woman who carries one of three particularly dangerous *BRCA* mutations has an 80 percent chance of developing breast cancer before the age of seventy.

BRCA genes are master genes whose protein products help maintain the structure and number of chromosomes in a dividing cell. The multiple functions of these proteins are still being unraveled. We do know they participate directly in DNA repair (Section 8.6), so any mutations that alter this function also alter the cell's capacity to repair damaged DNA. Other mutations are likely to accumulate, and that sets the stage for cancer.

The products of *BRCA* genes also bind to receptors for the hormones estrogen and progesterone, which are abundant on cells of breast and ovarian tissues. Binding suppresses transcription of growth factor genes in these cells. Among other effects, growth factors stimulate cells to divide during normal, cyclic renewals of breast and ovarian tissues. When a mutation alters a *BRCA* gene so that its product cannot bind to hormone receptors, the cells overproduce growth factors. Cell division goes out of control, and tissue growth becomes disorganized. In other words, cancer develops (Section 11.6 returns to this topic).

Researchers discovered that the RNA product of the *XIST* gene (Section 10.4) localizes abnormally in breast cancer cells. In those cells, both X chromosomes are active. It makes sense that two active X chromosomes would have something to do with abnormal gene expression, but why the RNA product of an unmutated *XIST* gene does not localize properly in cancer cells remains a mystery.

Mutations in the *BRCA1* gene may be part of the answer. The researchers found that the protein product of the *BRCA1* gene physically associates with the RNA product of the *XIST* gene. They were able to restore proper XIST RNA localization—and proper X chromosome inactivation—by restoring the function of the *BRCA1* gene product in breast cancer cells.

Inheritance of epigenetic modifications can adapt offspring to an environmental challenge much more quickly than evolution (we return to evolutionary processes in Chapter 17). Epigenetic modifications are not considered to be evolutionary because the underlying DNA sequence does not change. Even so, they may persist for generations after an environmental challenge has faded. For example, a recent study showed that grandsons of boys who endured a winter of famine (**FIGURE 10.11**) tend to outlive—by far—grandsons of boys who overate at the same age. The effect is presumed to be due to epigenetic modification because

these results were corrected for socioeconomic and genetic factors. In a similar study, nine-year-old boys whose fathers smoked cigarettes before age eleven were overweight compared with boys whose fathers did not smoke in childhood.

> ## TAKE-HOME MESSAGE 10.6
> ### Can gene expression patterns be inherited?
> ✔ Epigenetic modifications of chromosomal DNA, including DNA methylations acquired during an individual's lifetime, can be passed to offspring.

summary

Section 10.1 A complex interplay of controls over gene expression is a critical part of normal functioning of cells in a multicelled body. Cancer typically begins with mutations in master genes that govern cell division.

Section 10.2 Which genes a cell uses depends on the type of organism, the type of cell, conditions inside and outside the cell, and, in complex multicelled species, the organism's stage of development.

Gene expression control is necessary for differentiation during development in multicelled eukaryotes, and it also allows individual cells to respond appropriately to changes in their internal and external environments.

Different molecules and processes govern every step between transcription of a gene and delivery of the gene's product to its final destination. **Transcription factors** such as **activators** and **repressors** influence transcription by binding to chromosomal DNA at sites such as promoters, **enhancers**, and silencers.

Section 10.3 Control over gene expression drives embryonic development of complex, multicelled animal bodies. Various **master genes** are expressed locally in different parts of an embryo as it develops. Their products, which form gradients as they diffuse through the embryo, affect expression of other master genes, which in turn affect the expression of others, and so on. Cells differentiate according to where they fall in these gradients. Eventually, master gene expression induces the expression of **homeotic genes**, the products of which govern the development of body parts. The function of many homeotic genes was revealed by **knockouts** in fruit flies.

Section 10.4 In cells of female mammals, one of the two X chromosomes is condensed as a **Barr body**, rendering most of its genes permanently inaccessible. By the theory of **dosage compensation**, this **X chromosome inactivation** balances gene expression between the sexes. The inactivation occurs because the *XIST* gene's RNA product shuts down the one X chromosome that transcribes it. The *SRY* gene determines male sex in humans.

Overlapping expression of master genes guides expression of flower formation in plants. Cells differentiate and form sepals, petals, stamens, or carpels depending on which floral identity gene products they are exposed to.

Section 10.5 Prokaryotes, being single-celled, do not undergo development. Most of their gene control reversibly adjusts transcription rates in response to environmental conditions, especially nutrient availability. The *lac* **operon** governs expression of three genes, the three products of which allow a bacterial cell to metabolize lactose. Two **operators** that flank the promoter are binding sites for a repressor that blocks transcription. Binding to an mRNA's riboswitch affects its translation.

Section 10.6 **Epigenetic** refers to heritable modifications of DNA that affect gene expression but do not involve changes to the DNA sequence. DNA methylations and other epigenetic modifications acquired during an individual's lifetime can persist for generations.

self-quiz

Answers in Appendix VII

1. The expression of a gene may depend on _____ .
   a. the type of organism
   b. environmental conditions
   c. the type of cell
   d. all of the above

2. Gene expression in multicelled eukaryotic organisms changes in response to _____ .
   a. extracellular conditions
   b. master gene products
   c. operons
   d. a and b

3. Binding of _____ to _____ in DNA can increase the rate of transcription of specific genes.
   a. activators; repressors
   b. activators; enhancers
   c. repressors; operators
   d. both a and b

4. Proteins that influence RNA synthesis by binding directly to DNA are called _____ .
   a. promoters
   b. transcription factors
   c. operators
   d. enhancers

5. In eukaryotes, control over gene expression occurs at the level of _____ .
   a. transcription
   b. RNA processing
   c. RNA transport
   d. mRNA degradation
   e. translation
   f. protein modification
   g. a through e
   h. all of the above

6. Muscle cells differ from bone cells because _____ .
   a. they carry different genes
   b. they use different genes
   c. both a and b

7. Control over gene expression drives _____ in complex, multicelled eukaryotes.
   a. transcription factors
   b. nutrient availability
   c. development
   d. all of the above

8. Homeotic gene products _____ .
   a. flank a bacterial operon
   b. map out the overall body plan in embryos
   c. control the formation of specific body parts

9. A gene that is knocked out is _____ .
   a. deleted
   b. inactivated
   c. expressed
   d. either a or b

10. Which of the following includes all of the others?
    a. homeotic genes
    b. master genes
    c. *SRY* gene
    d. *PAX6*

11. The expression of *ABC* genes _____ .
    a. occurs in layers of an onion
    b. controls flower formation
    c. causes mutations in flowers

Effect of Paternal Grandmother's Food Supply on Infant Mortality Researchers are investigating long-reaching epigenetic effects of starvation, in part because historical data on periods of famine are widely available.

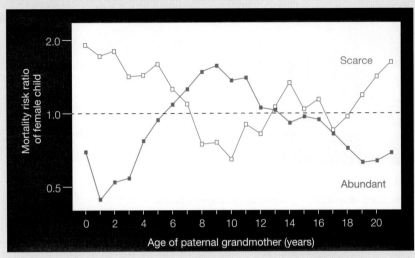

FIGURE 10.12 Graph showing the relative risk of early death of a female child, correlated with the age at which her paternal grandmother experienced a winter with a food supply that was scarce (blue) or abundant (red) during childhood. The dotted line represents no difference in risk of mortality. A value above the line means an increased risk; one below the line indicates a reduced risk.

Before the industrial revolution, a failed harvest in one autumn typically led to severe food shortages the following winter. A retrospective study has correlated female infant mortality at certain ages with the abundance of food during the paternal grandmother's childhood. **FIGURE 10.12** shows some of the results of this study.

1. Compare the mortality of girls whose paternal grandmothers ate well at age 2 with that of those who experienced famine at the same age. Which girl was more likely to die early? How much more likely was she to die?

2. Children have a period of slow growth around age 9. What trend in the data can you see around that age?

3. There was no correlation between early death of a male child and eating habits of his paternal grandmother, but there was a strong correlation with the eating habits of his paternal grandfather. What does this tell you about the probable location of epigenetic changes that gave rise to these data?

12. During X chromosome inactivation, _____ .
   a. female cells shut down
   b. RNA coats a chromosome
   c. pigments form
   d. both a and b

13. A cell with a Barr body is _____ .
   a. a bacterium   c. from a female mammal
   b. a sex cell   d. infected by the Barr virus

14. Operons _____ .
   a. only occur in bacteria
   b. have multiple genes
   c. involve selective gene expression

15. Match the terms with the most suitable description.
   ___ *SRY* gene   a. makes a man out of you
   ___ operator   b. binding site for repressor
   ___ Barr body   c. may be epigenetic
   ___ differentiation   d. inactivated X chromosome
   ___ mRNA zip code   e. controls multiple genes
   ___ methylation   f. localization mechanism
   ___ *eyeless*   g. works by binding DNA
   ___ homeodomain   h. required for eye formation
   ___ operon   i. driven by gene expression control

critical thinking

1. Why are some genes expressed and some not?

2. Do the same types of gene control operate in bacterial cells and eukaryotic cells?

3. Almost all calico cats (one is pictured in **FIGURE 10.7B**) are female. Why?

mutant   *wild-type*

4. The photos above show flowers from *Arabidopsis* plants. One plant is wild-type (unmutated); the other carries a mutation in one of its *ABC* floral identity genes. This mutation causes sepals and petals to form instead of stamens and carpels. Refer to Figure 10.8 to decide which gene (*A*, *B*, or *C*) has been inactivated by the mutation.

CREDITS: (12) Pembrey et al. *European Journal of Human Genetics* (2006) 14, 159–166; (in text CT #4) left, © Jose Luis Riechmann; right, Science Source.

LEARNING ROADMAP

Be sure you understand cell structure (Sections 4.2, 4.6, 4.10, 4.11), chromosomes (8.4), and DNA replication (8.5) before beginning. This chapter also revisits phosphorylation (5.6), membrane proteins (5.7), fermentation (7.6), mutations (8.6), animal cloning (8.7), cancer (10.1), gene control (10.2), and knockouts (10.3).

THE CELL CYCLE

A cell cycle starts when a new cell forms by division of a parent cell, and ends when the cell completes its own division. Built-in checkpoints control the timing and rate of the cycle.

MITOSIS

Mitosis, a mechanism by which a cell's nucleus divides, maintains the chromosome number. Four sequential stages parcel the cell's duplicated chromosomes into two new nuclei.

CYTOPLASMIC DIVISION

After nuclear division, the cytoplasm may divide, so one nucleus ends up in each of two new cells. Cytoplasmic division proceeds by different mechanisms in animal and plant cells.

MITOTIC CLOCKS

Built into eukaryotic chromosomes are DNA sequences that protect the cell's genetic information. Degradation of these sequences is associated with cell death and aging.

THE CELL CYCLE GONE AWRY

On rare occasions, cell cycle checkpoint mechanisms fail, and cell division becomes uncontrollable. Tumor formation and cancer are outcomes.

Mitosis is the basis of reproduction in single-celled eukaryotic organisms (Chapters 21 and 23). In multicelled eukaryotes, it has a role in reproduction and development (Chapters 29 and 42), as well as growth and tissue repair (Chapter 31). We compare mitosis with meiosis in Chapter 12. The HPV virus causes a sexually transmitted disease (Section 41.9) and cervical cancer (Section 37.1). Other cancers are discussed in Chapters 31, 32, and 34.

11.1 Henrietta's Immortal Cells

Each human starts out as a fertilized egg. By the time of birth, that cell has given rise to about a trillion cells, all organized as a human body. Even in an adult, billions of cells divide every day as new cells replace worn-out ones. However, despite the ability of human cells to continue dividing as part of a body, they tend to divide a limited number of times and die within weeks when grown in the laboratory.

Researchers had been trying to coax human cells to keep dividing outside of the body as early as the mid-1800s. Immortal cell lineages—cell lines—would allow the researchers to study human diseases (and potential cures for them) without experimenting on people. The quest to create a human cell line continued unsuccessfully until 1951. By this time, George and Margaret Gey had been trying to culture human cells for nearly thirty years. Then their assistant, Mary Kubicek, prepared a new sample of human cancer cells. Mary named the cells HeLa, after the first and last names of the patient from whom the cells had been taken. The HeLa cells began to divide, again and again. The cells were astonishingly vigorous, quickly coating the inside of their test tube and consuming their nutrient broth. Four days later, there were so many cells that the researchers had to transfer them to more tubes. The cell populations increased at a phenomenal rate. The cells were dividing every twenty-four hours and coating the inside of the tubes within days.

Sadly, cancer cells in the patient were dividing just as fast. Only six months after she had been diagnosed with cervical cancer, malignant cells had invaded tissues throughout her body. Two months after that, Henrietta Lacks, a young African American woman from Baltimore, was dead.

Although Henrietta passed away, her cells lived on in the Geys' laboratory (**FIGURE 11.1**). The Geys were able to grow poliovirus in HeLa cells, a practice that enabled them to determine which strains of the virus cause polio. That work was a critical step in the development of polio vaccines, which have since saved millions of lives.

Henrietta Lacks was just thirty-one, a wife and mother of five, when runaway cell divisions of cancer killed her. Her cells, however, are still dividing, again and again, more than fifty years after she died. Frozen away in tiny tubes and packed in Styrofoam boxes, HeLa cells continue to be shipped among laboratories all over the world. They are still widely used to investigate cancer, viral growth, protein synthesis, the effects of radiation, and countless other processes important in medicine and research. HeLa cells helped several researchers win Nobel Prizes, and some even traveled into space for experiments on satellites.

Understanding why cancer cells are immortal—and why we are not—begins with understanding the structures and mechanisms that cells use to divide.

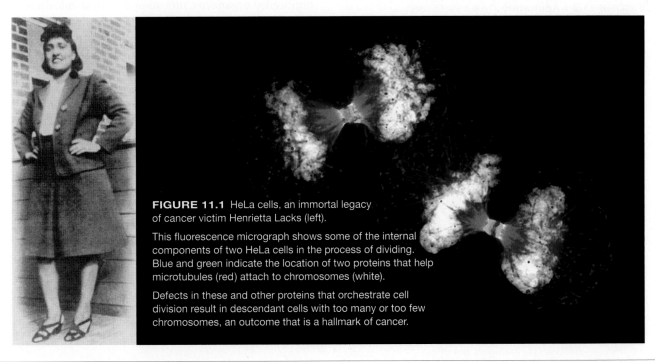

FIGURE 11.1 HeLa cells, an immortal legacy of cancer victim Henrietta Lacks (left).

This fluorescence micrograph shows some of the internal components of two HeLa cells in the process of dividing. Blue and green indicate the location of two proteins that help microtubules (red) attach to chromosomes (white).

Defects in these and other proteins that orchestrate cell division result in descendant cells with too many or too few chromosomes, an outcome that is a hallmark of cancer.

CREDITS: (opposite) © Carolina Biological Supply Company/Phototake; (1) left, Courtesy of the family of Henrietta Lacks; right, Dr. Paul D. Andrews/University of Dundee.

11.2 Multiplication by Division

✔ A cell's life occurs in a recognizable series of intervals and events collectively called the cell cycle.

✔ Division of a eukaryotic cell typically occurs in two steps: nuclear division followed by cytoplasmic division.

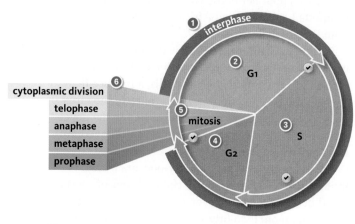

1 A cell spends most of its life in interphase, which includes three stages: G1, S, and G2.

2 G1 is the phase of growth before DNA replication. The cell's chromosomes are unduplicated.

3 S is the phase of synthesis, during which the cell makes copies of its chromosome(s) by DNA replication.

4 G2 is the phase after DNA replication and before mitosis. The cell prepares to divide during this stage.

5 The nucleus divides during mitosis, the four stages of which are detailed in the next section.

6 After mitosis, the cytoplasm may divide. Each descendant cell begins the cycle anew, in interphase.

✔ Built-in checkpoints stop the cycle from proceeding until certain conditions are met.

FIGURE 11.2 ▶**Animated** The eukaryotic cell cycle. The length of the intervals differs among cells. G1, S, and G2 are part of interphase.

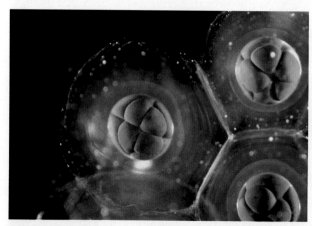

FIGURE 11.3 A multicelled eukaryote develops by repeated cell divisions. This photo shows early frog embryos, each a product of three mitotic divisions of one fertilized egg.

FIGURE IT OUT Each of these embryos has how many cells?

Answer: Eight

A life cycle is a sequence of recognizable stages that occur during an organism's lifetime, from the first cell of the new individual until its death. Multicelled organisms and free-living cells have life cycles, but what about cells that make up a multicelled body? Biologists consider such cells to be individually alive, each with its own lifetime. A cell's life passes through a series of recognizable intervals and events collectively called the **cell cycle** (**FIGURE 11.2**).

A typical cell spends most of its life in the interval called interphase **1**. During **interphase**, the cell increases its mass, roughly doubles the number of its cytoplasmic components, and replicates its DNA in preparation for division. Interphase comprises three major stages: G1, S, and G2. G1 and G2 were named "Gap" phases because outwardly they seem to be periods of inactivity, but they are not. Most cells going about their metabolic business are in G1 **2**. Cells preparing to divide enter S **3**, the phase of DNA synthesis, when they copy their chromosomes by DNA replication (Section 8.5). During G2 **4**, cells make the proteins that will drive the division process.

The remainder of the cell cycle consists of the division process itself. When the cell divides, both of its two cellular offspring end up with a blob of cytoplasm and some DNA. Each of the offspring of a eukaryotic cell inherits its DNA packaged inside a nucleus. Thus, a eukaryotic cell's nucleus has to divide before its cytoplasm does. **Mitosis** is a nuclear division mechanism that maintains the chromosome number **5**. In multicelled organisms, mitosis and cytoplasmic division **6** are the basis of increases in body size and tissue remodeling during development (**FIGURE 11.3**), as well as ongoing replacements of damaged or dead cells. Mitosis and cytoplasmic division are also part of **asexual reproduction**, a reproductive mode by which offspring are produced by one parent only. Some multicelled eukaryotes and many single-celled ones use this mode of reproduction. (Prokaryotes do not have a nucleus and do not undergo mitosis. We discuss their reproduction in Section 20.6.)

When a cell divides by mitosis, it produces two descendant cells, each with the chromosome number of the parent. However, if only the total number of chromosomes mattered, then one of the descendant cells might get, say, two pairs of chromosome 22 and no chromosome 9. A cell cannot function properly without a full complement of DNA, which means it needs to have *a copy of each* chromosome. Thus, the two cells produced by mitosis have the same number and types of chromosomes as the parent.

CREDITS: (2) © Cengage Learning; (3) © Carolina Biological Supply Company/Phototake.

Your body's cells are diploid, which means their nuclei contain pairs of chromosomes—two of each type (Section 8.4). One chromosome of a pair was inherited from your father; the other, from your mother. Except for a pairing of nonidentical sex chromosomes (XY) in males, the chromosomes of each pair are homologous. **Homologous chromosomes** have the same length, shape, and genes (*hom*– means alike).

FIGURE 11.4 shows how homologous chromosomes are distributed into descendant cells when a diploid cell divides by mitosis. When a cell is in G1, each of its chromosomes consists of one double-stranded DNA molecule ❷. The cell replicates its DNA in S, so by G2, each of its chromosomes consists of two double-stranded DNA molecules ❹. These molecules stay attached to one another at the centromere as sister chromatids until mitosis is almost over, and then they are pulled apart and packaged into two separate nuclei (the next section details this process).

When sister chromatids are pulled apart, each becomes an individual chromosome that consists of one double-stranded DNA molecule. Thus, each of the two new nuclei that form in mitosis contains a full complement of (unduplicated) chromosomes. When the cytoplasm divides, these nuclei are packaged into separate cells ❻. Each new cell starts the cell cycle over again in G1 of interphase.

Controls Over the Cell Cycle

When a cell divides—and when it does not—is determined by mechanisms of gene expression control (Section 10.2). Like the accelerator of a car, some of these mechanisms cause the cell cycle to advance. Others are like brakes, preventing the cycle from proceeding. In the adult body, brakes on the cell cycle normally keep the vast majority of cells in G1. Most of your nerve cells, skeletal muscle cells, heart muscle cells, and fat-storing cells have been in G1 since you were born, for example.

Control over the cell cycle also ensures that a dividing cell's descendants receive intact copies of its chro-

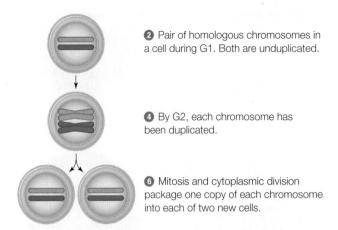

❷ Pair of homologous chromosomes in a cell during G1. Both are unduplicated.

❹ By G2, each chromosome has been duplicated.

❻ Mitosis and cytoplasmic division package one copy of each chromosome into each of two new cells.

FIGURE 11.4 How mitosis maintains chromosome number in a diploid cell. For clarity, only one pair of homologous chromosomes is shown. The maternal chromosome is shown in pink, the paternal one in blue. Numbered balls indicate cell cycle stages in **FIGURE 11.2**.

mosomes. Built-in checkpoints ensure the cell's DNA has been copied completely, that it is not damaged, and even that enough nutrients are available to support division. Protein products of "checkpoint genes" interact to carry out this control process. For example, at a checkpoint that operates in S, these molecules put the brakes on the cycle if the cell's chromosomes have been damaged during DNA replication (Section 8.6). Checkpoint proteins that function as sensors recognize damaged DNA and bind to it. Upon binding, they trigger other events that stall the cell cycle, and also enhance expression of genes involved in DNA repair. After the problem has been corrected, the brakes are lifted and the cell cycle proceeds. If the problem remains uncorrected, other checkpoint proteins may initiate a series of events that eventually cause the cell to self-destruct (you will read more about cell suicide, or apoptosis, in Section 42.4).

---

asexual reproduction Reproductive mode of eukaryotes by which offspring arise from a single parent only.
cell cycle The collective series of intervals and events of a cell's life, from the time it forms until it divides.
homologous chromosomes Chromosomes with the same length, shape, and genes.
interphase In a eukaryotic cell cycle, the interval during which a cell grows, roughly doubles the number of its cytoplasmic components, and replicates its DNA in preparation for division.
mitosis Nuclear division mechanism that maintains the chromosome number.

TAKE-HOME MESSAGE 11.2
How do eukaryotic cells reproduce?

✔ A cell cycle is the sequence of events and stages through which a cell passes during its lifetime. The cell cycle includes interphase, mitosis, and cytoplasmic division.

✔ A eukaryotic cell reproduces by division: nucleus first, then cytoplasm. Each descendant cell receives a set of chromosomes and some cytoplasm.

✔ When a nucleus divides by mitosis, each new nucleus has the same chromosome number as the parent cell.

✔ Gene expression controls advance, delay, or block the cell cycle in response to internal and external conditions. Checkpoints built into the cell cycle allow problems to be corrected before the cycle proceeds.

11.3 A Closer Look at Mitosis

Plant cell nucleus   **Animal cell nucleus**

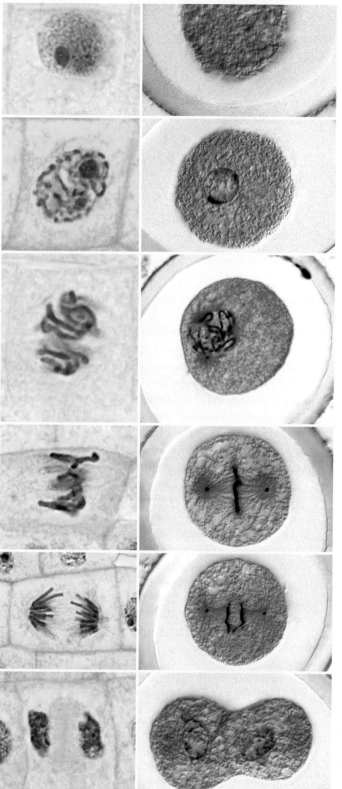

❶ Interphase
Interphase cells are shown for comparison, but interphase is not part of mitosis. The nuclear envelope is intact.

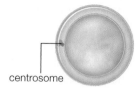

centrosome

❷ Early Prophase
Mitosis begins. Transcription stops, and the DNA begins to appear grainy as it starts to condense. The nuclear envelope begins to break up and the centrosome gets duplicated.

nuclear envelope breaking up

❸ Prophase
The duplicated chromosomes become visible as they condense. One of the two centrosomes moves to the opposite side of the cell as the nuclear envelope breaks up completely. Spindle microtubules assemble and bind to chromosomes at the centromere.

spindle microtubule

❹ Metaphase
All of the chromosomes are aligned midway between the spindle poles.

❺ Anaphase
Spindle microtubules separate the sister chromatids and move them toward opposite spindle poles. Each sister chromatid has now become an individual, unduplicated chromosome.

❻ Telophase
The chromosomes reach opposite sides of the cell and loosen up. Mitosis ends when a new nuclear envelope forms around each cluster of chromosomes.

FIGURE 11.5 ▶Animated Mitosis. Micrographs show nuclei of plant cells (onion root, left), and animal cells (fertilized eggs of a roundworm, right). A diploid (2n) animal cell with two chromosome pairs is illustrated.

CREDITS: (5) left, Michael Clayton/University of Wisconsin, Department of Botany; right, ISM/Phototake; far right, © Cengage Learning.

✔ The four main stages of mitosis are prophase, metaphase, anaphase, and telophase.

During interphase, a cell's chromosomes are loosened from histones to allow transcription and DNA replication. Loosening spreads out the chromosomes, so they are not easily visible under a light microscope (**FIGURE 11.5 ❶**). In preparation for nuclear division, the chromosomes begin to pack tightly ❷. Transcription and DNA replication stop as the chromosomes condense into their most compact "X" forms (Section 8.4). Tight condensation keeps the chromosomes from getting tangled and breaking during nuclear division.

Most animal cells have a structure called a centrosome, which typically consists of a pair of centrioles (Section 4.10) surrounded by a region of dense cytoplasm. The centrosome gets duplicated as mitosis begins. During **prophase**, the first stage of mitosis, the chromosomes condense so much that they become visible under a light microscope ❸. "Mitosis" is from *mitos*, the Greek word for thread, after the threadlike appearance of chromosomes during nuclear division. If the cell has centrosomes, one of them now moves to the opposite side of the cell. Microtubules begin to assemble and lengthen from the centrosomes (or from other structures in cells that have no centrosomes). The lengthening microtubules form a **spindle**, which is a temporary structure for moving chromosomes (**FIGURE 11.6**). The area from which the spindle originates on each side of the cell is called a spindle pole.

Spindle microtubules penetrate the nuclear region as the nuclear envelope breaks up. Some of the microtubules stop lengthening when they reach the middle of the cell. Others lengthen until they reach a chromosome and attach to it at the centromere. By the end of prophase, one sister chromatid of each chromosome has become attached to microtubules extending from one spindle pole, and the other sister chromatid has become attached to microtubules extending from the other spindle pole.

The opposing sets of microtubules then begin a tug-of-war by adding and losing tubulin subunits. As

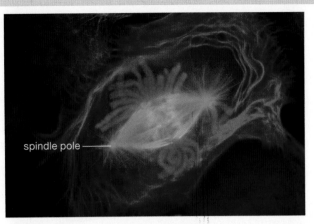

spindle pole

FIGURE 11.6 The spindle. In this dividing lung cell of a salamander, microtubules (green) have extended from two centrosomes to form the spindle, which has attached to and aligned the chromosomes (blue) midway between its two poles. Red shows actin microfilaments.

FIGURE IT OUT This cell is in which stage of mitosis?

Answer: Metaphase

the microtubules lengthen and shorten, they push and pull the chromosomes. When all the microtubules are the same length, the chromosomes are aligned midway between spindle poles ❹. The alignment marks **metaphase** (from *meta*, the ancient Greek word for between).

During **anaphase**, the spindle pulls the sister chromatids of each duplicated chromosome apart and moves them toward opposite spindle poles ❺. Each DNA molecule has now become a separate chromosome.

Telophase begins when one set of chromosomes reaches each spindle pole ❻. Each set consists of the same number and kinds of chromosomes as the parent cell nucleus had: two of each type of chromosome, if the parent cell was diploid. A new nuclear envelope forms around each set of chromosomes as they loosen up again. At this point, telophase—and mitosis—are over.

---

anaphase Stage of mitosis during which sister chromatids separate and move toward opposite spindle poles.
metaphase Stage of mitosis at which all chromosomes are aligned midway between spindle poles.
prophase Stage of mitosis during which chromosomes condense and become attached to a newly forming spindle.
spindle Temporary structure that moves chromosomes during nuclear division; consists of microtubules.
telophase Stage of mitosis during which chromosomes arrive at opposite spindle poles and decondense, and two new nuclei form.

TAKE-HOME MESSAGE 11.3
What is the sequence of events in mitosis?

✔ DNA replication occurs before mitosis. As mitosis begins, each chromosome consists of two DNA molecules attached as sister chromatids.

✔ In prophase, the chromosomes condense and a spindle forms. Spindle microtubules attach to the chromosomes as the nuclear envelope breaks up.

✔ At metaphase, the spindle has aligned all of the (still duplicated) chromosomes in the middle of the cell.

✔ In anaphase, sister chromatids separate and move toward opposite spindle poles. Each DNA molecule is now an individual chromosome.

✔ In telophase, two clusters of chromosomes reach opposite spindle poles. A new nuclear envelope forms around each cluster, so two new nuclei form.

11.4 Cytokinesis: Division of Cytoplasm

✔ The mechanism by which cytoplasm divides differs between plant and animal cells.

In most eukaryotes, the cell cytoplasm divides between late anaphase and the end of telophase, so two cells form, each with their own nucleus. The mechanism of

Animal cell cytokinesis

❶ In a dividing animal cell, the spindle disassembles as mitosis ends.

❷ At the midpoint of the former spindle, a ring of actin and myosin filaments attached to the plasma membrane contracts.

❸ This contractile ring pulls the cell surface inward, forming a cleavage furrow as it shrinks.

❹ The ring contracts until it pinches the cell in two.

Plant cell cytokinesis

❺ In a dividing plant cell, vesicles cluster at the future plane of division before mitosis ends.

❻ The vesicles fuse with each other, forming a cell plate along the plane of division.

❼ The cell plate expands outward along the plane of division. When it reaches and attaches to the plasma membrane, it partitions the cytoplasm.

❽ The cell plate matures as two new cell walls. These walls join with the parent cell wall, so each descendant cell becomes enclosed by its own wall.

FIGURE 11.7 ▶Animated Cytoplasmic division of animal cells (top) and plant cells (bottom).

cytoplasmic division, which is called **cytokinesis**, differs between plants and animals.

Typical animal cells pinch themselves in two after nuclear division ends (**FIGURE 11.7**). How? The spindle begins to disassemble during telophase ❶. The cell cortex, which is the mesh of cytoskeletal elements just under the plasma membrane (Section 4.10), includes a band of actin and myosin filaments that wraps around the cell, midway between the former spindle poles. The band is called a contractile ring because it contracts when its component proteins are energized by phosphate-group transfers from ATP. When the ring contracts, it drags the attached plasma membrane inward ❷. The sinking plasma membrane becomes visible on the outside of the cell as an indentation between the former spindle poles ❸. The indentation, which is called a **cleavage furrow**, deepens until the cytoplasm (and the cell) is pinched in two ❹. Each of the two cells formed by this division has its own nucleus and some of the parent cell's cytoplasm, and each is enclosed by a plasma membrane.

Dividing plant cells face a particular challenge because a stiff cell wall surrounds their plasma membrane (Section 4.11). Accordingly, plant cells have their own mechanism of cytokinesis. By the end of anaphase, a set of short microtubules has formed on either side of the future plane of division. These microtubules guide vesicles from Golgi bodies and the cell surface to the division plane ❺. These vesicles provide material for a new cell membrane and wall. After mitosis, the vesicles fuse into a disk-shaped **cell plate** ❻. The plate expands at its edges until it reaches the plasma membrane and attaches to it, thus partitioning the cytoplasm ❼. In time, the cell plate will develop into two new cell walls, so each of the descendant cells will be enclosed by its own plasma membrane and wall ❽.

---

cell plate A disk-shaped structure that forms during cytokinesis in a plant cell; matures as a cross-wall between the two new nuclei.
cleavage furrow In a dividing animal cell, the indentation where cytoplasmic division will occur.
cytokinesis Cytoplasmic division.

TAKE-HOME MESSAGE 11.4
How do eukaryotic cells divide?

✔ In most eukaryotes, the cell cytoplasm divides between late anaphase and the end of telophase. Two descendant cells form, each with its own nucleus.

✔ In animal cell cytokinesis, a contractile ring pinches the cytoplasm in two. In plant cell cytokinesis, a cell plate that forms in the middle of the cell partitions the cytoplasm when it reaches and connects to the parent cell wall.

11.5 Marking Time With Telomeres

✔ Telomeres protect eukaryotic chromosomes from losing genetic information at their ends.

Remember that Dolly, the first clone produced by somatic cell nuclear transfer (SCNT, Section 8.7), died early. The life expectancy of a sheep is normally about 10 to 12 years. By the time Dolly was five, however, she was as fat and arthritic as a twelve-year-old sheep. The following year, she contracted a lung disease typical of much older sheep, and had to be euthanized.

Dolly's early demise may have been the result of abnormally short telomeres. **Telomeres** are noncoding DNA sequences that occur at the ends of eukaryotic chromosomes (**FIGURE 11.8**). Vertebrate telomeres consist of a short DNA sequence, 5'-TTAGGG-3', repeated perhaps thousands of times. These "junk" repeats provide a buffer against the loss of more valuable DNA internal to the chromosomes.

A telomere buffer is particularly important because, under normal circumstances, a eukaryotic chromosome shortens by about 100 nucleotides with each DNA replication. When a cell's offspring receive chromosomes with too-short telomeres, checkpoint gene products halt the cell cycle, and the descendant cells die shortly thereafter. Most body cells can divide only a certain number of times before this happens. This cell division limit may be a fail-safe mechanism in case a cell loses control over the cell cycle and begins to divide again and again. A limit on the number of divisions keeps such cells from overrunning the body (an outcome that, as you will see in the next section, has dangerous consequences to health).

The cell division limit varies by species, and it may be part of the mechanism that sets an organism's life span. Dolly's DNA came from the nucleus of a mammary gland cell donated by an adult sheep. When Dolly was only two years old, her telomeres were as short as those of a six-year-old sheep—the exact age of the adult animal that had been her genetic donor.

A few normal cells in an adult retain the ability to divide indefinitely. Their descendants replace cell lineages that eventually die out when they reach their division limit. These cells are called stem cells, and they are immortal because they continue to make an enzyme called telomerase. Telomerase reverses the telomere shortening that normally occurs after DNA replication.

Mice that have had their telomerase enzyme knocked out age prematurely. Their tissues degener-

telomere Noncoding, repetitive DNA sequence at the end of chromosomes; protects the coding sequences from degradation.

FIGURE 11.8 ▶Animated Telomeres. The bright dots at the end of each DNA strand in these duplicated chromosomes are telomeres.

ate much more quickly than those of normal mice, and their life expectancy declines to about half that of a normal mouse. When one of these knockout mice is close to the end of its shortened life span, rescuing the function of its telomerase enzyme results in lengthened telomeres. The rescued mouse also regains vitality: Decrepit tissue in its brain and other organs repairs itself and begins to function normally, and the once-geriatric individual even begins to reproduce again.

Researchers are careful to point out that shortening telomeres may be an effect of aging rather than a cause. Also, while telomerase holds therapeutic promise for rejuvenating aged tissues, it can also be dangerous. Cancer cells—including the HeLa cells you learned about in Section 11.1—characteristically express high levels of this molecule, which is why, like stem cells, they can divide indefinitely.

TAKE-HOME MESSAGE 11.5
What is the function of telomeres?

✔ Telomeres are noncoding sequences at the end of eukaryotic chromosomes. These extra sequences prevent loss of genetic information.

✔ Telomeres shorten with every cell division in normal body cells. When they are too short, the cell stops dividing and dies. Thus, telomeres are associated with aging.

11.6 When Mitosis Becomes Pathological

✔ On rare occasions, controls over cell division are lost and a neoplasm forms.

✔ Cancer develops as cells of a neoplasm become malignant.

Sometimes a checkpoint gene mutates so that its protein product no longer works properly. In other cases, the controls that regulate its expression fail, and a cell makes too much or too little of its product. When enough checkpoint mechanisms fail, a cell loses control over its cell cycle. Interphase may be skipped, so division occurs over and over with no resting period. Signaling mechanisms that cause abnormal cells to die may stop working. The problem is compounded because checkpoint malfunctions are passed along to the cell's descendants, which form a **neoplasm**, an accu-

mulation of cells that lost control over how they grow and divide.

A neoplasm that forms a lump in the body is called a **tumor**, but the two terms are sometimes used interchangeably. Once a tumor-causing mutation has occurred, the gene it affects is called an oncogene. An **oncogene** is any gene that can transform a normal cell into a tumor cell (Greek *onkos*, or bulging mass). Oncogene mutations in reproductive cells can be passed to offspring, which is a reason that some types of tumors run in families.

Genes encoding proteins that promote mitosis are called **proto-oncogenes** because mutations can turn them into oncogenes. One proto-oncogene, *EGFR*, encodes a plasma membrane receptor for EGF (epidermal growth factor). EGF and other **growth factors** are molecules that stimulate a cell to divide and differentiate. When the EGF receptor binds to EGF, it becomes activated and triggers the cell to begin mitosis. When EGF is not present, the receptor is in an inactive form and does not trigger mitosis. Tumor-causing mutations in *EGFR* change the receptor so that it stimulates mitosis even in the absence of EGF. Most neoplasms carry mutations resulting in an overactivity or overabundance of this particular receptor (**FIGURE 11.9**).

Checkpoint gene products that inhibit mitosis are called tumor suppressors because tumors form when they are missing. The products of the *BRCA1* and *BRCA2* genes that you learned about in Chapter 10 are examples of tumor suppressors. These proteins regulate, among other things, the expression of DNA repair enzymes (**FIGURE 11.10**). Tumor cells often have mutations in their *BRCA* genes.

Viruses such as HPV (human papillomavirus) cause a cell to make proteins that interfere with its own tumor suppressors. Infection with HPV causes skin growths called warts, and some kinds are associated with neoplasms that form on the cervix.

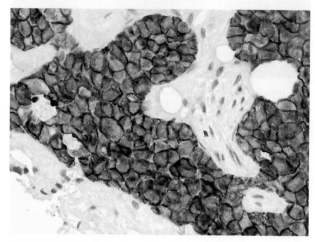

FIGURE 11.9 Effects of an oncogene. In this section of human breast tissue, a brown-colored tracer shows the active form of the EGF receptor. Normal cells are lighter in color. The dark cells have an overactive EGF receptor that is constantly stimulating mitosis; these cells have formed a neoplasm.

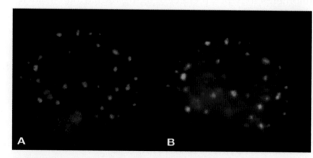

FIGURE 11.10 Checkpoint genes in action.

Radiation damaged the DNA inside this nucleus. (**A**) Red dots show the location of the *BRCA1* gene product. (**B**) Green dots pinpoint the location of the product of a gene called *53BP1*.

Both proteins have clustered around the same chromosome breaks in the same nucleus; both function to recruit DNA repair enzymes. The integrated action of these and other checkpoint gene products blocks mitosis until the DNA breaks are fixed.

Cancer

Benign neoplasms such as warts are not usually dangerous. They grow very slowly, and their cells retain the plasma membrane adhesion proteins that keep them properly anchored to the other cells in their home tissue (**FIGURE 11.11 ❶**).

A malignant neoplasm is one that gets progressively worse, and is dangerous to health. Malignant cells typically display the following three characteristics:

First, like cells of all neoplasms, malignant cells grow and divide abnormally. Controls that usually keep cells from getting overcrowded in tissues are lost in malig-

CREDITS: (9) From *Expression of the epidermal growth factor receptor (EGFR) and the phosphorylated EGFR in invasive breast carcinoma.* breast-cancer research.com/content/10/3/R49; (10) © Phillip B. Carpenter, Department of Biochemistry and Molecular Biology, University of Texas - Houston Medical School.

nant cells, so their populations may reach extremely high densities with cell division occurring very rapidly. The number of small blood vessels that transport blood to the growing cell mass also increases abnormally.

Second, the cytoplasm and plasma membrane of malignant cells are altered. Both are indications of cellular malfunction. The cytoskeleton may be shrunken, disorganized, or both. Malignant cells typically have an abnormal chromosome number, with some chromosomes present in multiple copies, and others missing or damaged. The balance of metabolism is often shifted, as in an amplified reliance on ATP formation by fermentation rather than aerobic respiration.

Altered or missing proteins impair the function of the plasma membrane of malignant cells. For example, these cells do not stay anchored properly in tissues because their plasma membrane adhesion proteins are defective or missing ❷. Malignant cells can slip easily into and out of vessels of the circulatory and lymphatic systems ❸. By migrating through these vessels, the cells can establish neoplasms elsewhere in the body ❹. The process in which malignant cells break loose from their home tissue and invade other parts of the body is called **metastasis**. Metastasis is the third hallmark of malignant cells.

The disease called **cancer** occurs when the abnormally dividing cells of a malignant neoplasm disrupt body tissues, both physically and metabolically. Unless chemotherapy, surgery, or another procedure eliminates malignant cells from the body, they can put an individual on a painful road to death. Each year, cancer causes 15 to 20 percent of all human deaths in developed countries. The good news is that mutations in multiple checkpoint genes are required to transform a normal cell into a malignant one, and such mutations may take a lifetime to accumulate. Lifestyle choices such as not smoking and avoiding exposure of unprotected skin to sunlight can reduce one's risk of acquiring mutations in the first place. Some neoplasms can be detected with periodic screening such as gynecology or dermatology exams (**FIGURE 11.12**). If detected early enough, many types of malignant neoplasms can be removed before metastasis occurs.

cancer Disease that occurs when a malignant neoplasm physically and metabolically disrupts body tissues.
growth factor Molecule that stimulates mitosis and differentiation.
metastasis The process in which malignant cells spread from one part of the body to another.
neoplasm An accumulation of abnormally dividing cells.
oncogene Gene that helps transform a normal cell into a tumor cell.
proto-oncogene Gene that, by mutation, can become an oncogene.
tumor A neoplasm that forms a lump.

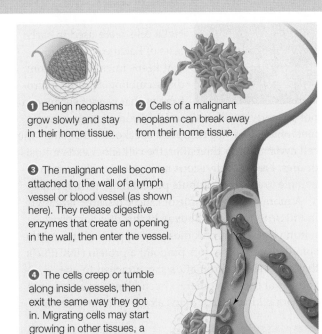

❶ Benign neoplasms grow slowly and stay in their home tissue.

❷ Cells of a malignant neoplasm can break away from their home tissue.

❸ The malignant cells become attached to the wall of a lymph vessel or blood vessel (as shown here). They release digestive enzymes that create an opening in the wall, then enter the vessel.

❹ The cells creep or tumble along inside vessels, then exit the same way they got in. Migrating cells may start growing in other tissues, a process called metastasis.

FIGURE 11.11 ▶**Animated** Neoplasms and malignancy.

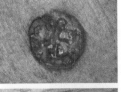

A Basal cell carcinoma is the most common type of skin cancer. This slow-growing, raised lump may be uncolored, reddish-brown, or black.

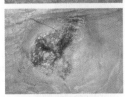

B Squamous cell carcinoma is the second most common form of skin cancer. This pink growth, firm to the touch, grows under the skin's surface.

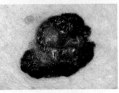

C Melanoma spreads fastest. Cells form dark, encrusted lumps that may itch or bleed easily.

FIGURE 11.12 Skin cancer can be detected with early screening.

TAKE-HOME MESSAGE 11.6

What is cancer?

✔ Neoplasms form when cells lose control over their cell cycle and begin dividing abnormally.

✔ Mutations in multiple checkpoint genes can give rise to a malignant neoplasm that gets progressively worse.

✔ Cancer is a disease that occurs when the abnormally dividing cells of a malignant neoplasm physically and metabolically disrupt body tissues.

✔ Although some mutations are inherited, lifestyle choices and early intervention can reduce one's risk of cancer.

CREDITS: (11) © Cengage Learning; (12A) Dr. Allan Harris/Phototake; (12B) Biophoto Associates/Science Source; (12C) James Stevenson/Science Source.

Henrietta's Immortal Cells (revisited)

HeLa cells were used in early tests of Paclitaxel, a drug that keeps microtubules from disassembling. Spindle microtubules that cannot shrink cannot properly position the cell's chromosomes during metaphase, and this triggers a checkpoint that stops the cell cycle. Shortly thereafter, the cell either exits mitosis or dies. Frequent divisions make cancer cells more vulnerable to this microtubule poison than normal cells.

A more recent example of cancer research is shown in **FIGURE 11.1** (and above). In this micrograph of mitotic HeLa cells, chromosomes appear white and the spindle is red. Blue dots pinpoint a protein (INCENP) that helps sister chromatids stay attached to one another at the centromere. Green identifies an enzyme (Aurora B kinase) that helps attach spindle microtu-

bules to centromeres. At this stage of telophase, these two proteins should be closely associated midway between the two clusters of chromosomes. The abnormal distribution means that the chromosomes are not properly attached to the spindle.

Defects in Aurora B kinase or its expression result in unequal distribution of chromosomes into descendant cells. Researchers recently correlated overexpression of Aurora B in cancer cells with shortened patient survival rates. Thus, drugs that inhibit Aurora B function are now being tested as potential cancer therapies.

Ongoing research with HeLa cells may one day allow researchers to identify drugs that target and destroy malignant cells or stop them from dividing. The research is far too late to have saved Henrietta Lacks, but it may one day yield drugs that put the brakes on cancer.

summary

Section 11.1 An immortal line of human cells (HeLa) is a legacy of cancer victim Henrietta Lacks. Researchers all over the world continue to work with these cells as they try to unravel the mechanisms of cancer.

Section 11.2 A **cell cycle** includes all the recognizable stages and events of a cell's lifetime; it starts when a new cell forms, and ends when the cell reproduces. Most of a cell's activities, including DNA replication that copies its **homologous chromosomes**, occur during **interphase**. A eukaryotic cell reproduces by dividing: nucleus first, then cytoplasm. **Mitosis** is a mechanism of nuclear division that maintains the chromosome number. It is the basis of growth, cell replacements, and tissue repair in multicelled species, and **asexual reproduction** in many species.

Section 11.3 Mitosis proceeds in four stages. The cell's (duplicated) chromosomes condensense during **prophase**. Microtubules assemble and form a **spindle**, and the nuclear envelope breaks up. Some microtubules that extend from one spindle pole attach to one chromatid of each chromosome; some that extend from the opposite spindle pole attach to its sister chromatid. These microtubules drag each chromosome toward the center of the cell.

At **metaphase**, all chromosomes are aligned at the spindle's midpoint.

During **anaphase**, the sister chromatids of each chromosome detach from each other, and the spindle microtubules move them toward opposite spindle poles.

During **telophase**, a complete set of chromosomes reaches each spindle pole. A nuclear envelope forms around each

cluster. Two new nuclei, each with the parental chromosome number, are the result.

Section 11.4 **Cytokinesis** typically follows nuclear division. In animal cells, a contractile ring of microfilaments pulls the plasma membrane inward, forming a **cleavage furrow** that pinches the cytoplasm in two. In plant cells, vesicles guided by microtubules to the future plane of division merge as a **cell plate**. The plate expands and fuses with the parent cell wall, thus becoming a cross-wall that partitions the cytoplasm.

Section 11.5 **Telomeres** that protect the ends of eukaryotic chromosomes shorten with every DNA replication. Cells that inherit too-short telomeres die, and in most cells this limits the number of divisions that can occur.

Section 11.6 The products of checkpoint genes work together to control the cell cycle. These molecules monitor the integrity of the cell's DNA, and can pause the cycle until breaks or other problems are fixed. When checkpoint mechanisms fail, a cell loses control over its cell cycle, and the cell's descendants form a **neoplasm**. Neoplasms may form lumps called **tumors**.

Genes encoding **growth factor** receptors are examples of **proto-oncogenes**, which means mutations can turn them into tumor-causing **oncogenes**. Mutations in multiple checkpoint genes can transform benign neoplasms into malignant ones. Cells of malignant neoplasms can break loose from their home tissues and colonize other parts of the body, a process called **metastasis**. **Cancer** occurs when malignant neoplasms physically and metabolically disrupt normal body tissues.

CREDIT: (in text) Dr. Paul D. Andrews/University of Dundee.

HeLa Cells Are a Genetic Mess HeLa cells can vary in chromosome number. Defects in proteins that orchestrate cell division result in descendant cells with too many or too few chromosomes, an outcome that is one of the hallmarks of cancer cells. The panel of chromosomes on the right, originally published in 1989, shows all of the chromosomes in a single metaphase HeLa cell.

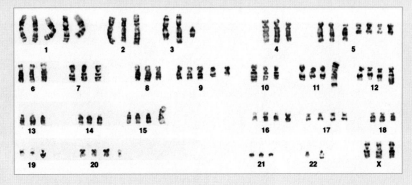

1. What is the chromosome number of this HeLa cell?

2. How many extra chromosomes does this cell have, compared to a normal human body cell?

3. Can you tell that this cell came from a female? How?

self-quiz

Answers in Appendix VII

1. Mitosis and cytoplasmic division function in _____ .
   a. asexual reproduction of single-celled eukaryotes
   b. growth and tissue repair in multicelled species
   c. asexual reproduction in prokaryotes
   d. both a and b   e. all of the above

2. A duplicated chromosome has how many chromatids?

3. Homologous chromosomes _____ .
   a. carry the same genes   c. are the same length
   b. are the same shape   d. all of the above

4. Most cells spend the majority of their lives in _____ .
   a. prophase   d. telophase
   b. metaphase   e. interphase
   c. anaphase   f. a and d

5. The spindle attaches to chromosomes at the _____ .
   a. centriole   c. centromere
   b. contractile ring   d. centrosome

6. Only _____ is not a stage of mitosis.
   a. prophase   c. interphase
   b. metaphase   d. anaphase

7. In intervals of interphase, G stands for _____ .
   a. gap   b. growth   c. Gey   d. gene

8. Interphase is the part of the cell cycle when _____ .
   a. a cell ceases to function
   b. a cell forms its spindle apparatus
   c. a cell grows and duplicates its DNA
   d. mitosis proceeds

9. After mitosis, the chromosome number of a descendant cell is _____ the parent cell's.
   a. the same as   c. rearranged compared to
   b. one-half of   d. double that of

10. A plant cell divides by the process of _____ .
    a. telekinesis   c. fission
    b. nuclear division   d. cytokinesis

11. *BRCA1* and *BRCA2* _____ .
    a. are checkpoint genes   c. encode tumor suppressors
    b. are proto-oncogenes   d. all of the above

12. _____ are characteristic of cancer.
    a. Malignant cells   b. Neoplasms   c. Tumors

13. Match each term with its best description.
    ___ cell plate   a. lump of cells
    ___ spindle   b. made of microfilaments
    ___ tumor   c. divides plant cells
    ___ cleavage furrow   d. organize(s) the spindle
    ___ contractile ring   e. dangerous metastatic cells
    ___ cancer   f. made of microtubules
    ___ centrosomes   g. indentation
    ___ telomere   h. shortens with age

14. Match each stage with the events listed.
    ___ metaphase   a. sister chromatids move apart
    ___ prophase   b. chromosomes condense
    ___ telophase   c. new nuclei form
    ___ interphase   d. all chromosomes are aligned
    ___ anaphase   midway between spindle poles
    ___ cytokinesis   e. DNA replication
                            f. cytoplasmic division

critical thinking

1. When a cell reproduces by mitosis and cytoplasmic division, does its life end?

2. The eukaryotic cell in the photo on the left is in the process of cytoplasmic division. Is this cell from a plant or an animal? How do you know?

3. Exposure to radioisotopes or other sources of radiation can damage DNA. Humans exposed to high levels of radiation face a condition called radiation poisoning. Why do you think that exposure to radiation is used as a therapy to treat some kinds of cancers?

4. Suppose you have a way to measure the amount of DNA in one cell during the cell cycle. You first measure the amount at the G1 phase. At what points in the rest of the cycle will you see a change in the amount of DNA per cell?

CENGAGE **brain**.com To access course materials, please visit www.cengagebrain.com.

LEARNING ROADMAP

This chapter will draw on your knowledge of eukaryotic chromosomes (Section 11.2), DNA replication (8.5), genes (9.2), cytoplasmic division (11.4), and cell cycle controls (11.6) as we compare meiosis with mitosis (11.3). You will also revisit clones (8.1, 8.7), and the effects of mutation (9.6).

ALLELES AND SEXUAL REPRODUCTION

Genes that vary a bit in sequence as alleles are the basis of variation in traits. In sexual reproduction, offspring inherit genes from two parents who usually differ in some number of alleles.

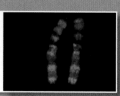

MEIOSIS IN THE LIFE CYCLE

Meiosis is a nuclear division process that reduces the chromosome number. It occurs only in cells that play a role in sexual reproduction in eukaryotes.

STAGES OF MEIOSIS

The chromosome number becomes reduced by the two nuclear divisions of meiosis. During this process, the chromosomes are sorted into four new nuclei.

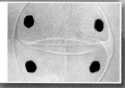

SHUFFLING PARENTAL DNA

During meiosis, homologous chromosomes swap segments, then are randomly sorted into separate nuclei. Both processes lead to novel combinations of alleles among offspring.

MITOSIS AND MEIOSIS COMPARED

Similarities between mitosis and meiosis suggest meiosis originated by evolutionary remodeling of mechanisms that already existed for mitosis and for repairing damaged DNA.

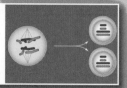

Meiosis is the basis of sexual reproduction, a topic covered in detail in Chapters 29 (plants) and 41 (animals). The variation in traits among individuals of sexually reproducing species arises from allele differences, a concept we revisit in context of inheritance in Chapters 13 and 14, and natural selection in Chapter 17.

A few species reproduce asexually, which means one individual gives rise to offspring that are identical to itself and to one another. By contrast, **sexual reproduction** involves two individuals and mixes their genetic material (**FIGURE 12.1**). Almost all species reproduce sexually. If the function of reproduction is the perpetuation of one's genes, then an asexual reproducer would seem to win the evolutionary race. When it reproduces, it passes all of its genes to every one of its offspring. Only about half of a sexual reproducer's genes are passed to each offspring. Yet most eukaryotes reproduce sexually, at least some of the time. Why?

Consider that all offspring of an asexual reproducer are clones: They have the same traits as their parent. Consistency is a good thing if an organism lives in a favorable, unchanging environment. Traits that help it survive and reproduce do the same for its descendants. However, most environments are constantly changing. All offspring of an asexual reproducer are identical to one another and to the parent, so all are equally vulnerable to a change that is unfavorable. In changing environments, sexual reproducers have the evolutionary edge. Their offspring vary in the details of their traits. Some may have a particular combination of traits that suits them perfectly to a change in their environment. As a group, their diversity offers them a better chance of surviving this challenge than clones.

Consider the interaction between a predatory species and its prey. In each generation, prey individuals with traits that allow them to hide from, fend off, or escape the predator will leave the most offspring. However, the predator is constantly changing too: In each generation, individuals best able to find, capture, and overcome prey leave the most descendants. Thus, predators and prey are locked in a constant race, with each genetic improvement in one species countered by a genetic improvement in the other. This idea is called the Red Queen hypothesis, a reference to Lewis Carroll's book *Through the Looking Glass*. In the book, the Queen of Hearts tells Alice, "It takes all the running you can do, to keep in the same place." Environmental challenges that constantly change favor sexual reproduction, because the genetic diversity it fosters is evolutionarily advantagous in a changing environment.

Perhaps the most important advantage of sexual reproduction involves the inevitable occurrence of harmful mutations. A population of sexual reproduc-

ers has a better chance of weathering the effects of such mutations. With asexual reproduction, individuals bearing a harmful mutation necessarily pass it to all of their offspring. This outcome would be rare in sexual reproduction, because each offspring of a sexual union has a 50 percent chance of inheriting a parent's mutation. Thus, all else being equal, harmful mutations accumulate in an asexually reproducing population more quickly than in a sexually reproducing one.

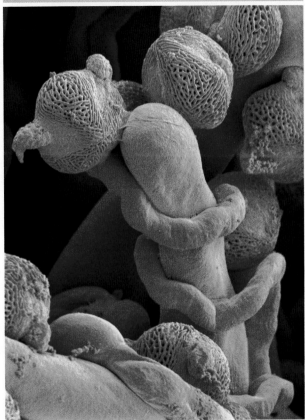

FIGURE 12.1 Moments in the stages of sexual reproduction. Sex mixes up the genetic material of two individuals.

Top, New Zealand mud snails can reproduce on their own (asexually) or with a partner (sexually). Natural populations of this species vary in their proportion of sexual and asexual individuals.

Bottom, flowering plants can reproduce asexually or sexually. The micrograph shows sexual reproduction, in which pollen grains (orange) germinate on flower carpels (yellow). Pollen tubes with male gametes inside grow from the grains down into tissues of the ovary, which house the flower's female gametes.

---

sexual reproduction Reproductive mode by which offspring arise from two parents and inherit genes from both.

CREDITS: (opposite) Reprinted from Fertility and Sterility, Vol 87/edition 3, Fei Sun, Paul Turek, Calvin Greene, Evelyn Ko, Alfred Rademaker, Renée H. Martin, Abnormal progression through meiosis in men with nonobstructive azoospermia, 565–571, copyright 2007, with permission from Elsevier; (1) top, © Bart Zijistra; bottom, Susumu Nishinaga/ Science Source.

✔ Small differences in shared genes are the basis of differences in shared traits.

✔ Sexual reproduction mixes up alleles from two parents.

✔ Meiosis, the basis of sexual reproduction, occurs in eukaryotic cells set aside for reproduction.

Introducing Alleles

Your **somatic** (body) cells and those of many other sexually reproducing eukaryotes are diploid, which means they contain pairs of chromosomes. One chromosome of each homologous pair is maternal, and the other is paternal (Section 11.2). Homologous chromosomes carry the same genes (**FIGURE 12.2A**). However, the corresponding genes on maternal and paternal chromosomes often vary—just a bit—in DNA sequence. Over evolutionary time, unique mutations accumulate in separate lines of descent, and some of those mutations occur in genes. Thus, the DNA sequence of any gene may differ from the corresponding gene on the homologous chromosome (**FIGURE 12.2B**). Different forms of the same gene are called **alleles**.

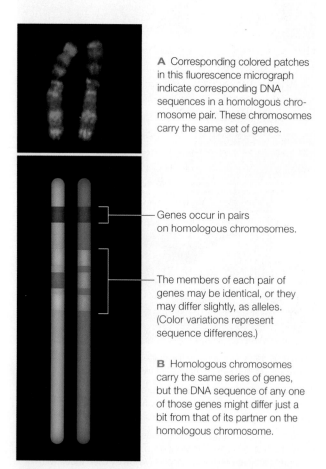

A Corresponding colored patches in this fluorescence micrograph indicate corresponding DNA sequences in a homologous chromosome pair. These chromosomes carry the same set of genes.

— Genes occur in pairs on homologous chromosomes.

— The members of each pair of genes may be identical, or they may differ slightly, as alleles. (Color variations represent sequence differences.)

B Homologous chromosomes carry the same series of genes, but the DNA sequence of any one of those genes might differ just a bit from that of its partner on the homologous chromosome.

FIGURE 12.2 ▶Animated Genes on chromosomes. Different forms of a gene are called alleles.

Alleles may encode slightly different forms of a gene's product, and such differences influence the details of traits shared by a species. Consider that one of the approximately 20,000 genes in human chromosomes encodes beta globin, a subunit of hemoglobin (Section 9.6). Like most human genes, the beta globin gene has multiple alleles—more than 700 in this case. A few beta globin alleles cause sickle-cell anemia, several cause beta thalassemia, and so on. Allele differences among individuals are one reason that the members of a sexually reproducing species are not identical. Offspring of sexual reproducers vary in the combinations of alleles they inherit; thus, they vary in their combinations of traits.

Meiosis Halves the Chromosome Number

Sexual reproduction involves the fusion of mature reproductive cells—**gametes**—from two parents. Gametes have a single set of chromosomes, so they are **haploid** (n): Their chromosome number is half of the diploid ($2n$) number (Section 8.4). **Meiosis**, the nuclear division mechanism that halves the chromosome number, is necessary for gamete formation. Meiosis also gives rise to new combinations of parental alleles.

Gametes arise by division of **germ cells**, which are immature reproductive cells that form in organs set aside for reproduction (**FIGURE 12.3**). Animals and plants make gametes somewhat differently. In animals, meiosis in diploid germ cells gives rise to eggs (female gametes) or sperm (male gametes). In plants, haploid germ cells form by meiosis. Gametes form when these cells divide by mitosis.

The first part of meiosis is similar to mitosis. A cell duplicates its DNA before either nuclear division process begins. As in mitosis, a spindle forms, and its microtubules move the duplicated chromosomes to opposite spindle poles. However, meiosis sorts the chromosomes into new nuclei not once, but twice, so it results in the formation of four haploid nuclei. The two consecutive nuclear divisions are called meiosis I and meiosis II:

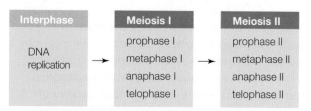

Interphase	Meiosis I	Meiosis II
DNA replication	prophase I metaphase I anaphase I telophase I	prophase II metaphase II anaphase II telophase II

In some cells, meiosis II occurs immediately after meiosis I. In others, a period of protein synthesis—but no DNA replication—intervenes between the divisions.

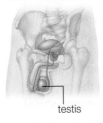
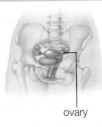

testis   ovary

A Reproductive organs of humans. Meiosis in germ cells inside testes and ovaries produces gametes (sperm and eggs).

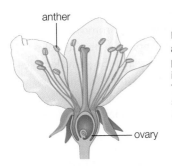

anther

ovary

B Reproductive organs of a flowering plant. Meiosis produces haploid germ cells inside anthers and ovaries. These cells divide by mitosis to give rise to gametes (sperm and eggs).

FIGURE 12.3 ▶**Animated** Examples of reproductive organs in (**A**) animals and (**B**) plants.

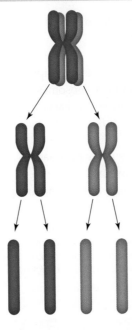

❶ Chromosomes are duplicated before meiosis begins. During meiosis I, each chromosome in the nucleus pairs with its homologous partner. The nucleus contains two of each chromosome, so it is diploid (2n).

❷ Homologous partners separate and are packaged into two new nuclei. Each new nucleus contains one of each chromosome, so it is haploid (n). The chromosomes are still duplicated.

❸ Sister chromatids separate in meiosis II and are packaged into four new nuclei. Each new nucleus contains one of each chromosome, so it is haploid (n). The chromosomes are now unduplicated.

FIGURE 12.4 How meiosis halves the chromosome number.

During meoisis I, every duplicated chromosome aligns with its homologous partner (**FIGURE 12.4** ❶). Then the homologous chromosomes are pulled away from one another and packaged into separate nuclei ❷. Each of the two new nuclei is haploid (n)—it has one copy of each chromosome—so the chromosome number is now half that of the parent.

The chromosomes are still duplicated (the sister chromatids remain attached to one another). During meiosis II, sister chromatids are pulled apart, and each becomes an individual, unduplicated chromosome ❸. The chromosomes are sorted into four new nuclei. Each new nucleus still has one copy of each chromosome, so it is haploid (n).

Thus, meiosis partitions the chromosomes of one diploid nucleus (2n) into four haploid (n) nuclei. The next section zooms in on the details of this process.

alleles Forms of a gene with slightly different DNA sequences; may encode slightly different versions of the gene's product.
fertilization Fusion of two gametes to form a zygote.
gamete Mature, haploid reproductive cell; e.g., an egg or a sperm.
germ cell Immature reproductive cell that gives rise to haploid gametes when it divides.
haploid Having one of each type of chromosome characteristic of the species.
meiosis Nuclear division process that halves the chromosome number. Basis of sexual reproduction.
somatic Relating to the body.
zygote Diploid cell that forms when two gametes fuse; the first cell of a new individual.

Fertilization Restores Chromosome Number

Haploid gametes form by meiosis. The diploid chromosome number is restored at **fertilization**, when two haploid gametes fuse to form a **zygote**, the first cell of a new individual. Thus, meiosis halves the chromosome number, and fertilization restores it.

If meiosis did not precede fertilization, then the chromosome number would double with every generation. As you will see in Chapter 14, chromosome number changes can have drastic consequences for health, particularly in animals. An individual's set of chromosomes is like a fine-tuned blueprint that must be followed exactly, page by page, in order to build a body that functions normally.

TAKE-HOME MESSAGE 12.2
Why is meiosis necessary for sexual reproduction?

✔ Paired genes on homologous chromosomes may vary in DNA sequence as alleles. Alleles arise by mutation.

✔ Alleles give rise to differences in shared traits. Offspring of sexual reproducers inherit new combinations of parental alleles—thus new combinations of traits.

✔ The nuclear division process of meiosis is the basis of sexual reproduction in plants and animals.

✔ Meiosis halves the diploid (2n) chromosome number, to the haploid number (n), for forthcoming gametes.

✔ When two gametes fuse at fertilization, the diploid chromosome number is restored in the resulting zygote.

12.3 Visual Tour of Meiosis

✔ During meiosis, chromosomes of one diploid nucleus become distributed into four haploid nuclei.

FIGURE 12.5 shows the stages of meiosis in a diploid (2*n*) cell, which contains two sets of chromosomes. DNA replication occurs before meiosis I, so each chromosome has two sister chromatids.

Meiosis I The first stage of meiosis I is prophase I ❶. During this phase, the chromosomes condense, and homologous chromosomes align tightly and swap

segments (more about segment-swapping in the next section). The nuclear envelope breaks up. A spindle forms, and by the end of prophase I, microtubules attach one chromosome of each homologous pair to one spindle pole, and the other to the opposite spindle pole. These microtubules grow and shrink, pushing and pulling the chromosomes as they do. At metaphase I ❷, all of the microtubules are the same length, and the chromosomes are aligned midway between the spindle poles. During anaphase I ❸, the spindle pulls the homologous chromosomes of each pair apart and toward opposite spindle poles. The two sets of chromosomes reach the spindle poles during telophase I ❹, and a new nuclear envelope forms around each cluster of chromosomes as the DNA loosens up. The two new nuclei are haploid (*n*);

FIGURE 12.5 ▶**Animated** Meiosis. Two pairs of chromosomes are illustrated in a diploid (2*n*) cell. Homologous chromosomes are indicated in blue and pink. Micrographs show meiosis in a lily plant cell (*Lilium regale*).

FIGURE IT OUT During which phase of meiosis does the chromosome number become reduced?

Answer: Anaphase I

MEIOSIS I One diploid nucleus to two haploid nuclei

① Prophase I
Homologous chromosomes condense, pair up, and swap segments. Spindle microtubules attach to them as the nuclear envelope breaks up.

② Metaphase I
Homologous chromosome pairs are aligned between spindle poles. Spindle microtubules attach the two chromosomes of each pair to opposite spindle poles.

③ Anaphase I
All of the homologous chromosomes separate and begin heading toward the spindle poles.

④ Telophase I
A complete set of chromosomes clusters at both ends of the cell. A nuclear envelope forms around each set, so two haploid (*n*) nuclei form.

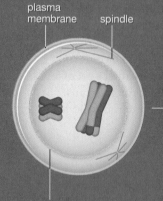

plasma membrane   spindle

nuclear envelope breaking up

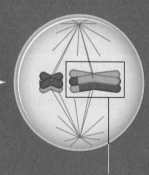

pair of homologous chromosomes

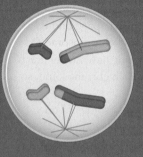

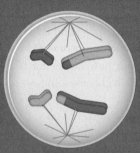

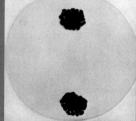

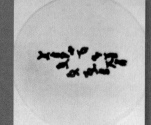

each contains one set of chromosomes. The cytoplasm often divides at this point. Each chromosome is still duplicated (it consists of two sister chromatids).

Meiosis may pause at this point, but no DNA replication occurs before meiosis II.

Meiosis II Meiosis II proceeds simultaneously in both nuclei that formed in meiosis I. During prophase II ❺, the chromosomes condense and the nuclear envelope breaks up. A new spindle forms. By the end of prophase II, spindle microtubules attach each chromatid to one spindle pole, and its sister chromatid to the opposite spindle pole. These microtubules push and pull the chromosomes, aligning them midway between spindle poles at metaphase II ❻. During anaphase II ❼, the spindle microtubules pull the sister

chromatids apart and toward opposite spindle poles. Each chromosome is now unduplicated (it consists of one molecule of DNA). During telophase II ❽, these chromosomes reach the spindle poles. New nuclear envelopes form around the four clusters of chromosomes as the DNA loosens up. Each of the four nuclei that form are haploid (*n*), with one set of (unduplicated) chromosomes. The cytoplasm often divides at this point.

TAKE-HOME MESSAGE 12.3
What happens to a cell during meiosis?

✔ During meiosis, the nucleus of a diploid (2*n*) cell divides twice. Four haploid (*n*) nuclei form, each with a full set of chromosomes—one of each type.

MEIOSIS II Two haploid nuclei to four haploid nuclei

❺ **Prophase II**
The chromosomes condense. Spindle microtubules attach to each sister chromatid as the nuclear envelope breaks up.

❻ **Metaphase II**
The (still duplicated) chromosomes are aligned midway between spindle poles.

❼ **Anaphase II**
Sister chromatids separate. The now unduplicated chromosomes head to the spindle poles.

❽ **Telophase II**
A complete set of chromosomes clusters at both ends of the cell. A new nuclear envelope forms around each set, so four haploid (*n*) nuclei form.

No DNA replication

✔ Crossovers and the random sorting of chromosomes into gametes result in new combinations of traits among offspring of sexual reproducers.

The previous section mentioned briefly that duplicated chromosomes swap segments with their homologous partners during prophase I. It also showed how spindle microtubules align and then separate homologous chromosomes during anaphase I. These events, along with fertilization, contribute to the variation in combinations of traits among the offspring of sexually reproducing species.

Crossing Over

Early in prophase I of meiosis, all chromosomes in the cell condense. When they do, each chromosome is drawn close to its homologous partner, so that the chromatids align along their length:

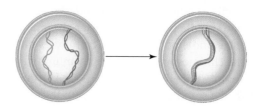

This tight, parallel orientation favors **crossing over**, a process by which a chromosome and its homologous partner exchange corresponding pieces of DNA during meiosis (**FIGURE 12.6A–C**). Homologous chromosomes may swap any segment of DNA along their length, although crossovers tend to occur more frequently in certain regions.

Swapping segments of DNA shuffles alleles between homologous chromosomes. It breaks up the particular combinations of alleles that occurred on the parental chromosomes, and makes new ones on the chromosomes that end up in gametes. Thus, crossing over introduces novel combinations of alleles—thus new combinations of traits—among offspring. It is a normal and frequent process in meiosis, but the rate of crossing over varies among species and among chromosomes. In humans, between 46 and 95 crossovers occur per meiosis, so on average each chromosome crosses over at least once (**FIGURE 12.6D**).

Chromosome Segregation

Normally, all of the new nuclei that form in meiosis I receive a complete set of chromosomes. However, whether a new nucleus ends up with the maternal or paternal version of a chromosome is entirely random. The chance that the maternal or the paternal version of any chromosome will end up in a particular nucleus is 50 percent. Why? The answer has to do with the way the spindle segregates the homologous chromosomes during meiosis I.

The process of chromosome segregation begins in prophase I. Imagine a germ cell undergoing meiosis. Crossovers have already made genetic mosaics of its chromosomes, but for simplicity let's put crossing over

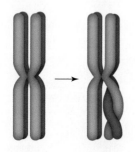

A Here, we focus on only two of the many genes on a chromosome. In this example, one gene has alleles *A* and *a*; the other has alleles *B* and *b*.

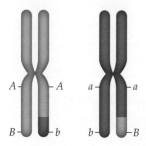

B Close contact between homologous chromosomes promotes crossing over between nonsister chromatids. Paternal and maternal chromatids exchange corresponding pieces.

C Crossing over mixes up paternal and maternal alleles on homologous chromosomes.

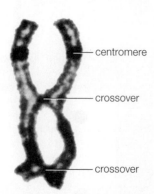

centromere

crossover

crossover

D Each pair of homologous chromosomes can cross over multiple times. This is a normal and common process of meiosis.

FIGURE 12.6 ▶Animated Crossing over. Blue signifies a paternal chromosome, and pink, its maternal homologue. For clarity, we show only one pair of homologous chromosomes.

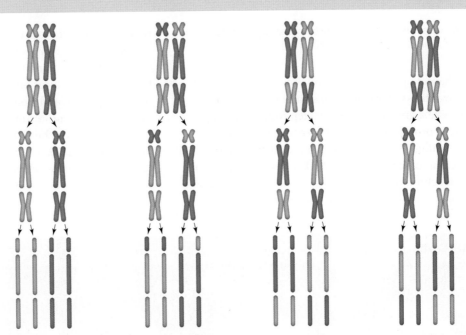

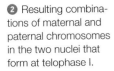 ① The four possible alignments of three pairs of chromosomes in a nucleus at metaphase I.

② Resulting combinations of maternal and paternal chromosomes in the two nuclei that form at telophase I.

③ Resulting combinations of maternal and paternal chromosomes in the four nuclei that form at telophase II. Eight different combinations are possible.

FIGURE 12.7 ▶Animated Hypothetical segregation of three pairs of chromosomes in meiosis I. Maternal chromosomes are pink; paternal, blue. Which chromosome of each pair gets packaged into which of the two new nuclei that form at telophase I is random. For simplicity, no crossing over occurs in this example, so all sister chromatids are identical.

aside for a moment. Just call the chromosomes inherited from the mother the maternal ones, and the ones from the father the paternal ones.

During prophase I, microtubules fasten the cell's chromosomes to the spindle poles. It is very unlikely that all of the maternal chromosomes will be attached to one pole, and all of the paternal chromosomes will be attached to the other. Microtubules extending from a spindle pole bind to the centromere of the first chromosome they contact, regardless of whether it is maternal or paternal. Though each homologous partner becomes attached to the opposite spindle pole, there is no pattern to the attachment of the maternal or paternal chromosomes to a particular pole.

Now imagine that the germ cell has three pairs of chromosomes (**FIGURE 12.7**). By metaphase I, those three pairs of maternal and paternal chromosomes have been divided up between the two spindle poles in one of four ways ①. In anaphase I, homologous chromosomes separate and are pulled toward opposite spindle poles. In telophase I, a new nucleus forms around the chromosomes that cluster at each spindle pole. Each nucleus contains one of eight possible combinations of maternal and paternal chromosomes ②.

In telophase II, each of the two nuclei divides and gives rise to two new haploid nuclei. The two new

nuclei are identical because no crossing over occurred in our hypothetical example, so all of the sister chromatids were identical. Thus, at the end of meiosis in this cell, two (2) spindle poles have divvied up three (3) chromosome pairs. The resulting four nuclei have one of eight (2^3) possible combinations of maternal and paternal chromosomes ③.

Cells that give rise to human gametes have twenty-three pairs of homologous chromosomes, not three. Each time a human germ cell undergoes meiosis, the four gametes that form end up with one of 8,388,608 (or 2^{23}) possible combinations of homologous chromosomes. That number does not even take into account crossing over, which mixes up the alleles on maternal and paternal chromosomes, or fusion with another gamete at fertilization.

TAKE-HOME MESSAGE 12.4

How does meiosis give rise to new combinations of parental alleles?

✔ Crossing over—recombination between nonsister chromatids of homologous chromosomes—occurs during prophase I.

✔ Homologous chromosomes can get attached to either spindle pole in prophase I, so each chromosome of a homologous pair can end up in either of the two new nuclei.

✔ Crossing over and random sorting of chromosomes into gametes give rise to new combinations of alleles—thus new combinations of traits—among offspring of sexual reproducers.

crossing over Process by which homologous chromosomes exchange corresponding segments during prophase I of meiosis.

CREDIT: (7) © Cengage Learning.

✔ Though they have different results, mitosis and meiosis are fundamentally similar processes.

This chapter opened with hypotheses about evolutionary advantages of asexual and sexual reproduction. It seems like a giant evolutionary step from producing clones to producing genetically varied offspring, but was it really? By mitosis and cytoplasmic division, one cell becomes two new cells that have copies of the parental chromosomes. Mitotic (asexual) reproduction produces clones of the parent. By contrast, in sexual

reproducers, meiosis gives rise to haploid gametes. Gametes of two parents fuse to form a zygote, which is a cell of mixed parentage. Meiotic (sexual) reproduction usually produces offspring that differ genetically from the parent, and from one another.

Though the end results differ, there are striking parallels between the four stages of mitosis and meiosis II (**FIGURE 12.8**). As one example, a spindle forms and separates chromosomes during both processes. There are many more similarities at the molecular level.

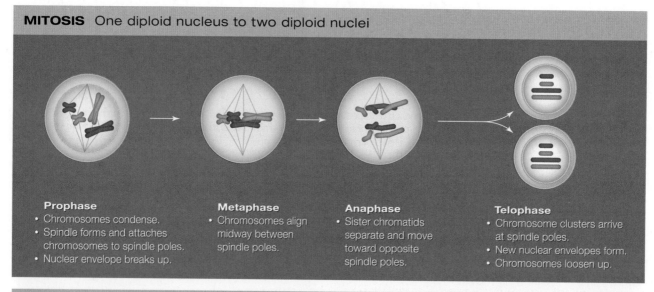

MITOSIS One diploid nucleus to two diploid nuclei

Prophase
- Chromosomes condense.
- Spindle forms and attaches chromosomes to spindle poles.
- Nuclear envelope breaks up.

Metaphase
- Chromosomes align midway between spindle poles.

Anaphase
- Sister chromatids separate and move toward opposite spindle poles.

Telophase
- Chromosome clusters arrive at spindle poles.
- New nuclear envelopes form.
- Chromosomes loosen up.

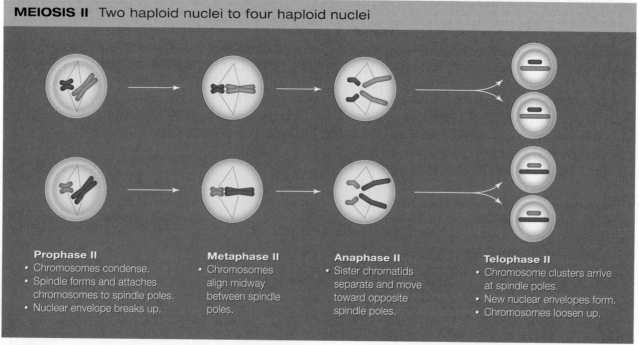

MEIOSIS II Two haploid nuclei to four haploid nuclei

Prophase II
- Chromosomes condense.
- Spindle forms and attaches chromosomes to spindle poles.
- Nuclear envelope breaks up.

Metaphase II
- Chromosomes align midway between spindle poles.

Anaphase II
- Sister chromatids separate and move toward opposite spindle poles.

Telophase II
- Chromosome clusters arrive at spindle poles.
- New nuclear envelopes form.
- Chromosomes loosen up.

FIGURE 12.8 ▶Animated Comparing meiosis II with mitosis.

Why do males exist? No male has ever been found among the tiny freshwater creatures called bdelloid rotifers (**FIGURE 12.10**). Females have been reproducing for 80 million years solely through cloning themselves. Bdelloids are one of the few animal groups to have completely abandoned sex.

Compared to sex, asexual reproduction is often seen as a poor long-term strategy because it lacks crossing over—the chromosomal shuffling that brings about genetic diversity thought to give species an adaptive edge in the face of new challenges. Bdelloids have contradicted this theory by being very successful; over 360 species are alive today.

A newly discovered ability may help to explain the success of the bdelloids despite their rejection of sex. These rotifers can apparently import genes from bacteria, fungi, protists, and even plants. If the main advantage of sex is that it promotes genetic diversity, why worry about it when you have the genes of entire kingdoms available to you?

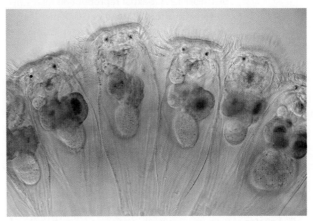

FIGURE 12.10 A common bdelloid rotifer (*Philodina rosea*). All of these tiny animals are female.

The direct swapping of genetic material is incredibly rare in animals, but bdelloids are bringing in external genes to an extent completely unheard of in complex organisms. Each rotifer is a genetic mosaic whose DNA spans almost all the major kingdoms of life: About 10 percent of its active genes have been pilfered from other organisms.

Long ago, the molecular machinery of mitosis may have been remodeled into meiosis. Evidence for this hypothesis includes a host of shared molecules, including the products of the *BRCA* genes (Section 11.6) that are made by all modern eukaryotes. By monitoring and fixing problems with the DNA—such as damaged or mismatched bases (Section 8.6)—these molecules actively maintain the integrity of a cell's chromosomes, particularly during DNA replication and mitosis. Many of the same molecules help homologous chromosomes cross over in prophase I of meiosis (**FIGURE 12.9**). As another example, consider that the same regulatory molecules involved in checkpoints of mitosis are also involved in checkpoints of meiosis.

During anaphase of mitosis, sister chromatids are pulled apart. What would happen if the connections between the sisters did not break? Each duplicated chromosome would be pulled to one or the other spindle pole—which is exactly what happens during anaphase I of meiosis. The shared molecules and mechanisms imply a shared evolutionary history; sexual reproduction probably originated with mutations that affected processes of mitosis. As you will see in later chapters, the remodeling of existing processes into new ones is a common evolutionary theme.

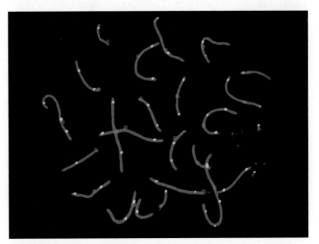

FIGURE 12.9 Example of a molecule that functions in mitosis and meiosis. This fluorescence micrograph shows homologous chromosome pairs (red) in the nucleus of a human cell during prophase I of meiosis. Centromeres are blue. Yellow pinpoints the location of a protein called MLH1 assisting with crossovers. MLH1 also helps repair mismatched bases during mitosis.

TAKE-HOME MESSAGE 12.5
Are the processes of mitosis and meiosis related?

✔ Meiosis may have evolved by the remodeling of existing mechanisms of mitosis.

CREDITS: (9) Reprinted from Fertility and Sterility, Vol 87/edition 3, Fei Sun, Paul Turek, Calvin Greene, Evelyn Ko, Alfred Rademaker, Renée H. Martin, Abnormal progression through meiosis in men with nonobstructive azoospermia, 565–571, copyright 2007, with permission from Elsevier; (in text revisited) © Bart Zijistra; (10) © Charles Krebs.

summary

Section 12.1 **Sexual reproduction** mixes up the genetic information of two parents. The offspring of sexual reproducers typically vary in shared, inherited traits. Particularly in environments where pressures change rapidly, this variation can offer an evolutionary advantage over genetically identical offspring produced by asexual reproduction.

Section 12.2 Sexual reproduction produces offspring whose **somatic** cells contain pairs of chromosomes, one of each homologous pair from the mother and the other from the father. The two chromosomes of a homologous pair carry the same genes. Paired genes on homologous chromosomes may vary in DNA sequence, in which case they are called **alleles**. Alleles are the basis of differences in shared, heritable traits. They arise by mutation.

Meiosis, the basis of sexual reproduction in eukaryotes, is a nuclear division mechanism that halves the chromosome number for forthcoming **gametes**. It occurs only in cells that are involved in sexual reproduction. **Haploid** (*n*) gametes are mature reproductive cells that form from division of **germ cells**. The fusion of two haploid gametes during **fertilization** restores the diploid parental chromosome number in the **zygote**, the first cell of the new individual.

Section 12.3 DNA replication occurs before meiosis begins, so each chromosome consists of two molecules of DNA (sister chromatids). The process of meiosis comprises two nuclear divisions, meiosis I and meiosis II.

The first nuclear division (meiosis I) begins when the chromosomes condense and align tightly with their homologous partners during prophase I. Microtubules then extend from the spindle poles, penetrate the nuclear region, and attach to one or the other chromosome of each homologous pair. At metaphase I, all chromosomes are lined up midway between the spindle poles. During anaphase I, homologous chromosomes separate and move to opposite spindle poles. Two nuclear envelopes form around the two sets of chromosomes during telophase I. The cytoplasm may divide at this point. There may be a resting period before meiosis resumes, but DNA replication does not occur.

The second nuclear division (meiosis II) occurs simultaneously in both haploid nuclei that formed during meiosis I. The chromosomes are still duplicated; each still consists of two sister chromatids. The chromosomes condense during prophase II, and align in metaphase II. The sister chromatids of each chromosome are pulled apart from each other during anaphase II, so at the end of meiosis each chromosome consists of one molecule of DNA. By the end of telophase II, four haploid nuclei have typically formed, each with a complete set of (unduplicated) chromosomes.

Section 12.4 Meiosis shuffles parental alleles, so offspring inherit nonparental combinations of them. During prophase I, homologous chromosomes exchange corresponding segments. This **crossing over** mixes up the alleles on maternal and paternal chromosomes, thus giving rise to combinations of alleles not present in either parental chromosome. Novel combinations of alleles are the basis of novel combinations of traits among offspring of sexual reproducing organisms. Meiosis also contributes to variation in traits by randomly segragating maternal and paternal chromosomes into gametes. Microtubules can attach the maternal or the paternal chromosome of each pair to one or the other spindle pole. Either chromosome may end up in any new nucleus, and in any gamete.

Section 12.5 The process of meiosis resembles that of mitosis, and may have evolved from it. Many of the same molecules function the same way in both processes. For example, a spindle forms, moves and sorts chromosomes, and disassembles during both mitosis and meiosis.

self-quiz
Answers in Appendix VII

1. One evolutionary advantage of sexual over asexual reproduction may be that it produces _____ .
   a. more offspring per individual
   b. more variation among offspring
   c. healthier offspring

2. Meiosis is a necessary part of sexual reproduction because it _____ .
   a. divides two nuclei into four new nuclei
   b. reduces the chromosome number for gametes
   c. produces clones that can cross over

3. Meiosis _____ .
   a. occurs in all eukaryotes
   b. supports growth and tissue repair in multicelled species
   c. gives rise to genetic diversity among offspring
   d. is part of the life cycle of all cells

4. Sexual reproduction in animals requires _____ .
   a. meiosis   c. germ cells
   b. fertilization   d. all of the above

5. Meiosis _____ the parental chromosome number.
   a. doubles   c. maintains
   b. halves   d. mixes up

6. Dogs have a diploid chromosome number of 78. How many chromosomes do their gametes have?
   a. 39   c. 156
   b. 78   d. 234

7. The cell in the diagram to the right is in anaphase I, not anaphase II. I know this because _____ .

8. Which of the following cells can undergo meiosis?
   a. the diploid body cells of an animal
   b. cells set aside for reproduction in eukaryotes
   c. haploid gametes
   d. all of the above

data analysis activities

BPA and Abnormal Meiosis In 1998, researchers at Case Western University were studying meiosis in mouse oocytes when they saw an unexpected and dramatic increase in abnormal meiosis events (**FIGURE 12.11**). The improper segregation of chromosomes during meiosis is one of the main causes of human genetic disorders, which we will discuss in Chapter 14.

The researchers discovered that the spike in meiotic abnormalities began immediately after the mouse facility started washing the animals' plastic cages and water bottles in a new, alkaline detergent. The detergent had damaged the plastic, which as a result was leaching bisphenol A (BPA). BPA is a synthetic chemical that mimics estrogen, the main female sex hormone in animals. BPA is still widely used to manufacture polycarbonate plastic items (such as water bottles) and epoxies (such as the coating on the inside of metal cans of food).

1. What percentage of mouse oocytes displayed abnormalities of meiosis with no exposure to damaged caging?

2. Which group of mice had the most meiotic abnormalities in their oocytes?

3. What is abnormal about metaphase I as it is occurring in the oocytes shown in **FIGURE 12.11B, C, and D**?

Caging materials	Total number of oocytes	Abnormalities
Control: New cages with glass bottles	271	5 (1.8%)
Damaged cages with glass bottles		
Mild damage	401	35 (8.7%)
Severe damage	149	30 (20.1%)
Damaged bottles	197	53 (26.9%)
Damaged cages with damaged bottles	58	24 (41.4%)

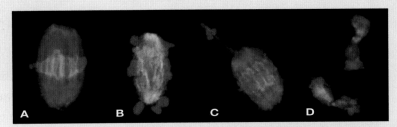

FIGURE 12.11 Meiotic abnormalities associated with exposure to damaged plastic caging.

Fluorescent micrographs show nuclei of single mouse oocytes in metaphase I. (**A**) Normal metaphase; (**B–D**) examples of abnormal metaphase. Chromosomes are stained red; spindle fibers, green.

---

9. The cell pictured to the right is in which stage of nuclear division?
   a. anaphase
   b. anaphase I
   c. anaphase II
   d. none of the above

10. Crossing over mixes up _____ .
    a. chromosomes
    b. alleles
    c. zygotes
    d. gametes

11. Crossing over happens during which phase of meiosis?
    a. prophase I
    b. prophase II
    c. anaphase I
    d. anaphase II

12. Which of the following is one of the very important differences between mitosis and meiosis?
    a. Chromosomes align midway between spindle poles only in meiosis.
    b. Homologous chromosomes pair up only in meiosis.
    c. DNA is replicated only in mitosis.
    d. Sister chromatids separate only in meiosis.
    e. Interphase occurs only in mitosis.

13. Match each term with the best description.
    ___ interphase
    ___ metaphase I
    ___ alleles
    ___ zygotes
    ___ gametes
    ___ males
    ___ prophase I

    a. different forms of a gene
    b. useful for varied offspring
    c. none between meiosis I and meiosis II
    d. chromosome lineup
    e. haploid
    f. form at fertilization
    g. mash-up time

14. _____ contributes to variation in traits among the offspring of sexual reproducers.
    a. Crossing over
    b. Random attachment of chromosomes to spindle poles
    c. Fertilization
    d. both a and b
    e. all are factors

critical thinking

1. The diploid chromosome number for the body cells of a frog is 26. What would that number be after three generations if meiosis did not occur before gamete formation?

2. In your own words, explain why sexual reproduction tends to give rise to greater genetic diversity among offspring in fewer generations than asexual reproduction.

3. Different populations of the tiny freshwater snails pictured in **FIGURE 12.1** reproduce sexually, or asexually. Individuals of the sexual populations are diploid; those in asexual populations are triploid ($3n$, having three sets of chromosomes). Huge populations of asexual snails are disrupting ecosystems worldwide.

   Fertilizers and detergents contain a lot of phosphorus. So does DNA. Explain why you might expect to find more sexual snail populations in an unpolluted river, and more asexual ones in a river polluted by agricultural and urban runoff.

4. Make a simple sketch of meiosis in a cell with a diploid chromosome number of 4. Now try it when the chromosome number is 3.

CENGAGE To access course materials, please visit **brain**.com www.cengagebrain.com.

CREDITS: (in text S-Q #9) Michael Clayton/University of Wisconsin, Department of Biology; (11A–D), Reprinted from *Current Biology*, Vol 13, (Apr 03), Authors Hunt, Koehler, Susiarjo, Hodges, Ilagan, Voight, Thomas, Thomas and Hassold, Bisphenol A Exposure Causes Meiotic Aneuploidy in the Female Mouse, pp. 546–553, © 2003 Cell Press. Published by Elsevier Ltd. With permission from Elsevier.

LEARNING ROADMAP

You may want to review traits (Section 1.5), chromosomes (8.4), genes and gene expression (9.2), sexual reproduction (12.1), alleles (12.2), and meiosis (12.3, 12.4). This chapter revisits probability and sampling error (1.8), laws of nature (1.9), protein structure (3.6), pigments (6.2), clones (8.7), gene expression control (10.2, 11.6), and epigenetics (10.6).

WHERE MODERN GENETICS STARTED

Gregor Mendel discovered that inherited traits are specified in units. The units, which are distributed into gametes in predictable patterns, were later identified as genes.

MONOHYBRID CROSSES

Tracking inheritance patterns of single traits led to the discovery that during meiosis, pairs of genes on homologous chromosomes separate and end up in different gametes.

DIHYBRID CROSSES

Tracking inheritance patterns of two unrelated traits led to the discovery that in most cases, genes of a pair segregate into gametes independently of other gene pairs.

VARIATIONS ON MENDEL'S THEME

An allele may be partly dominant over a nonidentical partner, or codominant with it. Multiple genes may influence a trait; some genes influence many traits.

COMPLEX VARIATIONS IN TRAITS

Environmental factors can alter the expression of genes that influence a trait. Many traits appear in a continuous range of forms.

We return to human skin color in Section 14.1, and human genetic disorders in the rest of Chapter 14. Chapter 17 explores the interplay between genes and the environment in an evolutionary context. Signaling pathways that affect gene expression comprise stimulus, perception, and response (Section 34.2). Chapter 29 details flowering plant reproduction. Neurons and how they work are the topic of Chapter 32, with neurological disorders in Section 32.7.

In 1988, researchers discovered a gene that, when mutated, causes cystic fibrosis (CF). Cystic fibrosis is the most common fatal genetic disorder in the United States. The gene, *CFTR*, encodes an active transport protein that moves chloride ions out of epithelial cells. Sheets of epithelial cells form the skin and also line the passageways and ducts of the lungs, liver, pancreas, intestines, and reproductive system. When the CFTR protein pumps chloride ions out of these cells, water follows the ions by osmosis. The two-step process maintains a thin film of water on the surface of epithelial cell sheets. Mucus slides easily over the wet sheets of cells.

The mutation most commonly associated with cystic fibrosis is a 3-base-pair deletion in the *CFTR* gene. The mutation is called $\Delta F508$ because it encodes a CFTR protein missing the phenylalanine (F) that is normally the 508th amino acid (Δ means deleted). This defect interferes with membrane trafficking of the newly synthesized CFTR protein. Normally, a new CFTR polypeptide moves from endoplasmic reticulum (ER) to a Golgi body, which packages it in vesicles routed to the plasma membrane. CFTR polypeptides with the $\Delta F508$ deletion misfold in a tiny domain, and this change is recognized by a quality control system in the endoplasmic reticulum. Most of the altered proteins are left stranded in the ER; the few that make it to the plasma membrane are quickly taken back into the cell by endocytosis and destroyed.

Epithelial cell membranes that lack the CFTR protein cannot transport chloride ions. Too few chloride ions leave these cells. Not enough water leaves them either, so the surfaces of epithelial cell sheets are too dry. Mucus that normally slips through the body's tubes sticks to their walls instead. Thick globs of mucus accumulate and clog passageways and ducts throughout the body. Breathing becomes difficult as the mucus obstructs the smaller airways of the lungs. Digestive problems arise as ducts that lead to the gut become clogged with mucus. Males are typically infertile because their sperm flow is hampered.

CFTR also helps alert the immune system to the presence of disease-causing bacteria in the lungs. The CFTR protein functions as a receptor: It binds directly to bacteria and triggers endocytosis. Endocytosis of bacteria into epithelial cells lining the respiratory tract initiates an immune response. When the cells lack CFTR, this early alert system fails, so bacteria have time to multiply before being detected by the immune system. Thus, chronic bacterial infections of the lungs are a hallmark of cystic fibrosis. Antibiotics help control infections, but there is no cure for the disorder. Most affected people die before age thirty, when their tormented lungs fail (**FIGURE 13.1**).

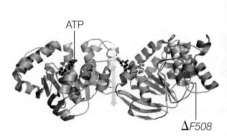

ATP

$\Delta F508$

Cody, 23   Jeff, 21   Lindsay, 22

Ben, 23   Savannah, 19   Brandon, 18

FIGURE 13.1 Cystic fibrosis.

Above, model of the CFTR protein. The parts shown here are a pair of ATP-driven motors that widen or narrow a channel (gray arrow) across the plasma membrane. The tiny part of the protein that is deleted in most people with cystic fibrosis is shown in green.

Right, a few of the many young victims of cystic fibrosis, which occurs most often in people of northern European ancestry. At least one young person dies every day in the United States from complications of this incurable disease.

CREDITS: (opposite) Jean M. Labat/Ardea London; (1) art, © Cengage Learning; photos, top row from left, Courtesy of © The Cody Dieruf Benefit Foundation, www.breathinisbelievin.org; Courtesy of © Bobby Brooks and The Family of Jeff Baird; Courtesy of © Steve & Ellison Widener and Breathe Hope, breathehope.tamu.edu; bottom row from left, Courtesy of The Family of Benjamin Hill, reprinted with permission of © Chappell/Marathonfoto; Courtesy of © The Family of Savannah Brooke Snider; Courtesy of © the family of Brandon Herriott.

✔ Recurring patterns of inheritance offer observable evidence of how heredity works.

In the nineteenth century, people thought that hereditary material must be some type of fluid, with fluids from both parents blending at fertilization like milk into coffee. However, the idea of "blending inheritance" failed to explain what people could see with their own eyes. Children sometimes have traits such as freckles that do not appear in either parent. A cross between a black horse and a white one does not produce gray offspring.

The naturalist Charles Darwin doubted the idea of blending inheritance, but he could not come up with an alternative hypothesis even though inheritance was central to his theory of natural selection. (We return to Darwin and this theory in Chapter 16.) At the time, no one knew that hereditary information is divided into discrete units (genes, Section 9.2), an insight that is critical to understanding how traits are inherited. However, even before Darwin presented his theory, someone had been gathering data that would support it. Gregor Mendel (left), an Austrian monk, had been carefully breeding thousands of pea plants. By keeping detailed records of how traits passed from one generation to the next, Mendel had been collecting evidence of how inheritance works.

Mendel's Experiments

Mendel cultivated the garden pea plant (**FIGURE 13.2**). This species is naturally self-fertilizing, which means its flowers produce male and female gametes ❶ that form viable embryos when they meet up. In order to study inheritance, Mendel had to carry out controlled matings between individuals with specific traits, then observe and document the traits of their offspring. To prevent an individual pea plant from self-fertilizing, Mendel removed the pollen-bearing parts (anthers) from its flowers. He then cross-fertilized the flowers by brushing their egg-bearing parts (carpels) with pollen from another plant ❷. He collected seeds ❸ from the cross-fertilized individual, and recorded the traits of the new pea plants that grew from them ❹.

Many of Mendel's experiments started with plants that "bred true" for particular traits such as white flowers or purple flowers. Breeding true for a trait means that, new mutations aside, all offspring have the same form of the trait as the parent(s), generation after gen-

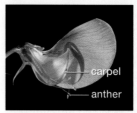

❶ In the flowers of garden pea plants, pollen grains that form in anthers produce male gametes; female gametes form in carpels.

❷ Experimenters control the transfer of hereditary material from one pea plant to another by snipping off a flower's pollen-producing anthers (to prevent it from self-fertilizing), then brushing pollen from another flower onto its egg-producing carpel.

In this example, pollen from a plant with purple flowers is brushed onto the carpel of a white-flowered plant.

❸ Later, seeds develop inside pods of the cross-fertilized plant. An embryo in each seed develops into a mature pea plant.

❹ Every plant that arises from the cross has purple flowers. Predictable patterns such as this are evidence of how inheritance works.

FIGURE 13.2 ▶Animated Breeding garden pea plants.

eration. For example, all offspring of two pea plants that breed true for white flowers also have white flowers. As you will see in the next section, Mendel cross-fertilized pea plants that breed true for different forms of a trait, and discovered that the traits of the offspring often appear in predictable patterns. Mendel's meticulous work tracking pea plant traits led him to conclude (correctly) that hereditary information passes from one generation to the next in discrete units.

Inheritance in Modern Terms

DNA was not proven to be hereditary material until the 1950s (Section 8.2), but Mendel discovered its units, which we now call genes, almost a century before then. Today, we know that individuals of a species share certain traits because their chromosomes carry the same genes.

Each gene occurs at a specific location, or **locus** (plural, loci), on a particular chromosome (**FIGURE 13.3**). The somatic cells of humans and other animals are diploid, so their nuclei contain pairs of genes, on pairs of homologous chromosomes. In most cases, both genes of a pair are expressed. Genes at the same locus on a pair of homologous chromosomes may be identical, or they may vary as alleles (Section 12.2).

Organisms breed true for a trait because they carry identical alleles of genes governing that trait. An individual with two identical alleles of a gene is **homozygous** for the allele. By contrast, an individual with two different alleles of a gene is **heterozygous** (*hetero–* means mixed). A **hybrid** is a heterozygous individual, such as an offspring of a cross or mating between individuals that breed true for different forms of a trait.

When we say that an individual is homozygous or heterozygous, we are discussing its **genotype**, the particular set of alleles it carries. Genotype is the basis of **phenotype**, which refers to the individual's observable traits. "White-flowered" and "purple-flowered" are examples of pea plant phenotypes that arise from differences in genotype.

The phenotype of a heterozygous individual depends on how the products of its two different alleles interact. In many cases, the effect of one allele influences the effect of the other, and the outcome of this interaction is reflected in the individual's phenotype. An allele is **dominant** when its effect masks that of a **recessive** allele paired with it. Usually, a dominant allele is represented by an italic capital letter such as *A*; a recessive allele, with a lowercase italic letter such as *a*. Consider the purple- and white-flowered pea plants that Mendel studied. In these plants, the allele that specifies purple flowers (let's call it *P*) is dominant over the allele that specifies white flowers (*p*). Thus, a pea plant homozygous

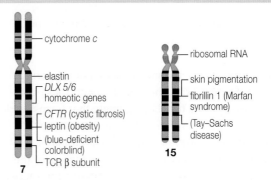

FIGURE 13.3 Loci of a few human genes. Genetic disorders arising from mutations in the genes are shown in parentheses. The number or letter below a chromosome is its name; characteristic banding patterns appear after staining. Appendix IV has a similar map of all 23 human chromosomes.

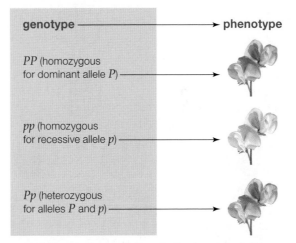

FIGURE 13.4 Genotype gives rise to phenotype. In this example, the dominant allele *P* specifies purple flowers; the recessive allele *p*, white flowers. **FIGURE IT OUT** Which individual is a hybrid?

Answer: The heterozygous one

for the dominant allele (*PP*) has purple flowers; one homozygous for the recessive allele (*pp*) has white flowers (**FIGURE 13.4**). A heterozygous plant (*Pp*) has purple flowers.

---

dominant Refers to an allele that masks the effect of a recessive allele paired with it in heterozygous individuals.
genotype The particular set of alleles that is carried by an individual's chromosomes.
heterozygous Having two different alleles of a gene.
homozygous Having identical alleles of a gene.
hybrid A heterozygous individual.
locus Location of a gene on a chromosome.
phenotype An individual's observable traits.
recessive Refers to an allele with an effect that is masked by a dominant allele on the homologous chromosome.

TAKE-HOME MESSAGE 13.2

How do alleles contribute to traits?

✔ Gregor Mendel indirectly discovered the role of alleles in inheritance by carefully breeding pea plants and tracking traits of their offspring.

✔ Genotype refers to the particular set of alleles that an individual carries. Genotype is the basis of phenotype, which refers to the individual's observable traits.

✔ A homozygous individual has two identical alleles of a gene. A heterozygous individual has two nonidentical alleles.

✔ A dominant allele masks the effect of a recessive allele paired with it in a heterozygous individual.

✔ Pairs of genes on homologous chromosomes separate during meiosis, so they end up in different gametes.

When homologous chromosomes separate during meiosis (Section 12.3), the gene pairs on those chromosomes separate too. Each gamete that forms carries only one of the two genes of a pair (**FIGURE 13.5**). Thus, plants homozygous for a dominant allele (*PP*, for example) can only make gametes that carry the dominant allele *P* ❶. Plants homozygous for a recessive allele (*pp*) can only make gametes that carry the recessive allele *p* ❷. If these homozygous plants are

crossed (*PP* × *pp*), only one outcome is possible: A gamete carrying allele *P* meets up with a gamete carrying allele *p* ❸. All offspring of this cross will have both alleles—they will be heterozygous (*Pp*). A grid called a **Punnett square** is helpful for predicting the outcomes of such crosses (**FIGURE 13.6**).

Our example illustrated a pattern so predictable that it can be used as evidence of a dominance relationship between alleles. In a **testcross**, an individual that has a dominant trait (but an unknown genotype) is crossed with an individual known to be homozygous for the recessive allele. The pattern of traits among the offspring of the cross can reveal whether the tested individual is heterozygous or homozygous. If all of the offspring of the testcross have the dominant trait (as occurred in our example above), then the parent with the unknown genotype is homozygous for the dominant allele. If some of the offspring have the recessive trait, then the parent is heterozygous.

Dominance relationships between alleles determine the phenotypic outcome of a **monohybrid cross**, in which individuals identically heterozygous for one gene (they have the same two alleles: *Pp*, for example)

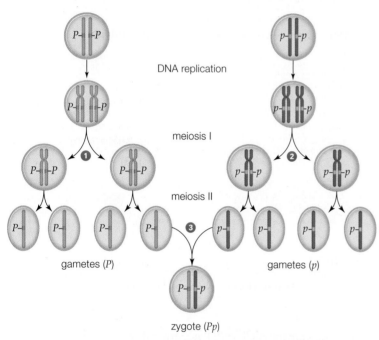

FIGURE 13.5 ▶Animated Segregation of genes into gametes. Homologous chromosomes separate during meiosis, so the pairs of genes they carry separate too. Each of the resulting gametes carries one of the two members of each gene pair. For clarity, only one set of chromosomes is illustrated.

❶ All gametes made by a parent homozygous for a dominant allele carry that allele.

❷ All gametes made by a parent homozygous for a recessive allele carry that allele.

❸ If these two parents are crossed, the union of any of their gametes at fertilization produces a zygote with both alleles. All offspring of this cross will be heterozygous.

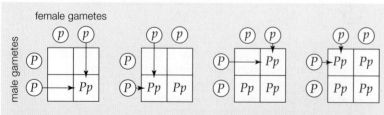

FIGURE 13.6 Making a Punnett square. Parental gametes are listed in circles on the top and left sides of a grid. Each square is filled with the combination of alleles that would result if the gametes in the corresponding row and column met up.

Table 13.1 Mendel's Seven Pea Plant Traits		
Trait	**Dominant Form**	**Recessive Form**
Seed Shape	Round	Wrinkled
Seed Color	Yellow	Green
Pod Texture	Smooth	Wrinkled
Pod Color	Green	Yellow
Flower Color	Purple	White
Flower Position	Along Stem	At Tip
Stem Length	Tall	Short

are bred together or self-fertilized ($Pp \times Pp$). The frequency at which the phenotype associated with each allele appears among offspring depends on whether one of the alleles is dominant over the other.

To do a monohybrid cross, we would start with two individuals that breed true for two distinct forms of a trait. In garden pea plants, flower color (purple or white) is one example of a trait with two distinct forms, but there are many others. Mendel investigated seven of them: stem length (tall or short), seed color (yellow or green), pod texture (smooth or wrinkled), and so on (**TABLE 13.1**). A cross between individuals that breed true for different forms of the trait yields offspring identically heterozygous for the alleles that govern the trait. A cross between these F_1 (first generation) hybrids is the monohybrid cross. The frequency at which the two traits appear in the F_2 (second generation) offspring offers information about a dominance relationship between the alleles. F is an abbreviation for filial, which means offspring.

A cross between two purple-flowered heterozygous individuals (Pp) is an example of a monohybrid cross. Each of these plants can make two types of gametes: ones that carry a P allele, and ones that carry a p allele (**FIGURE 13.7A**). So, in a monohybrid cross between two Pp plants ($Pp \times Pp$), the two types of gametes can meet up in four possible ways at fertilization:

Possible Event	Outcome
Sperm P meets egg P $\longrightarrow$	zygote genotype is PP
Sperm P meets egg p $\longrightarrow$	zygote genotype is Pp
Sperm p meets egg P $\longrightarrow$	zygote genotype is Pp
Sperm p meets egg p $\longrightarrow$	zygote genotype is pp

Three out of four possible outcomes of this cross include at least one copy of the dominant allele P. Each time fertilization occurs, there are 3 chances in 4 that the resulting offspring will inherit a P allele, and have purple flowers. There is 1 chance in 4 that it will inherit two recessive p alleles, and have white flowers. Thus, the probability that a particular offspring of this cross will have purple or white flowers is 3 purple to 1 white, which we represent as a ratio

law of segregation The two members of each pair of genes on homologous chromosomes end up in different gametes during meiosis.
monohybrid cross Cross between two individuals identically heterozygous for one gene; for example, $Aa \times Aa$.
Punnett square Diagram used to predict the genotypic and phenotypic outcomes of a cross.
testcross Method of determining genotype of an individual with a dominant phenotype: a cross between the individual and another individual known to be homozygous recessive.

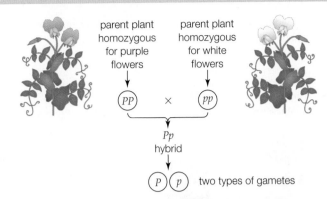

A All of the F_1 offspring of a cross between two plants that breed true for different forms of a trait are identically heterozygous (Pp). These offspring make two types of gametes: P and p.

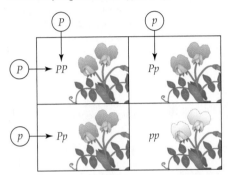

B A monohybrid cross is a cross between these F_1 offspring. In this example, the phenotype ratio in F_2 offspring is 3:1 (3 purple to 1 white).

FIGURE 13.7 ▶**Animated** An example of a monohybrid cross.
FIGURE IT OUT In this example, how many possible genotypes are there in the F_2 generation? Answer: Three: PP, Pp, and pp

of 3:1 (**FIGURE 13.7B**). The 3:1 pattern is an indication that purple and white flower color are specified by alleles with a clear dominance relationship: Purple is dominant; white, recessive. If the probability of an individual inheriting a particular genotype is difficult to imagine, think about it in terms of many offspring: In this example, there will be roughly three purple-flowered plants for every white-flowered one.

The phenotype ratios in the F_2 offspring of Mendel's monohybrid crosses were all close to 3:1. These results became the basis of his **law of segregation**, which we state here in modern terms: Diploid cells carry pairs of genes, on pairs of homologous chromosomes. The two genes of each pair are separated from each other during meiosis, so they end up in different gametes.

TAKE-HOME MESSAGE 13.3
How are alleles distributed into gametes?

✔ Homologous chromosomes carry pairs of genes. The two genes of each pair are separated from each other during meiosis, so they end up in different gametes.

13.4 Mendel's Law of Independent Assortment

✔ In many cases, pairs of genes are distributed into gametes independently of one another during meiosis.

parent plant homozygous for purple flowers and long stems

parent plant homozygous for white flowers and short stems

$PPTT$

$pptt$

❶ In each individual that is homozygous for two genes, meiosis results in only one type of gamete.

PT × pt

$PpTt$ dihybrid

❷ A cross between the two homozygous individuals yields offspring heterozygous for two genes (dihybrids).

four types of gametes

PT Pt pT pt

❸ Meiosis in dihybrid individuals results in four kinds of gametes.

PT Pt pT pt

	PT	Pt	pT	pt
PT	$PPTT$	$PPTt$	$PpTT$	$PpTt$
Pt	$PPTt$	$PPtt$	$PpTt$	$Pptt$
pT	$PpTT$	$PpTt$	$ppTT$	$ppTt$
pt	$PpTt$	$Pptt$	$ppTt$	$pptt$

❹ If two of the dihybrid individuals are crossed, the four types of gametes can meet up in 16 possible ways. Of 16 possible offspring genotypes, 9 will result in plants that are purple-flowered and tall; 3, purple-flowered and short; 3, white-flowered and tall; and 1, white-flowered and short. Thus, the ratio of phenotypes is 9:3:3:1.

FIGURE 13.8 ▶Animated A dihybrid cross between plants that differ in flower color and plant height. P and p stand for dominant and recessive alleles for flower color; T and t, dominant and recessive alleles for height.

FIGURE IT OUT What do the flowers inside the boxes represent?

Answer: Phenotypes of the F_2 offspring

A monohybrid cross allows us to study a dominance relationship between alleles of one gene. What about alleles of two genes? Dihybrids are individuals that have two alleles at two loci ($AaBb$, for example). In a **dihybrid cross**, individuals identically heterozygous for two genes are crossed ($AaBb \times AaBb$). As with a monohybrid cross, the pattern of traits seen among the offspring of the cross depends on the dominance relationships between the alleles.

Let's use a gene for flower color (P, purple; p, white) and one for plant height (T, tall; t, short) in an example. **FIGURE 13.8** shows a dihybrid cross starting with one parent plant that breeds true for purple flowers and tall stems ($PPTT$), and one that breeds true for white flowers and short stems ($pptt$). The $PPTT$ plant only makes gametes with the dominant alleles (PT); the $pptt$ plant only makes gametes with the recessive alleles (pt) ❶. So, all offspring from a cross between these parent plants ($PPTT \times pptt$) will be dihybrids ($PpTt$) with purple flowers and tall stems ❷.

Four combinations of alleles are possible in the gametes of $PpTt$ dihybrids ❸. If two $PpTt$ plants are crossed (a dihybrid cross, $PpTt \times PpTt$), the four types of gametes can combine in sixteen possible ways at fertilization ❹. Nine of the sixteen genotypes would give rise to tall plants with purple flowers; three, to short plants with purple flowers; three, to tall plants with white flowers; and one, to short plants with white flowers. Thus, the ratio of phenotypes among the offspring of this dihybrid cross would be 9:3:3:1.

Mendel discovered the 9:3:3:1 ratio of phenotypes among the offspring of his dihybrid crosses, but he had no idea what it meant. He could only say that "units" specifying one trait (such as flower color) are inherited independently of "units" specifying other traits (such as plant height). In time, Mendel's hypothesis became known as the **law of independent assortment**, which we state here in modern terms: During meiosis, the two genes of a pair tend to be sorted into gametes independently of how other gene pairs are sorted into gametes.

Mendel published his results in 1866, but apparently his work was read by few and understood by no one at the time. In 1871 he was promoted, and his pioneering experiments ended. When he died in 1884, he did not know that his work with pea plants would be the starting point for modern genetics.

The Contribution of Crossovers

How two pairs of genes get distributed into gametes depends partly on whether the genes are on the same chromosome. When homologous chromosomes sepa-

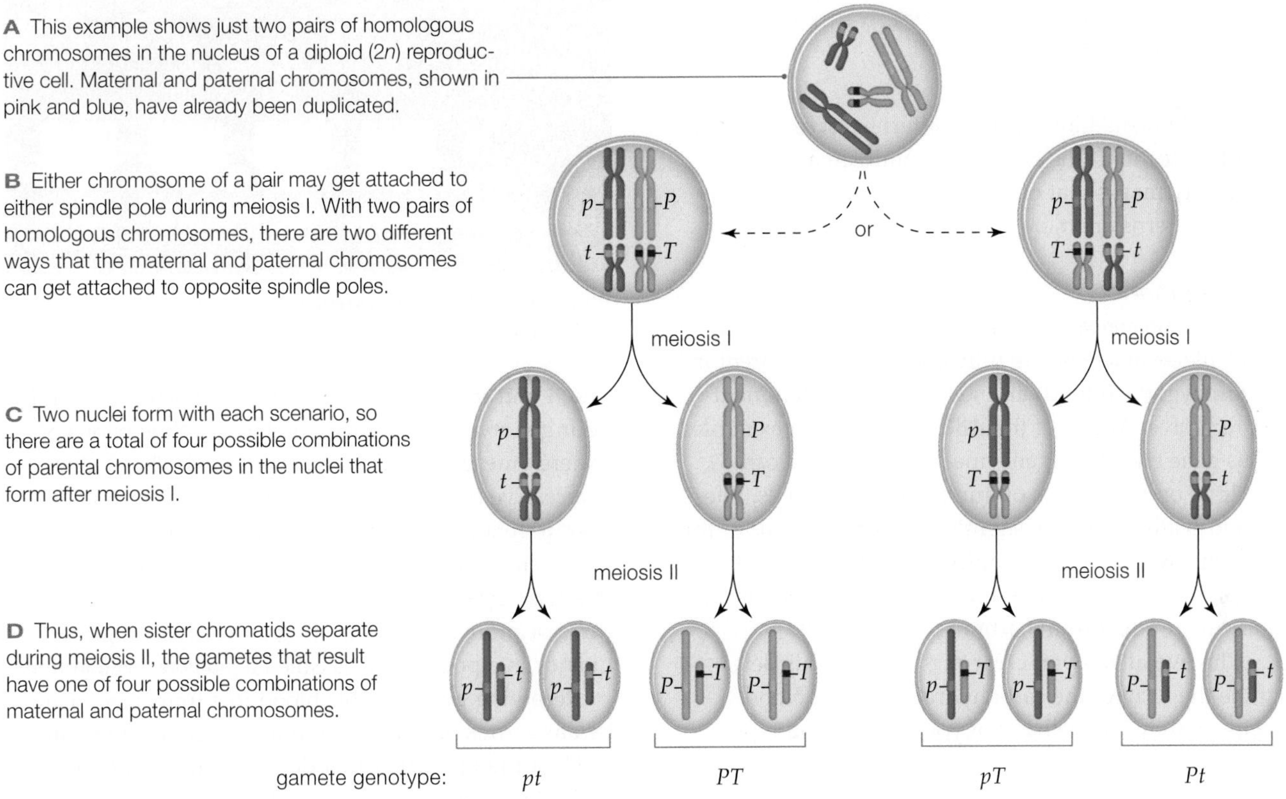

A This example shows just two pairs of homologous chromosomes in the nucleus of a diploid (2*n*) reproductive cell. Maternal and paternal chromosomes, shown in pink and blue, have already been duplicated.

B Either chromosome of a pair may get attached to either spindle pole during meiosis I. With two pairs of homologous chromosomes, there are two different ways that the maternal and paternal chromosomes can get attached to opposite spindle poles.

C Two nuclei form with each scenario, so there are a total of four possible combinations of parental chromosomes in the nuclei that form after meiosis I.

D Thus, when sister chromatids separate during meiosis II, the gametes that result have one of four possible combinations of maternal and paternal chromosomes.

gamete genotype: *pt*   *PT*   *pT*   *Pt*

FIGURE 13.9 ▶**Animated** Independent assortment of genes on different chromosomes. Genes that are far apart on the same chromosome usually assort independently too, because crossovers typically separate them.

rate during meiosis, either member of a gene pair can end up in either of the two new nuclei that form. This random assortment happens independently for each pair of homologous chromosomes in the cell. Thus, genes on one chromosome pair assort into gametes independently of genes on the other chromosome pairs (**FIGURE 13.9**).

Pea plants have seven chromosomes. Mendel studied seven pea genes, and all of them assorted into gametes independently of one another. Was he lucky enough to choose one gene on each of those chromosomes? As it turns out, some of the genes Mendel studied *are* on the same chromosome. These genes are far enough apart that crossing over occurs between them very frequently—so frequently that they tend to assort into gametes independently, just as if they were

on different chromosomes. By contrast, genes that are very close together on a chromosome usually do not assort independently into gametes, because crossing over does not happen between them very often. Thus, gametes usually end up with parental combinations of alleles of these genes.

Genes that do not assort independently into gametes are said to be linked. A **linkage group** comprises all genes on a chromosome. Peas have 7 different chromosomes, so they have 7 linkage groups. Humans have 23 different chromosomes, so they have 23 linkage groups.

---

dihybrid cross Cross between two individuals identically heterozygous for two genes; for example *AaBb* × *AaBb*.
law of independent assortment During meiosis, members of a pair of genes on homologous chromosomes tend to be distributed into gametes independently of other gene pairs.
linkage group All of the genes on a chromosome.

TAKE-HOME MESSAGE 13.4

How are gene pairs distributed into gametes?

✔ During meiosis, gene pairs on homologous chromosomes tend to be distributed into gametes independently of how other gene pairs are distributed.

✔ Independent assortment of genes on the same chromosome depends on proximity. Genes that are closer together get separated less frequently by crossovers, so gametes often receive parental combinations of alleles of these genes.

13.5 Beyond Simple Dominance

✔ Mendel studied traits with distinct forms arising from alleles that have a clear dominant–recessive relationship. Most inheritance patterns are less straightforward.

In the Mendelian inheritance patterns discussed in the last two sections, the effect of a dominant allele on a trait fully masks that of a recessive one. Other inheritance patterns are more common, and more complex.

Codominance

With **codominance**, traits associated with two nonidentical alleles of a gene are fully and equally apparent in heterozygous individuals; neither allele is dominant or recessive. Alleles of the *ABO* gene offer an example. This gene encodes an enzyme that modifies a carbohydrate on the surface of human red blood cells. The *A* and *B* alleles encode slightly different versions of this enzyme, which in turn modify the carbohydrate differently. The *O* allele has a mutation that prevents its enzyme product from becoming active at all.

The two alleles you carry at the *ABO* locus determine the form of the carbohydrate on your blood cells, so they are the basis of your ABO blood type.

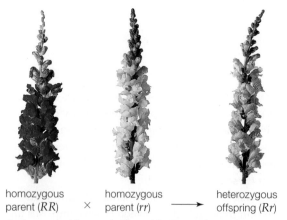

homozygous parent (*RR*) × homozygous parent (*rr*) ⟶ heterozygous offspring (*Rr*)

A Cross a red-flowered with a white-flowered snapdragon plant, and all of the offspring will have pink flowers.

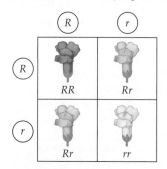

B If two of the pink-flowered snapdragons are crossed, the phenotypes of their offspring will occur in a 1:2:1 ratio.

FIGURE 13.11 ▶**Animated** Incomplete dominance in heterozygous (pink) snapdragons. One allele (*R*) results in the production of a red pigment; the other (*r*) results in no pigment.

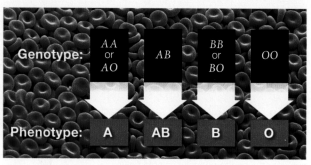

FIGURE 13.10 ▶**Animated** Combinations of alleles that are the basis of human ABO blood type.

The *A* and the *B* allele are codominant when paired. If your genotype is *AB*, then you have both versions of the enzyme, and your blood type is AB. The *O* allele is recessive when paired with either the *A* or *B* allele. If your genotype is *AA* or *AO*, your blood type is A. If your genotype is *BB* or *BO*, it is type B. If you are *OO*, it is type O (**FIGURE 13.10**). A gene such as *ABO*, with three or more alleles persisting at relatively high frequency in a population, is called a **multiple allele system**.

Receiving incompatible blood in a transfusion can be dangerous because the immune system attacks any cell bearing molecules that do not occur in one's own body. An immune system attack causes red blood cells to clump or burst, with potentially fatal results. Almost everyone makes the type O carbohydrate, so type O blood does not usually trigger an immune response in transfusion recipients. People with type O blood are called universal donors because they can donate blood to anyone. However, because their body is unfamiliar with the carbohydrates made by people with type A or B blood, they can receive type O blood only. People with type AB blood can receive a transfusion of any blood type, so they are called universal recipients.

Incomplete Dominance

With **incomplete dominance**, one allele is not fully dominant over the other, so the heterozygous phenotype is an intermediate blend of the two homozygous phenotypes. A gene that affects flower color in snapdragons is an example (**FIGURE 13.11**). One allele of the gene (*R*) encodes an enzyme that makes a red pigment. The enzyme encoded by a mutated allele (*r*) cannot make any pigment. Plants homozygous for the *R* allele (*RR*) make a lot of red pigment, so they have red flowers. Plants homozygous for the *r* allele (*rr*) make no pigment, so their flowers are white. Heterozygous plants (*Rr*) make only enough red pigment to tint their flowers pink. A cross between two heterozygous plants yields red-, pink-, and white-flowered offspring in a 1:2:1 ratio.

CREDITS: (10) photo, Annie Cavanagh/Wellcome Images; art, © Cengage Learning; (11A) © JupiterImages Corporation; (11B) © Cengage Learning.

FIGURE 13.12 ▶Animated

An example of epistasis. Interactions among products of two gene pairs affect coat color in Labrador retrievers. Dogs with alleles *E* and *B* have black fur. Those with an *E* and two recessive *b* alleles have brown fur. Dogs homozygous for the recessive *e* allele have yellow fur.

	EB	*Eb*	*eB*	*eb*
EB	*EEBB*	*EEBb*	*EeBB*	*EeBb*
Eb	*EEBb*	*EEbb*	*EeBb*	*Eebb*
eB	*EeBB*	*EeBb*	*eeBB*	*eeBb*
eb	*EeBb*	*Eebb*	*eeBb*	*eebb*

Epistasis

Some traits are affected by multiple genes, an effect called polygenic inheritance or **epistasis**. Coat color in Labrador retriever dogs, which depends on pigments called melanins, is an example of a trait affected by multiple genes (**FIGURE 13.12**). A dark brown melanin gives rise to brown or black fur; a reddish melanin, to yellow fur. Melanin synthesis and deposition in fur requires several genes. The product of one gene (*TYRP1*) helps make the brown melanin. A dominant allele (*B*) of this gene results in a higher production of brown melanin than the recessive allele (*b*). A different gene (*MC1R*) affects which type of melanin is produced. A dominant allele (*E*) of this gene triggers production of the brown melanin; its recessive partner (*e*) carries a mutation that results in production of the reddish form. Dogs homozygous for the *e* allele are yellow because they make only the reddish melanin.

Pleiotropy

In many cases, a single gene influences multiple traits, an effect called **pleiotropy**. Mutations that affect the gene's product or its expression affect all of the traits at once. Many complex genetic disorders, including sickle-cell anemia (Section 9.6) and cystic fibrosis, are caused by mutations in single genes. Another example, Marfan syndrome, is a result of mutations that affect fibrillin. Long fibers of this protein impart elasticity to tissues of the heart, skin, blood vessels, tendons, and other body parts. Mutations can cause tissues to form with defective fibrillin or none at all. The largest blood

vessel leading from the heart, the aorta, is particularly affected. The aorta's thick wall is not as elastic as it should be, and it eventually stretches and becomes leaky. Thinned and weakened, the aorta can rupture during exercise—an abruptly fatal outcome.

About 1 in 5,000 people have Marfan syndrome, and there is no cure. Its effects—and risks—are manageable with early diagnosis, but symptoms are easily missed. Affected people are tall and loose-jointed, but there are plenty of tall, loose-jointed people who do not have the syndrome. Thus, people with Marfan syndrome may die suddenly and early without ever knowing they had the disorder (**FIGURE 13.13**).

FIGURE 13.13 A heartbreaker: Marfan syndrome.

In 2006, 21-year-old basketball star Haris Charalambous collapsed and died suddenly during warm-up exercises. An autopsy revealed that his aorta had burst, an effect of the Marfan syndrome that Charalambous did not realize he had.

Assistant trainer Brian Jones says, "Haris was just the nicest, funniest kid in the world. With his size, he was sort of lovably goofy. He was everybody's best friend."

TAKE-HOME MESSAGE 13.5
Are all genes inherited in a Mendelian pattern?

✔ Some alleles are not dominant or recessive when paired.

✔ With incomplete dominance, one allele is not fully dominant over another, so the heterozygous phenotype is an intermediate blend of the two homozygous phenotypes.

✔ In codominance, two alleles have full and separate effects, so the phenotype of a heterozygous individual comprises both homozygous phenotypes.

✔ In some cases, one gene influences multiple traits. In other cases, multiple genes influence the same trait.

---

codominance Effect in which the full and separate phenotypic effects of two alleles are apparent in heterozygous individuals.
epistasis Polygenic inheritance, in which a trait is influenced by multiple genes.
incomplete dominance Effect in which one allele is not fully dominant over another, so the heterozygous phenotype is an intermediate blend between the two homozygous phenotypes.
multiple allele system Gene for which three or more alleles persist in a population at relatively high frequency.
pleiotropy Effect in which a single gene affects multiple traits.

13.6 Nature and Nurture

✔ Variations in traits are not always the result of differences in alleles. Many traits are also influenced by environmental factors.

The phrase "nature versus nurture" refers to a centuries-old debate about whether human behavioral traits arise from one's genetics (nature) or from environmental factors (nurture). It turns out that both play a role. The environment affects the expression of many genes, which in turn affects phenotype. We can summarize this thinking with an equation:

genotype + environment ⟶ phenotype

Epigenetics research is revealing that the environment has an even greater contribution to this equation than most biologists had suspected (Section 10.6).

Environmental cues initiate cell-signaling pathways that trigger changes in gene expression (you will learn more about such pathways in later chapters). Some of these cell-signaling pathways methylate or demethylate particular regions of DNA, so they suppress or enhance gene expression in those regions (Section 10.2). In humans, research has shown that DNA methylation patterns can be permanently and heritably affected by diet, stress, and exercise, and also by exposure to drugs and toxins such as tobacco, alcohol, arsenic, and asbestos.

Some Environmental Effects

Mechanisms that adjust phenotype in response to external cues are part of an individual's normal ability to adapt to a changing environment, as the following examples illustrate.

Alternative Phenotypes in Water Fleas

Water fleas (*Daphnia*) are tiny aquatic inhabitants of seasonal ponds, ditches, and other standing bodies of fresh water. Conditions such as temperature, oxygen content, and salinity vary dramatically over time and in different areas of these habitats. For example, water at the top of a still pond is typically warmer than water at the bottom, and it also contains more light and dissolved oxygen. Individual water fleas acclimate to such environmental differences by adjusting their gene expression. The adjustment provides an appropriate set of proteins to maintain cellular function in current environmental conditions.

Consider that *Daphnia* has a lot of genes—far more than other animals—and the abundance offers a striking plasticity of phenotype. For example, eleven *Daphnia* genes encode hemoglobin subunits; seven are

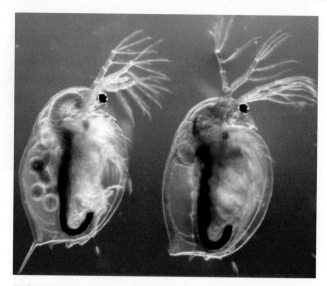

FIGURE 13.14 An environmental effect on phenotype of the water flea (*Daphnia*). Under low-oxygen conditions, water fleas can switch on their hemoglobin genes. Increased production of this protein enhances the individual's ability to take up oxygen from the water. The flea on the left has been living in water with a normal oxygen content; the one on the right, in water with a low oxygen concentration. The red pigment is hemoglobin.

differentially expressed depending on the temperature and oxygen content of the water. A flea survives low oxygen conditions by turning on synthesis of these genes—and turning red (**FIGURE 13.14**). The newly-produced hemoglobin improves the individual's ability to absorb oxygen from the water.

Other environmental factors also affect water flea phenotype. The presence of insect predators causes water fleas to form a protective pointy helmet and lengthened tail spine, for example. Individual water fleas also switch between asexual and sexual modes of reproduction. During early spring, food and space are typically abundant, and competition for these resources is scarce. Under these conditions, water fleas reproduce rapidly by asexual means, giving birth to large numbers of female offspring that quickly fill the ponds. Later in the season, competition intensifies as the pond water becomes warmer, saltier, and more crowded. Then, some of the water fleas start giving birth to males, and the population begins to reproduce sexually. The increased genetic diversity of sexually produced offspring may offer an advantage in the more challenging environment.

Seasonal Changes in Coat Color

Seasonal changes in temperature and the length of day affect the production of melanin and other pigments that color the skin and fur of many animals. These spe-

CREDIT: (14) From *Science* 4 February 2011: Vol 331 no.6017 pp. 555–561, Reprinted with permission from AAAS.

cies have different color phases in different seasons (**FIGURE 13.15A**). Hormonal signals triggered by the seasonal changes cause fur to be shed, and new fur grows back with different types and amounts of pigments deposited in it. The resulting change in phenotype provides these animals with seasonally appropriate camouflage from predators.

Effect of Altitude on Yarrow In plants, a flexible phenotype gives immobile individuals an ability to thrive in diverse habitats. For example, genetically identical yarrow plants grow to different heights at different altitudes (**FIGURE 13.15B**). More challenging temperature, soil, and water conditions are typically encountered at higher altitudes. Differences in altitude are also correlated with changes in the reproductive mode of yarrow: Plants at higher altitude tend to reproduce asexually, and those at lower altitude tend to reproduce sexually.

Psychiatric Disorders Researchers recently discovered that mutations in four human gene regions are associated with five psychiatric disorders: autism, depression, schizophrenia, bipolar disorder, and attention deficit hyperactivity disorder (ADHD). However, there must be environmental components to these disorders too, because one person with the mutations might get one type of disorder, while a relative with the same mutations might get another: two different results from the same genetic underpinnings. Moreover, the majority of people who carry these mutations never end up with a psychiatric disorder.

Recent discoveries in animal models are beginning to unravel some of the mechanisms by which environment can influence mental state in humans. For example, we now know that learning and memory are associated with dynamic and rapid DNA modifications in brain cells. Mood is, too. Stress-induced depression causes methylation-based silencing of a particular nerve growth factor gene; some antidepressants work by reversing this methylation. As another example, rats whose mothers are not very nurturing end up anxious and having a reduced resilience for stress as adults. The difference between these rats and ones who had nurturing maternal care is traceable to epigenetic DNA modifications that result in a lower than normal level of another nerve growth factor. Drugs can reverse these modifications—and their effects.

We do not yet know all of the genes that influence human mental state, but the implication of such research is that future treatments for many psychiatric

A The color of the snowshoe hare's fur varies by season. In summer, the fur is brown (left); in winter, white (right). Both forms offer seasonally appropriate camouflage from predators.

B The height of a mature yarrow plant (*Achillea millefolium*) depends on the elevation at which it grows.

FIGURE 13.15 ▶**Animated** Examples of environmental effects on phenotype.

disorders will involve deliberate modification of methylation patterns in an individual's DNA.

TAKE-HOME MESSAGE 13.6

Does an individual's environment affect its phenotype?

✔ The environment influences gene expression, and therefore can alter phenotype.

✔ Cell-signaling pathways link environmental cues with changes in gene expression.

FIGURE 13.16 Face length varies continuously in dogs. A gene with 12 alleles influences this trait; all arose by the spontaneous insertion of short tandem repeats. The longer the alleles, the longer the face.

63 64 65 66 67 68 69 70 71 72 73 74 75 76 77

A To see if human height varies continuously, male biology students at the University of Florida were divided into categories of one-inch increments in height and counted.

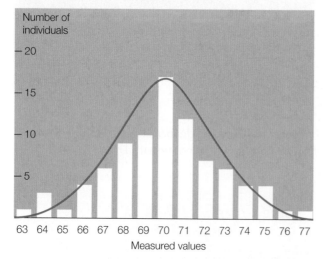

B Graphing the data that resulted from the experiment in (**A**) produces a bell-shaped curve, which is an indication that height does vary continuously in humans.

FIGURE 13.17 ►Animated Continuous variation.

✔ Individuals of most species vary in some of their shared traits. In some cases, the variation occurs in a continuous range.

The pea plant phenotypes that Mendel studied appeared in two or three forms, which made them easy to track through generations. However, many other traits do not appear in distinct forms. Such traits are often the result of complex genetic interactions—multiple genes, multiple alleles, or both—with added environmental influences. Tracking traits with complex variation presents a special challenge, which is why the genetic basis of many of them has not yet been completely unraveled.

Continuous Variation

Some traits occur in a range of small differences that is called **continuous variation**. Continuous variation can be an outcome of epistasis, in which multiple genes affect a single trait. The more genes that influence a trait, the more continuous is its variation. Traits that arise from genes with a lot of alleles may also vary continuously. Some genes have regions of DNA in which a series of 2 to 6 nucleotides is repeated hundreds or thousands of times in a row. These **short tandem repeats** can spontaneously expand or contract very quickly compared with the typical rate of mutation, and the resulting changes in the gene's DNA sequence may be preserved as alleles. For example, short tandem repeats have given rise to 12 alleles of a homeotic gene that influences the length of the face in dogs, with longer repeats associated with longer faces (**FIGURE 13.16**).

How do we know if a particular trait varies continuously? Let's use a human trait, height, in an example. First, the total range of phenotypes is divided into measurable categories—inches, in this case (**FIGURE 13.17A**). Next, the individuals in each category are counted; these counts reveal the relative frequencies of phenotypes across the range of values. Finally, this data is plotted as a bar chart (**FIGURE 13.17B**). A graph line around the top of the bars shows the distribution of values for the trait. If the line is a bell-shaped curve, or **bell curve**, then the trait varies continuously.

Human skin color varies continuously, as does human eye color (**FIGURE 13.18**). The colored part of the eye is a doughnut-shaped structure called the iris. Iris color, like skin color, results from interactions

bell curve Bell-shaped curve; typically results from graphing frequency versus distribution for a trait that varies continuously.
continuous variation Range of small differences in a shared trait.
short tandem repeat In chromosomal DNA, sequences of a few nucleotides repeated multiple times in a row.

Menacing Mucus (revisited)

The Δ*F508* allele that causes cystic fibrosis is at least 50,000 years old and very common: In some populations, 1 in 25 people are heterozygous for it. Why does the allele persist if it is so harmful?

The Δ*F508* allele is eventually lethal in homozygous individuals, but not in those who are heterozygous. It is codominant with the normal allele. Heterozygous individuals typically have no symptoms of cystic fibrosis because their cells have plasma membranes with enough CFTR to transport chloride ions normally.

Researchers think that the Δ*F508* allele has persisted because it offers heterozygous individuals an advantage in surviving certain deadly infectious diseases. CFTR's receptor function is an essential part of the immune response to bacteria in the respiratory tract. However, the same function allows bacteria to enter cells of the gastrointestinal tract, where they can be deadly. Thus, people who carry Δ*F508* are probably less susceptible to dangerous bacterial diseases that begin in the intestinal tract.

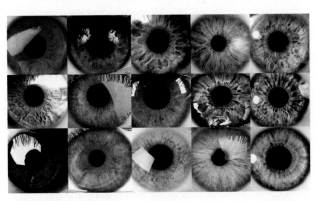

FIGURE 13.18 A small sampling of human eye color, a trait that varies continuously.

among several gene products that make and distribute melanins (an example of epistasis). The more melanin deposited in the iris, the less light is reflected from it. Dark irises have dense melanin deposits that absorb almost all light, and reflect almost none. Green and blue irises have the least amount of melanin, so they reflect the most light.

TAKE-HOME MESSAGE 13.7

Do all traits occur in distinct forms?

✔ The more genes and other factors that influence a trait, the more continuous is its range of variation.

---

summary

Section 13.1 Symptoms of cystic fibrosis are pleiotropic effects of mutations in the *CFTR* gene. The allele associated with most cases persists at high frequency despite its devastating effects in homozygous people. Carrying the allele may offer heterozygous individuals protection from dangerous gastrointestinal tract infections.

Section 13.2 Gregor Mendel indirectly discovered the role of alleles in inheritance by breeding pea plants and carefully tracking traits of the offspring over many generations.

Each gene occurs at a **locus**, or location, on a chromosome. Individuals with identical alleles are **homozygous** for the allele. **Heterozygous** individuals, or **hybrids**, have two nonidentical alleles.

A **dominant** allele masks the effect of a **recessive** allele partnered with it on the homologous chromosome. **Genotype** (an individual's particular set of alleles) gives rise to **phenotype** (an individual's observable traits).

Section 13.3 Crossing two individuals that breed true for different forms of a trait yields identically heterozygous offspring. A cross between such offspring is called a **monohybrid cross**. The frequency at which traits appear in offspring of a **testcross** can reveal the genotype of an individual with a dominant phenotype. **Punnett squares** are useful for determining the probability of offspring genotype and phenotype.

Mendel's monohybrid cross results led him to formulate his **law of segregation**, which we state here in modern terms: Diploid cells have pairs of genes on homologous chromosomes. The two genes of a pair become separated from each other during meiosis, so they end up in different gametes.

Section 13.4 Crossing two individuals that breed true for different forms of two traits yields F_1 offspring identically heterozygous for alleles governing those traits. A cross between such offspring is a **dihybrid cross**. The frequency at which the two traits appear in F_2 offspring can reveal dominance relationships between alleles associated with those traits. Mendel's dihybrid cross results led to his **law of independent assortment**, which we state here in modern terms: Pairs of genes on homologous chromosomes tend to sort into gametes independently of other gene pairs during meiosis. Crossovers can break up **linkage groups**.

Section 13.5 With **incomplete dominance**, the phenotype of heterozygous individuals is an intermediate blend of the two homozygous phenotypes. With **codominant** alleles, heterozygous individuals have both homozygous phenotypes. Codominance may occur in **multiple allele systems** such as the one underlying ABO blood typing. With **epistasis**, two or more genes affect the same trait. With **pleiotropy**, one gene affects two or more traits.

CREDITS: (18) top row from left, © Aaron Amat/Shutterstock; © jayfish/Shutterstock; © J. Helgason/Shutterstock; © Tishenko Irina/Shutterstock; © Tatiana Makotra/Shutterstock; second row from left, © Villedieu Christophe/Shutterstock; © Vaaka/Shutterstock; © evantravels/Shutterstock; © rawcaptured/Shutterstock; © Anemone/Shutterstock; third row from left, © szefei/Shutterstock; © Audrey Armyagov/Shutterstock; © Tatiana Makatra/ Shutterstock; © lightpoet/Shutterstock; © Anemone/Shutterstock; (in text revisited) Courtesy of © Bobby Brooks and the Family of Jeff Baird.

Section 13.6 An individual's phenotype is influenced by environmental factors. Environmental cues alter gene expression by way of cell signaling pathways that ultimately affect gene expression control.

Section 13.7 A trait that is influenced by multiple genes often occurs in a range of small increments of phenotype called **continuous variation**. Continuous variation typically occurs as a **bell curve** in the range of values. Multiple alleles such as those that arise in regions of **short tandem repeats** can give rise to continuous variation.

self-quiz

Answers in Appendix VII

1. A heterozygous individual has a _____ for a trait being studied.
   a. pair of identical alleles
   b. pair of nonidentical alleles
   c. haploid condition, in genetic terms

2. An organism's observable traits constitute its _____ .
   a. phenotype   c. genotype
   b. variation   d. pedigree

3. In genetics, independent assortment means _____ .
   a. genes of a pair end up in different gametes.
   b. gene pairs separate independently of other gene pairs

4. The second-generation offspring of a cross between individuals who are homozygous for different alleles of a gene are called the _____ .
   a. F_1 generation   c. hybrid generation
   b. F_2 generation   d. none of the above

5. The F_1 offspring of the cross $AA \times aa$ are _____ .
   a. all AA c. all Aa
   b. all aa d. half are AA; half are aa

6. Refer to question 5. Assuming complete dominance, the F_2 generation will show a phenotypic ratio of _____ .
   a. 3:1   b. 9:1   c. 1:2:1   d. 9:3:3:1

7. A testcross is a way to determine _____ .
   a. phenotype   b. genotype   c. both a and b

8. Assuming complete dominance, crosses between two dihybrid F_1 pea plants, which are offspring from a cross $AABB \times aabb$, result in F_2 phenotype ratios of _____ .
   a. 1:2:1   b. 3:1   c. 1:1:1:1   d. 9:3:3:1

9. The probability of a crossover occurring between two genes on the same chromosome _____ .
   a. is unrelated to the distance between them
   b. decreases with increasing distance between them
   c. increases with the distance between them

10. True or false? All traits are inherited in a Mendelian pattern.

11. A gene that affects three traits is _____ .
    a. epistatic   c. pleiotropic
    b. a multiple allele system   d. dominant

12. The phenotype of individuals heterozygous for _____ alleles comprises both homozygous phenotypes.
    a. epistatic   c. pleiotropic
    b. codominant   d. hybrid

13. _____ in a trait is indicated by a bell curve.
    a. Epigenetic effects   c. Incomplete dominance
    b. Pleiotropy   d. Continuous variation

14. Match the terms with the best description.
    ___ dihybrid cross   a. bb
    ___ monohybrid cross   b. $AaBb \times AaBb$
    ___ homozygous condition c. Aa
    ___ heterozygous condition d. $Aa \times Aa$

genetics problems

Answers in Appendix VII

1. Assuming that independent assortment occurs during meiosis, what type(s) of gametes will form in individuals with the following genotypes?
   a. $AABB$ b. $AaBB$ c. $Aabb$ d. $AaBb$

2. Refer to problem 1. Determine the frequencies of each genotype among offspring from the following matings:
   a. $AABB \times aaBB$ c. $AaBb \times aabb$
   b. $AaBB \times AABb$ d. $AaBb \times AaBb$

3. Refer to problem 2. Assume a third gene has alleles C and c. For each genotype listed, what allele combinations will occur in gametes, assuming independent assortment?
   a. $AABBCC$ c. $AaBBCc$
   b. $AaBBcc$ d. $AaBbCc$

4. Heterozygous individuals perpetuate some alleles that have lethal effects in homozygous individuals. A mutated allele (M^L) associated with taillessness in Manx cats is an example (left). Cats homozygous for this allele ($M^L M^L$) typically die before birth due to severe spinal cord defects. In a case of incomplete dominance, cats heterozygous for the M^L allele and the normal, unmutated allele (M) have a short, stumpy tail or none at all. Two $M^L M$ cats mate. What is the probability that any one of their surviving kittens will be heterozygous?

5. Suppose you identify a new gene in mice. One of its alleles specifies white fur, another specifies brown. You want to see if these alleles are inherited in a Mendelian pattern, or with incomplete dominance. What crosses would give you the answer?

6. Mendel crossed a true-breeding pea plant with green pods and a true-breeding pea plant with yellow pods. All the F_1 plants had green pods. Which color is recessive?

7. Several alleles affect traits of roses, such as plant form and bud shape. Alleles of one gene govern whether a plant will be a climber (dominant) or shrubby (recessive). All F_1 offspring from a cross between a true-breeding climber and a shrubby plant are climbers. If an F_1 plant is crossed with a shrubby plant, about 50 percent of the offspring will be shrubby; 50 percent will be climbers. Using symbols A and a

data analysis activities

Carrying the Cystic Fibrosis Allele Offers Protection from Typhoid Fever

Epithelial cells that lack the CFTR protein cannot take up bacteria by endocytosis. Endocytosis is an important part of the respiratory tract's immune defenses against common *Pseudomonas* bacteria, which is why *Pseudomonas* infections of the lungs are a chronic problem in cystic fibrosis patients. Endocytosis is also the way that *Salmonella typhi* bacteria (shown at right) enter cells of the gastrointestinal tract, where internalization of this bacteria can result in typhoid fever.

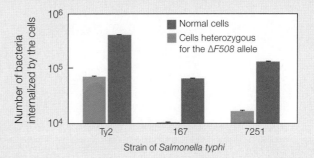

Typhoid fever is a common worldwide disease. Its symptoms include extreme fever and diarrhea, and the resulting dehydration causes delirium that may last several weeks. If untreated, it kills up to 30 percent of those infected. Around 600,000 people, most of whom are children, die annually from typhoid fever.

In 1998, Gerald Pier and his colleagues compared the uptake of *S. typhi* by different types of epithelial cells: those homozygous for the normal allele, and those heterozygous for the Δ*F508* allele associated with CF. (Cells that are homozygous for the mutation do not take up any *S. typhi* bacteria.) Some of the results are shown in **FIGURE 13.19**.

FIGURE 13.19 Effect of the Δ*F508* mutation on the uptake of three different strains of *Salmonella typhi* bacteria by epithelial cells.

1. Regarding the Ty2 strain of *S. typhi*, about how many more bacteria were able to enter normal cells (those heterozygous for the normal allele) than cells heterozygous for the Δ*F508* allele?

2. Which strain of bacteria entered normal epithelial cells most easily?

3. Entry of all three *S. typhi* strains into the heterozygous epithelial cells was inhibited. Is it possible to tell from this graph which strain was most inhibited?

---

for the dominant and recessive alleles, make a Punnett-square diagram of the expected genotypes and phenotypes in the cross between the F$_1$ offspring and the shrubby plant.

8. Mutations in the *TYR* gene may render its enzyme product—tyrosinase—nonfunctional. Individuals homozygous for such mutations cannot make the pigment melanin. Albinism, the absence of melanin, results. Humans and many other organisms can have this phenotype (left). Mutated tyrosinase alleles are recessive when paired with the normal allele in heterozygous individuals. In the following situations, what are the probable genotypes of the father, the mother, and their children?

a. Both parents have normal phenotypes; some of their children have the albino phenotype and others are unaffected.

b. Both parents and children have the albino phenotype.

c. The mother and three children are unaffected; the father and one child have the albino phenotype.

9. In sweet pea plants, an allele for purple flowers (*P*) is dominant when paired with a recessive allele for red flowers (*p*). An allele for long pollen grains (*L*) is dominant when paired with a recessive allele for round pollen grains (*l*). Bateson and Punnett crossed a plant having purple flowers

and long pollen grains with one having white flowers and round pollen grains. All F$_1$ offspring have purple flowers and long pollen grains. Among the F$_2$ generation, the researchers observed the following phenotypes:

    296 purple flowers/long pollen grains
      19 purple flowers/round pollen grains
      27 red flowers/long pollen grains
      85 red flowers/round pollen grains

What is the best explanation for these results?

10. Red-flowering snapdragons are homozygous for allele R^1. White-flowering snapdragons are homozygous for a different allele (R^2). Heterozygous plants (R^1R^2) bear pink flowers. What phenotypes should appear among first-generation offspring of the crosses listed? What are the expected proportions for each phenotype?

  a. $R^1R^1 \times R^1R^2$ c. $R^1R^2 \times R^1R^2$
  b. $R^1R^1 \times R^2R^2$ d. $R^1R^2 \times R^2R^2$

(Note that alleles inherited in a pattern of incomplete dominance are designated by superscript numerals, as shown, rather than by upper- and lowercase letters.)

11. A single allele gives rise to the HbS form of hemoglobin. Individuals who are homozygous for the allele (*HbS*/*HbS*) develop sickle-cell anemia (Section 9.6). Heterozygous individuals (*HbA*/*HbS*) have few symptoms. A couple who are both heterozygous for the *HbS* allele plan to have children. For each of the pregnancies, state the probability that they will have a child who is:

a. homozygous for the *HbS* allele

b. homozygous for the normal allele (*HbA*)

c. heterozygous: *HbA*/*HbS*

CREDITS: (in text Genetics Problems #8) © Rick Guidotti, Positive Exposure; (data analysis activities inset) © Gary Gaugler/The Medical file/Peter Arnold, Inc.; (19) © Cengage Learning.

14 Chromosomes and Human Inheritance

LEARNING ROADMAP

This chapter revisits proteins (Section 3.6); cell components (4.6, 4.8, 4.10, 4.11, 5.7); metabolism (5.5); pigments (6.2); chromosomes (8.4); DNA replication and repair (8.5); mutations (8.6); gene expression (9.2, 9.3) and control (10.4, 10.6); telomeres (11.5); oncogenes (11.6); meiosis (12.3); and inheritance patterns (13.2, 13.5, and 13.7).

TRACKING TRAITS IN HUMANS

Inheritance patterns in humans are revealed by following traits through generations of a family. Tracked traits are often genetic abnormalities or syndromes associated with a genetic disorder.

AUTOSOMAL INHERITANCE

Traits associated with dominant alleles on autosomes appear in every generation. Traits associated with recessive alleles on autosomes can skip generations.

SEX-LINKED INHERITANCE

Traits associated with alleles on the X chromosome tend to affect more men than women. Men cannot pass such alleles to a son; carrier mothers bridge affected generations.

CHROMOSOME CHANGES

Some genetic disorders arise after large-scale change in chromosome structure. With few exceptions, a change in the number of autosomes is fatal in humans.

GENETIC TESTING

Genetic testing provides information about the risk of passing a harmful allele to offspring. Prenatal testing can reveal a genetic abnormality or disorder in a developing fetus.

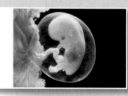

Genetic disorders are discussed in later chapters, in the context of the systems that they affect. Chapter 15 returns to human chromosomes as part of genomics and genetic engineering. Chapter 17 explores evolutionary adaptations and factors that influence the frequency of alleles in a population. The cells and other structural components of human skin are covered in detail in Section 31.8. Chapters 41 and 42 return to human reproduction and development.

The color of human skin begins with melanosomes, which are organelles that make melanin pigments. Most people have about the same number of melanosomes in their skin cells. Variations in skin color arise from differences in the size, shape, and cellular distribution of melanosomes in the skin, as well as in the kinds and amounts of melanins they make.

Human skin color variation may have evolved as a balance between vitamin production and protection against harmful ultraviolet (UV) radiation in the sun's rays. Dark skin is beneficial under the intense sunlight of African savannas where humans first evolved. Melanin is a natural sunscreen: It prevents UV radiation from breaking down folate, a vitamin essential for normal sperm formation and embryonic development.

Early human groups that migrated to regions with cold climates were exposed to less sunlight. In these regions, lighter skin color is beneficial. Why? UV radiation stimulates skin cells to make a molecule the body converts to vitamin D. Where sunlight exposure is minimal, UV radiation is less of a risk than vitamin D deficiency, which has serious health consequences for developing fetuses and children. People with dark, UV-shielding skin have a high risk of this deficiency in regions with long, dark winters.

Skin color, like most other human traits, has a genetic basis; at least 100 gene products are involved in pigmentation. The evolution of regional variations in human skin color began with mutations in these genes. Consider a gene on chromosome 15, *SLC24A5*, that encodes a transport protein in melanosome membranes. Nearly all people of African, Native American, or east Asian descent carry the same allele of this gene. Between 6,000 and 10,000 years ago, a mutation gave rise to a different allele. The mutation, a single base-pair substitution (Section 9.6), changed the 111th amino acid of the transport protein from alanine to threonine. The change results in less melanin—and lighter skin color—than the original African allele does. Today, nearly all people of European descent are homozygous for the mutated allele.

A person of mixed ethnicity may make gametes that contain different combinations of alleles for dark and light skin. It is fairly rare that one of those gametes contains all of the alleles for dark skin, or all of the alleles for light skin, but it happens (**FIGURE 14.1**). Skin color is only one of many human traits that vary as a result of single nucleotide mutations. The small scale of such changes offers a reminder that all of us share the genetic legacy of common ancestry.

FIGURE 14.1 Variation in human skin color (left) begins with differences in alleles inherited from parents. Above, twins Kian and Remee with their parents. Both of the children's grandmothers are of European descent, and have pale skin. Both of their grandfathers are of African descent, and have dark skin. The twins inherited different alleles of some genes that affect skin color from their parents, who, given the appearance of their children, must be heterozygous for those alleles.

CREDITS: (opposite) Ciarra, photo by © Michelle Harmon; (1) left, Richard A. Sturm, Molecular genetics of human pigmentation diversity, *Human Molecular Genetics*, 2009 Apr 15;18(R1):R9-17, by permission of Oxford University Press; right, © Gary Roberts/worldwidefeatures.com.

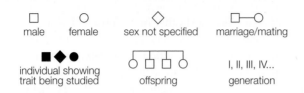

A Standard symbols used in pedigrees.

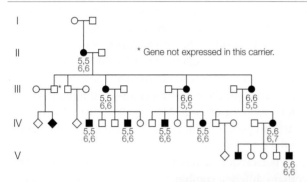

B Above, a pedigree for poly-dactyly, which is characterized by extra fingers (right), toes, or both. The black numbers signify the number of fingers on each hand; the red numbers signify the number of toes on each foot. Polydactyly that appears as part of a syndrome (such as Ellis–van Creveld syndrome) can be inherited in an autosomal recessive pattern. It also occurs on its own, in which case it is typically inherited in an autosomal dominant pattern.

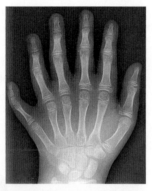

C For more than 30 years, researcher Nancy Wexler has studied the genetic basis of Huntington's disease, an inherited disorder that causes progressive degeneration of the nervous system. Wexler and her team constructed an extended family tree for nearly 10,000 Venezuelans. Their analysis of relationships among unaffected and affected individuals revealed that a dominant allele on human chromosome 4 is the culprit. Wexler has a special interest in the disorder: It runs in her family.

FIGURE 14.2 Pedigrees.

✔ Geneticists study inheritance patterns in humans by tracking the appearance of genetic disorders and abnormalities through generations of families.

✔ Charting these genetic connections with pedigrees can reveal patterns of inheritance for certain traits.

Some organisms, including pea plants and fruit flies, are ideal for genetic studies. They have relatively few chromosomes, they reproduce quickly under controlled conditions, and breeding them poses few ethical problems. It does not take long to follow a trait through many generations. Humans, however, are a different story. Unlike flies grown in laboratories, we humans live under variable conditions, in different places, and we live as long as the geneticists who study our inheritance patterns. Most of us select our own mates and reproduce if and when we want to. Our families tend to be on the small side, so sampling error (Section 1.8) is a major factor in studying them.

Because of these and other challenges, geneticists often use historical records to track traits through many generations of a family. They use standardized charts called **pedigrees** to illustrate the phenotypes of family members and genetic connections among them (**FIGURE 14.2**). Analysis of a pedigree can reveal whether a trait is associated with a dominant or recessive allele, and whether the allele is on an autosome or a sex chromosome. Pedigree analysis also allows geneticists to determine the probability that a trait will recur in future generations of a family or a population.

Types of Genetic Variation

Some easily observed human traits follow Mendelian inheritance patterns. Like the flower color of Mendel's pea plants, these traits are controlled by a single gene with alleles that have a clear dominance relationship. Consider the *MC1R* gene (Section 13.5), which encodes a protein that triggers production of the brownish melanin. Mutations can result in a defective protein; an allele with one of these loss-of-function mutations is recessive when paired with an unmutated allele. A person who is homozygous for a mutated allele does not make the brownish melanin—only the reddish type—so this individual has red hair.

Single genes on autosomes or sex chromosomes also govern more than 6,000 genetic abnormalities and disorders. **TABLE 14.1** lists a few examples. A genetic abnormality is a rare or uncommon version of a trait, such as having six fingers on a hand or having a web between two toes. Genetic abnormalities are not inherently life-threatening, and how you view them is a matter of opinion. By contrast, a genetic

CREDITS: (2A, B above) © Cengage Learning; (2B right) Courtesy of Irving Buchbinder, DPM, DABPS, Community Health Services, Hartford CT; (2C) Acey Harper/Time & Life Pictures/Getty Images.

disorder sooner or later causes medical problems that may be severe. A genetic disorder is often characterized by a specific set of symptoms (a syndrome). In general, much more research focuses on genetic disorders than on other human traits, because what we learn helps us develop treatments for affected people.

The next two sections of this chapter focus on inheritance patterns of human single-gene disorders, which collectively affect about 1 in 200 people. Keep in mind that these inheritance patterns are the least common kind. Most human traits, including skin color, are polygenic (influenced by multiple genes, Section 13.5), and some have epigenetic contributions or causes (Section 10.6). Environmental effects (Section 13.6) make these traits even harder to study. Many genetic disorders, including diabetes, asthma, obesity, cancers, heart disease, and multiple sclerosis, are inherited in patterns so complex that our understanding of the genetics behind them remains incomplete despite intense research. For example, mutations associated with an increased risk of autism (a developmental disorder) have been found on almost every chromosome, but most people who carry these mutations do not have autism. Appendix IV shows a map of human chromosomes with the locations of some alleles known to play a role in genetic disorders and other human traits.

Alleles that give rise to severe genetic disorders are generally rare in populations because they compromise the health and reproductive ability of their bearers. Why do they persist? Mutations periodically reintroduce them. In some cases, a codominant allele offers a survival advantage in a particular environment. You learned about one example, the Δ*F508* allele that causes cystic fibrosis, in Chapter 13: People heterozygous for this allele are protected from infection by bacteria that cause typhoid fever. You will see additional examples in later chapters.

---

pedigree Chart of family connections that shows the appearance of a trait through generations.

TAKE-HOME MESSAGE 14.2
How do we study inheritance patterns in humans?

✔ Human inheritance patterns are often studied by tracking genetic abnormalities or disorders through family trees.

✔ A genetic disorder is an inherited condition that causes medical problems. A genetic abnormality is a rare but harmless version of an inherited trait.

✔ A few genetic disorders are governed by single genes inherited in a Mendelian fashion. Most human traits are polygenic, and some have epigenetic contributions.

Table 14.1 Examples of Genetic Abnormalities and Disorders in Humans

Disorder/Abnormality	Main Symptoms
Autosomal dominant inheritance pattern	
Achondroplasia	One form of dwarfism
Aniridia	Defects of the eyes
Camptodactyly	Rigid, bent fingers
Familial hypercholesterolemia	High cholesterol level; clogged arteries
Huntington's disease	Degeneration of the nervous system
Marfan syndrome	Abnormal or missing connective tissue
Polydactyly	Extra fingers, toes, or both
Progeria	Drastic premature aging
Neurofibromatosis	Tumors of nervous system, skin
Autosomal recessive inheritance pattern	
Albinism	Absence of pigmentation
Hereditary methemoglobinemia	Blue skin coloration
Cystic fibrosis	Difficulty breathing; chronic lung infections
Ellis–van Creveld syndrome	Dwarfism, heart defects, polydactyly
Fanconi anemia	Physical abnormalities, marrow failure
Galactosemia	Brain, liver, eye damage
Hereditary hemochromatosis	Joints, organs damaged by iron overload
Phenylketonuria (PKU)	Mental impairment
Sickle-cell anemia	Anemia, pain, swelling, frequent infections
Tay–Sachs disease	Deterioration of mental and physical abilities; early death
X-linked recessive inheritance pattern	
Androgen insensitivity syndrome	XY individual but having some female traits; sterility
Red–green color blindness	Inability to distinguish red from green
Hemophilia	Impaired blood clotting ability
Muscular dystrophies	Progressive loss of muscle function
X-linked anhidrotic dysplasia	Mosaic skin (patches with or without sweat glands); other ill effects
X-linked dominant inheritance pattern	
Fragile X syndrome	Intellectual, emotional disability
Incontinentia pigmenti	Abnormalities of skin, hair, teeth, nails, eyes; neurological problems
Changes in chromosome number	
Down syndrome	Mental impairment; heart defects
Turner syndrome (XO)	Sterility; abnormal ovaries, sexual traits
Klinefelter syndrome	Sterility; mild mental impairment
XXX syndrome	Minimal abnormalities
XYY condition	Mild mental impairment or no effect
Changes in chromosome structure	
Chronic myelogenous leukemia (CML)	Overproduction of white blood cells; organ malfunctions
Cri-du-chat syndrome	Mental impairment; abnormal larynx

✔ An allele is inherited in an autosomal dominant pattern if the trait it specifies appears in heterozygous people.

✔ An allele is inherited in an autosomal recessive pattern if the trait it specifies appears only in homozygous people.

The Autosomal Dominant Pattern

A trait associated with a dominant allele on an autosome appears in people who are heterozygous for it as well as those who are homozygous. Such traits appear in every generation of a family, and they occur with equal frequency in both sexes. When one parent is heterozygous, and the other is homozygous for the recessive allele, each of their children has a 50 percent chance of inheriting the dominant allele and having the associated trait (**FIGURE 14.3A**).

Achondroplasia A form of hereditary dwarfism called achondroplasia offers an example of an autosomal dominant disorder (one caused by a dominant allele on an autosome). Mutations associated with achondroplasia occur in a gene for a growth factor receptor. The mutations cause the receptor, which normally slows bone development, to be overly active. About 1 in 10,000 people is heterozygous for one of these mutations. As adults, affected people are, on average, about 4 feet 4 inches (1.3 meters) tall, with arms and legs that are short relative to torso size (**FIGURE 14.3B**). An allele that causes achondroplasia can be passed to children because its expression does not interfere with reproduction, at least in heterozygous people. The homozygous condition results in severe skeletal malformations that cause early death.

Huntington's Disease Alleles associated with Huntington's disease are also inherited in an autosomal dominant pattern. Mutations that cause this disorder alter a gene for a cytoplasmic protein whose function is still unknown. The mutations are insertions caused by expansion of a short tandem repeat (Section 13.7), in which the same three nucleotides become repeated many times in the gene's sequence. The oversized protein product of the altered gene gets chopped into pieces inside nerve cells of the brain. The pieces accumulate in cytoplasm as large clumps that eventually prevent the cells from functioning properly. Brain cells involved in movement, thinking, and emotion are particularly affected. Dramatic, involuntary jerking and writhing movements that are symptoms of the most common form of Huntington's appear after age thirty. Affected people die during their forties or fifties. With this and other late-onset disorders, people may reproduce before symptoms appear, so the allele can be passed unknowingly to children.

Hutchinson–Gilford Progeria Hutchinson–Gilford progeria is an autosomal dominant disorder characterized by drastically accelerated aging. It is usually caused by a mutation that affects lamin A, a

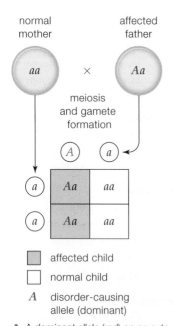

normal mother   affected father

aa × *Aa*

meiosis and gamete formation

Ⓐ   ⓐ

	A	*a*
a	*Aa*	*aa*
a	*Aa*	*aa*

▢ affected child

▢ normal child

A   disorder-causing allele (dominant)

A A dominant allele (red) on an autosome affects all heterozygous people.

B Achondroplasia affects Ivy Broadhead (left), her brother, father, and grandfather.

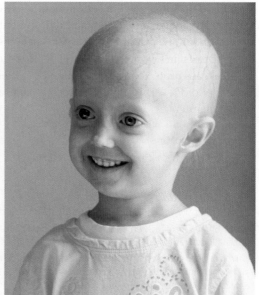

C Five-year-old Megan is already showing symptoms of Hutchinson–Gilford progeria.

FIGURE 14.3 ▶Animated Autosomal dominant inheritance.

fibrous protein component of intermediate filaments of the nuclear lamina (Section 4.10). The mutation, a base-pair substitution, adds a signal for an alternative splice site (Section 9.3). The resulting protein is defective, and so is the nuclear lamina. In cells that carry this mutation, the nucleus is grossly abnormal, with improperly assembled nuclear pore complexes and membrane proteins localized to the wrong side of the nuclear envelope. The function of the nucleus as protector of chromosomes and gateway for transcription is severely impaired, and DNA damage accumulates quickly. The effects are pleiotropic. Outward symptoms begin to appear before age two, as skin that should be plump and resilient starts to thin, muscles weaken, and bones soften. Premature baldness is inevitable (**FIGURE 14.3C**). Most people with the disorder die in their early teens as a result of a stroke or heart attack brought on by hardened arteries, a condition typical of advanced age. Progeria does not run in families because affected people do not live long enough to reproduce.

The Autosomal Recessive Pattern

A recessive allele on an autosome is expressed only in homozygous individuals, so traits associated with the allele tend to skip generations. Both sexes are equally affected. Heterozygous individuals are called carriers because they have the allele but not the trait. Any child of two carriers has a 25 percent chance of inheriting the allele from both parents—and developing the trait (**FIGURE 14.4A**).

Tay-Sachs Disease Alleles associated with Tay–Sachs disease are inherited in an autosomal recessive pattern. In the general population, about 1 in 300 people is a carrier for one of these alleles, but the incidence is ten times higher in some groups, such as Jews of eastern European descent. The gene altered in Tay–Sachs encodes a lysosomal enzyme responsible for breaking down a particular type of lipid. Mutations result in an enzyme that misfolds and becomes destroyed, so cells make the lipid but cannot break it down. Typically, newborns homozygous for a Tay–Sachs allele seem normal, but within three to six months they become irritable, listless, and may have seizures as the lipid accumulates in their nerve cells. Blindness, deafness, and paralysis follow. Affected children usually die by age five (**FIGURE 14.4B**).

Albinism Albinism, a phenotype characterized by an abnormally low level of the pigment melanin, is also inherited in an autosomal recessive pattern. Mutations

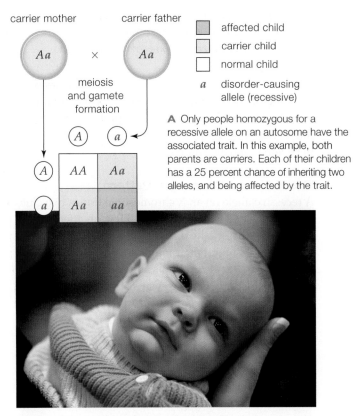

A Only people homozygous for a recessive allele on an autosome have the associated trait. In this example, both parents are carriers. Each of their children has a 25 percent chance of inheriting two alleles, and being affected by the trait.

B Conner Hopf was diagnosed with Tay–Sachs disease at age 7½ months. He died before his second birthday.

FIGURE 14.4 ▶Animated Autosomal recessive inheritance.

associated with albinism affect proteins involved in melanin synthesis. Skin, hair, or eye pigmentation may be reduced or missing. In the most dramatic form, the skin is very white and does not tan, and the hair is white. The irises of the eyes appear red because the lack of pigment allows underlying blood vessels to show through. Melanin also plays a role in the retina, so vision problems are typical. In skin, melanin acts as a sunscreen; without it, the skin is defenseless against UV radiation. Thus, people with the albino phenotype have a very high risk of skin cancer.

TAKE-HOME MESSAGE 14.3

How do we know when a trait is affected by an allele on an autosome?

✔ With an autosomal dominant inheritance pattern, anyone with the allele, homozygous or heterozygous, has the associated trait. The trait typically appears in every generation.

✔ With an autosomal recessive inheritance pattern, only persons who are homozygous for an allele have the associated trait. The trait tends to skip generations.

✔ Traits associated with recessive alleles on the X chromosome appear more frequently in men than in women.

✔ A man cannot pass an X chromosome allele to a son.

Many genetic disorders are associated with alleles on the X chromosome (**FIGURE 14.5**). Almost all of them are inherited in a recessive pattern, probably because those caused by dominant X chromosome alleles tend to be lethal in male embryos.

The X-Linked Recessive Pattern

A recessive allele on an X chromosome leaves two clues when it causes a genetic disorder. First, an affected father never passes the disorder to a son, because all children who inherit their father's X chromosome are female (**FIGURE 14.6A**). Thus, a heterozygous female is always the bridge between an affected male and his affected grandson. Second, the disorder appears in males more often than in females. This is because all males who carry the allele have the disorder, but not all heterozygous females do. Remember that one of the two X chromosomes in each cell of a female is inactivated as a Barr body (Section 10.4). As a result, only about half of a heterozygous female's cells express the recessive allele. The other half of her cells express the dominant, normal allele that she carries on her other

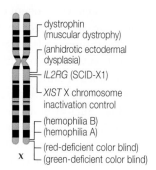

FIGURE 14.5 The human X chromosome.

This chromosome carries about 2,000 genes—almost 10 percent of the total. Most X chromosome alleles that cause genetic disorders are inherited in a recessive pattern. A few disorders are listed (in parentheses).

- dystrophin (muscular dystrophy)
- (anhidrotic ectodermal dysplasia)
- *IL2RG* (SCID-X1)
- *XIST* X chromosome inactivation control
- (hemophilia B)
- (hemophilia A)
- (red-deficient color blind)
- (green-deficient color blind)

X chromosome, and this expression can mask the phenotypic effects of the recessive allele.

Red–Green Color Blindness Color blindness refers to a range of conditions in which an individual cannot distinguish among colors in the spectrum of visible light. These conditions are typically inherited in an X-linked recessive pattern, because most of the genes involved in color vision are on the X chromosome.

Humans can sense the differences among 150 colors, and this perception depends on pigment-containing receptors in the eyes. Mutations that result in altered or missing receptors affect color vision. For example, people who have red–green color blindness see fewer than 25 colors because receptors that respond to the red

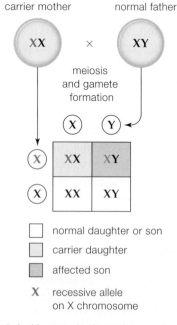

carrier mother × normal father

meiosis and gamete formation

☐ normal daughter or son
▨ carrier daughter
▩ affected son
X recessive allele on X chromosome

A In this example, the mother carries a recessive allele on one of her two X chromosomes (red).

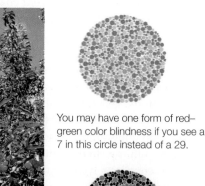

You may have one form of red–green color blindness if you see a 7 in this circle instead of a 29.

You may have another form of red–green color blindness if you see a 3 instead of an 8 in this circle.

B A view of color blindness. The photo on the left shows how a person with red–green color blindness sees the photo on the right. The perception of blues and yellows is normal; red and green appear similar. The circle diagrams are part of a standardized test for color blindness. A set of 38 of these diagrams is commonly used to diagnose deficiencies in color perception.

FIGURE 14.6 ▶**Animated** X-linked recessive inheritance.

CREDITS: (5, 6A) © Cengage Learning; (6B) left, Photos by Gary L. Friedman, www.FriedmanArchives.com; right, Life Nature Library, The Primates, 1965, Sarel Eimerl and Irven DeVore.

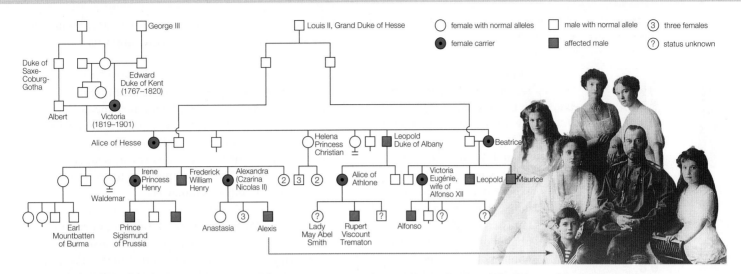

FIGURE 14.7 ▶Animated A classic case of X-linked recessive inheritance: a partial pedigree of the descendants of Queen Victoria of England. At one time, the recessive X-linked allele that resulted in hemophilia was present in eighteen of Victoria's sixty-nine descendants, who sometimes intermarried. Of the Russian royal family members shown in the photo, the mother (Alexandra Czarina Nicolas II) was a carrier.

FIGURE IT OUT How many of Alexis's siblings were affected by hemophilia? Answer: None

and green wavelengths of light are weakened or absent (**FIGURE 14.6B**). Some confuse red and green; others see green as gray.

Duchenne Muscular Dystrophy A genetic disorder called Duchenne muscular dystrophy (DMD) is characterized by progressive muscle degeneration. It is caused by mutations in the X chromosome gene for dystrophin, a cytoskeletal protein that links actin microfilaments in cytoplasm to a complex of proteins in the plasma membrane. This complex structurally and functionally links the cell to extracellular matrix. When dystrophin is absent, the entire protein complex is unstable. Muscle cells, which are subject to stretching, are particularly affected. Their plasma membrane is easily damaged, and they become flooded with calcium ions. Eventually, the cells die and become replaced by fat cells and connective tissue.

DMD affects about 1 in 3,500 people; almost all are boys. Symptoms begin between ages three and seven. Anti-inflammatory drugs can slow the progression of DMD, but there is no cure. When an affected boy is about twelve years old, he will begin to use a wheelchair and his heart will start to fail. Even with the best care, he will probably die before the age of thirty, from a heart disorder or respiratory failure (suffocation).

Hemophilia Hemophilias are genetic disorders in which the blood does not clot properly. Most of us have a blood clotting mechanism that quickly stops bleeding from minor injuries. That mechanism involves two proteins, clotting factor VIII and IX, both products of X chromosome genes. Mutations in these two genes cause two type of hemophilia (A and B, respectively). Males who carry one of these mutations have prolonged bleeding, as do homozygous females (heterozygous females make enough clotting protein to have a clotting time that is close to normal). Affected people bruise easily, but internal bleeding is their most serious problem. Repeated bleeding inside the joints disfigures them and causes chronic arthritis.

In the nineteenth century, the incidence of hemophilia A was relatively high in royal families of Europe and Russia, probably because the common practice of inbreeding kept the allele in their family trees (**FIGURE 14.7**). Today, about 1 in 7,500 people in the general population is affected. That number may be rising because the disorder is now treatable. More affected people now live long enough to transmit a mutated allele to children.

TAKE-HOME MESSAGE 14.4

How do we know when a trait is affected by an allele on an X chromosome?

✔ Men who have an X-linked allele have the associated trait, but not all heterozygous women do. Thus, the trait appears more often in men.

✔ Men transmit an X-linked allele to their daughters, but not to their sons.

14.5 Heritable Changes in Chromosome Structure

✔ Chromosome structure rarely changes, but when it does, the outcome can be severe or lethal.

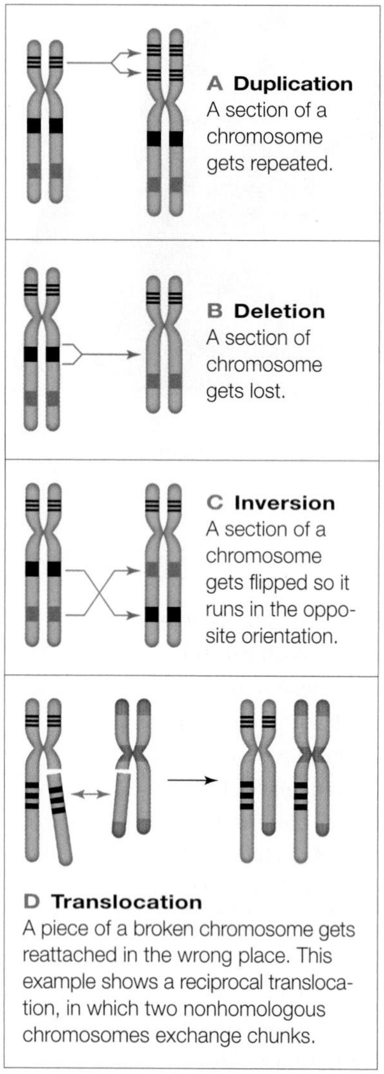

A Duplication
A section of a chromosome gets repeated.

B Deletion
A section of chromosome gets lost.

C Inversion
A section of a chromosome gets flipped so it runs in the opposite orientation.

D Translocation
A piece of a broken chromosome gets reattached in the wrong place. This example shows a reciprocal translocation, in which two nonhomologous chromosomes exchange chunks.

FIGURE 14.8 ▶**Animated** Major changes in chromosome structure.

Mutation is a term that usually refers to small-scale changes in DNA sequence—one or a few nucleotides. Chromosome changes on a larger scale also occur. Like mutations, these changes may be induced by exposure to chemicals or radiation. Others are an outcome of faulty crossing over during prophase I of meiosis. For example, nonhomologous chromosomes sometimes align and swap segments at spots where the DNA sequence is similar. Homologous chromosomes also may misalign along their length. In both cases, crossing over results in the exchange of segments that are not equivalent. The activity of transposable elements also alters chromosome structure. A **transposable element** is a segment of DNA hundreds or thousands of nucleotides long that can move spontaneously within or between chromosomes. Repeated DNA sequences at the ends allow the element to "jump" during mitosis or meiosis. Transposable elements are common in the DNA of all species; about 45 percent of human DNA consists of them and their evolutionary remnants.

Types of Chromosomal Change

Large-scale changes in chromosome structure can be categorized into several groups (**FIGURE 14.8**). In most cases, these changes drastically affect health; about half of all miscarriages are due to chromosome abnormalities of the developing embryo.

Duplication Even normal chromosomes have DNA sequences that are repeated two or more times. These repetitions are called **duplications** (**FIGURE 14.8A**). Some newly occurring duplications, such as the expansion mutations that cause Huntington's disease, cause

genetic abnormalities or disorders. Others, as you will soon see, have been evolutionarily important.

Deletion Large-scale deletions (**FIGURE 14.8B**) often have severe consequences. Duchenne muscular dystrophy most often arises from X chromosome deletions. A chromosome 5 deletion causes cri-du-chat syndrome, in which lifespan is shortened, mental functioning is impaired, and the larynx is abnormally shaped. Cri-du-chat (French for "cat's cry") refers to the sound made by affected infants when they cry.

Inversion With an **inversion**, a segment of chromosomal DNA becomes oriented in the reverse direction, with no loss of nucleotides (**FIGURE 14.8C**). An inversion may not affect a carrier's health if it does not interrupt a gene or gene control region, because the individual's cells still contain their full complement of genetic material. However, fertility may be compromised because a chromosome with an inversion does not pair properly with its homologous partner during meiosis. Crossovers may occur between the mispaired chromosomes, producing other chromosome abnormalities that reduce the viability of forthcoming embryos. People who carry an inversion may not know about it until they are diagnosed with infertility and their karyotype is checked.

Translocation If a chromosome breaks, the broken part may become attached to a different chromosome, or to a different part of the same one. This type of structural change is called a **translocation**. Most translocations are reciprocal, in which two nonhomologous chromosomes exchange broken parts (**FIGURE 14.8D**). A reciprocal translocation between chromosomes 8 and 14 is the usual cause of Burkitt's lymphoma, an aggressive cancer of the immune system. The translocation moves a proto-oncogene to a region that is vigorously transcribed in immune cells, with the result being uncontrolled cell divisions that are characteristic of cancer (Section 11.6).

Many other reciprocal translocations have no adverse effects on health, but, like inversions, they can compromise fertility. During meiosis, translocated chromosomes pair abnormally and segregate improperly; about half of the resulting gametes carry major duplications or deletions. If one of these gametes unites with a normal gamete at fertilization, the resulting embryo almost always dies. As with inversions, people who carry a translocation may not know about it until they have difficulty with fertility.

CREDIT: (8) © Cengage Learning.

(autosome pair)

Y X
SRY—

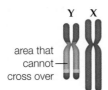
Y X
area that
cannot—
cross over

Y X

Y X Y X Y X

Ancestral reptiles
>350 mya

Ancestral reptiles
350 mya

Monotremes
320–240 mya

Marsupials
170–130 mya

Monkeys
130–80 mya

Humans
50–30 mya

A Before 350 mya, sex was determined by temperature, not by chromosome differences.

B The *SRY* gene begins to evolve 350 mya. The DNA sequences of the chromosomes diverge as other mutations accumulate.

C By 320–240 mya, the DNA sequences of the chromosomes are so different that the pair can no longer cross over in one region. The Y chromosome begins to shorten.

D Three more times, the pair stops crossing over in yet another region. Each time, the DNA sequences of the chromosomes diverge, and the Y chromosome shortens. Today, the pair crosses over only at a small region near the ends.

FIGURE 14.9 Evolution of the Y chromosome. Today, the *SRY* gene determines male sex. Homologous regions of the chromosomes are shown in pink; mya, million years ago. Monotremes are egg-laying mammals; marsupials are pouched mammals.

Chromosome Changes in Evolution

There is evidence of major structural alterations in the chromosomes of all known species. For example, duplications have often allowed a copy of a gene to mutate while the original carried out its unaltered function. The multiple and strikingly similar globin chain genes of mammals apparently evolved by this process. Globin chains, remember, associate to form molecules of hemoglobin (Section 9.6). Two identical genes for the alpha chain—and five other slightly different versions of it—form a cluster on chromosome 16. The gene for the beta chain clusters with four other slightly different versions on chromosome 11.

As another example, X and Y chromosomes were once homologous autosomes in ancient, reptilelike ancestors of mammals (**FIGURE 14.9**). Ambient temperature probably determined the gender of those organisms, as it still does in turtles and some other modern reptiles. About 350 million years ago, a gene on one of the two homologous chromosomes mutated. The mutation, which interfered with crossing over during meiosis, was the beginning of the male sex determination gene *SRY* (Section 10.4). A reduced frequency of crossovers allowed the chromosomes to diverge around the changed region as mutations began to accumulate separately in the two chromosomes. Over evolutionary time, the chromosomes became so different that they no longer crossed over at all in the changed region, so they diverged even more. Today, the Y chromosome is much smaller than the X, and is homologous with it only in a tiny part. The Y crosses over mainly with itself—by translocating duplicated regions of its own DNA.

Some chromosome structure changes contributed to differences among closely related organisms, such as apes and humans. Human somatic cells have twenty-three pairs of chromosomes, but cells of chimpanzees, gorillas, and orangutans have twenty-four. Thirteen human chromosomes are almost identical with chimpanzee chromosomes. Nine more are similar, except for some inversions. One human chromosome matches up with two in chimpanzees and the other great apes (**FIGURE 14.10**). During human evolution, two chromosomes evidently fused end to end and formed our chromosome 2. How do we know? The region where the fusion occurred contains remnants of a telomere (Section 11.5).

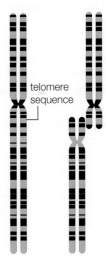
telomere
sequence

human chimpanzee

FIGURE 14.10
Human chromosome 2 compared with chimpanzee chromosomes 2A and 2B.

---

duplication Repeated section of a chromosome.
inversion Structural rearrangement of a chromosome in which part of the DNA becomes oriented in the reverse direction.
translocation Structural change of a chromosome in which a broken piece gets reattached in the wrong location.
transposable element Segment of DNA that can move spontaneously within or between chromosomes.

TAKE-HOME MESSAGE 14.5
How does chromosome structure change?

✔ A segment of a chromosome may be duplicated, deleted, inverted, or translocated. Any of these changes are usually harmful or lethal, but may be conserved in the rare circumstance that it has a neutral or beneficial effect.

✔ Occasionally, abnormal events occur before or during meiosis, and new individuals end up with the wrong chromosome number. Consequences range from minor to lethal changes in form and function.

A **polyploid** individual has three or more complete sets of chromosomes. About 70 percent of flowering plant species are polyploid, as are some insects, fishes, and other animals—but not humans. In our species, inheriting more than two full sets of chromosomes is invariably fatal, although some somatic cells are normally polyploid in adult tissues.

An **aneuploid** individual has too many or too few copies of a particular chromosome. Less than 1 percent of children are born with a diploid chromosome number that differs from the normal 46. Changes in chromosome number are usually an outcome of **nondisjunction**, the failure of chromosomes to separate properly during mitosis or meiosis. Nondisjunction during meiosis (**FIGURE 14.11**) can affect chromosome number at fertilization. For example, if a normal gamete (*n*) fuses with a gamete that has an extra chromosome (*n*+1), the resulting zygote will have three copies of one type of chromosome and two of every other type (2*n*+1), an aneuploid condition called trisomy. If a normal gamete (*n*) fuses with a gamete missing a chromosome (*n*−1), the new individual will have one copy of one chromosome and two of every other type (2*n*−1), an aneuploid condition called monosomy.

Autosomal Aneuploidy and Down Syndrome

In most cases, autosomal aneuploidy in humans is fatal before birth or shortly thereafter. An important excep-

tion is trisomy 21. A person born with three chromosomes 21 has Down syndrome and a high likelihood of surviving infancy. Mild to moderate mental impairment and health problems such as heart disease are hallmarks of this disorder. Other effects may include a somewhat flattened facial profile, a fold of skin that starts at the inner corner of each eyelid, white spots on the iris (**FIGURE 14.12**), and one deep crease (instead of two shallow creases) across each palm. The skeleton grows and develops abnormally, so older children have short body parts, loose joints, and misaligned bones of the fingers, toes, and hips. Muscles and reflexes are weak, and motor skills such as speech develop slowly. With medical care, affected individuals live about fifty-five years. Early training can help these individuals learn to care for themselves and to take part in normal activities. Down syndrome occurs in about 1 of 700 births, and the risk increases with maternal age.

Sex Chromosome Aneuploidy

Nondisjunction also causes alterations in the number of X and Y chromosomes, with a frequency of about 1 in 400 live births. Most often, such alterations lead to mild difficulties in learning and impaired motor skills such as a speech delay. These problems may be very subtle.

Turner Syndrome Individuals with Turner syndrome have an X chromosome and no corresponding X or Y chromosome (XO). The syndrome is thought to arise most frequently as an outcome of inheriting an unstable Y chromosome from the father. The zygote starts out being genetically male, with an X and a Y chromosome. Sometime during early development,

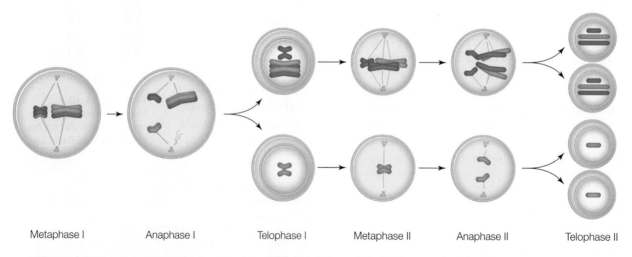

| Metaphase I | Anaphase I | Telophase I | Metaphase II | Anaphase II | Telophase II |

FIGURE 14.11 ▸Animated An example of nondisjunction during meiosis. Of the two pairs of homologous chromosomes shown here, one fails to separate during anaphase I. The chromosome number is altered in the resulting gametes.

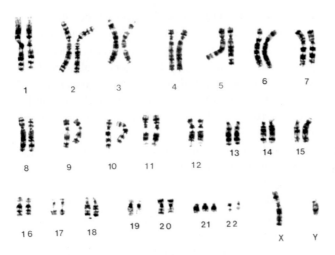

A Example of a Down syndrome genotype.

B Example of a Down syndrome phenotype. Excess tissue deposits on the iris give rise to a ring of starlike white speckles, a lovely effect of the chromosome number change that causes Down syndrome.

FIGURE 14.12 Down syndrome. **FIGURE IT OUT** Is the karyotype from an individual who is male or female? Answer: Male (XY)

the Y chromosome breaks up and is lost, so the embryo continues to develop as a female.

There are fewer people affected by Turner syndrome than other chromosome abnormalities: Only about 1 in 2,500 newborn girls has it. XO individuals grow up well proportioned but short, with an average height of 4 feet 8 inches (1.4 meters). Their ovaries do not develop properly, so they do not make enough sex hormones to become sexually mature and do not develop secondary sexual traits such as enlarged breasts.

XXX Syndrome A female may inherit multiple X chromosomes, a condition called XXX syndrome. This syndrome occurs in about 1 of 1,000 births. As with Down syndrome, the risk increases with maternal age. Only one X chromosome is typically active in female cells, so having extra X chromosomes usually does not cause physical or medical problems, but mild mental impairment may occur.

Klinefelter Syndrome About 1 out of every 500 males has an extra X chromosome (XXY). The resulting disorder, Klinefelter syndrome, becomes apparent at puberty. As adults, XXY males tend to be overweight and tall, with mild mental impairment. They make more estrogen and less testosterone than normal males.

This hormone imbalance causes affected men to have small testes and a small prostate gland, a low sperm count, sparse facial and body hair, a high-pitched voice, and enlarged breasts. Testosterone injections during puberty can minimize these traits.

XYY Syndrome About 1 in 1,000 males is born with an extra Y chromosome (XYY), a result of nondisjunction of the Y chromosome during sperm formation. Adults tend to be taller than average and have mild mental impairment, but most are otherwise normal. XYY men were once thought to be predisposed to live a life of crime. This misguided view was based on sampling error (too few cases in narrowly chosen groups such as prison inmates) and bias (the researchers who gathered the karyotypes also took the personal histories of the participants). That view has since been disproven: Men with XYY syndrome are only slightly more likely to be convicted for crimes than unaffected men. Researchers believe this slight increase can be explained by poor socioeconomic conditions related to the effects of the syndrome.

TAKE-HOME MESSAGE 14.6

What are the effects of chromosome number changes in humans?

✔ Polyploidy is fatal in humans, but not in flowering plants and some other organisms.

✔ Aneuploidy can arise from nondisjunction during meiosis. In humans, most cases of aneuploidy are associated with some degree of mental impairment.

---

aneuploid Having too many or too few copies of a particular chromosome.
nondisjunction Failure of sister chromatids or homologous chromosomes to separate during nuclear division.
polyploid Having three or more of each type of chromosome characteristic of the species.

CREDITS: (12A) L. Willatt, East Anglian Regional Genetics Service/Science Source; (12B) Ciarra, photo by © Michelle Harmon.

✔ Our understanding of human inheritance can provide prospective parents with information about the health of their future children.

Studying human inheritance patterns has given us many insights into how genetic disorders arise and progress, and how to treat them. Some disorders can be detected early enough to start countermeasures before symptoms develop. For this reason, most hospitals in the United States now screen newborns for mutations that cause phenylketonuria, or PKU. The mutations affect an enzyme that converts one amino acid (phenylalanine) to another (tyrosine). Without this enzyme, the body becomes deficient in tyrosine, and phenylalanine accumulates to high levels. The imbalance inhibits protein synthesis in the brain, which in turn results in severe neurological symptoms. Restricting all intake of phenylalanine can slow the progression of PKU, so routine early screening has resulted in fewer individuals suffering from the symptoms of the disorder.

The probability that a child will inherit a genetic disorder can be estimated by testing prospective parents for alleles known to be associated with genetic disorders. Karyotypes and pedigrees are also useful in this type of screening, which can help the parents make decisions about family planning.

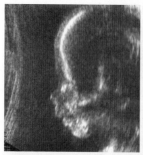

A Conventional ultrasound.

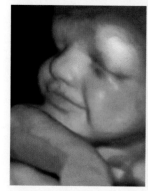

B 4D ultrasound.   C Fetoscopy.

FIGURE 14.13 Three ways of imaging a developing human fetus.

Genetic screening is also done post-conception, in which case it is called prenatal diagnosis (prenatal means before birth). Prenatal diagnosis checks an embryo or fetus for physical and genetic abnormalities. Early diagnosis of these conditions gives parents time to prepare for the birth of an affected child, and an opportunity to decide whether to continue with the pregnancy or terminate it. More than 30 conditions are detectable prenatally, including aneuploidy, hemophilia, Tay–Sachs disease, sickle-cell anemia, muscular dystrophy, and cystic fibrosis. If a disorder is treatable, early detection can allow the newborn to receive prompt and appropriate treatment. A few conditions are even surgically correctable before birth.

As an example of how prenatal diagnosis works, consider a woman who becomes pregnant at age thirty-five. Her doctor will probably perform a procedure called obstetric sonography, in which ultrasound waves directed across the woman's abdomen form images of the fetus's limbs and internal organs (**FIGURE 14.13A,B**). If the images reveal a physical defect that may be the result of a genetic disorder, a more invasive technique such as fetoscopy would be recommended for further diagnosis. With fetoscopy, sound waves pulsed from inside the mother's uterus yield images much higher in resolution than ultrasound (**FIGURE 14.13C**). Samples of tissue or blood are often taken at the same time, and some corrective surgeries can be performed.

Human genetics studies show that our thirty-five-year-old woman has about a 1 in 80 chance that her baby will be born with a chromosomal abnormality, a risk more than six times greater than when she was twenty years old. Thus, even if no abnormalities are detected by ultrasound, she probably will be offered an additional diagnostic procedure, amniocentesis, in which a small sample of fluid is drawn from the amniotic sac enclosing the fetus (**FIGURE 14.14**). The fluid contains cells shed by the fetus, and those cells can be tested for genetic disorders. Chorionic villus sampling (CVS) can be performed earlier than amniocentesis. With this technique, a few cells from the chorion are removed and tested for genetic disorders. (The chorion is a membrane that surrounds the amniotic sac and helps form the placenta, an organ that allows substances to be exchanged between mother and embryo.)

An invasive procedure often carries a risk to the fetus. The risks vary by the procedure. Amniocentesis has improved so much that, in the hands of a skilled physician, the procedure no longer increases the risk of miscarriage. CVS occasionally disrupts the placenta's

Individuals of European descent share several alleles that influence skin pigmentation with individuals of east Asian descent. However, most people of east Asian descent carry a particular mutation in their *OCA2* gene—a single base-pair substitution in which an adenine changed to a cytosine—that results in lightened skin color. The product of the *OCA2* gene is a protein of unknown function, but it is named after the condition that occurs when the protein is missing: oculocutaneous albinism type II.

The *OCA2* mutation that lightens east Asian skin is uncommon in people of European ancestry. The *SLC24A5* allele that lightens European skin is uncommon in people of east Asian ancestry. Taken together, the distribution of these alleles suggests that (1) an African population with dark skin was ancestral to both east Asians and Europeans, and (2) east Asian and European populations separated before their pigmentation genes mutated and their skin color changed.

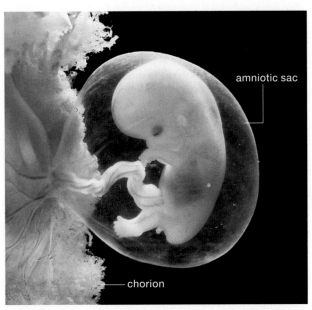

FIGURE 14.14 An 8-week-old fetus. With amniocentesis, a tiny bit of the fluid inside the amniotic sac is removed, and fetal cells that have been shed into the fluid are tested for genetic disorders. Chorionic villus sampling tests cells of the chorion, which is part of the placenta.

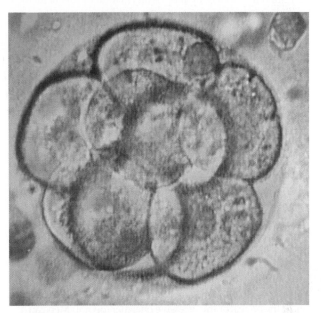

FIGURE 14.15 Clump of cells formed by three mitotic divisions after *in vitro* fertilization. All eight of the cells are identical; one can be removed for genetic analysis to determine whether the embryo carries any genetic defects. The remaining cells can continue development to form a viable embryo.

development and thus causes underdeveloped or missing fingers and toes in 0.3 percent of newborns. Fetoscopy raises the miscarriage risk by a whopping 2 to 10 percent.

Couples who discover they are at high risk of having a child with a genetic disorder may opt for reproductive interventions such as *in vitro* fertilization. With this procedure, sperm and eggs taken from prospective parents are mixed in a test tube. If an egg becomes fertilized, the resulting zygote will begin to divide. In about forty-eight hours, it will have become an embryo that consists of a ball of eight cells (**FIGURE 14.15**). All of the cells in this ball have the same genes, but none has yet committed to being specialized one way or another. Doctors can remove one of these undifferenti-

ated cells and analyze its genes, a procedure called preimplantation diagnosis. The withdrawn cell will not be missed. If the embryo has no detectable genetic defects, it is inserted into the woman's uterus to continue developing. Many of the resulting "test-tube babies" are born in good health.

TAKE-HOME MESSAGE 14.7

How do we use what we know about human inheritance?

✔ Studying inheritance patterns for genetic disorders has helped researchers develop treatments for some of them.

✔ Genetic testing can provide prospective parents with information about the health of their future children.

CREDITS: (14) © Lennart Nilsson/Bonnierforlagen AB; (15) Fran Heyl Associates © Jacques Cohen, computer-enhanced by © Pix Elation; (in text) Gary Roberts/worldwidefeatures.com.

Section 14.1 Like most other human traits, skin color has a genetic basis. Minor differences in the alleles that govern melanin production and the size, shape, and distribution of melanosomes affect skin color. Skin color differences probably evolved as a balance between vitamin production and protection against harmful UV radiation.

Section 14.2 Geneticists study inheritance patterns in humans by tracking genetic disorders and abnormalities through generations of families. A genetic abnormality is an uncommon version of a heritable trait that does not result in medical problems. A genetic disorder is a heritable condition that sooner or later results in mild or severe medical problems. Geneticists make **pedigrees** to reveal inheritance patterns for alleles that can be predictably associated with specific phenotypes.

Section 14.3 An allele is inherited in an autosomal dominant pattern if the trait it specifies appears in everyone who carries it, and both sexes are affected with equal frequency. Such traits appear in every generation of families that have the allele. An allele is inherited in an autosomal recessive pattern if the trait it specifies appears only in homozygous people. Such traits also appear in both sexes equally, but they can skip generations.

Section 14.4 An allele is inherited in an X-linked pattern when it occurs on the X chromosome. Most X-linked disorders are inherited in a recessive pattern, and these tend to appear in men more often than in women. Heterozygous women have a dominant, normal allele that can mask the effects of the recessive one; men do not. Men can transmit an X-linked allele to their daughters, but not to their sons. Only a woman can pass an X-linked allele to a son.

Section 14.5 Faulty crossovers and the activity of **transposable elements** can give rise to major changes in chromosome structure, including **duplications, inversions**, and **translocations**. Some of these changes are harmful or lethal in humans; others affect fertility. Even so, major structural changes have accumulated in the chromosomes of all species over evolutionary time.

Section 14.6 Occasionally, abnormal events occur before or during meiosis, and new individuals end up with the wrong chromosome number. Consequences of such changes range from minor to lethal alterations in form and function. Chromosome number change is usually an outcome of **nondisjunction**, in which chromosomes fail to separate properly during nuclear division. **Polyploid** individuals have three or more of each type of chromosome. Polyploidy is lethal in humans, but not in flowering plants and some insects, fishes, and other animals. **Aneuploid** individuals

have too many or too few copies of a chromosome. In humans, most cases of autosomal aneuploidy are lethal. Trisomy 21, which causes Down syndrome, is an exception. A change in the number of sex chromosomes usually results in some degree of impairment in learning and motor skills.

Section 14.7 Prospective parents can use genetic screening to estimate their risk of transmitting a harmful allele to offspring. The procedure involves analysis of parental pedigrees and genotype by a genetic counselor. Amniocentesis and other methods of prenatal genetic testing can reveal a genetic disorder before birth.

self-quiz
Answers in Appendix VII

1. Constructing a pedigree is particularly useful when studying inheritance patterns in organisms that _____ .
   a. produce many offspring per generation
   b. produce few offspring per generation
   c. have a very large chromosome number
   d. reproduce asexually
   e. have a fast life cycle

2. Pedigree analysis is necessary when studying human inheritance patterns because _____ .
   a. humans have more than 20,000 genes
   b. of ethical problems with experimenting on humans
   c. inheritance in humans is more complicated than it is in other organisms
   d. genetic disorders occur only in humans
   e. all of the above

3. A recognized set of symptoms that characterize a genetic disorder is a(n) _____ .
   a. syndrome   b. disease   c. abnormality

4. If one parent is heterozygous for a dominant allele on an autosome and the other parent does not carry the allele, any child of theirs has a _____ chance of being heterozygous.
   a. 25 percent   c. 75 percent
   b. 50 percent   d. no chance; it will die

5. True or false? A son can inherit an X-linked recessive allele from his father.

6. A trait that is present in a male child but not in either of his parents is characteristic of _____ inheritance.
   a. autosomal dominant   d. It is impossible to answer
   b. autosomal recessive   this question without
   c. X-linked recessive   more information.

7. Color blindness is a case of _____ inheritance.
   a. autosomal dominant   c. X-linked dominant
   b. autosomal recessive   d. X-linked recessive

8. A female child inherits one X chromosome from her mother and one from her father. What sex chromosome does a male child inherit from each of his parents?

Skin Color Survey of Native Peoples In 2000, researchers measured the average amount of UV radiation received in more than fifty regions of the world, and correlated it with the average skin reflectance of people native to those regions (reflectance is a way to measure the amount of melanin pigment in skin). Some of the results of this study are shown in **FIGURE 14.16**.

1. Which country receives the most UV radiation? The least?

2. The people native to which country have the darkest skin? The lightest?

3. According to these data, how does the skin color of indigenous peoples correlate with the amount of UV radiation incident in their native regions?

Country	Skin Reflectance	UVMED
Australia	19.30	335.55
Kenya	32.40	354.21
India	44.60	219.65
Cambodia	54.00	310.28
Japan	55.42	130.87
Afghanistan	55.70	249.98
China	59.17	204.57
Ireland	65.00	52.92
Germany	66.90	69.29
Netherlands	67.37	62.58

FIGURE 14.16 Skin color of indigenous peoples and regional incident UV radiation. Skin reflectance measures how much light of 685-nanometer wavelength is reflected from skin; UVMED is the annual average UV radiation received at Earth's surface.

9. Alleles for Tay–Sachs disease are inherited in an autosomal recessive pattern. Why would two parents with a normal phenotype have a child with Tay–Sachs?
   a. Both parents are homozygous for a Tay–Sachs allele.
   b. Both parents are heterozygous for a Tay–Sachs allele.
   c. A new mutation gave rise to Tay–Sachs in the child.
   d. b or c

10. The *SRY* gene gives rise to the male phenotype in humans (Sections 10.4 and 14.5). What do you think the inheritance pattern of *SRY* alleles is called?

11. Nondisjunction may occur during _____ .
   a. mitosis
   b. meiosis
   c. fertilization
   d. both a and b

12. Nondisjunction can result in _____ .
   a. duplications
   b. aneuploidy
   c. crossing over
   d. pleiotropy

13. True or false? An individual may inherit three or more of each type of chromosome characteristic of the species, a condition called polyploidy.

14. Klinefelter syndrome (XXY) can be easily diagnosed by _____ .
   a. pedigree analysis
   b. aneuploidy
   c. karyotyping
   d. phenotypic treatment

15. Match the chromosome terms appropriately.
   ____ polyploidy
   ____ deletion
   ____ aneuploidy
   ____ translocation
   ____ syndrome
   ____ transposable element
   a. symptoms of a genetic disorder
   b. chromosomal mashup
   c. extra sets of chromosomes
   d. gets around
   e. a chromosome segment lost
   f. one extra chromosome

genetics problems

Answers in Appendix VII

1. Duchenne muscular dystrophy (DMD), which is inherited in an X-linked recessive pattern, occurs almost exclusively in males. Suggest why.

2. Does the phenotype indicated by the red circles and squares in this pedigree show an inheritance pattern that is autosomal dominant, autosomal recessive, or X-linked?

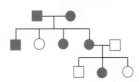

3. Human females have two X chromosomes (XX); males have one X and one Y chromosome (XY).
   a. With respect to X-linked alleles, how many different types of gametes can a male produce?
   b. If a female is homozygous for an X-linked allele, how many types of gametes can she produce with respect to that allele?
   c. If a female is heterozygous for an X-linked allele, how many types of gametes can she produce with respect to that allele?

4. A mutated allele responsible for Marfan syndrome (Section 13.5) is inherited in an autosomal dominant pattern. What is the chance that any child will inherit it if one parent does not carry the allele and the other is heterozygous for it?

5. Somatic cells of individuals with Down syndrome usually have an extra chromosome 21; they contain forty-seven chromosomes.
   a. At which stage(s) of meiosis could nondisjunction alter the chromosome number?
   b. A few individuals with Down syndrome have forty-six chromosomes: two normal-appearing chromosomes 21, and a longer-than-normal chromosome 14. Speculate on how this chromosome abnormality may arise.

6. Mutations in the genes for clotting factor VIII and IX cause hemophilia A and B, respectively. A woman may be heterozygous for mutations in both genes, with a mutated factor VIII allele on one X chromosome, and a mutated factor IX allele on the other. All of her sons should have either hemophilia A or B. However, on rare occasions, one of these women gives birth to a son who does not have hemophilia, and his one X chromosome does not have either mutated allele. Explain how this boy's X chromosome probably arises.

CREDITS: (16) © Cengage Learning, based on *Journal of Human Evolution* (2000)39, 57–106 doi: 10.1006/jhev.2000.0403 © 2000 Academic Press; (in text) © Cengage Learning.

15 Studying and Manipulating Genomes

LEARNING ROADMAP

This chapter builds on your understanding of DNA (Sections 8.3–8.5, 13.2, 14.5, 14.7). Clones (8.1), gene expression (9.2, 9.3), and knockouts (10.3) turn up in the context of human traits (13.7) and genetic disorders (Chapter 14). You will revisit tracers (2.2), triglycerides (3.5), denaturation (3.7), β-carotene (6.2), the lac operon (10.5), and cancer (11.6).

DNA CLONING

Researchers make recombinant DNA by cutting and pasting together DNA from different species. Plasmids and other vectors can carry foreign DNA into host cells.

FINDING NEEDLES IN HAYSTACKS

Genetic engineering, the directed modification of an organism's genes, relies on laboratory techniques for isolating and identifying targeted fragments of DNA.

DNA SEQUENCING

Sequencing reveals the linear order of nucleotides in DNA. Comparing genomes offers insights into human genes and evolution. DNA sequence can be used to identify individuals.

GENETIC ENGINEERING

Genetic engineering is now a routine part of research and industrial applications. Genetically modified organisms are used to produce food, medicines, and other products.

GENE THERAPY

The directed modification of human DNA continues to be tested in medical applications. It also continues to raise ethical questions about modifying the human genome.

Evolutionary biology relies heavily on DNA sequence comparisons (Section 18.4). The escape of transgenic genes into the environment is an example of gene flow (17.8). In nature, plasmids transfer genes among bacteria (20.5). Genetic engineering returns in the context of phytoremediation (28.1) and vaccines (37.12). Research using engineered animals is explained in relevant chapters. Heat-loving bacteria return in Section 20.7.

15.1 Personal Genetic Testing

About 99 percent of your DNA is exactly the same as everyone else's. The shared part is what makes you human; the differences make you a unique member of the species. If you compared your DNA with your neighbor's, about 2.97 billion nucleotides of the two sequences would be identical; the remaining 30 million nonidentical nucleotides are sprinkled throughout your chromosomes. The sprinkling is not entirely random because some regions of DNA vary less than others. These conserved regions are of particular interest because they are the ones most likely to have an essential function. If a conserved sequence does vary among people, the variation tends to be in single nucleotides at a particular location. A base-pair substitution that is carried by a measurable percentage of a population, usually above 1 percent, is called a **single-nucleotide polymorphism**, or **SNP** (pronounced "snip").

Alleles of most genes differ by single nucleotides, and differences in alleles are the basis of the variation in human traits that makes each individual unique (Section 12.2). Thus, SNPs account for many of the differences in the way humans look, and they also have a lot to do with differences in the way our bodies work—how we age, respond to drugs, weather assaults by pathogens and toxins, and so on.

Consider the lipoprotein particles that carry fats and cholesterol through our bloodstreams (Section 3.6). These particles consist of variable amounts and types of lipids and proteins. One of these proteins is called apolipoprotein E, and it is encoded by the *APOE* gene.

About one in four people carries an allele of this gene, *E4*, that has a cytosine instead of the more common thymine at a particular location in its sequence. The gene with this SNP encodes an apolipoprotein E with one amino acid substitution, an arginine instead of a cysteine in position 112. How this change affects the function of the protein is not yet clear, but we do know that having the *E4* allele increases one's risk of developing Alzheimer's disease later in life, particularly in people homozygous for it.

At this writing, about 73 million SNPs in human DNA have been identified, and that number grows every day. A few companies now offer to determine some of the SNPs you carry. The companies extract your DNA from the cells in a few drops of spit, then analyze it for SNPs. Such personalized genetic testing is now revolutionizing medicine, for example by allowing physicians to determine a patient's ability to respond to certain drugs before treatment begins. Cancer treatments are being tailored to fit the genetic makeup of individual patients and their tumor cells. People who discover they carry SNPs associated with a heightened risk of a medical condition are being encouraged to make lifestyle changes that could delay the condition's onset or prevent it entirely; preventive medical treatments based on these SNPs are becoming more common—and more mainstream (**FIGURE 15.1**).

single-nucleotide polymorphism (**SNP**) One-nucleotide DNA sequence variation carried by a measurable percentage of a population.

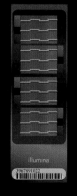

Only about 1 percent of the 3 billion bases in a person's DNA are unique to the individual. Personal genetic testing companies use chips like this one to analyze their customers' chromosomes for SNPs. This chip reveals which versions of 1,140,419 SNPs occur in the DNA of four individuals at a time.

FIGURE 15.1 Celebrity Angelina Jolie chose preventive treatment after genetic testing showed she had a very high risk of breast cancer. She carries a *BRCA1* mutation associated with an 87% lifetime risk of developing breast cancer. Even though Jolie did not yet have cancer, she underwent a double mastectomy, thereby reducing her risk of breast cancer to 5%.

CREDITS: (opposite) Courtesy of © Dr. Jean Levit. The Brainbow technique was developed in the laboratories of Jeff W. Lichtman and Joshua R. Sanes at Harvard University. This image has received the Bioscape imaging competition 2007 prize; (l) left, © Oli Scarff/Getty Images; right, Courtesy of © Illumina, Inc., www.illumina.com.

15.2 Cloning DNA

✔ Researchers cut up DNA from different sources, then paste the resulting fragments together.

✔ Cloning vectors can carry foreign DNA into host cells.

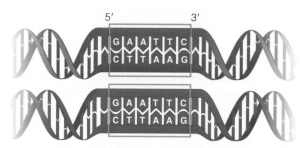

① The restriction enzyme *Eco*RI (named after the *E. coli* bacteria from which it was isolated) recognizes a specific base sequence (GAATTC) in DNA from two different sources.

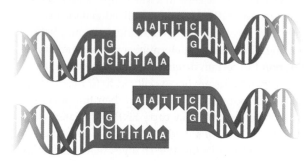

② The enzyme cuts the DNA into fragments. *Eco*RI leaves single-stranded tails ("sticky ends") where it cuts DNA.

③ When the DNA fragments from the two sources are mixed together, matching sticky ends base-pair with each other.

④ DNA ligase joins the base-paired DNA fragments to produce molecules of recombinant DNA.

FIGURE 15.2 ▶**Animated** Making recombinant DNA.

FIGURE IT OUT Why did the enzyme cut both strands of DNA?

Answer: Because the recognition sequence occurs on both strands.

Cut and Paste

In the 1950s, excitement over the discovery of DNA's structure (Section 8.3) gave way to frustration: No one could determine the order of nucleotides in a molecule of DNA. Identifying a single base among thousands or millions of others turned out to be a huge technical hurdle. Research in a seemingly unrelated field yielded a solution when Werner Arber, Hamilton Smith, and their coworkers discovered how some bacteria resist infection by bacteriophage (Section 8.2). These bacteria have enzymes that chop up any injected viral DNA before it has a chance to integrate into the bacterial chromosome. The enzymes restrict viral growth; hence their name, restriction enzymes. A **restriction enzyme** cuts DNA wherever a specific nucleotide sequence occurs (**FIGURE 15.2 ①**).

The discovery of restriction enzymes allowed researchers to cut chromosomal DNA into manageable chunks. It also allowed them to combine DNA fragments from different organisms. How? Many restriction enzymes leave single-stranded tails on DNA fragments **②**. Researchers realized that complementary tails will base-pair, regardless of the source of DNA **③**. The tails are called "sticky ends" because two DNA fragments stick together when their matching tails base-pair. The enzyme DNA ligase (Section 8.5) can be used to seal the gaps between base-paired sticky ends, so continuous DNA strands form **④**. Thus, using appropriate restriction enzymes and DNA ligase, researchers can cut and paste DNA from different sources. The result, a hybrid molecule that consists of genetic material from two or more organisms, is called **recombinant DNA**.

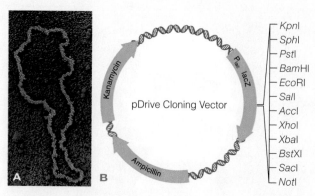

FIGURE 15.3 Plasmid cloning vectors. (**A**) Micrograph of a plasmid. (**B**) Commercial plasmid cloning vector. Restriction enzyme recognition sequences are indicated (right) by the name of the enzyme that cuts them. Researchers insert foreign DNA into the vector at these sequences. Bacterial genes (gold) help them identify host cells that take up a vector with inserted DNA. This vector carries two antibiotic resistance genes and the *lac* operon (Section 10.5).

CREDITS: (2) © Cengage Learning; (3A) Professor Stanley Cohen/Science Source; (3B) Taken from QIAGEN, Showing a reduced pDrive Cloning Vector.

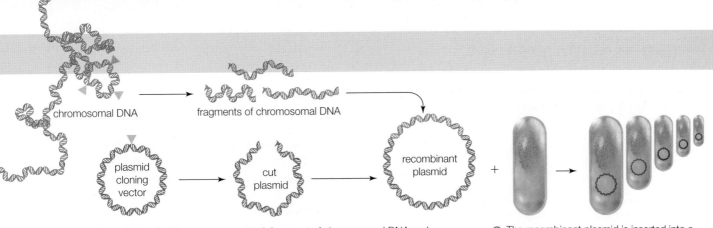

A A restriction enzyme (gold triangles) cuts a specific nucleotide sequence in chromosomal DNA and in a plasmid cloning vector.

B A fragment of chromosomal DNA and the cut plasmid base-pair at their sticky ends. DNA ligase joins the two pieces of DNA, so a recombinant plasmid forms.

C The recombinant plasmid is inserted into a host bacterial cell. When the cell reproduces, it copies the plasmid along with its chromosome. Each descendant cell receives a plasmid.

FIGURE 15.4 ▶Animated An example of cloning. Here, a fragment of chromosomal DNA is inserted into a plasmid.

Making recombinant DNA is the first step in **DNA cloning**, a set of laboratory methods that uses living cells to mass-produce specific DNA fragments. Researchers clone a fragment of DNA by inserting it into a **cloning vector**, which is a molecule that can carry foreign DNA into host cells. Bacterial plasmids (Section 4.4) may be used as cloning vectors (**FIGURE 15.3**). A bacterium copies all of its DNA before it divides, so its offspring inherit plasmids along with chromosomes. If a plasmid carries a fragment of foreign DNA, that fragment gets copied and distributed to descendant cells along with the plasmid DNA (**FIGURE 15.4**).

A host cell into which a cloning vector has been inserted can be grown in the laboratory (cultured) to yield a huge population of genetically identical cells, or clones (Section 8.7). Each clone contains a copy of the vector and the inserted DNA fragment. The hosted DNA fragment can be harvested in large quantities from the clones.

cDNA Cloning

Remember from Section 9.3 that eukaryotic DNA contains introns. Unless you are a eukaryotic cell, it is not very easy to determine which parts of eukaryotic DNA encode gene products. Thus, researchers who study gene expression in eukaryotes often start with mature mRNA. Post-transcriptional processing removes introns from an mRNA, so just the coding sequence remains.

An mRNA cannot be cut with restriction enzymes or pasted with DNA ligase, because these enzymes work only on double-stranded DNA. Thus, cloning with mRNA requires **reverse transcriptase**, a replication enzyme that uses an RNA template to assemble a strand of complementary DNA, or **cDNA**:

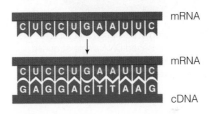

DNA polymerase is used to copy the cDNA into a second strand of DNA. The outcome is a double-stranded DNA version of the original mRNA:

*Eco*RI recognition site

Like any other double-stranded DNA, this fragment may be cut with restriction enzymes and pasted into a cloning vector using DNA ligase.

cDNA Complementary strand of DNA synthesized from an RNA template by the enzyme reverse transcriptase.
cloning vector A DNA molecule that can accept foreign DNA and be replicated inside a host cell.
DNA cloning Set of methods that uses living cells to mass-produce targeted DNA fragments.
recombinant DNA A DNA molecule that contains genetic material from more than one organism.
restriction enzyme Type of enzyme that cuts DNA at a specific nucleotide sequence.
reverse transcriptase An enzyme that uses mRNA as a template to make a strand of cDNA.

TAKE-HOME MESSAGE 15.2
What is DNA cloning?

✔ DNA cloning uses living cells to mass-produce targeted DNA fragments. Restriction enzymes cut DNA into fragments, then DNA ligase seals the fragments into cloning vectors. Recombinant DNA molecules result.

✔ A cloning vector that holds foreign DNA can be introduced into a living cell. When the host cell divides, it gives rise to huge populations of genetically identical cells (clones), each with a copy of the foreign DNA.

✔ DNA libraries and the polymerase chain reaction (PCR) help researchers find and isolate targeted DNA fragments.

A Individual bacterial cells from a DNA library are spread over the surface of a solid growth medium. The cells divide repeatedly and form colonies—clusters of millions of genetically identical descendant cells.

B Special paper is pressed onto the surface of the growth medium. Some cells from each colony stick to the paper.

C The paper is soaked in a solution that ruptures the cells and makes the released DNA single-stranded. The DNA clings to the paper in spots mirroring the distribution of colonies.

D A radioactive probe is added to the liquid bathing the paper. The probe hybridizes with any spot of DNA that contains a complementary sequence.

E The paper is pressed against x-ray film. The radioactive probe darkens the film in a spot where it has hybridized. The spot's position is compared to the positions of the original bacterial colonies. Cells from the colony that corresponds to the spot are cultured, and their DNA is harvested.

FIGURE 15.5 ▶Animated Nucleic acid hybridization. In this example, a radioactive probe helps identify a colony of bacteria that contain a targeted fragment of DNA.

DNA Libraries

The entire set of genetic material—the **genome**—of most organisms consists of thousands of genes. To study or manipulate a single gene, researchers must first find it, and then separate it from all of the other genes in a genome. They often begin by cutting an organism's DNA into fragments, and then cloning all the fragments. The result is a genomic library, a set of clones that collectively contain all of the DNA in a genome. Researchers may also harvest mRNA, make cDNA copies of it, and then clone the cDNA. The resulting cDNA library represents only those genes being expressed at the time the mRNA was harvested.

Genomic and cDNA libraries are **DNA libraries**, sets of cells that host various cloned DNA fragments. In such libraries, a clone that contains a targeted DNA fragment of interest is mixed up with thousands or millions of others that do not—a needle in a genetic haystack. One way to find that clone among the others involves the use of a **probe**, which is a fragment of DNA or RNA labeled with a tracer (Section 2.2).

For example, to find a targeted gene, researchers may use radioactive nucleotides to synthesize a short strand of DNA complementary in sequence to a similar gene. Because the nucleotide sequences of the probe and the gene are complementary, the two can hybridize. (Remember from Section 8.5 that nucleic acid hybridization is the establishment of base pairing between nucleic acid strands.) When the probe is mixed with DNA from a library, it will hybridize with the gene, but not with other DNA (**FIGURE 15.5**). Researchers can pinpoint a cell that hosts the gene by detecting the label on the probe. That cell is isolated and cultured, and DNA can be extracted in bulk from the cultured cells for research or other purposes.

PCR

The **polymerase chain reaction (PCR)** is a technique used to mass-produce copies of a particular section of DNA without having to clone it in living cells (**FIGURE 15.6**). The reaction can transform a needle in a haystack—that one-in-a-million fragment of DNA— into a huge stack of needles with a little hay in it.

The starting material for PCR is any sample of DNA with at least one molecule of a targeted sequence. It might be extracted from a mixture of 10 million different clones, a sperm, a hair left at a crime scene, or a mummy—essentially any sample that has DNA in it.

The PCR reaction is similar to DNA replication (Section 8.5). It requires two primers; each base-pairs with one end of the section of DNA to be amplified,

CREDIT: (5) © Cengage Learning.

targeted section

FIGURE 15.6 ▶Animated Two rounds of PCR. Each cycle of this reaction can double the number of copies of a targeted sequence of DNA. Thirty cycles can make a billion copies.

❶ DNA template (blue) is mixed with primers (pink), nucleotides, and heat-tolerant *Taq* DNA polymerase.

or mass-produced **❶** . Researchers mix these primers with the starting (template) DNA, nucleotides, and DNA polymerase, then expose the reaction mixture to repeated cycles of high and low temperatures. A few seconds at high temperature disrupts the hydrogen bonds that hold the two strands of a DNA double helix together (Section 8.3), so every molecule of DNA unwinds and becomes single-stranded. As the temperature of the reaction mixture is lowered, the single DNA strands hybridize with the primers **❷** .

The DNA polymerases of most organisms denature at the high temperature required to separate DNA strands. The kind that is used in PCR reactions, *Taq* polymerase, is from *Thermus aquaticus*. This bacterial species lives in hot springs and hydrothermal vents, so its DNA polymerase necessarily tolerates heat. *Taq* polymerase, like other DNA polymerases, recognizes hybridized primers as places to start DNA synthesis **❸** . Synthesis proceeds along the template strand until the temperature rises and the DNA separates into single strands **❹** . The newly synthesized DNA is a copy of the targeted section. When the mixture is cooled, the primers rehybridize, and DNA synthesis begins again. Each cycle of heating and cooling takes only a few minutes, but it can double the number of copies of the targeted section of DNA **❺** . Thirty PCR cycles may amplify that number a billionfold.

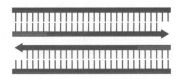

❷ When the mixture is heated, the double-stranded DNA separates into single strands. When the mixture is cooled, some of the primers base-pair with the DNA at opposite ends of the targeted sequence.

❸ *Taq* polymerase begins DNA synthesis at the primers, so it produces complementary strands of the targeted DNA sequence.

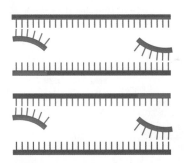

❹ The mixture is heated again, so all double-stranded DNA separates into single strands. When it is cooled, primers base-pair with the targeted sequence in the original template DNA and in the new DNA strands.

DNA library Collection of cells that host different fragments of foreign DNA, often representing an organism's entire genome.
genome An organism's complete set of genetic material.
polymerase chain reaction (**PCR**) Method that rapidly generates many copies of a specific section of DNA.
probe Short fragment of DNA designed to hybridize with a nucleotide sequence of interest and labeled with a tracer.

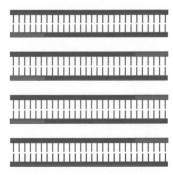

TAKE-HOME MESSAGE 15.3
How do researchers study one gene in the context of many?

✔ A DNA library can be made from cDNA or genomic DNA. Probes are used to identify one clone that hosts a targeted DNA fragment among many other clones in a DNA library.

✔ PCR quickly mass-produces copies of a targeted section of DNA.

❺ Each cycle of heating and cooling can double the number of copies of the targeted DNA section.

CREDIT: (6) © Cengage Learning.

15.4 DNA Sequencing

✔ DNA sequencing reveals the order of nucleotide bases in a section of DNA.

Once a fragment of DNA has been isolated (for example by cloning or PCR), researchers can use a technique called **sequencing** to determine the order of nucleotides in it. The most common method uses DNA polymerase (Section 8.5). This enzyme is mixed with a primer, nucleotides, and the DNA to be sequenced (the template). Starting at the primer, the polymerase joins the nucleotides into a new strand of DNA, in the order dictated by the sequence of the template (**FIGURE 15.7**).

The sequencing reaction mixture includes all four kinds of nucleotides, and also four kinds of dideoxynucleotides—DNA nucleotides that lack the hydroxyl group on their 3′ carbon ❶. Each dideoxynucleotide base (A, C, G, or T) is labeled with a different colored pigment. During the sequencing reaction, DNA polymerase randomly adds either a regular nucleotide or a dideoxynucleotide to the 3′ end of a growing DNA strand. If it adds a regular nucleotide, synthesis can continue because the 3′ carbon of the strand will have

a hydroxyl group on it. If it adds a dideoxynucleotide, the 3′ carbon will not have a hydroxyl group, so synthesis of the strand ends there ❷. (Remember that DNA synthesis proceeds only in the 5′ to 3′ direction.)

The reaction produces millions of DNA fragments of different lengths—incomplete, complementary copies of the starting DNA ❸. Each fragment of a given length ends with the same dideoxynucleotide base. For example, if the tenth base in the template DNA was thymine, then any newly synthesized fragment that is 10 nucleotides long ends with an adenine.

The DNA fragments are then separated by length. Using a technique called **electrophoresis**, an electric field pulls the fragments through a semisolid gel. Fragments of different sizes move through the gel at different rates. The shorter the fragment, the faster it moves, because shorter fragments slip through the tangled molecules of the gel faster than longer fragments do. All fragments of the same length move through the gel at the same speed, so they gather into bands. Because all fragments in a given band are the same length, all have the same dideoxynucleotide base at

DNA template strand

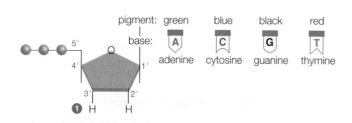

pigment: green   blue   black   red
base: A   C   G   T
adenine cytosine guanine thymine

❶ H   H

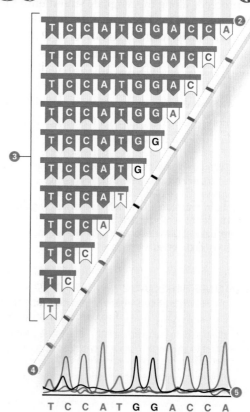

FIGURE 15.7 ▶**Animated** A method of sequencing DNA, in which DNA polymerase is used to incompletely replicate a section of DNA.

❶ This sequencing method depends on dideoxynucleotides, which are nucleotides that have a hydrogen atom instead of a hydroxyl group on their 3′ carbon (compare the structure with those in **FIGURE 9.2**). Each base (G, A, T, or C) is labeled with a different colored pigment.

❷ DNA polymerase uses a section of DNA as a template to synthesize new strands of DNA. Synthesis of each new strand stops when a dideoxynucleotide is added.

❸ At the end of the reaction, there are many incomplete copies of the template DNA in the mixture.

❹ Electrophoresis separates the copied DNA fragments into bands according to their length. All of the DNA strands in each band end with the same base, so each band is the color of the base's tracer pigment.

❺ A computer detects and records the color of successive bands on the gel (see **FIGURE 15.8** for an example). The order of colors of the bands represents the sequence of the template DNA.

T C C A T G G A C C A

CREDIT: (7) © Cengage Learning.

their ends, and the pigment labels now impart distinct colors to the bands ❹. Each color designates one of the four bases, so the order of colored bands in the gel represents the DNA sequence ❺.

The Human Genome Project

The sequencing method we have just described was invented in 1975. Ten years later, it had become so routine that scientists began to consider sequencing the entire human genome—all 3 billion nucleotides. Proponents of the idea said it could provide huge payoffs for medicine and research. Opponents said this daunting task would divert attention and funding from more urgent research. It would require 50 years to sequence the human genome given the techniques of the time. However, the techniques continued to improve rapidly, and with each improvement more nucleotides could be sequenced in less time. Automated (robotic) DNA sequencing and PCR had just been invented. Both were still too cumbersome and expensive to be useful in routine applications, but they would not be so for long. Waiting for faster, cheaper technologies seemed the most efficient way to sequence the genome, but just how fast did they need to be before the project should begin?

A few privately owned companies decided not to wait, and started sequencing. One of them intended to determine the genome sequence in order to patent it. The idea of patenting the human genome provoked widespread outrage, but it also spurred commitments in the public sector. In 1988, the National Institutes of Health (NIH) essentially took over the project by hiring James Watson (of DNA structure fame) to head an official Human Genome Project, and providing $200 million per year to fund it. A partnership formed between the NIH and international institutions that were sequencing different parts of the genome. Watson set aside 3 percent of the funding for studies of ethical and social issues arising from the work. He later resigned over a patent disagreement, and geneticist Francis Collins took his place.

Amid ongoing squabbles over patent issues, Celera Genomics formed in 1998. With biologist Craig Venter at its helm, the company intended to commercialize human genetic information. Celera invented faster techniques for sequencing genomic DNA, because the first to have the complete sequence had a legal basis for patenting it. The competition motivated the international

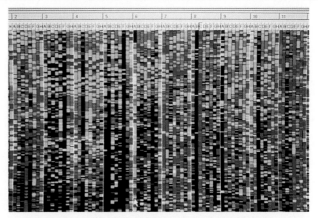

FIGURE 15.8 Human DNA sequence data. The order of colors in each vertical lane reveals one part of the DNA sequence.

FIGURE 15.9 Today's automated DNA sequencing machines can sequence an individual's entire genome in 2–4 hours, for a cost of about $1,000.

partnership to accelerate its efforts. Then, in 2000, U.S. President Bill Clinton and British Prime Minister Tony Blair jointly declared that the sequence of the human genome could not be patented. Celera kept sequencing anyway, and, in 2001, the competing governmental and corporate teams published about 90 percent of the sequence. In 2003, fifty years after the discovery of the structure of DNA, the sequence of the human genome was officially completed (**FIGURE 15.8**). The technology has improved so much that a human genome can now be sequenced in a few hours (**FIGURE 15.9**).

TAKE-HOME MESSAGE 15.4
How is DNA sequence determined?

✔ In the most common method of DNA sequencing, a strand of DNA is partially replicated. Electrophoresis is used to separate the resulting fragments by length.

✔ Improved sequencing techniques and worldwide efforts allowed the human genome sequence to be determined.

electrophoresis Technique that separates DNA fragments by size.
sequencing Method of determining the order of nucleotides in DNA.

✔ Comparing the human genome sequence with that of other species is helping us understand how the human body works.

✔ Unique sequences of genomic DNA can be used to distinguish an individual from all others.

Despite our ability to determine the sequence of an individual's genome, it will be a long time before we understand all the information coded within that sequence. The human genome contains a massive amount of seemingly cryptic data. We can decipher some of this data by comparing genomes of different species, the premise being that all organisms are descended from shared ancestors, so all genomes are related to some extent. We see evidence of such genetic relationships simply by comparing the raw sequence data, which, in some regions, is extremely similar across many species (**FIGURE 15.10**).

The study of genomes is called **genomics**. This broad field encompasses whole-genome comparisons, structural analysis of gene products, and surveys of small-scale variations in sequence. Genomics is providing powerful insights into evolution. For example, comparing primate genomes revealed a change in chromosome structure that occurred during the evolution of our species (Section 14.5). Comparing genomes also showed us that structural changes in chromosomes are not entirely random; some specific alterations are much more common than others. This is because chromosomes that break tend to do so in particular spots. Human, mouse, rat, cow, pig, dog, cat, and horse chromosomes have undergone several translocations at these breakage hot spots during evolution. In humans, chromosome abnormalities that contribute to the progression of cancer also occur at the very same breakage hot spots.

Comparing the coding regions of genomes offers medical benefits. We have learned the function of many human genes by studying their counterpart genes in other species. For instance, researchers comparing human and mouse genomes discovered a human version of a mouse gene, *APOA5*, that encodes a lipopro-

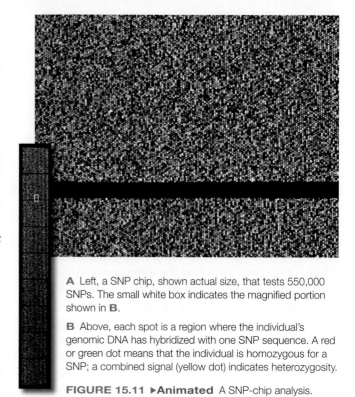

A Left, a SNP chip, shown actual size, that tests 550,000 SNPs. The small white box indicates the magnified portion shown in **B**.

B Above, each spot is a region where the individual's genomic DNA has hybridized with one SNP sequence. A red or green dot means that the individual is homozygous for a SNP; a combined signal (yellow dot) indicates heterozygosity.

FIGURE 15.11 ▶**Animated** A SNP-chip analysis.

tein. Mice with an *APOA5* knockout have four times the normal level of triglycerides in their blood. The researchers then looked for—and found—a correlation between *APOA5* mutations and high triglyceride levels in humans. High triglycerides are a risk factor for coronary artery disease.

DNA Profiling

As you learned in Section 15.1, about 99% of the human genome sequence is identical in every member of the species. The differences you carry in your DNA make you unique. In fact, those differences are so unique that they can be used to identify you. Identifying an individual by his or her DNA is a method called **DNA profiling**. One DNA profiling method uses SNP-chips (**FIGURE 15.11**). A SNP-chip

```
758  GATAATCCTGTTTTGAACAAAAGGTCAAATTGCTGAATAGAAA-GTCTTGATTAACTAAAAGATGTACAAAGTGGAATTA 836  Human
752  GATAATCCTGTTTTGAACAAAAGGTCAAATTGCTGAATAGAAA-GTCTTGATTAACTAAAAGATGTACAAAGTGGAATTA 830  Mouse
751  GATAATCCTGTTTTGAACAAAAGGTCAAATTGCTGAATAGAAA-GTCTTGATTAACTAAAAGATGTACAAAGTGGAATTA 829  Rat
754  GATAATCCTGTTTTGAACAAAAGGTCAAATTGCTGAATAGAAA-GTCTTGATTAACTAAAAGATGTACAAAGTGGAATTA 832  Dog
782  GATAATCCTGTTTTGAACAAAAGGTCAAATTGCTGAATAGAAA-GTCTTGATTAACTAAAAGATGTACAAAGTGGAATTA 860  Chicken
758  GATAATCCTGTTTTGAACAAAAGGTCAAATTGCTGAATAGAAA-GTCTTGATTAAGTAAAAGATGTACAAAGTGGAATTA 836  Frog
823  GATAATCCTGTTTTGAACAAAAGGTCAGATTGCTGAATAGAAAAGGCTTGATTAAAGCAGAGATGTACAAAGTGGACGCA 902  Zebrafish
763  GATAATCCTGTTTTGAACAAAAGGTCAAATTGTTGAATAGACGCTTTGATAAAGCGGAGGAGGTACAAAGTGGGACC- 841  Pufferfish
```

FIGURE 15.10 Genomic DNA alignment. This is a region of the gene for a DNA polymerase. Nucleotides that differ from those in the human sequence are highlighted. The chance that any two of these sequences would randomly match is about 1 in 10^{46}.

CREDITS: (10) © Cengage Learning; (11A) The Sanger Institute. Wellcome Images; (11B) Wellcome Trust Sanger Institute.

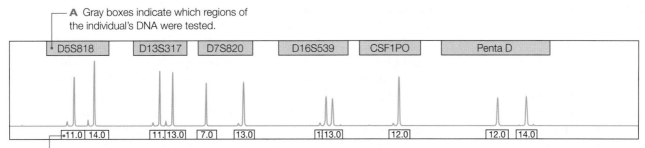

A Gray boxes indicate which regions of the individual's DNA were tested.

| D5S818 | D13S317 | D7S820 | D16S539 | CSF1PO | Penta D |

| •11.0 | 14.0 | | 11. | 13.0 | 7.0 | | 13.0 | | 1 | 13.0 | | 12.0 | | 12.0 | 14.0 |

B The number of repeats is shown in a box below each peak. A peak's location on the x-axis corresponds to the length of the DNA fragment amplified (a measure of the number of repeats). Peak size reflects the amount of DNA.

FIGURE 15.12 ▶Animated An individual's (partial) short tandem repeat profile. Remember, human body cells are diploid. Double peaks appear on a profile when the two members of a chromosome pair carry a different number of repeats.

is a tiny glass plate with microscopic spots of DNA stamped on it. The DNA sample in each spot is a short, synthetic single strand with a unique SNP sequence. When an individual's genomic DNA is washed over a SNP-chip, it hybridizes only with DNA spots that have a matching SNP sequence. Probes reveal where the genomic DNA has hybridized—and which of the SNPs are carried by the individual.

Another method of DNA profiling involves analysis of short tandem repeats in an individual's chromosomes (Section 13.7). Short tandem repeats usually occur in the same location in human chromosomes, but the number of times a sequence is repeated in each location differs among individuals. For example, one person's DNA may have fifteen repeats of the nucleotides TTTTC at a certain spot on one chromosome. Another person's DNA may have four repeats of this sequence in the same location. Short tandem repeats slip spontaneously into DNA during replication, and their numbers grow or shrink over generations. Unless two people are identical twins, the chance that they have identical short tandem repeats in even three regions of DNA is 1 in a quintillion (10^{18}), which is far more than the number of people who have ever lived. Thus, an individual's array of short tandem repeats is, for all practical purposes, unique.

Analyzing a person's short tandem repeats begins with PCR, which is used to copy ten to thirteen particular regions of chromosomal DNA known to have repeats. The lengths of the copied DNA fragments differ among most individuals, because the number

of tandem repeats in those regions also differs. Thus, electrophoresis can be used to reveal an individual's unique array of short tandem repeats (**FIGURE 15.12**).

Short tandem repeat analysis will soon be replaced by full genome sequencing, but for now it continues to be a common DNA profiling method. Geneticists compare short tandem repeats on Y chromosomes to determine relationships among male relatives, and to trace an individual's ethnic heritage. They also track mutations that accumulate in populations over time by comparing DNA profiles of living humans with those of ancient ones. Such studies are allowing us to reconstruct population dispersals that happened in the ancient past.

Short tandem repeat profiles are routinely used to resolve kinship disputes, and as evidence in criminal cases. Within the context of a criminal or forensic investigation, DNA profiling is called DNA fingerprinting. As of January 2014, the database of DNA fingerprints maintained by the Federal Bureau of Investigation (the FBI) contained the short tandem repeat profiles of 10.7 million convicted offenders, and had been used in more than 200,000 criminal investigations. DNA fingerprinting is also used to identify human remains, including the individuals who died in the World Trade Center on September 11, 2001.

DNA profiling Identifying an individual by analyzing the unique parts of his or her DNA.
genomics The study of genomes.

TAKE-HOME MESSAGE 15.5
How do we use what researchers discover about genomes?

✔ Analysis of the human genome sequence is yielding new information about our genes and how they work.

✔ DNA profiling identifies individuals by the unique parts of their DNA.

CREDIT: (12) Raw STR data courtesy of © Orchid Cellmark, www.orchidcellmark.com.

15.6 Genetic Engineering

✔ Bacteria and yeast are the most common genetically engineered organisms.

Traditional cross-breeding methods can alter genomes, but only if individuals with the desired traits will interbreed. Genetic engineering takes gene-swapping to an entirely different level. **Genetic engineering** is a process by which an individual's genome is deliberately modified. A gene from one species may be transferred to another to produce an organism that is **transgenic**, or a gene may be altered and reinserted into an individual of the same species. Both methods yield a **genetically modified organism**, or **GMO**.

The most common GMOs are yeast and bacteria (**FIGURE 15.13**). Both types of cells have the metabolic machinery to make complex organic molecules, and they are easily modified to produce, for example, medically important proteins. People with diabetes were among the first beneficiaries of such organisms. Insulin for their injections was once extracted from animals, but it provoked an allergic reaction in some people. Human insulin, which does not provoke allergic reactions, has been produced by transgenic *E. coli* since 1982. Slight modifications of the gene have yielded fast-acting and slow-release forms of human insulin.

Genetically modified microorganisms also make proteins used in foods. For example, enzymes produced by modified microorganisms improve the taste and clarity of beer and fruit juice, slow bread staling, or modify certain fats. Cheese is traditionally made with an enzyme, chymosin, extracted from calf stomachs. Today, almost all cheese is made with calf chymosin produced by transgenic fungi.

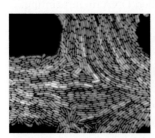

FIGURE 15.13 *E. coli* bacteria transgenic for a fluorescent jellyfish protein. Variation in fluorescence among the genetically identical cells reveals differences in gene expression that may help us understand why some bacteria become dangerously resistant to antibiotics, and others do not.

TAKE-HOME MESSAGE 15.6
What is genetic engineering?

✔ Genetic engineering is the deliberate alteration of an individual's genome, and it results in a genetically modified organism (GMO).

✔ A transgenic organism carries a gene from a different species. Transgenic bacteria and yeast are used in research, medicine, and industry.

15.7 Designer Plants

✔ Genetically engineered crop plants are widespread in the United States.

As crop production expands to keep pace with human population growth, it places unavoidable pressure on ecosystems everywhere. Irrigation leaves mineral and salt residues in soils. Tilled soil erodes, taking topsoil with it. Runoff clogs rivers, and fertilizer in it causes algae to grow so fast that fish suffocate. Pesticides can be harmful to humans and other animals, including beneficial insects such as bees.

Pressured to produce more food at lower cost and with less damage to the environment, many farmers have begun to rely on genetically modified crop plants. Genes can be introduced into plant cells by way of electric or chemical shocks, by blasting them with microscopic DNA-coated pellets, or by using *Agrobacterium tumefaciens* bacteria. *A. tumefaciens* carries a plasmid with genes that cause tumors to form on infected plants; hence the name Ti plasmid (for Tumor-inducing). Researchers replace the tumor-inducing genes with foreign or modified genes, then use the plasmid as a vector to deliver the desired genes into plant cells. Whole plants can be grown from plant cells that integrate a recombinant plasmid into their chromosomes (**FIGURE 15.14**).

Many genetically modified crops carry genes that impart resistance to devastating plant diseases. Others offer improved yields. GMO crops such as Bt corn and soy help farmers use smaller amounts of toxic pesticides. Organic farmers often spray their crops with spores of Bt (*Bacillus thuringiensis*), a bacterial species that makes a protein toxic only to some insect larvae. Researchers transferred the gene encoding the Bt protein into plants. The engineered plants produce the Bt protein, but otherwise they are essentially identical with unmodified plants. Larvae die shortly after eating their first and only GMO meal. Farmers can use much less pesticide on crops that make their own (**FIGURE 15.15**).

Transgenic crop plants are also being developed for impoverished regions of the world. Genes that confer drought tolerance, insect resistance, and enhanced nutritional value are being introduced into plants such as corn, rice, beans, sugarcane, cassava, cowpeas, banana, and wheat. The resulting GMO crops may help

genetic engineering Process by which deliberate changes are introduced into an individual's genome.
genetically modified organism (**GMO**) Organism whose genome has been modified by genetic engineering.
transgenic Refers to a genetically modified organism that carries a gene from a different species.

CREDIT: (13) Photo Courtesy of Systems Biodynamics Lab, P. I. Jeff Hasty, UCSD Department of Bioengineering, and Scott Cookson.

people in these regions who rely mainly on agriculture for food and income.

Genetic modifications can make food plants more nutritious. For example, rice plants have been engineered to make β-carotene, an orange photosynthetic pigment that is remodeled by cells of the small intestine into vitamin A. These rice plants carry two genes in the β-carotene synthesis pathway: one from corn, the other from bacteria. One cup of the engineered rice seeds—grains of Golden Rice—has enough β-carotene to satisfy a child's daily need for vitamin A.

The USDA Animal and Plant Health Inspection Service (APHIS) regulates the introduction of GMOs into the environment. At this writing, APHIS has deregulated ninety-two crop plants, which means the plants are approved for unrestricted use in the United States. Worldwide, more than 330 million acres are currently planted in GMO crops, the majority of which are corn, sorghum, cotton, soy, canola, and alfalfa engineered for resistance to the herbicide glyphosate. Rather than tilling the soil to control weeds, farmers can spray their fields with glyphosate, which kills the weeds but not the engineered crops.

Crops genetically engineered to resist glyphosate have been used in conjunction with the herbicide since the mid 1970s. Genes that confer glyphosate resistance are now appearing in weeds and other wild plants, as well as in nonengineered crops—which means that recombinant DNA can (and does) escape into the environment. Glyphosate resistance genes are probably being transferred from transgenic plants to nontransgenic ones via pollen carried by wind or insects.

Many people are opposed to any GMO. Some worry that our ability to tinker with genetics has surpassed our ability to understand the impact of the tinkering. Controversy raised by GMO use invites you to read the research and form your own opinions. The alternative is to be swayed by media hype (the term "Frankenfood," for instance), or by reports from possibly biased sources (such as herbicide manufacturers).

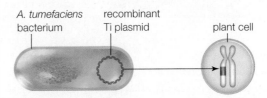

A Ti plasmid carrying a foreign gene is inserted into an *Agrobacterium tumefaciens* bacterium.

B The bacterium infects a plant cell and transfers the Ti plasmid into it. The plasmid DNA, along with the foreign gene, becomes integrated into one of the cell's chromosomes.

C The infected plant cell divides, and its descendants form an embryo, then a plant (left). Cells of the transgenic plant carry and express the foreign gene.

FIGURE 15.14 ▶Animated Using the Ti plasmid to make a transgenic plant.

FIGURE 15.15 Farmers can use much less pesticide on crops that make their own. The genetically modified plants that produced the row of corn on the top carry a gene from the bacteria *Bacillus thuringiensis* (Bt) that conferred insect resistance. Compare the corn from unmodified plants, bottom. No pesticides were used on either crop.

TAKE-HOME MESSAGE 15.7

> Are genetically modified plants used as commercial crops?

✔ Genetically modified crop plants can help farmers be more productive while reducing overall costs.

✔ The widespread use of GMO crops has had unintended environmental effects. Herbicide resistant weeds are now common, and recombinant genes have spread to wild plants and non-GMO crops.

CREDITS: (14A, B) © Cengage Learning; (14C) Pascal Goetgheluck/Science Source; (15) The Bt and Non-Bt corn photos were taken as part of field trial conducted on the main campus of Tennessee State University at the Institute of Agricultural and Environmental Research. The work was supported by a competitive grant from the CSREES, USDA titled *Southern Agricultural Biotechnology Consortium for Underserved Communities*, (2000–2005). Dr. Fisseha Tegegne and Dr. Ahmad Aziz served as Principal and Co-principal Investigators respectively to conduct the portion of the study in the State of Tennessee.

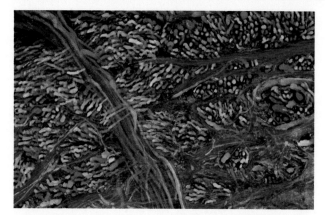

A Mice transgenic for multiple pigments ("brainbow mice") are allowing researchers to map the complex neural circuitry of the brain. Individual nerve cells in the brain stem of a brainbow mouse are visible in this fluorescence micrograph.

B Zebrafish engineered to glow in places where BPA, an endocrine-disrupting chemical, is present. The fish are literally illuminating where this pollutant acts in the body—and helping researchers discover what it does when it gets there.

C Transgenic goats produce human antithrombin, an anticlotting protein. Antithrombin harvested from their milk is used as a drug during surgery or childbirth to prevent blood clotting in people with hereditary antithrombin deficiency. This genetic disorder carries a high risk of life-threatening clots.

FIGURE 15.16 Examples of genetically modified animals.

✔ Genetically engineered animals are invaluable in medical research and in other applications.

Traditional cross-breeding can produce animals so unusual that transgenic animals may appear mundane by comparison. Consider featherless chickens (right) that were bred to survive in deserts where cooling systems are not an option. Cross-breeding is a form of genetic manipulation, but many transgenic animals would probably never have occurred without laboratory intervention.

The first genetically modified animals were mice. Today, such mice are commonplace, and they are invaluable in research (**FIGURE 15.16A**). For example, we have discovered the function of human genes (including the *APOA5* gene discussed in Section 15.5) by inactivating their counterparts in mice. Genetically modified mice are also used as models of human diseases. For example, researchers inactivated the molecules involved in the control of glucose metabolism, one by one. Studying the effects of the knockouts in mice has resulted in much of our current understanding of how diabetes works in humans.

Genetically modified animals other than mice are also useful in research (**FIGURE 15.16B**), and some make molecules that have medical and industrial applications (**FIGURE 15.16C**). Various transgenic goats produce proteins used to treat cystic fibrosis, heart attacks, blood clotting disorders, and even nerve gas exposure. Milk from goats transgenic for lysozyme, an antibacterial protein in human milk, may protect infants and children in developing countries from acute diarrheal disease. Goats transgenic for a spider silk gene produce the silk protein in their milk; researchers can spin this protein into nanofibers that have medical and electronics applications. Rabbits make human interleukin-2, a protein that triggers immune cells to divide and is used as a cancer drug.

We also engineer food animals. Genetic engineering has given us pigs with heart-healthy fat and environmentally friendly low-phosphate feces, muscle-bound trout, chickens that do not transmit bird flu, and cows that do not get mad cow disease, among other examples. Many people think that genetically engineering livestock is unconscionable. Others see it as an extension of thousands of years of acceptable animal husbandry practices. The techniques have changed, but not the intent: We humans continue to have an interest

CREDITS: (16A) Courtesy of © Dr. Jean Levit. The Brainbow technique was developed in the laboratories of Jeff W. Lichtman and Joshua R. Sanes at Harvard University. This image has received the Bioscape imaging competition 2007 prize; (16B) © Charles Taylor/University of Exeter; (16C) © CTC Biotherapeutics, Inc.; (in text) © Adi Nes, Dvir Gallery Ltd.

in improving our livestock. Either way, tinkering with the genes of animals raises a host of ethical dilemmas. Consider animals genetically modified to carry mutations associated with human diseases—multiple sclerosis, cystic fibrosis, diabetes, cancer, or Huntington's disease, for example. Researchers study these animals in order to understand the diseases, and to test potential treatments, without experimenting on humans. However, the modified animals often suffer the same terrible symptoms of the condition as humans do.

Knockouts and Organ Factories

Millions of people have a heart, kidney, or other organ that has been damaged and cannot heal. In any given year, more than 80,000 of them are on waiting lists for an organ transplant in the United States alone. Human donors are in such short supply that illegal organ trafficking is now a common problem.

Pigs are considered a potential source of organs for transplantation, because pig and human organs are about the same in both size and function. However, the human immune system battles anything it recognizes as nonself. It rejects a pig organ at once, because it recognizes proteins and carbohydrates on the plasma membrane of pig cells. Within a few hours, blood coagulates inside the organ's vessels and dooms the transplant. Drugs can suppress the immune response, but they also render organ recipients particularly vulnerable to infection. Researchers have produced genetically modified pigs that lack the offending molecules on their cells. The human immune system may not reject tissues or organs transplanted from these pigs.

Pig-to-human transplants are an example of **xenotransplantation**, the transfer of an organ from one species into another. Critics of xenotransplantation are concerned that, among other things, such transplants would invite pig viruses to infect humans, perhaps with catastrophic results. Their concerns are not unfounded: All human pandemics have been caused by animal viruses that adapted to replicate in people.

xenotransplantation Transplantation of an organ from one species into another.

TAKE-HOME MESSAGE 15.8
Why do we genetically engineer animals?

✔ Genetic engineering creates animals that would be impossible to produce using traditional cross-breeding methods.

✔ Most engineered animals are used for research and medical applications.

✔ The technique also offers a way to improve livestock.

15.9 Safety Issues

✔ The first transfer of foreign DNA into bacteria ignited an ongoing debate about potential dangers of transgenic organisms that enter the environment.

When James Watson and Francis Crick presented their model of DNA in 1953, they ignited a global blaze of optimism. The very book of life seemed to be open for scrutiny, but in reality, no one could read it. New techniques would have to be invented before that book would be readable. Twenty years later, Paul Berg and his coworkers discovered how to make recombinant organisms by fusing DNA from two species of bacteria, providing the tools to be able to study DNA sequence in detail. They began to clone DNA from many different organisms. The technique of genetic engineering was born, and suddenly everyone was worried about it. Researchers knew that DNA itself was not toxic, but they could not predict with certainty what would happen each time they fused genetic material from different organisms. Would they accidentally make a new, dangerous form of life by fusing DNA of two harmless organisms? What if an engineered organism escaped the laboratory and transformed other organisms?

In a remarkably quick act of self-regulation, scientists reached a consensus on new safety guidelines for DNA research. Adopted at once by the NIH, these guidelines included precautions for laboratory procedures. They covered the design and use of host organisms that could survive only under a narrow range of conditions inside the laboratory. Researchers stopped using DNA from pathogenic or toxic organisms for recombinant DNA experiments until proper containment facilities were developed.

Today, all genetic engineering should be done under these laboratory guidelines, but the rules are not a guarantee of safety. We are still learning about escaped GMOs and their effects, and enforcement is a problem. For example, the expense of deregulating a GMO is prohibitive for endeavors in the public sector. Thus, most commercial GMOs were produced by large, private companies—the same ones that typically wield tremendous political influence over the very government agencies charged with regulating them.

TAKE-HOME MESSAGE 15.9
Is genetic engineering safe?

✔ Guidelines for DNA research have been in place for decades in the United States and other countries. Researchers are expected to comply, but the guidelines are not a guarantee of safety.

✔ We as a society continue to work our way through the ethical implications of applying new DNA technologies.

✔ The manipulation of individual genomes continues even as we are weighing the risks and benefits of this research.

Gene Therapy

We know of more than 15,000 serious genetic disorders. Collectively, they cause 20 to 30 percent of infant deaths each year, and account for half of all mentally impaired patients and a fourth of all hospital admissions. They also contribute to many age-related disorders, including cancer, Parkinson's disease, and diabetes. Drugs and other treatments can minimize the symptoms of some genetic disorders, but gene therapy is the only cure. **Gene therapy** is the transfer of recombinant DNA into an individual's body cells, with the intent to correct a genetic defect or treat a disease. Typically, the transfer inserts an unmutated gene into the individual's chromosomes.

DNA can be introduced into human cells in many ways, for example by direct injection, electrical pulses, lipid clusters, nanoparticles, or genetically engineered viruses. Viruses have molecular machinery that delivers their genomes into cells they infect. Those used as vectors have DNA that splices itself into the infected cells' chromosomes, along with foreign DNA that has been inserted into it.

Human gene therapy is a compelling reason to embrace genetic engineering research. It is now being tested as a treatment for AIDS, muscular dystrophy, heart attack, sickle-cell anemia, cystic fibrosis, hemophilia A, Parkinson's disease, Alzheimer's disease, several types of cancer, and inherited diseases of the eye, the ear, and the immune system.

People have already benefited from gene therapy. Consider Rhys Evans (**FIGURE 15.17**), who was born with SCID-X1, a severe genetic disorder that stems from a mutated allele of the *IL2RG* gene. The gene encodes a receptor for an immune signaling molecule. Without treatment, people affected by this disorder can survive only in germ-free isolation tents because they cannot fight infections (a diminished life in a sterile isolation tent was the source of the term "bubble boy"). In the late 1990s, researchers used a genetically engineered virus to insert unmutated copies of *IL2RG* into cells taken from the bone marrow of twenty boys with SCID-X1. Each child's modified cells were infused back into his bone marrow. Within months of their treatment, eighteen of the boys left their isolation tents for good. Rhys was one of them. Gene therapy had permanently repaired their immune systems.

FIGURE 15.17
Rhys Evans, shown here at age 10, was born with SCID-X1. His immune system has been permanently repaired by gene therapy.

Recently, gene therapy has been used to treat acute lymphoblastic leukemia, a typically fatal cancer of bone marrow cells. A viral vector was used to insert a gene into immune cells extracted from patients. When the engineered cells were reintroduced into the patients' bodies, the inserted gene directed the destruction of the cancer cells. The therapy worked astonishingly well: In one patient, all traces of the leukemia vanished in eight days.

Despite the successes, manipulating a gene within the context of a living individual is unpredictable even when we know its sequence and location on a chromosome. No one, for example, can predict with absolute certainty where a virus-injected gene will become integrated into a chromosome. Its insertion might disrupt other genes. If it interrupts a gene that is part of the controls over cell division, then cancer might be the outcome. Consider that five of the twenty boys first treated with gene therapy for SCID-X1 developed a type of bone marrow cancer called leukemia, and one of them died. Developers of the gene therapy had wrongly predicted that cancer related to it would be rare. Research now implicates the very gene targeted for repair, especially when combined with the virus that delivered it. The viral DNA preferentially inserted itself into the children's chromosomes at a site near a proto-oncogene (Section 11.6). The insertion activated the gene by triggering its transcription, and that is how the leukemia began. Since that time, researchers used PCR to detect viral integration sites and improve the design of the vector. The development of more efficient and specific viral vectors has reduced the risk associated with all types of gene therapy.

Personal Genetic Testing (revisited)

Results of a personal genetic test may include estimated risks of developing conditions associated with your particular set of SNPs. For example, the test will probably determine whether you are homozygous for certain alleles of the *MC1R* gene (Section 13.5). If you are, the company's report may tell you that you have red hair. Few SNPs have such a clear effect, however. Most human traits are polygenic, and many are also influenced by environmental factors (Section 13.6). Thus, although a DNA test can reliably determine the SNPs in an individual's genome, it cannot reliably predict the effect of those SNPs on the individual.

For example, if you carry one *E4* allele of the *APOE* gene, a DNA testing company cannot tell you whether you will develop Alzheimer's disease later in life.

However, it may report your lifetime risk of developing the disease, which is about 29 percent, as compared with about 9 percent for someone who has no *E4* allele.

What, exactly, does a 29 percent lifetime risk of Alzheimer's mean? The number is a probability statistic; it means, on average, 29 of every 100 people who have the *E4* allele eventually get the disease. However, a risk is just that. Not everyone who has the allele develops Alzheimer's, and not everyone who develops the disease has the allele. Other unknown factors, including epigenetic modifications of DNA, contribute to the disease. We still have a limited understanding of how genes contribute to many health conditions, particularly age-related ones such as Alzheimer's. Geneticists believe that it will be at least five to ten more years before genotyping can be used to accurately predict an individual's future health problems.

Eugenics

The idea of selecting the most desirable human traits, **eugenics**, is an old one. It has been used as a justification for some of the most horrific episodes in human history, including the genocide of 6 million Jews during World War II; thus, it continues to be a hotly debated social issue. For example, using gene therapy to cure human genetic disorders seems like an acceptable goal to most people, but imagine taking this idea a bit further. Would it also be acceptable to engineer the genome of an individual who is within a normal range of phenotype in order to modify a particular trait? Researchers have already produced mice that have improved memory, enhanced learning ability, bigger muscles, and longer lives. Why not people?

Given the pace of genetics research, the debate is no longer about how we would engineer desirable traits, but how we would choose the traits that are desirable. Realistically, cures for many severe but rare genetic disorders will not be found, because the financial return would not cover the cost of the research. Eugenics, however, may be profitable. How much would potential parents pay to be sure that their child will be tall or blue-eyed, with breathtaking strength or intelligence? What about a treatment that can help you lose that extra weight, and keep it off permanently? The gray area between interesting and abhorrent can be very different depending on who is asked. In a survey con-

ducted in the United States, more than 40 percent of those interviewed said they would be fine with using gene therapy to make smarter and cuter babies. In one poll of British parents, 10 percent would use it to keep a child from growing up to be homosexual, and 18 percent would be willing to use it to keep a child from being aggressive.

Some people are concerned that gene therapy puts us on a slippery slope that may result in irreversible damage to ourselves and to the biosphere. We as a society may not have the wisdom to know how to stop once we set foot on that slope; one is reminded of our peculiar human tendency to leap before we look. And yet, something about the human experience allows us to dream of such things as wings of our own making, a capacity that carried us into space. In this brave new world, the questions before you are these: What do we stand to lose if serious risks are not taken? And, do we have the right to impose the potential consequences on people who would choose not to take those risks?

---

eugenics Idea of deliberately improving the genetic qualities of the human race.
gene therapy Treating a genetic defect or disorder by transferring a normal or modified gene into the affected individual.

TAKE-HOME MESSAGE 15.10
Can people be genetically modified?

✔ Genes can be transferred into a person's cells to correct a genetic defect or treat a disease. However, the outcome of altering a person's genome can be unpredictable.

summary

Section 15.1 Personal genetic testing involves identifying a person's unique array of **single-nucleotide polymorphisms (SNPs)**. This type of test is revolutionizing the way medicine is practiced.

Section 15.2 In **DNA cloning**, researchers use **restriction enzymes** to cut DNA into pieces, and then use DNA ligase to splice the fragments into plasmids or other **cloning vectors**. The resulting molecules of **recombinant DNA** are inserted into host cells such as bacteria. Division of host cells produces huge populations of genetically identical descendant cells (clones), each with a copy of the cloned DNA fragment. The enzyme **reverse transcriptase** is used to transcribe RNA into **cDNA** for cloning.

Section 15.3 A **DNA library** is a collection of cells that host different fragments of DNA. A genomic library represents an organism's entire **genome**. Researchers can use **probes** to identify cells in a library that carry a targeted fragment of DNA. The **polymerase chain reaction (PCR)** uses primers and a heat-resistant DNA polymerase to rapidly increase the number of copies of a targeted section of DNA.

Section 15.4 Advances in **sequencing**, which reveals the order of nucleotides in DNA, allowed the DNA sequence of the entire human genome to be determined. DNA polymerase is used to partially replicate a DNA template. The reaction produces a mixture of DNA fragments of different lengths; **electrophoresis** separates the fragments by length into bands.

Section 15.5 **Genomics** provides insights into the function of the human genome. Similarities between genomes of different organisms are evidence of evolutionary relationships, and can be used as a predictive tool in research. **DNA profiling** identifies a person by the unique parts of his or her DNA. An example is the determination of an individual's array of short tandem repeats or single-nucleotide polymorphisms. Within the context of a criminal investigation, a DNA profile is called a DNA fingerprint.

Sections 15.6–15.9 Recombinant DNA technology is the basis of **genetic engineering**, the directed modification of an organism's genetic makeup with the intent to change its phenotype. A gene from one species is inserted into an individual of a different species to make a **transgenic** organism, or a gene is modified and reinserted into an individual of the same species. The result of either process is a **genetically modified organism (GMO)**.

Bacteria and yeast, the most common genetically engineered organisms, produce proteins that have medical value. Most transgenic crop plants, which are now in widespread use worldwide, were created to help farmers produce food more efficiently. Engineered animals, which are mainly used for research and medical applications, may one day provide organs and tissues for **xenotransplantation** into humans.

Section 15.10 With **gene therapy**, a gene is transferred into body cells to correct a genetic defect or treat a disease. Potential benefits of genetically modifying humans must be weighed against potential risks. The practice raises ethical issues such as whether **eugenics** is desirable in some circumstances.

self-quiz

Answers in Appendix VII

1. _____ cut(s) DNA molecules at specific sites.
   a. DNA polymerase   c. Restriction enzymes
   b. DNA probes   d. Reverse transcriptase

2. A _____ is a molecule that can be used to carry a fragment of DNA into a host organism.
   a. cloning vector   c. GMO
   b. chromosome   d. cDNA

3. Reverse transcriptase assembles a(n) _____ on a(n) _____ template.
   a. mRNA; DNA   c. DNA; ribosome
   b. cDNA; mRNA   d. protein; mRNA

4. For each species, all _____ in the complete set of chromosomes is the _____ .
   a. genomes; library   c. mRNA; start of cDNA
   b. DNA; genome   d. cDNA; start of mRNA

5. A set of cells that host various DNA fragments collectively representing an organism's entire set of genetic information is a _____ .
   a. genome   c. genomic library
   b. clone   d. GMO

6. _____ is a technique to determine the order of nucleotide bases in a fragment of DNA.
   a. PCR   c. Electrophoresis
   b. Sequencing   d. Nucleic acid hybridization

7. Fragments of DNA can be separated by electrophoresis according to _____ .
   a. sequence   b. length   c. species

8. PCR can be used _____ .
   a. to increase the number of specific DNA fragments
   b. in DNA fingerprinting
   c. to modify a human genome
   d. a and b are correct

9. An individual's set of unique _____ can be used in DNA profiling.
   a. DNA sequences   c. SNPs
   b. short tandem repeats d. all of the above

10. A transgenic organism _____ .
    a. carries a gene from another species
    b. has been genetically modified
    c. both a and b

Enhanced Spatial Learning in Mice With an Autism Mutation Autism is a neurobiological disorder with symptoms that include impaired social interactions and stereotyped patterns of behavior. Around 10 percent of autistic people have an extraordinary skill or talent such as greatly enhanced memory.

Mutations in neuroligin 3, an adhesion protein that connects brain cells to one another, have been associated with autism. One mutation changes amino acid 451 from arginine to cysteine. In 2007, Katsuhiko Tabuchi and his colleagues genetically modified mice to carry the same arginine-to-cysteine substitution in their neuroligin 3. Mice with the mutation had impaired social behavior.

To test spatial learning ability, the mice were placed in a water maze: a deep pool of warm water in which a platform is submerged a few millimeters below the surface. The platform is not visible to swimming mice. Mice do not particularly enjoy swimming, so they locate a hidden platform as fast as they can. When tested again, they can remember its location by checking visual cues around the edge of the pool. How quickly they remember the platform's location is a measure of spatial learning ability (**FIGURE 15.18**).

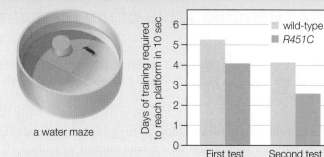

a water maze

FIGURE 15.18 Spatial learning ability in mice with a mutation in neuroligin 3 (*R451C*), compared with unmodified (wild-type) mice.

1. In the first test, how many days did it take unmodified mice to learn to find the location of the hidden platform within 10 seconds?

2. Did the modified or the unmodified mice learn the location of the platform faster in the first test?

3. Which mice learned faster the second time around?

4. Which mice showed the greatest improvement in memory between the first and the second test?

---

11. True or false? A transgenic organism can pass a foreign gene to offspring.

12. Which of the following can be used to carry foreign DNA into host cells? Choose all correct answers.

a. RNA	e. lipid clusters
b. viruses	f. blasts of pellets
c. PCR	g. xenotransplantation
d. plasmids	h. nanoparticles

13. True or false? Some humans are genetically modified.

14. Match the method with the appropriate enzyme.

___ PCR	a. *Taq* polymerase
___ cutting DNA	b. DNA ligase
___ cDNA synthesis	c. reverse transcriptase
___ DNA sequencing	d. restriction enzyme
___ pasting DNA	e. DNA polymerase (not *Taq*)

15. Match each term with the most suitable description.

___ DNA profile	a. GMO with a foreign gene
___ Ti plasmid	b. alleles commonly have them
___ eugenics	c. a person's unique collection
___ SNPs	of short tandem repeats
___ transgenic	d. selecting "desirable" traits
___ GMO	e. genetically modified
	f. used in plant gene transfers

critical thinking

1. In 1918, an influenza pandemic that originated with avian flu killed 50 million people. Researchers isolated samples of that virus from bodies of infected people preserved in Alaskan permafrost since 1918. From the samples, they sequenced the viral genome, then reconstructed the virus. The reconstructed virus is 39,000 times more infectious than modern influenza strains, and 100 percent lethal in mice.

Understanding how this virus works can help us defend ourselves against other strains that may arise. For example, discovering what makes it so infectious and deadly would help us design more effective vaccines. Critics of the research are concerned: If the virus escapes the containment facilities (even though it has not done so yet), it might cause another pandemic. Worse, the published DNA sequence and methods to make the virus could be used for criminal purposes. Do you think this research makes us more or less safe?

2. The results of a paternity test using short tandem repeats are listed in the table below. Who's the daddy? How sure are you?

	Mother	Baby	Alleged Father #1	Alleged Father #2
CSF1PO	15, 17	17, 23	23, 27	17, 15
FGA	9, 9	9, 9	9, 12	9, 12
THO1	29, 29	29, 27	27, 28	29, 28
TPOX	14, 18	18, 20	15, 20	17, 22
VWA	14, 14	14, 14	14, 14	14, 16
D3S1358	11, 14	14, 16	12, 16	14, 20
D5S818	11, 13	10, 13	8, 10	18, 18
D7S820	7, 13	13, 13	13, 19	13, 13
D8S1179	13, 13	13, 15	12, 15	10, 12
D13S317	12, 12	10, 12	8, 10	12, 17
D16S539	12, 14	14, 12	14, 14	18, 25
D18S51	5, 6	6, 22	22, 6	5, 22
D21S11	15, 17	17, 22	15, 22	22, 22

Appendix I. Periodic Table of the Elements

The symbol for each element is an abbreviation of its name. Some symbols for elements are abbreviations for their Latin names. For instance, Pb (lead) is short for *plumbum*; the word "plumbing" is related—ancient Romans made their water pipes with lead.

Elements in each vertical column of the table behave in similar ways. For instance, all of the elements in the far right column of the table are inert gases; they do not interact with other atoms. In nature, such elements occur only as solitary atoms.

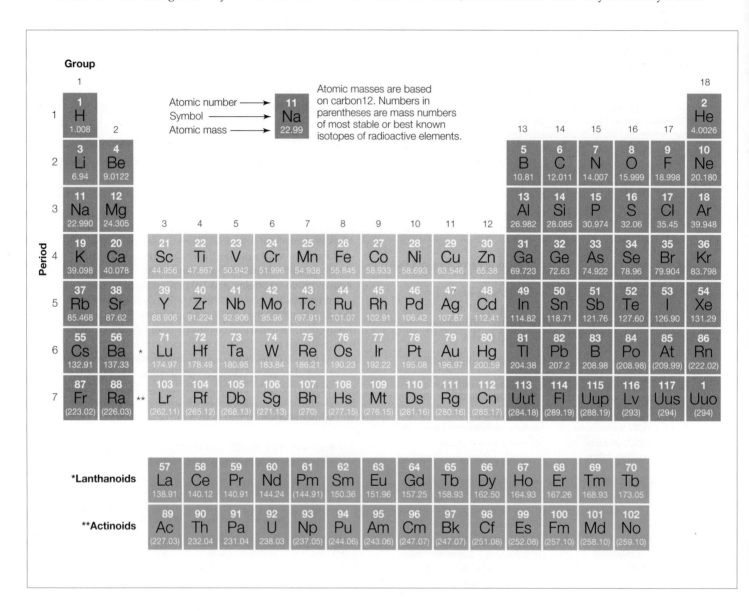

Appendix II. The Amino Acids

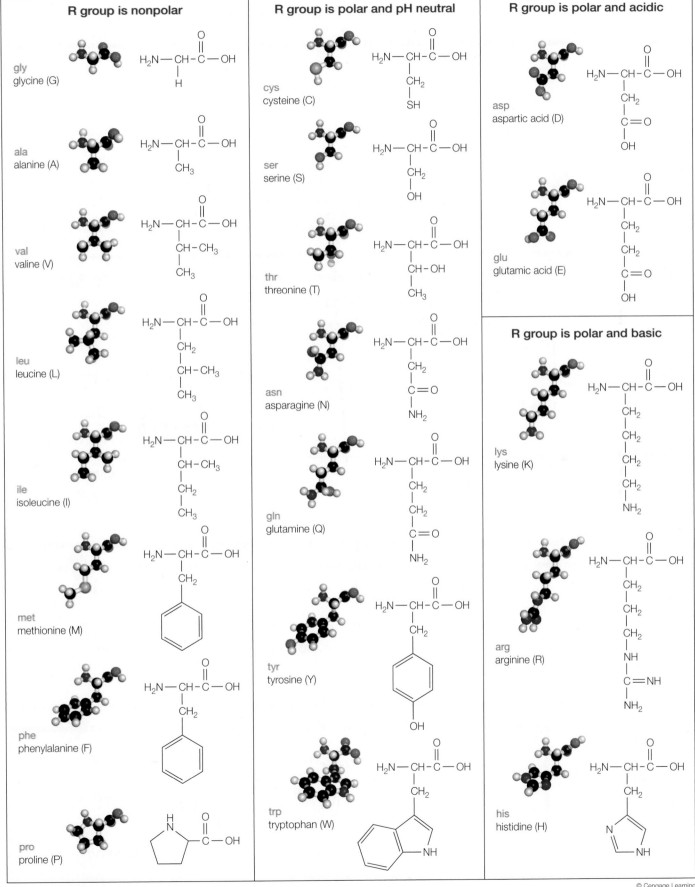

R group is nonpolar

gly
glycine (G)

$H_2N-CH-C-OH$, with $=O$ above C and H below CH

ala
alanine (A)

$H_2N-CH-C-OH$, with $=O$ above C and CH_3 below

val
valine (V)

$H_2N-CH-C-OH$, with $=O$ above C and $CH-CH_3$ / CH_3 below

leu
leucine (L)

$H_2N-CH-C-OH$, with $=O$ above C and CH_2 / $CH-CH_3$ / CH_3 below

ile
isoleucine (I)

$H_2N-CH-C-OH$, with $=O$ above C and $CH-CH_3$ / CH_2 / CH_3 below

met
methionine (M)

$H_2N-CH-C-OH$, with $=O$ above C and CH_2 below and phenyl ring

phe
phenylalanine (F)

$H_2N-CH-C-OH$, with $=O$ above C and CH_2 below and phenyl ring

pro
proline (P)

ring structure with N-H, C, $=O$, OH

R group is polar and pH neutral

cys
cysteine (C)

$H_2N-CH-C-OH$, with $=O$ above C and CH_2 / SH below

ser
serine (S)

$H_2N-CH-C-OH$, with $=O$ above C and CH_2 / OH below

thr
threonine (T)

$H_2N-CH-C-OH$, with $=O$ above C and $CH-OH$ / CH_3 below

asn
asparagine (N)

$H_2N-CH-C-OH$, with $=O$ above C and CH_2 / $C=O$ / NH_2 below

gln
glutamine (Q)

$H_2N-CH-C-OH$, with $=O$ above C and CH_2 / CH_2 / $C=O$ / NH_2 below

tyr
tyrosine (Y)

$H_2N-CH-C-OH$, with $=O$ above C and CH_2 below and phenyl ring with OH

trp
tryptophan (W)

$H_2N-CH-C-OH$, with $=O$ above C and CH_2 below and indole ring with NH

R group is polar and acidic

asp
aspartic acid (D)

$H_2N-CH-C-OH$, with $=O$ above C and CH_2 / $C=O$ / OH below

glu
glutamic acid (E)

$H_2N-CH-C-OH$, with $=O$ above C and CH_2 / CH_2 / $C=O$ / OH below

R group is polar and basic

lys
lysine (K)

$H_2N-CH-C-OH$, with $=O$ above C and CH_2 / CH_2 / CH_2 / CH_2 / NH_2 below

arg
arginine (R)

$H_2N-CH-C-OH$, with $=O$ above C and CH_2 / CH_2 / CH_2 / NH / $C=NH$ / NH_2 below

his
histidine (H)

$H_2N-CH-C-OH$, with $=O$ above C and CH_2 below and imidazole ring with N and NH

Appendix III. A Closer Look at Some Major Metabolic Pathways

Glycolysis

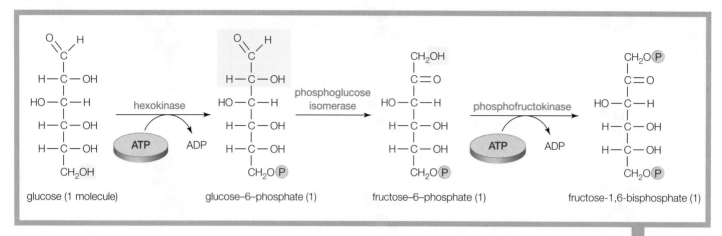

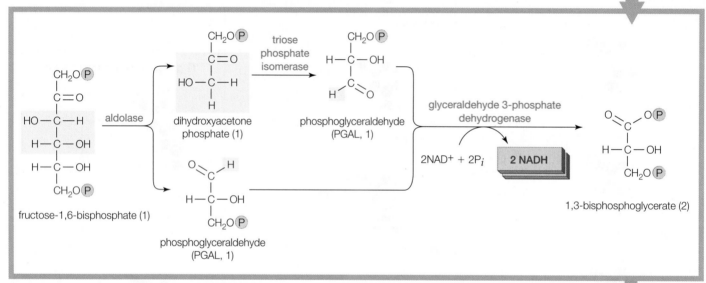

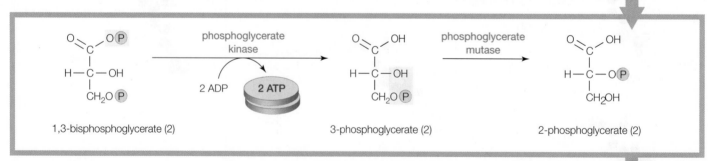

FIGURE A Glycolysis breaks down one glucose molecule into two 3-carbon pyruvate molecules for a net yield of two ATP. Enzyme names are indicated in green; parts of substrate molecules undergoing chemical change are highlighted blue.

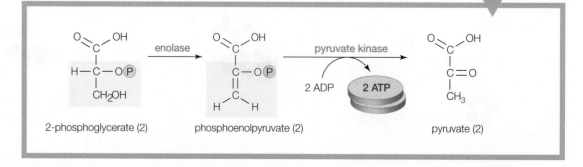

© Cengage Learning

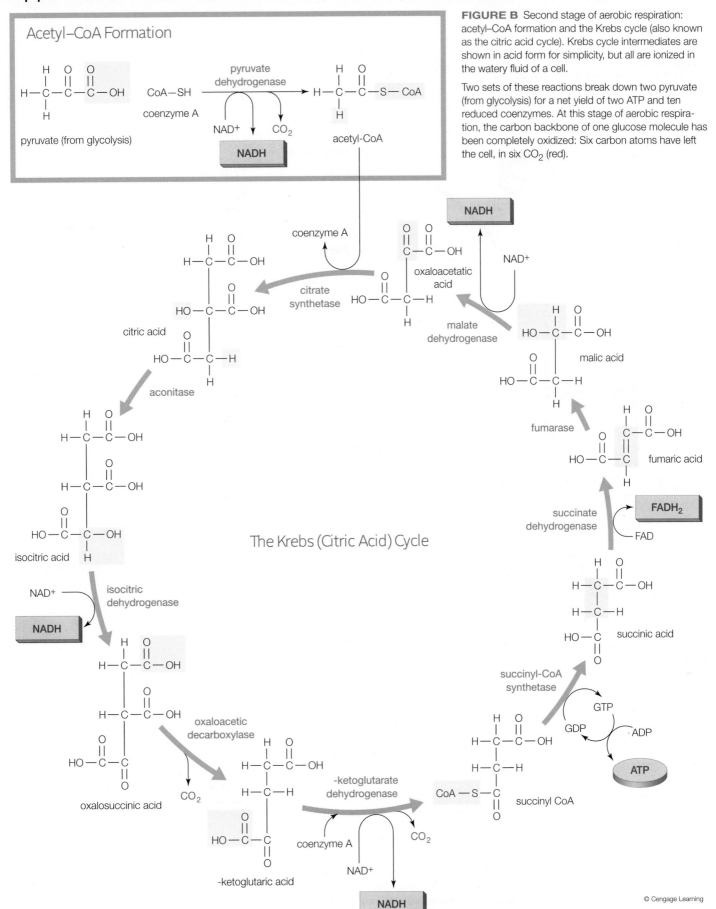

Acetyl–CoA Formation

pyruvate (from glycolysis)

coenzyme A

pyruvate dehydrogenase

NAD^+ CO_2

NADH

acetyl-CoA

FIGURE B Second stage of aerobic respiration: acetyl–CoA formation and the Krebs cycle (also known as the citric acid cycle). Krebs cycle intermediates are shown in acid form for simplicity, but all are ionized in the watery fluid of a cell.

Two sets of these reactions break down two pyruvate (from glycolysis) for a net yield of two ATP and ten reduced coenzymes. At this stage of aerobic respiration, the carbon backbone of one glucose molecule has been completely oxidized: Six carbon atoms have left the cell, in six CO_2 (red).

The Krebs (Citric Acid) Cycle

NADH

coenzyme A

oxaloacetatic acid

NAD^+

citrate synthetase

citric acid

malate dehydrogenase

malic acid

aconitase

fumarase

fumaric acid

isocitric acid

FADH₂

succinate dehydrogenase

FAD

NAD^+

isocitric dehydrogenase

NADH

succinic acid

oxaloacetic decarboxylase

succinyl-CoA synthetase

GTP

GDP

ADP

ATP

oxalosuccinic acid

CO_2

-ketoglutarate dehydrogenase

CoA—S—C

succinyl CoA

CO_2

coenzyme A

NAD^+

-ketoglutaric acid

NADH

FIGURE C Details of the Calvin–Benson cycle. These light-independent reactions of photosynthesis use ATP and NADPH to fix carbon from carbon dioxide. The enzyme rubisco catalyzes the attachment of CO_2 to RuBP. The resulting PGA molecules are converted to PGAL, and the complex series of reactions that follow shuffle carbon atoms among sugar molecules to regenerate RuBP. One molecule of glucose is produced for six CO_2 molecules that enter the reactions. Water and some of the molecular participants are not shown, for clarity.

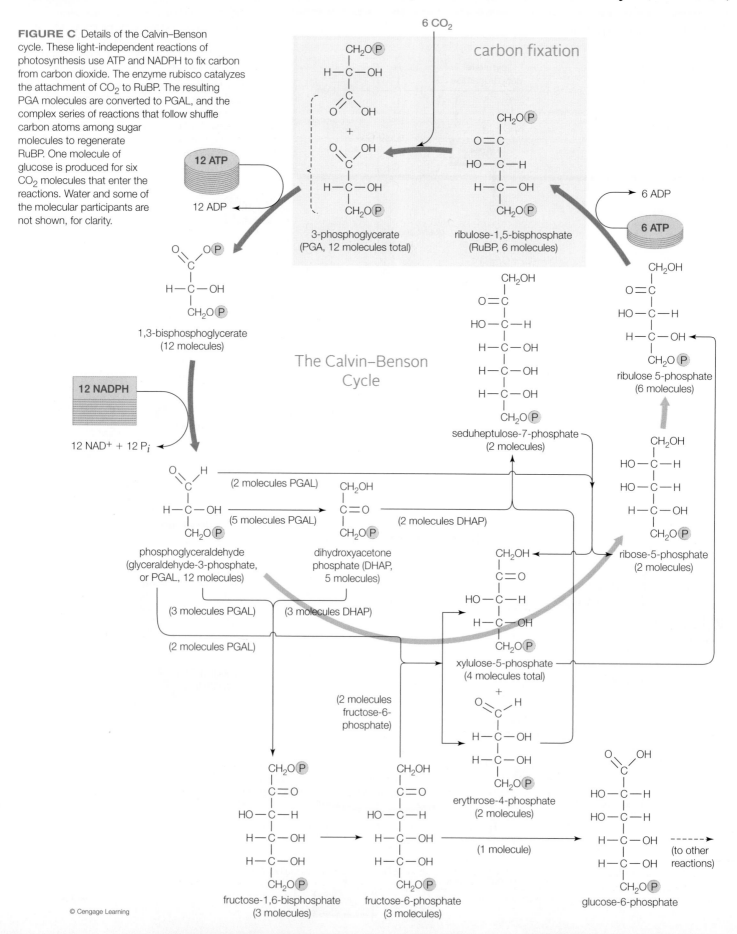

© Cengage Learning

Appendix IV. A Plain English Map of the Human Chromosomes

Haploid set of human chromosomes. The banding patterns characteristic of each type of chromosome appear after staining with a reagent called Giemsa. The locations of some of the 20,065 known genes (as of November, 2005) are indicated. Also shown are locations that, when mutated, cause some of the genetic diseases discussed in the text.

Appendix VI. Units of Measure

LENGTH

1 kilometer (km) = 0.62 miles (mi)
1 meter (m) = 39.37 inches (in)
1 centimeter (cm) = 0.39 inches

To convert	multiply by	to obtain
inches	2.25	centimeters
feet	30.48	centimeters
centimeters	0.39	inches
millimeters	0.039	inches

AREA

1 square kilometer = 0.386 square miles
1 square meter = 1.196 square yards
1 square centimeter = 0.155 square inches

VOLUME

1 cubic meter = 35.31 cubic feet
1 liter = 1.06 quarts
1 milliliter = 0.034 fluid ounces = 1/5 teaspoon

To convert	multiply by	to obtain
quarts	0.95	liters
fluid ounces	28.41	milliliters
liters	1.06	quarts
milliliters	0.03	fluid ounces

WEIGHT

1 metric ton (mt) = 2,205 pounds (lb) = 1.1 tons (t)
1 kilogram (kg) = 2.205 pounds (lb)
1 gram (g) = 0.035 ounces (oz)

To convert	multiply by	to obtain
pounds	0.454	kilograms
pounds	454	grams
ounces	28.35	grams
kilograms	2.205	pounds
grams	0.035	ounces

TEMPERATURE

Celsius (°C) to Fahrenheit (°F): $°F = 1.8 \, (°C) + 32$

Fahrenheit (°F) to Celsius: $°C = \dfrac{(°F - 32)}{1.8}$

	°C	°F
Water boils	100	212
Human body temperature	37	98.6
Water freezes	0	32

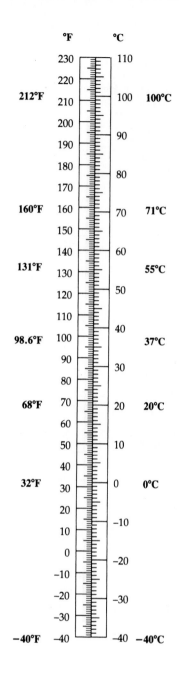

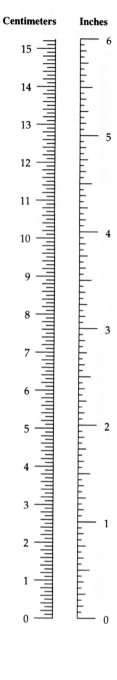

Appendix VII. Answers to Self-Quizzes and Genetics Problems

CHAPTER 1

1. a — 1.2
2. c — 1.2
3. c — 1.3
4. homeostasis — 1.3
5. d — 1.3
6. a — 1.3
7. d — 1.3
8. a, d, e — 1.2–1.5
9. Animals — 1.4
10. a, b — 1.2, 1.4
11. domains — 1.5
12. b — 1.6
13. b — 1.8
14. b — 1.9
15. c — 1.2
    e — 1.8
    b — 1.5
    d — 1.6
    a — 1.6
    f — 1.3

CHAPTER 2

1. a — 2.3
2. b — 2.2
3. d — 2.3
4. d — 2.2
5. b — 2.3
6. a — 2.4
7. a — 2.4
8. c — 2.4
9. c — 2.5
10. c — 2.5
11. d — 2.3, 2.6
12. a — 2.6
13. c — 2.6
14. b — 2.5
15. c — 2.5
    b — 2.2
    d — 2.5
    a — 2.2–2.4
    f — 2.5
    e — 2.2–2.4

CHAPTER 3

1. c — 3.2
2. four — 3.2
3. b — 3.2, 3.4, 3.5
4. e — 3.4, 3.8
5. a — 3.4
6. c — 3.5
7. False — 3.1, 3.5
8. b — 3.5
9. e — 3.5
10. d — 3.6, 3.8
11. d — 3.7
12. d — 3.8
13. a: amino acid — 3.6
    b: carbohydrate — 3.4
    c: polypeptide — 3.6
    d: fatty acid — 3.5
14. c — 3.5
    e — 3.4
    f — 3.5
    d — 3.8
    a — 3.6
    b — 3.8
15. g — 3.6
    a — 3.5
    b — 3.6
    c — 3.5
    d — 3.8
    j — 3.5
    f — 3.8
    i — 3.6
    h — 3.4
    e — 3.4

CHAPTER 4

1. c — 4.2
2. any of the following are correct: all organisms consist of one or more cells; or the cell is the smallest unit of life; or each new cell arises from another, preexisting cell; or a cell passes hereditary material to its offspring. — 4.2
3. b — 4.2
4. False — 4.2
5. b — 4.2
6. False — 4.4
7. c — 4.2, 4.4, 4.5
8. a — 4.7
9. c, b, d, a — 4.7
10. a — 4.7
11. False (some cells have a cell wall and/or other ECM) — 4.11
12. plasmodesmata — 4.11
13. b — 4.7, 4.8, 4.9
14. d — 4.11
15. h — 4.8
    g — 4.9
    b — 4.11
    f — 4.7
    e — 4.7
    a — 4.7
    d — 4.7
    c — 4.9
    i — 4.4, 4.10

CHAPTER 5

1. b — 5.2
2. c — 5.2
3. d — 5.2
4. a — 5.3
5. c — 5.3
6. c — 5.4, 5.6
7. temperature, pH, salt, pressure — 5.4
8. d — 5.2, 5.5, 5.6
9. c — 5.5
10. a — 5.6
11. more/less — 5.8
12. c — 5.7–5.9
13. b — 5.9
14. a — 5.8
15. d — 5.10
16. c — 5.3
    e — 5.10
    f — 5.2
    b — 5.3
    a — 5.6
    g — 5.8
    h — 5.9
    d — 5.9
    k — 5.5
    j — 5.5
    i — 5.7

CHAPTER 6

1. weed=autotroph, all others are heterotrophs — 6.1
2. a — 6.1
3. b — 6.1, revisited
4. c — 6.2
5. a — 6.4
6. d — 6.5
7. b — 6.4, 6.5
8. b — 6.5
9. c — 6.5
10. c — 6.4, 6.6
11. b — 6.6
12. e — 6.6
13. a — 6.7
14. c — 6.7
15. f — 6.6
    h — 6.6
    g — 6.5
    d — 6.5
    e — 6.7
    b — 6.2, 6.5
    a — 6.2
    c — 6.1

CHAPTER 7

1. False — 7.2
2. d — 7.2, 7.3
3. a — 7.2, 7.6
4. c — 7.3
5. b — 7.2, 7.4, 7.5
6. e — 7.4
7. b — 7.4
8. c — 7.5
9. c — 7.6
10. b — 7.6
11. c — 7.5
12. d — 7.7
13. f — 7.6
14. b — 7.3
    c — 7.6
    a — 7.4, 7.6
    d — 7.5
15. b — 7.4
    d — 7.3
    a — 7.3, 7.6
    c — 7.2, 7.5
    e — 7.2–7.4
    f — 7.1

CHAPTER 8

1. c — 8.3
2. c — 8.3
3. b — 8.3
4. b — 8.3
5. a — 8.4
6. b — 8.4
7. b — 8.4
8. a — 8.5
9. d — 8.5
10. c — 8.5
11. b — 8.5
12. d — 8.4, 8.5
13. d — 8.6
14. d — 8.7
15. d — 8.2
    b — 8.1
    a — 8.3
    g — 8.4
    e — 8.5
    h — 8.5
    c — 8.4
    f — 8.6

CHAPTER 9

1. c — 9.2
2. b — 9.3
3. a — 9.2
4. c — 9.2
5. a — 9.2
6. b — 9.2, 9.4
7. c — 9.4
8. b — 9.3
9. a — 9.4
10. a — 9.4
11. a — 9.3
12. a — 9.3, 9.5
13. b — 9.5
14. c — 9.5
15. c — 9.4
    b — 9.3
    e — 9.5
    a — 9.3
    f — 9.4
    d — 9.3

CHAPTER 10

1. d — 10.2
2. d — 10.2, 10.3
3. b — 10.2
4. b — 10.2
5. h — 10.2
6. b — 10.2
7. c — 10.3
8. c — 10.3
9. d — 10.3
10. b — 10.3
11. b — 10.4
12. b — 10.4
13. c — 10.4
14. b — 10.5
15. a — 10.4
    b — 10.2, 10.5
    d — 10.4
    i — 10.2
    f — 10.2
    c — 10.6
    h — 10.3
    g — 10.3
    e — 10.5

CHAPTER 11

1. d — 11.2
2. two — 11.2
3. d — 11.2
4. e — 11.2
5. c — 11.2, 11.3
6. c — 11.2
7. a — 11.2
8. c — 11.2
9. a — 11.2
10. d — 11.4
11. d — 11.6
12. a — 11.6
13. c — 11.4
    f — 11.3
    a — 11.6
    g — 11.4
    b — 11.4
    e — 11.6
    d — 11.3
    h — 11.5
14. d — 11.3
    b — 11.3
    c — 11.3
    e — 11.2
    a — 11.3
    f — 11.4

CHAPTER 12

1. b — 12.1
2. b — 12.2
3. c — 12.1, 12.4
4. d — 12.2
5. b — 12.2
6. a — 12.2
7. Sister chromatids are still attached — 12.3
8. b — 12.2
9. c — 12.3
10. b — 12.4
11. a — 12.4, 12.5
12. b — 12.4, 12.5
13. c — 12.2, 12.3
    d — 12.3
    a — 12.2
    f — 12.2
    e — 12.2
    b — 12.1
    g — 12.3, 12.4
14. e — 12.4

CHAPTER 13

1. b — 13.2
2. a — 13.2
3. b — 13.4
4. b — 13.3
5. c — 13.3
6. a — 13.3
7. b — 13.3
8. d — 13.4
9. c — 13.4
10. False — 13.5
11. c — 13.5
12. b — 13.5
13. d — 13.7
14. b — 13.4
    d — 13.3
    a — 13.2
    c — 13.2

CHAPTER 14

1. b — 14.2
2. b — 14.2
3. a — 14.2
4. b — 14.3
5. False — 14.4
6. d — 14.3, 14.4 (could be due to both parents carrying an autosomal recessive allele, or the mom carrying an x-linked recessive allele, or a new mutation)
7. d — 14.4
8. X from mom, Y from dad — 14.4
9. d — 14.3
10. Y-linked — 14.4
11. d — 14.6
12. b — 14.6
13. True — 14.6
14. c — 14.6
15. c — 14.6
    e — 14.5
    f — 14.6
    b — 14.5
    a — 14.2
    d — 14.5

CHAPTER 15

1. c — 15.2
2. a — 15.2
3. b — 15.2
4. b — 15.3
5. c — 15.3
6. b — 15.4
7. b — 15.4
8. d — 15.3, 15.5
9. d — 15.5
10. c — 15.6
11. True — 15.7–15.9
12. b — 15.8, 15.10
    d — 15.2, 15.7
    e — 15.10
    f — 15.7
    h — 15.10
13. True — 15.10
14. a — 15.3
    d — 15.2
    c — 15.2
    b — 15.4
    b — 15.2
15. c — 15.5
    f — 15.7
    d — 15.10
    b — 15.1
    a — 15.6
    e — 15.6

CHAPTER 13: GENETICS PROBLEMS

1. **a.** *AB*
    b. *AB, aB*
    c. *Ab, ab*
    d. *AB, Ab, aB, ab*

2. **a.** All offspring will be *AaBB*.
    b. 1/4 *AABB*   (25% each genotype)
    1/4 *AABb*
    1/4 *AaBB*
    1/4 *AaBb*

    c. 1/4 *AaBb*   (25% each genotype)
    1/4 *Aabb*
    1/4 *aaBb*
    1/4 *aabb*

    d. 1/16 *AABB* (6.25% of genotype)
    1/8 *AaBB*   (12.5%)
    1/16 *aaBB* (6.25%)
    1/8 *AABb*   (12.5%)
    1/4 *AaBb*   (25%)
    1/8 *aaBb*   (12.5%)
    1/16 *AAbb* (6.25%)
    1/8 *Aabb*   (12.5%)
    1/16 *aabb* (6.25%)

3. **a.** *ABC*
    b. *ABC, aBC*
    c. *ABC, aBC, ABc, aBc*
    d. *ABC, aBC, AbC, abC, ABc, aBc, Abc, abc*

4. A mating of two M^L cats yields 1/4 *MM*, 1/2 M^LM, and 1/4 M^LM^L. Because M^LM^L is lethal, the probability that any one kitten among the survivors will be heterozygous is 2/3.

5. A mating between a mouse from a true-breeding, white-furred strain and a mouse from a true-breeding, brown-furred strain would provide you with the most direct evidence. Because true-breeding strains typically are homozygous for a trait being studied, all F_1 offspring from this mating should be heterozygous. Record the phenotype of each F_1 mouse, then let them mate with one another. Assuming only one gene locus is involved, these are possible outcomes for the F_1 offspring:

    a. All F_1 mice are brown, and their F_2 offspring segregate: 3 brown : 1 white. *Conclusion*: Brown is dominant to white.

    b. All F_1 mice are white, and their F_2 offspring segregate: 3 white : 1 brown. *Conclusion*: White is dominant to brown.

    c. All F_1 mice are tan, and the F_2 offspring segregate: 1 brown : 2 tan : 1 white. *Conclusion*: The alleles at this locus show incomplete dominance.

6. Yellow is recessive. Because F_1 plants have a green phenotype and must be heterozygous, green must be dominant over the recessive yellow.

7. Possible outcomes of a testcross between an F_1 rose plant heterozygous for height (*Aa*) and a shrubby rose plant:

    Gametes F_1 hybrid:

	Ⓐ	ⓐ	
Gametes shrubby plant:	ⓐ	*Aa* climber	*aa* shrubby
	ⓐ	*Aa* climber	*aa* shrubby

    1:1 possible ratio of genotypes and phenotypes in F_2 generation

8. **a.** Both parents are heterozygous (*Aa*). Their children may be albino (*aa*) or unaffected (*AA* or *Aa*).
    b. All are homozygous recessive (*aa*).
    c. Homozygous recessive (*aa*) father, and heterozygous (*Aa*) mother. The albino child is *aa*, the unaffected children *Aa*.

9. The data reveal that these genes do not assort independently because the observed ratio is very far from the 9:3:3:1 ratio expected with independent assortment. Instead, the results can be explained if the genes are located close to each other on the same chromosome.

10. **a.** 1/2 red   1/2 pink   0 white
    b. 0 red   All pink   0 white
    c. 1/4 red   1/2 pink   1/4 white
    d. 0 red   1/2 pink   1/2 white

11. Because both parents are heterozygous (*HbA/HbS*), the following are the probabilities for each child:
    a. 1/4 *HbS/HbS*
    b. 1/4 *HbA/HbA*
    c. 1/2 *HbA/HbS*

CHAPTER 14: GENETICS PROBLEMS

1. A daughter could develop DMD only if she inherited two mutated alleles, one from each parent, on each of her two X chromosomes. However, males who have the allele are unlikely to father children because they develop the disorder and die early in life.

2. Autosomal recessive. If the allele was inherited in a dominant pattern, individuals in the last generation would all have the phenotype. If it was X-linked, offspring of the first generation would all have the phenotype.

3. **a.** A male with an X-linked allele produces two kinds of gametes: one with an X chromosome (and the X-linked allele), and the other with a Y chromosome.
    b. A female homozygous for an X-linked allele produces one type of gamete, which carries the X-linked allele.
    c. A female heterozygous for an X-linked allele produces two types of gametes: one that carries the X-linked allele, and another that carries the partnered allele on the homologous chromosome.

4. 50 percent.

5. **a.** Anaphase I or anaphase II.
    b. As a result of translocation, chromosome 21 may get attached to the end of chromosome 14. The new individual's chromosome number would still be 46, but his or her somatic cells would have the translocated chromosome 21 in addition to two normal chromosomes 21.

6. A crossover between the two genes during meiosis generates an X chromosome that carries neither mutated allele.

Glossary of Biological Terms

acid Substance that releases hydrogen ions in water. **32**

activation energy Minimum amount of energy required to start a chemical reaction. **80**

activator Regulatory protein that increases the rate of transcription when it binds to a promoter or enhancer. **164**

active site Pocket in an enzyme where substrates bind and a chemical reaction occurs. **82**

active transport Energy-requiring mechanism in which a transport protein pumps a solute across a cell membrane against the solute's concentration gradient. **93**

adhering junction Cell junction composed of adhesion proteins that connect to cytoskeletal elements. Fastens animal cells to each other or to basement membrane. **71**

adhesion protein Plasma membrane protein that helps cells stick together in animal tissues. Some types form adhering junctions and tight junctions. **89**

aerobic Involving or occurring in the presence of oxygen. **117**

aerobic respiration Oxygen-requiring metabolic pathway that breaks down sugars to produce ATP. Includes glycolysis, acetyl–CoA formation, the Krebs cycle, and electron transfer phosphorylation. **118**

alcoholic fermentation Anaerobic sugar breakdown pathway that produces ATP, CO_2, and ethanol. **126**

alleles Forms of a gene with slightly different DNA sequences; may encode slightly different versions of the gene's product. **190**

allosteric regulation Control of enzyme activity by a regulatory molecule or ion that binds to a region outside the enzyme's active site. **84**

alternative splicing Post-translational RNA modification process in which some exons are removed or joined in different combinations. **153**

amino acid Small organic compound that is a subunit of proteins. Consists of a carboxyl group, an amine group, and a characteristic side group (R), all typically bonded to the same carbon atom. **46**

anaerobic Occurring in the absence of oxygen. **117**

anaphase Stage of mitosis during which sister chromatids separate and move toward opposite spindle poles. **181**

aneuploid Having too many or too few copies of a particular chromosome. **226**

animal Multicelled heterotroph that has unwalled cells, develops through a series of stages, and moves about during part or all of its life. **8**

anticodon In a tRNA, set of three nucleotides that base-pairs with an mRNA codon. **155**

antioxidant Substance that prevents oxidation of other molecules. **86**

archaea Group of single-celled prokaryotic organisms that are more closely related to eukaryotes than to bacteria. **8**

asexual reproduction Reproductive mode of eukaryotes by which offspring arise from a single parent only. **178**

atom Fundamental building block of all matter. Consists of varying numbers of protons, neutrons, and electrons. **4**

atomic number Number of protons in the atomic nucleus; determines the element. **24**

ATP Adenosine triphosphate. Nucleotide that consists of an adenine base, a ribose sugar, and three phosphate groups. Functions as a subunit of RNA and as a coenzyme in many reactions. Important energy carrier in cells. **49**

ATP/ADP cycle Process by which cells regenerate ATP. ADP forms when ATP loses a phosphate group, then ATP forms again as ADP gains a phosphate group. **87**

autosome A chromosome that is the same in males and females. **139**

autotroph Producer. Organism that makes its own food using energy from the environment and carbon from inorganic molecules such as CO_2. **101**

bacteria (singular **bacterium**) The most diverse and well-known group of prokaryotes. **8**

bacteriophage Virus that infects bacteria. **135**

Barr body Inactivated X chromosome in a cell of a female mammal. The other X chromosome is active. **168**

basal body Organelle that develops from a centriole. **69**

base Substance that accepts hydrogen ions when it dissolves in water. **32**

base-pair substitution Type of mutation in which a single base pair changes. **158**

bell curve Bell-shaped curve; typically results from graphing frequency versus distribution for a trait that varies continuously. **212**

biodiversity Of a region, the genetic variation within its species, variety of species, and variety of ecosystems. **8**

biofilm Community of microorganisms living within a shared mass of secreted slime. **59**

biology The scientific study of life. **3**

biosphere All regions of Earth where organisms live. **5**

buffer Set of chemicals that can keep the pH of a solution stable by alternately donating and accepting ions that contribute to pH. **33**

C3 plant Type of plant that uses only the Calvin–Benson cycle to fix carbon. **110**

C4 plant Type of plant that minimizes photorespiration by fixing carbon twice, in two cell types. **110**

Calvin–Benson cycle Cyclic carbon-fixing pathway that builds sugars from CO_2; the light-independent reactions of photosynthesis. **109**

CAM plant Type of C4 plant that minimizes photorespiration by fixing carbon twice, at different times of day. **111**

cancer Disease that occurs when a malignant neoplasm physically and metabolically disrupts body tissues. **185**

carbohydrate Molecule that consists primarily of carbon, hydrogen, and oxygen atoms in a 1:2:1 ratio. Complex kinds (e.g., cellulose, starch, glycogen) are polymers of simple kinds (sugars). **42**

carbon fixation Process by which carbon from an inorganic source such as carbon dioxide becomes incorporated (fixed) into an organic molecule. **109**

catalysis The acceleration of a chemical reaction by a molecule that is unchanged by participating in the reaction. **82**

cDNA Complementary strand of DNA synthesized from an RNA template by the enzyme reverse transcriptase. **235**

cell Smallest unit of life; at minimum, consists of plasma membrane, cytoplasm, and DNA. **4**

cell cortex Reinforcing mesh of cytoskeletal elements under a plasma membrane. **68**

cell cycle The collective series of intervals and events of a cell's life, from the time it forms until it divides. **178**

cell junction Structure that connects a cell to another cell or to extracellular matrix; e.g., tight junction, adhering junction, or gap junction (of animals); plasmodesmata (of plants). **70**

cell plate A disk-shaped structure that forms during cytokinesis in a plant cell; matures as a cross-wall between the two new nuclei. **182**

cell theory Theory that all organisms consist of one or more cells, which are the basic unit of life;

all cells come from division of preexisting cells; and all cells pass hereditary material to offspring. **54**

cellulose Tough, insoluble polysaccharide that is the major structural material in plants. **42**

cell wall Rigid but permeable layer of extracellular matrix structure that surrounds the plasma membrane of some cells. **59**

central vacuole Large fluid-filled organelle in many plant cells. **64**

centriole Barrel-shaped organelle from which microtubules grow. **69**

centromere Of a duplicated eukaryotic chromosome, constricted region where sister chromatids attach to each other. **138**

charge Electrical property. Opposite charges attract, and like charges repel. **24**

chemical bond An attractive force that arises between two atoms when their electrons interact; joins atoms as molecules. *See also* covalent bond, ionic bond. **28**

chlorophyll *a* Main photosynthetic pigment in plants. **102**

chloroplast Organelle of photosynthesis in the cells of plants and photosynthetic protists. Has two outer membranes enclosing semifluid stroma. Light-dependent reactions occur at its inner thylakoid membrane; light-independent reactions, in the stroma. **67**

chromatin Collective term for all of the DNA and associated proteins in a cell nucleus. **62**

chromosome A molecule of DNA together with associated proteins; carries part or all of a cell's genetic information. **138**

chromosome number The total number of chromosomes in a cell of a given species. **139**

cilium Short, movable structure that projects from the plasma membrane of some eukaryotic cells. **68**

cleavage furrow In a dividing animal cell, the indentation where cytoplasmic division will occur. **182**

clone Genetically identical copy of an organism. **133**

cloning vector A DNA molecule that can accept foreign DNA and be replicated inside a host cell. **235**

codominance Effect in which the full and separate phenotypic effects of two alleles are apparent in heterozygous individuals. **208**

codon In an mRNA, a nucleotide base triplet that codes for an amino acid or stop signal during translation. **154**

coenzyme An organic cofactor; e.g., NAD. **86**

cofactor A coenzyme or metal ion that associates with an enzyme and is necessary for its function. **86**

cohesion Property of a substance that arises from the tendency of its molecules to resist separating from one another. **31**

community All populations of all species in a given area. **5**

compound Molecule that has atoms of more than one element. **28**

concentration Amount of solute per unit volume of solution. **30**

condensation Chemical reaction in which an enzyme builds a large molecule from smaller subunits; water also forms. **41**

consumer Organism that gets energy and nutrients by feeding on tissues, wastes, or remains of other organisms; a heterotroph. **6**

continuous variation Range of small differences in a shared trait. **212**

control group Group of individuals identical to an experimental group except for the independent variable under investigation. **13**

covalent bond Type of chemical bond in which two atoms share a pair of electrons. **28**

critical thinking The act of evaluating information before accepting it. **12**

crossing over Process by which homologous chromosomes exchange corresponding segments during prophase I of meiosis. **194**

cuticle Secreted covering at a body surface. **70**

cytokinesis Cytoplasmic division; process in which a eukaryotic cell divides in two after mitosis or meiosis. **182**

cytoplasm Jellylike mixture of water and solutes enclosed by a cell's plasma membrane. **54**

cytoskeleton Network of interconnected protein filaments that support, organize, and move eukaryotic cells and their internal structures. *See also* microtubule, microfilament, intermediate filament. **68**

data Experimental results. **13**

deductive reasoning Using a general idea to make a conclusion about a specific case. **12**

deletion Mutation in which one or more nucleotides are lost from DNA. **158**

denature To unravel the shape of a protein or other large biological molecule. **48**

deoxyribonucleic acid *See* DNA.

dependent variable In an experiment, a variable that is presumably affected by an independent variable being tested. **13**

development Multistep process by which the first cell of a new multicelled organism gives rise to an adult. **7**

differentiation Process by which cells become specialized during development; occurs as different cells in an embryo begin to use different subsets of their DNA. **144**

diffusion Spontaneous spreading of molecules or ions. **90**

dihybrid cross Cross between two individuals identically heterozygous for two genes; for example, *AaBb* × *AaBb*. **206**

diploid Having two of each type of chromosome characteristic of the species (2*n*). **139**

disaccharide Carbohydrate that is a polymer of two monosaccharides. **42**

DNA Deoxyribonucleic acid. Nucleic acid that carries hereditary information. Consists of two chains of nucleotides twisted into a double helix. **7, 49**

DNA cloning Set of methods that uses living cells to mass-produce targeted DNA fragments. **235**

DNA library Collection of cells that host different fragments of foreign DNA, often representing an organism's entire genome. **236**

DNA ligase Enzyme that seals gaps in double-stranded DNA. **141**

DNA polymerase DNA replication enzyme. Uses a DNA template to assemble a complementary strand of DNA. **140**

DNA profiling Identifying an individual by analyzing the unique parts of his or her DNA. **240**

DNA replication Process by which a cell duplicates its DNA before it divides. **140**

DNA sequence Order of nucleotides in a strand of DNA. **137**

DNA sequencing *See* sequencing.

dominant Refers to an allele that masks the effect of a recessive allele paired with it in heterozygous individuals. **203**

dosage compensation Mechanism in which X chromosome inactivation equalizes gene expression between males and females. **168**

duplication Repeated section of a chromosome. **224**

ECM *See* extracellular matrix.

ecosystem A community interacting with its environment. **5**

electron Negatively charged subatomic particle. **24**

electron transfer chain Array of membrane-bound enzymes and other molecules that accept and give up electrons in sequence, thus releasing the energy of the electrons in small, usable steps. **85**

electron transfer phosphorylation Process in which electron flow through electron transfer chains sets up a hydrogen ion gradient that drives ATP formation. **106**

electronegativity Measure of the ability of an atom to pull electrons away from other atoms. **28**

electrophoresis Technique that separates DNA fragments by size. **238**

element A pure substance that consists only of atoms with the same number of protons. **24**

emergent property A characteristic of a system that does not appear in any of the system's component parts. **4**

endergonic Describes a reaction that requires a net input of free energy to proceed. **80**

endocytosis Process by which a cell takes in a small amount of extracellular fluid (and its contents) by the ballooning inward of the plasma membrane. **94**

endomembrane system Series of interacting organelles (endoplasmic reticulum, Golgi bodies, vesicles) between nucleus and plasma membrane; produces lipids, proteins. **64**

endoplasmic reticulum (ER) Membrane-enclosed organelle that is a continuous system of sacs and tubes extending from the nuclear envelope. Smooth ER makes lipids and breaks down carbohydrates and fatty acids; rough ER modifies polypeptides made by ribosomes on its surface. **64**

energy The capacity to do work. **78**

enhancer Binding site in DNA for proteins that enhance the rate of transcription. **164**

entropy Measure of how much the energy of a system is dispersed. **78**

enzyme Protein or RNA that speeds up a reaction without being changed by it. **41**

epigenetic Refers to heritable changes in gene expression that are not the result of changes in DNA sequence. **172**

epistasis Polygenic inheritance, in which a trait is influenced by multiple genes. **209**

ER *See* endoplasmic reticulum.

eugenics Idea of deliberately improving the genetic qualities of the human race. **247**

eukaryote Organism whose cells characteristically have a nucleus; a protist, fungus, plant, or animal. **8**

evaporation Transition of a liquid to a gas. **31**

exergonic Describes a reaction that ends with a net release of free energy. **80**

exocytosis Process by which a cell expels a vesicle's contents to extracellular fluid. **94**

exon Nucleotide sequence that remains in an RNA after post-transcriptional modification. **153**

experiment A test designed to support or falsify a prediction. **13**

experimental group In an experiment, a group of individuals who have a certain characteristic or receive a certain treatment as compared with a control group. **13**

extracellular matrix (ECM) Complex mixture of cell secretions; its composition and function vary by cell type. **70**

facilitated diffusion Passive transport mechanism in which a solute follows its concentration gradient across a membrane by moving through a transport protein. **92**

fat Lipid that consists of a glycerol molecule with one, two, or three fatty acid tails. Saturated fats have three saturated fatty acid tails. Unsaturated fats have one or more unsaturated fatty acid tails. **44**

fatty acid Organic compound that consists of an acidic carboxyl group "head" and a long hydrocarbon "tail." *See also* saturated fatty acid, unsaturated fatty acid. **44**

feedback inhibition Regulatory mechanism in which a change that results from some activity decreases or stops the activity. **84**

fermentation A metabolic pathway that breaks down sugars to produce ATP and does not require oxygen. **119**

fertilization Fusion of two gametes to form a zygote; part of sexual reproduction. **191**

first law of thermodynamics Energy cannot be created or destroyed. **78**

flagellum Long, slender cellular structure used for motility. **59**

fluid mosaic Model of a cell membrane as a two-dimensional fluid of mixed composition. **88**

free radical Atom with an unpaired electron; most are highly reactive and can damage biological molecules. **27**

functional group An atom (other than hydrogen) or a small molecular group bonded to a carbon of an organic compound; imparts a specific chemical property. **40**

fungus Single-cell or multicelled eukaryotic organism with cell walls of chitin; obtains nutrients by extracellular digestion and absorption. **8**

gamete Mature, haploid reproductive cell; e.g., an egg or a sperm. **190**

gap junction Cell junction that forms a closable channel across the plasma membranes of adjoining animal cells. **71**

gene A part of a chromosome that encodes an RNA or protein product in its DNA sequence. **150**

gene expression Process by which the information in a gene guides assembly of an RNA or protein product. **150**

gene therapy Treating a genetic defect or disorder by transferring a normal or modified gene into the affected individual. **246**

genetic code Complete set of sixty-four mRNA codons. **154**

genetic engineering Process by which deliberate changes are introduced into an individual's genome. **242**

genetically modified organism (GMO) Organism whose genome has been modified by genetic engineering. **242**

genome An organism's complete set of genetic material. **236**

genomics The study of genomes. **240**

genotype The particular set of alleles that is carried by an individual's chromosomes. **203**

genus (plural **genera**) A group of species that share a unique set of traits; also the first part of a species name. **10**

germ cell Immature reproductive cell that gives rise to haploid gametes when it divides. **190**

glycogen Polysaccharide that serves as an energy reservoir in animal cells. **43**

glycolysis Set of reactions in which a six-carbon sugar (such as glucose) is converted to two pyruvate for a net yield of two ATP. **120**

GMO *See* genetically modified organism.

Golgi body Membrane-enclosed organelle that modifies proteins and lipids, then packages the finished products into vesicles. **65**

growth In multicelled species, an increase in the number, size, and volume of cells. **7**

growth factor Molecule that stimulates mitosis and differentiation. **184**

haploid Having one of each type of chromosome characteristic of the species. **190**

heterotroph Organism that obtains carbon from organic compounds assembled by other organisms. **101**

heterozygous Having two different alleles of a gene. **203**

histone Type of protein that associates with DNA and structurally organizes eukaryotic chromosomes. **138**

Glossary of Biological Terms (continued)

homeostasis Process in which an organism keeps its internal conditions within tolerable ranges by sensing and responding to change. **7**

homeotic gene Type of master gene with a homeodomain; its expression directs formation of a specific body part during development. **166**

homologous chromosomes Chromosomes that have the same length, shape, and genes. In sexual reproducers, one member of a homologous pair is paternal and the other is maternal. **179**

homozygous Having identical alleles of a gene. **203**

hybrid A heterozygous individual. **203**

hydrocarbon Compound or region of one that consists only of carbon and hydrogen atoms. **40**

hydrogen bond Attraction between a covalently bonded hydrogen atom and another atom taking part in a separate covalent bond. **30**

hydrogenosome Organelle that produces ATP and hydrogen gas by an anaerobic pathway; evolved from mitochondria. **340**

hydrolysis Water-requiring chemical reaction in which an enzyme breaks a molecule into smaller subunits. **41**

hydrophilic Describes a substance that dissolves easily in water. **30**

hydrophobic Describes a substance that resists dissolving in water. **30**

hypertonic Describes a fluid that has a high solute concentration relative to another fluid separated by a semipermeable membrane. **90**

hypothesis Testable explanation of a natural phenomenon. **12**

hypotonic Describes a fluid that has a low solute concentration relative to another fluid from which it is separated by a semipermeable membrane. **90**

incomplete dominance Effect in which one allele is not fully dominant over another, so the heterozygous phenotype is an intermediate blend between the two homozygous phenotypes. **208**

independent variable Variable that is controlled by an experimenter in order to explore its relationship to a dependent variable. **13**

induced-fit model Substrate binding to an active site improves the fit between the two. **82**

inductive reasoning Drawing a conclusion based on observation. **12**

inheritance Transmission of DNA to offspring. **7**

insertion Mutation in which one or more nucleotides become inserted into DNA. **158**

intermediate filament Stable cytoskeletal element that structurally supports cell membranes

and tissues; also forms external structures such as hair. **68**

interphase In a eukaryotic cell cycle, the interval during which a cell grows, roughly doubles the number of its cytoplasmic components, and replicates its DNA in preparation for division. **178**

intron Nucleotide sequence that intervenes between exons and is removed during post-transcriptional modification. **153**

inversion Structural rearrangement of a chromosome in which part of the DNA becomes oriented in the reverse direction. **224**

ion Charged atom. **27**

ionic bond Type of chemical bond in which a strong mutual attraction links ions of opposite charge. **28**

isotonic Describes two fluids with identical solute concentrations and separated by a semipermeable membrane. **90**

isotopes Forms of an element that differ in the number of neutrons their atoms carry. **24**

karyotype Image of an individual's set of chromosomes arranged by size, length, shape, and centromere location. **139**

kinetic energy The energy of motion. **78**

knockout An experiment in which a gene is deliberately inactivated in a living organism; also, an organism that has a knocked-out gene. **166**

Krebs cycle Cyclic pathway that, along with acetyl–CoA formation, breaks down pyruvate to carbon dioxide in aerobic respiration's second stage. **122**

lactate fermentation Anaerobic sugar breakdown pathway that produces ATP and lactate. **126**

law of independent assortment During meiosis, members of a pair of genes on homologous chromosomes tend to be distributed into gametes independently of other gene pairs. **206**

law of nature Generalization that describes a consistent natural phenomenon for which there is incomplete scientific explanation. **18**

law of segregation The two members of each pair of genes on homologous chromosomes end up in different gametes during meiosis. **205**

light-dependent reactions First stage of photosynthesis; convert light energy to chemical energy. **105**

light-independent reactions Second stage of photosynthesis; use ATP and NADPH to assemble sugars from water and CO_2. A noncyclic pathway produces oxygen; a cyclic pathway does not. **105**

lignin Material that stiffens cell walls of vascular plants. **70**

linkage group All of the genes on a chromosome. **207**

lipid A fat, steroid, or wax. **44**

lipid bilayer Double layer of lipids arranged tail-to-tail; structural foundation of cell membranes. **45**

locus Location of a gene on a chromosome. **203**

lysosome Enzyme-filled vesicle that breaks down cellular wastes and debris. **64**

mass number Of an isotope, the total number of protons and neutrons in the atomic nucleus. **24**

master gene Gene encoding a product that affects the expression of many other genes. **166**

meiosis Nuclear division process that halves the chromosome number. Basis of sexual reproduction. **190**

messenger RNA (mRNA) Type of RNA that carries a protein-building message. **150**

metabolic pathway Series of enzyme-mediated reactions by which cells build, remodel, or break down an organic molecule. **84**

metabolism All of the enzyme-mediated chemical reactions by which cells build and break down organic molecules. **41**

metaphase Stage of mitosis at which all chromosomes are aligned midway between spindle poles. **181**

metastasis The process in which malignant cells spread from one part of the body to another. **185**

microfilament Cytoskeletal element composed of actin subunits. Reinforces cell membranes; functions in movement and muscle contraction. **68**

microtubule Hollow cytoskeletal element composed of tubulin subunits. Involved in movement of a cell or its parts. **68**

mitochondrion Double-membraned organelle that produces ATP by aerobic respiration in eukaryotes. **66**

mitosis Nuclear division mechanism that maintains the chromosome number. Basis of body growth and tissue repair in multicelled eukaryotes; also asexual reproduction in some multicelled eukaryotes and many single-celled ones. **178**

model Analogous system used for testing hypotheses. **13**

molecule Two or more atoms joined by chemical bonds. **4**

monohybrid cross Cross between two individuals identically heterozygous for one gene; for example, $Aa \times Aa$. **204**

monomers Molecules that are subunits of polymers. **41**

monosaccharid Simple sugar; carbohydrate that is a monomer of polysaccharides. **42**

motor protein Type of energy-using protein that interacts with cytoskeletal elements to move the cell's parts or the whole cell. **68**

mRNA *See* messenger RNA.

multiple allele system Gene for which three or more alleles persist in a population at relatively high frequency. **208**

mutation Permanent change in the DNA sequence of a chromosome. **142**

neoplasm An accumulation of abnormally dividing cells. **184**

neutron Uncharged subatomic particle in the atomic nucleus. **24**

nondisjunction Failure of sister chromatids or homologous chromosomes to separate during nuclear division. **226**

nuclear envelope A double membrane that constitutes the outer boundary of the nucleus. Nuclear pores in the membrane control the entry and exit of large molecules. **62**

nucleic acid Polymer of nucleotides; DNA or RNA. **49**

nucleic acid hybridization Spontaneous establishment of base-pairing between two nucleic acid strands. **140**

nucleoid Of a bacterium or archaeon, region of cytoplasm where the DNA is concentrated. **59**

nucleolus In a cell nucleus, a dense, irregularly shaped region where ribosomal subunits are being produced. **63**

nucleoplasm Viscous fluid enclosed by the nuclear envelope. **62**

nucleosome A length of DNA wound twice around a spool of histone proteins. **138**

nucleotide Monomer of nucleic acids; has a ribose or deoxyribose sugar, a nitrogen-containing base, and one, two, or three phosphate groups. E.g., adenine, guanine, cytosine, thymine, uracil. **49**

nucleus Of an atom; core area occupied by protons and neutrons. **24** Of a eukaryotic cell, organelle with a double membrane that holds, protects, and controls access to the cell's DNA. **55**

nutrient Substance that an organism needs for growth and survival but cannot make for itself. **6**

oncogene Gene that helps transform a normal cell into a tumor cell. **184**

operator Part of an operon; a DNA binding site for a repressor. **170**

operon Group of genes together with a promoter–operator DNA sequence that controls their transcription. **170**

organ In multicelled organisms, a structure that consists of tissues engaged in a collective task. **4**

organ system In multicelled organisms, set of organs engaged in a collective task that keeps the body functioning properly. **5**

organelle Structure that carries out a specialized metabolic function inside a cell. **55**

organic Describes a compound that consists mainly of carbon and hydrogen atoms. **38**

organism Individual that consists of one or more cells. **4**

osmosis Diffusion of water across a selectively permeable membrane; occurs in response to a difference in solute concentration between the fluids on either side of the membrane. **90**

osmotic pressure Amount of turgor that prevents osmosis into cytoplasm or other hypertonic fluid. **91**

oxidation–reduction reaction *See* redox reaction.

passive transport Membrane-crossing mechanism that requires no energy input. **92**

PCR *See* polymerase chain reaction.

pedigree Chart of family connections that shows the appearance of a trait through generations. **218**

peptide Short chain of amino acids linked by peptide bonds. **46**

peptide bond A bond between the amine group of one amino acid and the carboxyl group of another. Joins amino acids in proteins. **46**

periodic table Tabular arrangement of all known elements by their atomic number. **24**

peroxisome Enzyme-filled vesicle that breaks down amino acids, fatty acids, and toxic substances. **64**

pH Measure of the number of hydrogen ions in a fluid. Decreases with increasing acidity. **32**

phagocytosis "Cell eating"; an endocytic pathway by which a cell engulfs large particles such as microbes or cellular debris. **94**

phenotype An individual's observable traits. **203**

phospholipid A lipid with a phosphate group in its hydrophilic head, and two nonpolar fatty acid tails; main constituent of eukaryotic cell membranes. **45**

phosphorylation A phosphate-group transfer. **87**

photolysis Process by which light energy breaks down a molecule. **106**

photorespiration Inefficient sugar-production pathway initiated when rubisco attaches oxygen instead of carbon dioxide to RuBP (ribulose bisphosphate). **110**

photosynthesis Metabolic pathway by which most autotrophs use light energy to make sugars from carbon dioxide and water. **6**

photosystem Cluster of pigments and proteins that converts light energy to chemical energy in photosynthesis. **106**

pigment An organic molecule that can absorb light of certain wavelengths. Reflected light imparts a characteristic color. **102**

pilus Protein filament that projects from the surface of some bacterial and archaeal cells. **59**

pinocytosis Endocytic pathway by which fluid and materials in bulk are brought into the cell. **94**

plants Lineage of multicelled, typically photosynthetic eukaryotes adapted to life on land. **8**

plasma membrane A cell's outermost membrane; controls movement of substances into and out of the cell. **54**

plasmid Of many bacteria and archaea, a small ring of DNA replicated independently of the chromosome. **59**

plasmodesma Cell junction that forms an open channel between the cytoplasm of adjacent plant cells. **71**

plastid Double-membraned organelle that functions in photosynthesis, pigmentation, or storage in plants and algal cells; for example, a chloroplast, chromoplast, or amyloplast. **67**

pleiotropy Effect in which a single gene affects multiple traits. **209**

polarity Separation of charge into positive and negative regions. **28**

polygenic inheritance *See* epistasis.

polymer Molecule that consists of multiple monomers. **41**

polymerase chain reaction (PCR) Laboratory method that mass-produces copies of a specific section of DNA. **236**

polypeptide Long chain of amino acids linked by peptide bonds. **46**

polyploid Having three or more of each type of chromosome characteristic of the species. **226**

polysaccharide Carbohydrate that is a polymer of hundreds or thousands of monosaccharides. **42**

population A group of organisms of the same species who live in a specific location and breed with one another more often than they breed with members of other populations. **5**

potential energy Stored energy. **79**

prediction Statement, based on a hypothesis, about a condition that should exist if the hypothesis is correct. **12**

Glossary of Biological Terms (continued)

primary wall The first cell wall of young plant cells. **70**

primer Short, single strand of DNA or RNA that base-pairs with a specific DNA sequence in DNA synthesis. **140**

prion Infectious protein. **48**

probability The chance that a particular outcome of an event will occur; depends on the total number of outcomes possible. **17**

probe Short fragment of DNA designed to hybridize with a nucleotide sequence of interest and labeled with a tracer. **236**

producer Autotroph. Organism that makes its own food using energy and nonbiological raw materials from the environment. **6**

product A molecule that is produced by a reaction. **80**

prokaryote Informal name for a single-celled organism without a nucleus; a bacterium or archaeon. **8**

promoter In DNA, a sequence to which RNA polymerase binds; site where transcription begins. **152**

prophase Stage of mitosis during which chromosomes condense and become attached to a newly forming spindle. **181**

protein Organic molecule that consists of one or more polypeptides. **46**

protist General term for member of one of the eukaryotic lineages that is not a fungus, animal, or plant. **8**

proton Positively charged subatomic particle that occurs in the nucleus of all atoms. **24**

proto-oncogene Gene that, by mutation, can become an oncogene. **184**

pseudopod A temporary protrusion that helps some eukaryotic cells move and engulf prey. **69**

Punnett square Diagram used to predict the genetic and phenotypic outcomes of a cross. **204**

pupil Adjustable opening that allows light into a camera eye. **570**

pyruvate Three-carbon end product of glycolysis. **120**

radioactive decay Process by which atoms of a radioisotope emit energy and/or subatomic particles when their nucleus spontaneously breaks up. **25**

radioisotope Isotope with an unstable nucleus. **25**

reactant A molecule that enters a reaction and is changed by participating in it. **80**

reaction Process of molecular change. **41**

receptor protein Membrane protein that triggers a change in cell activity after binding to a particular substance. **89**

recessive Refers to an allele with an effect that is masked by a dominant allele on the homologous chromosome. **203**

recombinant DNA A DNA molecule that contains genetic material from more than one organism. **234**

redox reaction Oxidation–reduction reaction in which one molecule accepts electrons (it becomes reduced) from another molecule (which becomes oxidized). Also called electron transfer. **85**

repressor Regulatory protein that blocks transcription. **164**

reproduction Processes by which parents produce offspring. *See also* sexual reproduction, asexual reproduction. **7**

reproductive cloning Any of several technologies that produce genetically identical individuals. **144**

restriction enzyme Type of enzyme that cuts DNA at a specific nucleotide sequence. **234**

reverse transcriptase An enzyme that uses mRNA as a template to make a strand of cDNA. **235**

ribonucleic acid *See* RNA.

ribosomal RNA (rRNA) RNA that becomes part of ribosomes. **150**

ribosome Organelle of protein synthesis. An intact ribosome has two subunits, each composed of rRNA and proteins. **59**

RNA Ribonucleic acid. Nucleic acid with roles in gene expression; consists of a single-stranded chain of nucleotides (adenine, guanine, cytosine, and uracil). *See also* messenger RNA, transfer RNA, ribosomal RNA. **49**

RNA polymerase Enzyme that carries out transcription. **152**

rRNA *See* ribosomal RNA.

rubisco Ribulose bisphosphate carboxylase. Carbon-fixing enzyme of the Calvin–Benson cycle. **109**

salt Ionic compound that releases ions other than H^+ and OH^- when it dissolves in water. **30**

sampling error Difference between results derived from testing an entire group of events or individuals, and results derived from testing a subset of the group. **16**

saturated fatty acid Fatty acid with only single bonds linking the carbons in its tail. **44**

science Systematic study of the observable world. **12**

scientific method Making, testing, and evaluating hypotheses about the natural world. **13**

scientific theory Hypothesis that has not been disproven after many years of rigorous testing. **18**

SCNT *See* somatic cell nuclear transfer.

second law of thermodynamics Energy tends to disperse spontaneously. **78**

secondary wall Lignin-reinforced wall that forms inside the primary wall of a plant cell. **70**

semiconservative replication Describes the process of DNA replication, which produces two copies of a DNA molecule: one strand of each copy is new, and the other is conserved (parental). **141**

sequencing Laboratory method of determining the order of nucleotides in DNA. **238**

sex chromosome Member of a pair of chromosomes that differs between males and females. **139**

sexual reproduction Reproductive mode by which offspring arise from two parents and inherit genes from both. **189**

shell model Model of electron distribution in an atom. **26**

short tandem repeat In chromosomal DNA, sequences of a few nucleotides repeated multiple times in a row. Used in DNA profiling. **212**

single-nucleotide polymorphism (SNP) One-nucleotide DNA sequence variation carried by a measurable percentage of a population. **233**

sister chromatids The two DNA molecules of a duplicated eukaryotic chromosome, attached at the centromere. **138**

SNP *See* single-nucleotide polymorphism.

solute A dissolved substance. **30**

solution Uniform mixture of solute completely dissolved in solvent. **30**

solvent Liquid in which other substances dissolve. **30**

somatic Relating to the body. **190**

somatic cell nuclear transfer (SCNT) Reproductive cloning method in which the DNA of an adult donor's body cell is transferred into an unfertilized egg. **144**

species Unique type of organism designated by genus name and specific epithet. Of sexual reproducers, often defined as one or more groups of individuals that can potentially interbreed, produce fertile offspring, and do not interbreed with other groups. **10**

specific epithet Second part of a species name. **10**

spindle Temporary structure that moves chromosomes during nuclear division; consists of microtubules. **181**

starch Polysaccharide that serves as an energy reservoir in plant cells. **42**

statistically significant Refers to a result that is statistically unlikely to have occurred by chance alone. **17**

steroid Type of lipid with four carbon rings and no fatty acid tails. **45**

stomata (singular **stoma**) Closable gaps defined by guard cells on plant surfaces; when open, they allow water vapor and gases to diffuse across the epidermis. **110**

stroma Cytoplasm-like fluid between the thylakoid membrane and the two outer membranes of a chloroplast. Site of light-independent reactions of photosynthesis. **105**

substrate Of an enzyme, a reactant that is specifically acted upon by the enzyme. **82**

substrate-level phosphorylation The formation of ATP by the direct transfer of a phosphate group from a substrate to ADP. **120**

surface-to-volume ratio A relationship in which the volume of an object increases with the cube of the diameter, and the surface area increases with the square. Limits cell size. **55**

taxon (plural **taxa**) A rank of organisms that share a unique set of traits. **10**

taxonomy The science of naming and classifying species. **10**

telomere Noncoding, repetitive DNA sequence at the end of chromosomes; protects the coding sequences from degradation. **183**

telophase Stage of mitosis during which chromosomes arrive at opposite spindle poles and decondense, and two new nuclei form. **181**

temperature Measure of molecular motion. **31**

testcross Method of determining genotype of an individual with a dominant phenotype: a cross between the individual and another individual known to be homozygous recessive. **204**

theory, scientific See scientific theory.

therapeutic cloning The use of SCNT to produce human embryos for research purposes. **145**

thylakoid membrane A chloroplast's highly folded inner membrane system; forms a continuous compartment in the stroma. Site of light reactions of photosynthesis. **105**

tight junction Cell junction that fastens together the plasma membrane of adjacent animal cells; collectively prevent fluids from leaking between the cells. Composed of adhesion proteins. **71**

tissue In multicelled organisms, specialized cells organized in a pattern that allows them to perform a collective function. **4**

tracer A substance that can be traced via its detectable component. **25**

trait An inherited characteristic of an organism or species. **10**

transcription Process by which enzymes assemble an RNA using the nucleotide sequence of a gene as a template. **150**

transcription factor Regulatory protein that influences transcription by binding directly to DNA; e.g., an activator or repressor. **164**

transfer RNA (**tRNA**) RNA that delivers amino acids to a ribosome during translation. **150**

transgenic Refers to a genetically modified organism that carries a gene from a different species. **242**

transition state Point during a reaction at which substrate bonds will break and the reaction will run spontaneously. **82**

translation Process by which a polypeptide chain is assembled from amino acids in the order specified by an mRNA. **150**

translocation Structural change of a chromosome in which a broken piece gets reattached in the wrong location. **224**

transport protein Protein that passively or actively assists specific ions or molecules across a membrane. **89**

transposable element Segment of DNA that can move spontaneously within or between chromosomes. **224**

triglyceride A fat with three fatty acid tails. **44**

tRNA See transfer RNA.

tumor A neoplasm that forms a lump. **184**

turgor Pressure that a fluid exerts against a wall, membrane, or other structure that contains it. **91**

unsaturated fatty acid Fatty acid with one or more carbon–carbon double bonds in its tail. **44**

vacuole A membrane-enclosed organelle filled with fluid; isolates or disposes of waste, debris, or toxic materials. **64**

variable In an experiment, a characteristic or event that differs among individuals or over time. See also dependent variable, independent variable. **13**

vector See cloning vector.

vesicle Small, membrane-enclosed organelle; different kinds store, transport, or break down their contents. **64**

wavelength Distance between the crests of two successive waves. **102**

wax Water-repellent mixture of lipids with fatty acid tails bonded to long-chain alcohols or carbon rings. **45**

X chromosome inactivation Developmental shutdown of one of the two X chromosomes in the cells of female mammals. See also Barr body. **168**

xenotransplantation Transplantation of an organ from one species into another. **245**

zygote Diploid cell formed by fusion of two gametes; the first cell of a new individual. **191**

Index

Page numbers followed by an f or t indicate figures and tables. ■ indicates human health topics; ■ indicates environmental topics. Bold terms indicate major topics.

A

■ ABO blood typing, 208, 208f
■ Abortion, spontaneous (miscarriage), 224, 228–229
Absorption spectrum, of pigments, 104, 104f
Accessory pigments, 102, 103, 106
Acetaldehyde, 96, 126, 126f
■ Acetyl-CoA, 122, 122f, 123, 123f, 124f–125f, 128f, 129
Acetyl groups, 40, 40t, 164
■ Achondroplasia, 219t, 220, 220f
Acid(s), 32–33
■ Acid deposition, 33
Actin, 47, 68, 68f, 181f, 182, 182f, 223
Activation energy, 80–81, 81f, 82–83, 83f
Activator(s), 164
Active site, 82, 82f, 84
Active transport, 93, 93f, 201
Adaptation, evolutionary. See also Selective advantage
carbon-fixing pathways in plants as, 110–111, 110f, 111f
to extreme environments, 8, 8f
human skin color variation and, 217
of viruses, to human host, 245
Adaptive traits. See Adaptation, evolutionary; Selective advantage
Adenine (A), 49f, 136–137, 136f, 137f, 150, 151f, 152, 154, 154f, 160, 229
Adenosine monophosphate. See AMP
Adenosine triphosphate. See ATP
■ ADHD (Attention deficit hyperactivity disorder), 211
Adhering junctions, 71, 71f
Adhesion proteins, 88t, 89, 89f, 185, 249
ADP (adenosine diphosphate)
ATP/ADP cycle, 87, 87f, 93, 93f
in ATP synthesis, 124–125, 124f–125f
Aerobic respiration
ATP yield, 118, 119, 124–125, 124f–125f
electron transfer chains in, 85, 118, 118f, 124–125, 124f–125f
equation for, 118
evolution of, 112, 117
fuel sources, alternative, 128–129, 128f
mitochondria and, 66
in muscle, 127
overview, 118–119, 118f, 124f–125f
steps in, 118–119, 118f, 120–125
Africa
genome, and evidence of evolution, 229
■ **Agriculture**. See also Crops (food); Fertilizers; Pesticides
irrigation, 242
Agrobacterium, 242, 243f
■ **Air pollution**
and acid rain, 33
and mercury contamination, 23, 23f, 33, 35, 35f
Alanine, 128f, 217
Albinism, 219t, 221
Albumins, 48
Alcohol(s), functional groups, 40t
■ **Alcohol (ethanol)**
and alcoholism, 96–97, 96f
and DNA methylation, 210
hangovers, 96
and liver, 96f, 97
metabolism of, 64, 77, 77f, 96–97
Alcohol dehydrogenase (ADH), 77, 77f, 96–97

Alcoholic fermentation, 126, 126f
Aldehyde dehydrogenase (ALDH), 96
Aldehyde group, 40, 40f, 40t
Alfalfa, 243
Algae, pigments in, 104, 104f
Allele(s)
codominant, 208, 208f
defined, 190, 190f
development of, 194–195, 194f, 195f
dominant
autosomal, 218f, 219t, 220–221, 220f
defined, 203, 203f
X-linked, 219t
formation of, 190
incompletely dominant, 208, 208f
multiple allele systems, 208
notation for, 203
recessive
autosomal, 218f, 219t, 221, 221f
defined, 203, 203f
X-linked, 219t, 222–223, 222f, 223f
single-nucleotide polymorphisms, 233
■ X-linked genetic disorders, 219t, 222–223, 222f, 223f
Allosteric regulation, 84, 84f
■ Alzheimer's disease, 86, 233, 246, 247
Amine groups, 40, 40t, 46, 46f
Amino acid(s)
defined, 46
as fuel for aerobic respiration, 128f, 129
functional groups, 40, 40t
genetic code for, 154–155, 154f, 155f, 158
structure, 46, 46f
Ammonia (NH_3)
in early atmosphere, 112
as metabolic waste, 129
■ Amniocentesis, 228, 229f
Amoebas, 64, 69, 95
AMP (adenosine monophosphate), 87f
Amyloplasts, 67
Anaerobes
on early Earth, 117
Anaerobic respiration, 119
Anaphase (mitosis), 178f, 180f, 181, 182, 196f, 197
Anaphase I (meiosis), 190, 192, 192f, 194, 195
Anaphase II (meiosis), 190, 193, 193f, 196f
Androgen insensitivity syndrome, 219t
■ Anemia(s). See also Sickle-cell anemia
beta thalassemia, 158f, 159, 160
causes of, 158–159
Aneuploidy, 226–227, 227f, 228
Angiosperm(s). See also Flower(s)
polyploidy in, 226
reproductive structures, 202, 202f
■ Anhidrotic ectodermal dysplasia, 222f
Animal(s)
cell junctions, 71, 71f
cloning of, 133, 133f, 144–145, 144f, 145f
defined, 8
development
disease. See Disease, animal
gamete formation, 190, 191f
■ genetically modified, 244–245, 244f, 245f
genomes, comparison of, 240, 240f
as kingdom, 11f
polyploidy in, 226
tissue. See specific types

Animal cells
cytokinesis in, 182, 182f
mitosis in, 180f, 181
structure and components, 54–55, 54f, 60f, 61f
■ Aniridia, 167f, 219t
Antarctica
atmospheric record in ice of, 113, 113f
■ Antelope Valley Poppy Reserve, 5, 5f
Antennapedia gene, 166, 167f, 300f
Anther, 191f, 202, 202f
Anthocyanins, 103, 103f
■ **Antibiotics**, resistance to, 58f, 59, 234f, 242f
Anticodons, 155, 155f, 156
■ Antidepressants, 211
Antigen(s). See Immune system
■ Antioxidants, 86, 103
Aorta, 209, 209f
Apes, chromosomes, 225
Apoptosis, 179
Apple (Malus domestica), 10f
Aptostichus stephencolberti, 19f
Arabidopsis, 169, 169f
Arber, Werner, 234
Archaea. See also Prokaryotes
acid-loving, 99, 99f
cell membrane, 88
characteristics, 8, 8f
defined, 8
as domain, 11f, 11t
gene expression in, 157, 170
genetic code in, 155
as kingdom, 11f
structure, 58–59, 58f
Arginine, 233, 249
Aristotle, 252
Arsenic
■ exposure, and DNA methylation, 210
Arthropod(s) (Arthropoda)
exoskeleton/cuticle of, 70
Artificial twinning, 144
■ Asbestos, as mutagen, 210
■ Ascorbic acid (Vitamin C), 42, 86t
Asexual reproduction, 178
vs. sexual reproduction, 189, 197
Asia
genome, and evidence of evolution, 229
Asians
■ alcohol metabolism, 96
■ Asthma, cause of, 219
■ Atherosclerosis, 86
Atmosphere. See also Air pollution
■ carbon dioxide levels, 112–113, 112f, 113f
composition
history of, 112–113, 112f, 117
Atom(s). See also Ion(s)
defined, 4, 4f, 24
interaction
atomic structure and, 26–27
bonds, types of, 28–29, 28f, 29f
as level of organization, 4, 4f
structure, 24, 26–27, 26f, 27f
Atomic number, 24, 24f
Atomic theory, 18, 18t
ATP (adenosine triphosphate)
ATP/ADP cycle, 87, 87f, 93, 93f
as coenzyme, 86–87, 87f
as energy carrier, 49, 86–87
functions
cross-membrane transport, 93, 93f
cytokinesis, 182
glycolysis, 120, 121f
photorespiration, 110, 110f

photosynthesis, 108–109, 109f
vesicle formation, 94
phosphorylation, 87, 87f
structure, 49, 49f, 86–87, 87f
ATP/ADP cycle, 87, 87f, 93, 93f
ATP synthases, 106, 106f, 107f, 124, 124f–125f
ATP synthesis
in aerobic respiration, 66, 85, 118–119, 118f, 122, 122f, 123, 123f, 124–125, 124f–125f, 128–129, 128f
in alcoholic fermentation, 126, 126f
■ in cancer cells, 185
in chloroplasts, 67
■ disorders, 117, 117f
in fermentation, 118f, 119
in glycolysis, 120, 121f
in lactate fermentation, 126–127, 127f
in mitochondria, 66
in muscle, 127, 127f
in photosynthesis, 105, 106–107, 106f, 107f
■ Attention deficit hyperactivity disorder (ADHD), 211
■ Autism, 211, 219, 249, 249f
■ **Autoimmune disorders**
multiple sclerosis, 219, 245
Autosomal aneuploidy, 226–227
Autosomal dominant inheritance, 218f, 219t, 220–221, 220f
Autosomal recessive inheritance, 218f, 219t, 221, 221f
Autosome(s), 139, 139f
Autotrophs, 101. See also Producers
Avery, Oswald, 134

B

B cell. See B lymphocyte(s)
B lymphocyte(s) (B cells)
antigen receptors, 89f
Bacillus thuringiensis (Bt), 242, 243f
Bacteriophage
in DNA research, 135, 135f, 147, 147f, 234
Bacterium (bacteria). See also Cyanobacteria; Prokaryotes
■ antibiotic resistance in, 58f, 59, 234f, 242f
bacteriophage resistance, 234
biofilms, 59, 59f, 75
characteristics, 8, 8f
defined, 8
■ in dental plaque, 8f, 59, 59f
in DNA cloning, 234–235, 234f, 235f
in DNA research, 134, 134f, 135, 135f
as domain, 11f, 11t
gene expression, 157, 170–171, 170f
genetic code in, 155
■ in intestines, 53, 53f, 96, 170–171
as kingdom, 11f
magnetotactic, 8f
■ as pathogen, 53, 53f
photosynthetic. See Cyanobacteria
pigments produced by, 103f
reproduction, 59
size and shape, 56f
structure, 54f, 58–59, 58f
transgenic, 242, 243f
Baker's yeast (Saccharomyces cerevisiae), 126, 126f
Ball-and-stick models, 38f, 39
Bamboo, 110–111
Barley (Hordeum vulgare), 111f, 126
Barr bodies, 168, 168f, 222
Basal body, 69, 69f
■ Basal cell carcinoma, 185f
Base(s), characteristics of, 32–33

Base-pair substitutions, 158–159, 158f, 159f, 221, 233
Base pairing
in DNA, 137, 137f, 151f
in DNA replication, 140–141, 140f, 141f, 152, 152f
in RNA, 151f
in transcription, 152–153, 152f
in translation, 156–157, 157f
Base sequence, in DNA, 137, 137f
Bdelloid rotifers, 197, 197f
Beehler, Bruce, 3
Beer, 126, 126f, 242
Behavioral traits, 10, 11, 219t
Bell curve, phenotype variation and, 212, 212f
Benson, Andrew, 25
Berg, Paul, 245
Beta-carotene, 103, 103f, 104f, 243
Beta globin, 158, 158f, 159f, 190
Beta thalassemia, 158f, 159, 190
Bias in data interpretation, 17, 19
Bicarbonate (HCO₃-)
as buffer, 33
Big Bang theory, 18t
Bile, 45
Binge drinking, 77, 97
Biochemical traits, 10, 11
Biodiversity, 8, 8f, 9f
DNA and, 137
in humans, 217–225
Biofilms, 59, 59f, 75
Biofuels, 101, 101f, 113, 115, 115f
Biogeochemical cycles
carbon cycle, 118, 118f
Biological weapons, 149
Biology
defined, 3
Bioluminescence, 85f
Biosphere
defined, 5, 5f
as level of organization, 5, 5f
Biotin (vitamin B₇), 86t
Bipolar disorder, 211
Bird(s)
sex determination in, 139
Birth defects
incidence, maternal age and, 228
screening for, 228–229, 228f, 229f
Bisphenol A (BPA), 199, 199f
Blair, Tony, 239
Blending inheritance, 202
Blood
components of, 525, 525f, 628–629, 628f, 629f
as connective tissue, 525, 525f, 628
disorders. See Sickle-cell anemia
pH levels, 33
transfusions, 208
typing, 208, 208f
Blood cells. See Red blood cells; White blood cells
Blood clotting
disorders, 223, 223f, 244, 244f. See also Hemophilia
Blue tit, 15f
Body height, human, variation in, 212, 212f
Boivin, André, 135
Bond(s). See Chemical bonds
Bond, Jason, 19f
Bone(s)
cancer of, 246
as extracellular matrix, 70
Bovine spongiform encephalopathy (BSE), 48
BPA. See Bisphenol A
BRCA genes, 173, 184, 184f, 197, 233

Breast
neoplasms in, 184f
tissue, growth factor regulation, 173
Breast cancer
gene mutations and, 173, 233f
incidence, 163
treatment, 163
Broadhead, Ivy, 220f
BSE. See Bovine spongiform encephalopathy
Bt (Bacillus thuringiensis), 242, 243f
Bt gene, 242
Buffer system, 33
Bundle-sheath cells, 110f, 111, 111f
Burkitt's lymphoma, 224
Heliconius, 11f
and natural selection, 14–15, 15f, 15t, 17f, 21
sex determination in, 139

C
C3 plants, 110–111, 110f, 111f
C4 plants, 110–111, 110f, 111f
Cactus
water-conserving adaptations, 111
Calcium, in muscle function, 223
Calcium pumps/channels, 92, 93, 93f
California poppy (Eschscholzia californica), 5, 5f
Calvin, Melvin, 25
Calvin-Benson cycle, 105f, 108–109, 109f, 110–111, 110f, 111f
CAM plants, 111, 111f
Camphor tree seeds, 149f
Camptodactyly, 219t
Cancer. See also Tumor(s)
bone marrow, 246
gene mutations and, 173, 233f
incidence, 163
treatment, 163
carcinomas, 185f
cause of, 219
cervical, 177
gene therapy and, 246
genetic controls loss and, 163, 173
genetic mutation and, 142, 143, 163, 173, 177f, 184–185, 246
of immune system, 224
leukemias, 219t, 246
lymphomas, 224
metastasis, 185, 185f
ovarian, 143, 173
oxidative damage and, 86
research on, 160, 161, 161f, 186, 187, 245
risk factors, 185
screening for, 185
skin, 143, 185, 185f, 221
treatment, 163, 173, 185, 186, 233, 244, 246
Cancer cells, 163
chromosome number in, 185, 187, 187f
immortality of, 177
telomerase levels, 183
Cannabis. See Marijuana
Capsule, bacterial, 58f, 59
Carbohydrates. See also Monosaccharides; Polysaccharides
breakdown pathways, 42–43, 118–119, 118f. See also specific pathways
conversion to fat, 129
digestion of, 128–129, 128f
metabolism of, 42–43, 118–119, 118f
photosynthesis and, 101
structure, 42
types, 42–43, 43f
Carbon. See also Organic molecules
atomic number, 24, 24f

as basis of life, 101
bonding behavior, 38, 38f, 40f, 44, 44f
in human body, 112
isotopes, 25
structure, 26f, 38
Carbon cycle, 118, 118f
atmospheric carbon dioxide and, 112–113, 112f
human disruption of, 112–113
photosynthesis and, 112–113
Carbon dioxide
and acid-base balance, 33
in aerobic respiration, 118, 118f, 122, 122f, 123, 123f, 124f–125f
atmospheric, 112–113, 112f, 113f
in fermentation, 126, 126f
in photorespiration, 110
in photosynthesis, 105, 105f, 109, 109f, 110, 110f, 111, 111f
Carbon fixation, 109, 109f
variations, in C3, C4, and CAM plants, 110–111, 110f, 111f
Carbonic acid, 33
Carcinomas, 185f
Cardiovascular disease
atherosclerosis, 86
Cardiovascular system. See Circulatory system
Carotenes, 104f
Carotenoids, 67, 103f
Carpel, plant 169, 169f, 189, 202, 202f
Carroll, Lewis, 189
Carrot (Daucus carota), 10f, 103
Casein, 127
Castor-oil plant (Ricinus communis), 149, 149f
Cat(s)
calico, 168f
Catalase, 86
Catalysis, 82–83, 82f
cDNA (complementary DNA)
assembly of, 235
cloning of, 235
and DNA libraries, 236
Celera Genomics, 239
Cell(s). See also Animal cells; Cell membrane; Nucleus, cellular; Plant cells
cytoskeleton, 68–69, 68f
defined, 4, 4f, 54
as defining feature of life, 72
differentiation. See Differentiation
discovery of, 54
eukaryotic
cell junctions, 70–71, 71f
structure and components, 54–55, 54f, 60, 60f, 60t, 61f, 62–69
functions of, 54
human
division outside body, 177
research on, 145, 177, 177f
as level of organization, 4, 4f
prokaryotic, 54–55, 54f, 56f, 58–59, 58f
Cell cycle
checkpoints and control mechanisms, 178f, 179, 183, 184–185, 184f, 186, 197
overview, 178–179, 178f
Cell division. See Meiosis; Mitosis
Cell junctions, types, 70–71, 71f. See also specific types
Cell membrane, 88–89. See also Lipid bilayer
archaeans, 88
fluid mosaic model, 88
membrane trafficking, 94–95, 94f, 95f

of organelle, 60
proteins, 54, 55f, 63, 63f, 88–89, 88f–89f, 92–93, 92f, 93f, 95, 95f
selective permeability in, 90–91, 90f, 91f
structure, 88–89, 88f–89f
transport across, 92–93, 92f, 93f
Cell plate, 182, 182f
Cell theory, 18t, 54, 54t
Cell wall
eukaryotes, 60f, 61f, 70, 70f
prokaryotes, 58f, 59
Cellulose
in plant cells, 42, 80, 81f
structure, 42, 43f
Central vacuole, 61f, 64
Centriole, 60t, 61f, 69, 69f, 181
Centromeres, 138, 138f, 140, 179, 186, 194f, 197f
Centrosomes, 180f, 181, 181f
Cervical cancer, 177
Cervix, 184
Charalambous, Haris, 209f
Chargaff, Erwin, 136
Chargaff's rules, 136–137
Charge
of molecule, and diffusion rate, 90
of subatomic particles, 24, 24f
Chase, Martha, 135, 135f, 147, 147f
Checkpoint(s), in cell cycle, 178f, 179, 184, 186, 197
Cheese, manufacture of, 242
Chemical bonds
covalent bond, 28–29, 29f, 29t
defined, 28
energy storage and release in, 78–79, 80–81, 80f, 81f, 101
hydrogen bond, 30–31, 30f
ionic bond, 28, 28f, 29
notation, 28–29, 29t, 38–39, 38f, 39f
Chemical equation, 80, 80f
Chemical formula, 28t
Chemical name, of molecule, 28t
Chemical reaction
activation energy, 80–81, 81f
defined, 41, 80
endergonic, 80–81, 80f, 81f, 87
equation for, 80, 80f
exergonic, 80–81, 80f, 81f, 87
transition state, 82–83, 83f
Chemical signals. See Hormone(s), animal
Chemoautotrophs, 112
Cheung, Melissa, 161, 161f
Chicken(s)
genome, vs. other species, 240f
muscle fibers in, 127
sex chromosomes, 139f
transgenic, 244
Chimpanzee(s) (Pan troglodytes)
chromosomes, 225, 225f
Chinn, Mari, 101f
Chitin
in arthropods, 43, 43f, 70
structure, 43, 43f
Cladophora (green algae), 104, 104f
Chloride ions, 201, 213
Chlorine
atomic structure, 26f, 27, 27f
in early atmosphere, 112
Chlorophylls
chlorophyll a, 102–103, 103f, 104f, 106
chlorophyll b, 104f
in light-harvesting complex, 106
in photosynthesis, 67, 106–107, 108f
types and characteristics, 103, 103f

Chloroplast(s), 111, 111*f*
  defined, 105
  in eukaryotic cell, 60*f*, 61*f*
  functions, 60*t*, 67, 67*f*, 105, 109, 109*f*
  genetic code in, 155
  structure, 105, 105*f*, 106, 106*f*
Cholesterol
  in cell membrane, 88
  ■ dietary, 37, 51, 51*f*
  ■ HDL (high density lipoprotein), 47, 47*f*, 51, 51*f*
  ■ LDL (low density lipoprotein), 51, 51*f*, 94
  structure, 45
  Chorion, 228, 229*f*
  Chorionic villi sampling (CVS), 228–229, 229*f*
Chromatid(s), sister
  in DNA replication, 138, 138*f*, 140
  in meiosis, 191, 191*f*, 192–193, 192*f*–193*f*, 195, 196–197, 196*f*, 207*f*
  in mitosis, 179, 179*f*, 180*f*, 181, 186, 196–197, 196*f*
  Chromatin, 62, 62*f*, 62*t*, 63*f*, 164
  Chromoplasts, 67, 67*f*
Chromosome(s)
  in DNA replication, 140
  eukaryotic, structure of, 138–139, 138*f*, 139*f*
  evolution of, 225, 225*f*
  in HELA cells, 177*f*
  homologous, 179, 190, 190*f*, 192, 197*f*, 203, 204, 204*f*, 206–207
  human, 218–219
    vs. ape or monkey chromosomes, 225, 225*f*
  number of, 138, 225, 226–227, 226*f*, 227*f*
  number of nucleotides, 158, 233, 239
  sex chromosomes, 168–169
  karyotyping, 224
  locus of gene on, 203, 203*f*
  in meiosis, 192–193, 192*f*–193*f*, 196–197, 196*f*
  in mitosis, 179, 179*f*, 180*f*, 181, 181*f*, 186, 196–197, 196*f*
  ■ mutations in. *See also* Mutation(s)
    and allele creation, 190
    chromosome number changes, 219*t*, 226–227
    deletions, 201, 224, 224*f*
    duplications, 224, 224*f*
    inversions, 224, 224*f*
    translocations, 224, 224*f*, 240
    types of, 158, 158*f*
  polytene, 164
  radiation damage to, 142*f*
  segregation into gametes, 194–195, 195*f*, 204, 204*f*, 206–207, 207*f*
  structure
    evolution of, 225, 225*f*, 227, 227*f*
    heritable changes in, 219*t*, 224, 224*f*
    telomeres, 183, 183*f*
Chromosome number, 139
  ■ abnormalities, 191, 219*t*, 226–227, 226*f*, 227*f*
  of cancer cells, 185, 187, 187*f*
  cell division and, 179, 179*f*
  changes in, 219*t*
    and diversity, 219*t*
  diploid, 139, 179, 179*f*, 190, 191, 191*f*, 192, 203
  fertilization and, 191
  haploid, 190–191, 191*f*, 192–193, 192*f*–193*f*, 195, 196
  meiosis and, 190–191, 191*f*

■ Chronic myelogenous leukemia (CML), 219*t*
Cilium (cilia)
  structure, 68–69, 69*f*
Circulatory system. *See* Heart
■ Cirrhosis, 77, 96*f*, 97
  Citrate, 123, 123*f*
  Citric acid cycle. *See* Krebs cycle
  CJD. *See* Creutzfeldt-Jakob disease
  Classification. *See* Taxonomy
  Cleavage furrow, 182, 182*f*
■ **Climate.** *See* Global warming
■ **Climate change, global**
  atmospheric carbon dioxide and, 113
  fossil fuels and, 112–113
  Clinton, William "Bill," 239
Cloning
  of animals, 133, 133*f*, 144–145, 144*f*, 145*f*
  of cDNA, 235
  conservation efforts and, 183
  of DNA, 234–235, 234*f*, 235*f*
    vectors, plasmids as, 234–235, 234*f*, 235*f*, 242, 243*f*
  as ethical issue, 145
  health problems in clones, 145, 183
  ■ of humans, 145
    potential benefits of, 133, 145
    reproductive, 144–145, 144*f*, 145*f*
    reprogramming of DNA in, 133, 144
    telomeres and, 183
  ■ therapeutic, 145
  Cloning vectors, 234–235, 234*f*, 235*f*
  Clotting. *See* Blood clotting
  Cloud forests, New Guinea, 3, 3*f*, 16, 16*f*
  CoA (coenzyme A), 86*t*, 122, 123*f*
  Coal
  ■ and air pollution, 23, 23*f*, 35, 35*f*
  ■ Cobalamin (Vitamin B$_{12}$), 171
  Codominance, 208, 208*f*
  Codon(s), 154–155, 154*f*, 155*f*, 158–159
  Coenzymes, 86–87, 86*f*, 86*t*, 87*f*
    ATP as, 86–87, 87*f*
    coenzyme A (CoA), 86*t*, 122, 123*f*
    coenzyme Q$_{10}$, 86*f*, 86*t*
    defined, 86
    functional groups in, 40*f*
  Cofactors, 86–87
  Cohesion
    defined, 31
    of water molecules, 31
  Collagen, in bone, 71
  Collins, Francis, 239
  ■ Color blindness, 203*f*, 219*t*, 222–223, 222*f*
  ■ **Community**
    defined, 5, 5*f*
    as level of organization, 5, 5*f*
  Complementary DNA. *See* cDNA
  Complex carbohydrates. *See* Polysaccharides
  Compound, defined, 28
  Concentration
    defined, 30
    and diffusion rate, 90
  Condensation reaction, 41, 41*f*, 42, 46*f*–47*f*
  Consumers
    defined, 6
    energy acquisition by, 101
    role in ecosystem, 6, 6*f*, 79, 79*f*
  Continuous variation, in traits, 212–213, 212*f*, 213*f*
  Contractile vacuoles, 57*f*
  Control group, 12, 14, 14*f*
  Copernicus, Nicolaus, 19
  Corn (Zea mays)
  ■ as biofuel, 101, 115, 115*f*

  genetically engineered, 242, 243, 243*f*
  Cotton, 243
  Covalent bond
    defined, 28–29, 29*f*
    notation, 28–29, 29*t*, 39
  Crab(s)
    cuticle of, 70
    as detritivores, 63
  Crabgrass, 111*f*
  Crassulacean acid metabolism. *See* CAM plants
  ■ Creutzfeldt-Jakob disease (CJD), 48, 48*f*
  ■ Cri-du-chat syndrome, 219*t*, 224
  Crick, Francis, 136–137, 137*f*, 143, 245
  Critical thinking, 12, 17
  Crops (food)
  ■ genetically engineered, 242–243, 243*f*
    pests of, 242–243, 243*f*,
  Crossing over, of chromosomes 194, 194*f*, 197, 206–207, 224, 225, 225*f*
  Cuticle
    arthropod, 70
    plant, 70, 110
  Cyanobacteria, 8*f*
    structure of, 58*f*
  Cyclic pathway, in photosynthesis, 106*f*, 107, 108, 112
  Cysteine, 249
  Cystic fibrosis (CF), 201, 201*f*, 203*f*, 209, 213, 215, 215*f*, 219, 219*t*, 228, 244, 245
  Cytochrome c, 203*f*
  Cytochrome P450, 89*f*
  Cytokinesis (cytoplasmic division), 178, 178*f*, 179, 179*f*, 182, 182*f*, 193
  Cytoplasm, 54–55, 54*f*, 63*f*
    of cancer cell, 185
    fermentation in, 118*f*, 119, 126
    glycolysis in, 118, 118*f*, 119, 120, 121*f*, 126
    prokaryotic cell, 58*f*, 59
  Cytoplasmic division. *See* Cytokinesis
  Cytosine (C), 136–137, 136*f*, 137*f*, 150, 151*f*, 152, 154, 154*f*, 172, 172*f*, 229, 233
  Cytoskeleton, 60*t*, 61*f*, 68–69, 68*f*, 69*f*, 185

D
Darwin, Charles, 202
  Data, in scientific process, 13, 13*t*, 16–17
■ **Decomposers**
  fungi as, 9*f*
  role of, 6
  Deductive reasoning, 12
■ Deletion mutation, 158, 158*f*, 159, 201, 224, 224*f*
  Δ*F508* mutation, 201, 213, 215, 215*f*, 219
  Denaturation, protein, 48, 83
  Deoxyribonucleic acid. *See* DNA
  Dependent variable, 13, 14
■ Depression, 211
  Deuterium, 25
  Development. *See also* Differentiation
    defined, 7
    as a feature of life, 7, 72
    fruit fly, 166–167, 166*f*, 167*f*
■ **Diabetes mellitus**
    cause of, 219
    genetic factors in, 246
    oxidative damage and, 86
    research on, 245
    treatment of, 242

■ **Diet.** *See also* Food; Nutrition
    cholesterol in, 37, 51, 51*f*
    and DNA methylation, 210
    and epigenetic inheritance, 172–173, 172*f*, 175, 175*f*
    fiber, 42
  Differentiation
    cloning and, 144
    control of, 166–167, 166*f*, 167*f*
    DNA methylation in, 172
    gene expression and, 164
  Diffusion, 90, 90*f*
    across cell membrane, 90–91, 91*f*, 92, 92*f*
  Digestion
    carbohydrates, 128–129, 128*f*
    fats, 64, 65
  Disaccharides, 42
  Disease. *See also* Autoimmune disorders; Cancer; Genetic disorders; Pathogen(s); *specific diseases*
    animal
      bovine spongiform encephalopathy (BSE), 48
    ■ human
      bacterial, 53, 53*f*
      mitochondrial, 117, 117*f*, 129, 131, 131*f*
      prion, 48, 48*f*
      research on, 177, 177*f*, 245
      xenotransplantation and, 245
    infectious. *See* Pathogen(s)
    oxidative damage and, 86
    plant
      GMO crops and, 242
  Diversity. *See* Biodiversity; Genetic diversity
  DNA (deoxyribonucleic acid), 138*f*. *See also* Chromosome(s)
    base pairings, 137, 137*f*, 151*f*
      in DNA cloning, 234, 234*f*
      in DNA replication, 140–141, 140*f*, 141*f*, 152, 152*f*
      in transcription, 152–153, 152*f*
      in translation, 156–157, 157*f*
    base sequence, 137, 137*f*
    breakage hot spots, 240
    chloroplast, 67
    cloning of, 234–235, 234*f*, 235*f*
    compared to RNA, 150, 151*f*
    conserved regions, 233
    as defining feature of life, 7, 72
    deletions, 224, 224*f*
    duplications, 224, 224*f*
    and epigenetic mechanisms, 172–173, 172*f*
    eukaryotic, 54, 55, 61*f*, 62, 62*t*, 63*f*.
    fragments, copying/multiplication of, 236–237, 237*f*
    functions, 7, 150, 151*f*
      discovery of, 134–135, 134*f*, 135*f*
    genetic code and, 154–155, 154*f*, 158
    genetic information, nature of, 150
    individual, 233, 240–241, 241*f*
    ■ inversions in, 224, 224*f*
    length of, 138
    ■ methylation of, 172–173, 172*f*, 210, 211
    mitochondrial, 66, 129
    prokaryotic, 54, 55*f*, 58*f*, 59
    recombinant, 234–235, 234*f*, 235*f*
    repair mechanisms, 142–143, 142*f*, 173, 184, 197, 197*f*
    reprogramming of, in cloning, 133, 144
    research on, 134–137, 147, 147*f*, 202, 234
    single-nucleotide polymorphisms, 233
    size of molecule, 62

"sticky ends" of, 234, 234f, 235f
structure
    base pairings, 137, 137f, 151f, 152, 152f
    discovery of, 136–137, 137f, 143
    double helix, 49, 49f, 136–137, 137f
    functional groups, 40, 40t
    nucleotides, 49, 49f, 136–137, 136f, 137f, 150, 150f, 151f
    sugar-phosphate backbone, 137, 137f, 151f
    transcription. See Transcription
    translocation of, 224, 224f, 240
    transposable elements in, 224
DNA chips. See SNP chip analysis
DNA fingerprinting, 241, 241f
DNA libraries, 236
DNA ligase, 140f, 141, 141f, 234, 234f, 235f
DNA polymerase, 140–141, 140f, 141f
    DNA repair mechanisms and, 142–143
    in DNA sequencing, 238, 238f
    genetic code variations by species, 240f
    genetic mutation and, 142–143
    in polymerase chain reaction, 237, 237f
    proofreading by, 142
    Taq polymerase, 237, 237f
DNA profiling, 240–241, 241f
DNA replication
    in cell cycle, 178, 178f, 179, 180f, 181
    cell cycle checkpoints and, 184
    in meiosis, 190, 192
    mutations and, 142–143, 142f
    overview, 138
    PCR and, 236–237, 237f
    process, 140–141, 140f, 141f
    speed of, 142
DNA sequencing, 238–239, 238f, 239f
    automated, 239, 239f
    Human Genome Project, 239, 239f
DNA testing. See Genetic testing
Dog(s)
    cloning of, 133, 133f, 145
    face length in, 212, 212f
    genome, vs. other species, 240f
Dog rose (Rosa canina), 10, 10f
Dolly (cloned sheep), 145, 183
Domain, as classification, 11f
Domain, of protein, 46, 47f
Dominant allele
    autosomal dominant inheritance, 218f, 219t, 220–221, 220f
    defined, 203, 203f
    X-linked disorders and abnormalities, 219t
Dosage compensation, 168
Down syndrome (Trisomy 21), 219t, 226, 227f
Drosophila melanogaster
    development, 166–167, 166f, 167f
    in genetic research, 166–167, 166f, 167f
Drought, GMO plants and, 242
Drug(s), antidepressants, 211
Drug abuse
    and DNA methylation, 210
Duchenne muscular dystrophy (DMD), 223, 224
Dunce gene, 166
Duplication, of DNA, 224, 224f
Dwarfism, hereditary, 220, 220f
Dynein, 68, 69f, 75
Dysentery, 140
Dystrophin, 222f

E
EcoRI enzyme, 234f
Ecosystem(s)
    defined, 5, 5f
    energy flow in, 6–7, 6f, 79, 79f
    as level of organization, 5, 5f
    nutrient cycling in, 6, 6f. See also Biogeochemical cycles
EGFR gene, 184
Egg. See Ovary(ies)
Elastin, 203f
Electromagnetic energy
    and genetic mutation, 142–143, 142f
Electromagnetic field
    animal's ability to detect, 8f
Electron(s)
    in atomic structure, 24, 24f, 26–27, 26f, 27f
Electron microscope, 56f–57f, 57
Electron transfer, 84–85, 85f
Electron transfer chains
    in aerobic respiration, 85, 118, 118f, 124–125, 124f–125f
    cell membrane and, 89, 89f
    defined, 85, 85f
    disorders of, 117
    in photosynthesis, 85, 106–107, 107f, 108, 108f
Electron transfer phosphorylation, 106–107, 118, 118f, 124–125, 124f–125f, 128f, 129
Electronegativity, 28–29
Electrophoresis, 238–239, 238f
Element(s). See also Periodic table
    atomic structure, and chemical reactivity, 27
    defined, 24
    isotopes, 24–25
    symbols for, 24, 24f
Ellis–van Creveld syndrome, 218f, 219t
Embryo
    bird, 27
    frog, 178f
    prenatal screening of, 228–229, 229f
    in vitro fertilization, 229, 229f
    splitting, 144
Emergent properties, 4
Encyclopedia of Life (website), 19
Endergonic chemical reaction, 80–81, 80f, 81f, 87
Endocrine system. See Hormone(s), animal
Endocytosis, 94–95, 94f, 95f, 201, 215
Endomembrane system, 64–65, 64f–65f
Endoplasmic reticulum (ER), 60, 60f, 62, 64–65, 71f
    functions, 60t, 95, 95f, 201
    rough, 61f, 64, 64f–65f
    smooth, 61f, 64–65, 64f–65f
Energy
    for cellular work, 84–85, 85f. See also Aerobic respiration; ATP; Fermentation; Oxidation-reduction (redox) reactions
    in chemical bonds, 78–79, 80–81, 80f, 81f, 101
    conservation of, 78
    defined, 78
    from dietary molecules, 128–129, 129f
    dispersal of, 78, 79, 79f
    energy metabolism, 80–81, 81f
    entropy and, 78, 78f
    first law of thermodynamics, 78
    flow, in photosynthesis, 108, 108f

flow through ecosystems, 6–7, 6f, 79, 79f
    forms of, 78
    free, 80, 83
    human use
        fossil fuels. See Fossil fuels
    kinetic, 78
    losses in transfer of, 78–79, 78f
    photosynthesis as source of, 101
    potential, 78, 79, 79f
    second law of thermodynamics, 78, 79
    sunlight as source of, 6f, 78, 79, 79f, 102–103, 102f, 103f, 104, 104f. See also Photosynthesis
    use of, as characteristic of life, 72
Engelmann, Theodor, 104, 104f, 349
Enhancers, 164, 164f–165f
Entropy, 78, 78f, 79, 90. See also Second law of thermodynamics
Environment. See also Epigenetic mechanisms
    and DNA methylation, 172
    and genetic mutation, 142–143, 142f
    human impact on. See Global warming
        agriculture and, 242. See also Agriculture
        transgene escape and, 243, 245
    and phenotype, 210–211, 210f, 211f
    and sex determination in animals, 210, 225, 225f
Environmental factors
    in phenotype, 210–211, 210f, 211f
    in psychiatric disorders, 211
Enzyme(s), 82–83
    action, 82, 82f, 84
    in cell membrane, 88t, 89, 89f
    defined, 41
    in DNA repair, 142, 142f
    environmental factors affecting, 83, 83f, 91
    genetic engineering and, 242
    induced-fit model, 82
    liver, 77
    metabolism and, 84–85, 84f
    in protein synthesis, 47
    as proteins, 46, 82
    regulation of, 84–85, 84f
    restriction, 234, 234f, 235f
    RNA and, 82
    stomach, 83, 83f
    structure of, 47, 82, 82f, 86
    synthesis of, 64–65
Epidermal growth factor (EGF), 184, 184f
Epigenetic mechanisms, 172–173, 172f, 175, 175f, 210–211, 210f, 211f, 219
    genetic testing and, 247
Epistasis, 209, 209f, 212, 213
Epithelial tissue
    cystic fibrosis and, 201
    Epithet, specific, 10
Equation, chemical, 80, 80f
Erosion
    agriculture and, 242
Error bars, 17, 17f
Erythrocytes. See Red blood cells
Escherichia coli, 58f, 83f, 149, 149f
    gene expression, control of, 170–171, 170f
    and lactose intolerance, 171
    pathogenic, 53, 53f, 149f
    transgenic, 242, 242f
Essential fatty acids, 44f
Estrogen(s)
    functions, 169
    as steroid, 45f

Estrogen receptors
    BRCA proteins and, 173
Ethanol. See Alcohol (ethanol)
Ethical issues
    biofuels vs. fossil fuels, 113
    extinctions, current rate of, 3
    genetic engineering, 243, 244–245, 246–247
    mercury contamination, 23, 33
    organ transplant screening, 97
    as outside of science's purview, 19
    patenting of Human Genome, 239
Ethyl alcohol. See Alcohol (ethanol)
Eugenics, 247
Eukaryotes
    defined, 8
    as domain, 11f, 11t
    overview, 8, 9f
Eukaryotic cells
    aerobic respiration and, 118
    cell junctions, 70–71, 71f
    division of, 178f, 182, 182f
    size of, 56f
    structure and components, 54f, 55, 60, 60f, 60t, 61f, 62–69
Evans, Rhys, 246, 246f
Evaporation, process, 31
Even-skipped gene, 166f
Evolution. See also Adaptation, evolutionary; Natural selection
    of aerobic respiration, 112, 117
    atmosphere and, 112, 117
    of chromosomes, 225, 225f
    defined, 18t
    evidence for, genetics, 225, 225f, 230, 240, 240f
    genetic code and, 155
    of homeotic genes, 167
    of humans, 225, 225f
    of mitochondria, 66
    mutations and, 225, 225f
    of photosynthesis, 108, 112
    of reptiles, 225f
    as scientific theory, 18, 18t
    of sexual reproduction, 196–197
    of viruses, 245
    of Y chromosome, 225, 225f
Evolutionary tree diagrams (cladograms), 295, 295f
Exergonic chemical reaction, 80–81, 80f, 81f, 87
Exocytosis, 94, 94f, 95
Exons, 153, 153f, 164f–165f
Experiment(s), in scientific process, 13, 13t
Experimental group, 12, 13, 14, 14f
Extinction(s)
    current rate of, 3
Extracellular matrix (ECM), 70, 70f, 71, 89f
Extreme thermophiles, 8, 8f
Eye(s). See also Vision
    human
        color variations in, 203f, 212–213, 213f
Eyeless gene, 167, 167f

F
Facilitated diffusion, 92, 92f. See also Passive transport
FAD (flavin adenine dinucleotide), 86t, 122, 123, 123f
Familial hypercholesterolemia, 219t
Famine, and epigenetic inheritance, 172–173, 172f, 175, 175f
Fanconi anemia, 219t
Fat(s)
    carbohydrate conversion to, 129
    cis, 37f, 44, 51f

Fats *(continued)*
- in diet
  - and blood cholesterol levels, 37, 51, 51*f*
  - health impact, 37
  - digestion of, 64, 65
  - as fuel for aerobic respiration, 128*f*, 129
- in liver of alcoholics, 77
  - saturation of, 44–45, 44*f*, 51*f*
  - structure of, 37, 44, 44*f*
- synthetic, 14, 14*f*
  - *trans*, 37, 37*f*, 44, 45, 49, 51*f*
Fatty acids, 37, 37*f*, 44, 44*f*, 45
- *cis*, 37*f*, 44, 51*f*
- essential, 44*f*
- as fuel for aerobic respiration, 128*f*, 129
- functional groups, 40, 40*t*
- saturation of, 44–45, 44*f*, 51*f*
- *trans*, 37, 37*f*, 44, 45, 49, 51*f*
Feedback mechanisms
- in enzymes, 84, 84*f*
- in gene expression, 171
Female(s)
- chromosomes, 168, 168*f*
Fermentation, 126–127, 126*f*, 127*f*
- alcoholic, 126, 126*f*
- ATP yield, 119
- lactate, 126–127, 127*f*
- in muscle, 127, 172*f*
- overview, 118*f*, 119
Ferroplasma acidarmanus, 99, 99*f*
Fertilization
- and chromosome number, 191
Fertilizers
- as pollutant, 242
Fetoscopy, 228, 228*f*
Fetus
- human
  - development, 229*f*
  - prenatal screening of, 228–229, 228*f*, 229*f*
Fiber(s)
- in diet, 42
Fibrillin, 209
First law of thermodynamics, 78
Fish
- mercury and, 23, 33
- polyploid species, 226
- sex determination in, 139
Flagellum (flagella)
- eukaryotes, 68–69, 69*f*
- prokaryotes, 58*f*, 59
- sperm, 75, 75*f*
Flavin adenine dinucleotide. *See* FAD
Floral identity genes, 169, 169*f*
Flower(s)
- floral identity genes, 169, 169*f*
- formation, 169, 169*f*
- structure, 191*f*
Flowering plants. *See* Angiosperm(s)
Fluid mosaic model, 88
Fluorescence microscopy, 56, 57, 57*f*
Fly (dipteran). *See Drosophila melanogaster*
Foja Mountains (New Guinea), new species in, 3, 3*f*, 16, 16*f*, 19
Folate (folic acid), and neural tube defects, 217
Folic acid. *See* Folate
Food. *See also* Crops (food); Diet
- fermentation and, 126–127, 126*f*
- genetically modified animals as, 244–245
- genetically modified plants as, 242–243, 243*f*
- Food poisoning
  - *E. coli*, 53, 53*f*
  - mercury contamination, 23, 33
Forest(s)
- cloud forest, New Guinea, 3, 3*f*, 16, 16*f*

rain forest
- in New Guinea, new species in, 19
Fossil fuels. *See also* Coal
- and acid rain, 33
- alternatives to, 113
- and carbon cycle, 112–113, 113*f*
- source of, 101
Fragile X syndrome, 219*t*
Frameshift mutations, 158, 158*f*, 159
Franklin, Rosalind, 136, 143
Free energy, 80, 83
Free radicals, 27
- antioxidants and, 86
- and DNA methylation, 172
- and early atmosphere, 117
- in mitochondria, 117, 131, 131*f*
- photosynthesis and, 107
- in tobacco smoke, 143
Frogs
- embryo, 178*f*
- genome, *vs.* other species, 240*f*
- sex determination in, 139
Fructose, 42, 42*f*, 120
Fructose-1,6-bisphosphate, 121*f*
Fruit fly. *See Drosophila melanogaster*
Fucoxanthin, 103*f*
Functional group(s), 40–41, 40*t*
Fungus (fungi), 375–385
- characteristics, 8, 9*f*
- as decomposer, 8
- defined, 8
- as kingdom, 11*f*
Fur, 211, 211*f*

G
Galactose, 42, 120, 170, 171
Galactosemia, 219*t*
Galileo, Galilei, 19
Gametes
- animal, 190, 191*f*
- chromosome number, 191
- defined, 190
- formation of, 191
- plant, 189*f*, 190, 191*f*, 202, 202*f*
Gamma rays, 102*f*, 142
Gap junctions, 71, 71*f*
Garden pea (*Pisum sativum*) 202–207, 202*f*, 203*f*
Gastric fluid, 32, 32*f*, 71
Gel electrophoresis. *See* Electrophoresis
Gene(s)
- alternative splicing, 153
- cloning of, 235, 235*f*
- converting to RNA, 150
- defined, 150
- exons, 153, 153*f*, 164*f*–165*f*
- genetic information in, 150
- and genetic inheritance, 203
- homeotic, 166–167, 167*f*, 168–169, 169*f*, 212
- human
  - *APOA5*, 240, 244
  - *APOE*, 233, 247
  - *BRCA*, 173, 184, 184*f*, 197, 233
  - *CFTR*, 201, 203*f*
  - *DLX 5/6*, 203*f*
  - *IL2RG*, 222*f*, 246
  - loci of, 203*f*
  - *MC1R* gene, 218
  - number of, 190
  - *OCA2* gene, 229
  - PAX6 gene, 167, 167*f*
  - *SLC24A5*, 217, 229
  - *SRY* gene, 225, 225*f*
  - *XIST*, 222*f*
- importation of, 197
- introns, 153, 153*f*, 164*f*–165*f*, 235
- isolation of, 236–237, 236*f*, 237*f*
- linked, 207
- loci of, 203, 203*f*

master. *See* Master genes
mutation. *See* Mutation(s)
- *SRY*, 225, 225*f*
- transfer of, methods, 242, 243*f*, 246
- on X chromosome, 168, 222*f*
- on Y chromosome, 168
Gene expression. *See also* Differentiation
- control of, 164–173
  - and cell cycle, 179
  - in eukaryotes, 164, 164*f*, 165, 167, 168–169, 168*f*, 169*f*, 170
  - failure of, 163, 173, 184–185
  - in prokaryotes, 170–171, 170*f*
- DNA methylation and, 172–173, 172*f*, 210, 211
- environmental factors in, 210
- overview, 150–151
Gene therapy, 246, 246*f*
Genetic abnormalities
- albinism, 219*t*, 221
- color blindness, 203*f*, 219*t*, 222–223, 222*f*
- defined, 218
- polydactyly, 218*f*, 219*t*
- research on, 218, 218*f*
Genetic code, 154–155, 154*f*, 155*f*, 158
Genetic disorders. *See also* Sickle-cell anemia; X-linked disorders
- achondroplasia, 219*t*, 220, 220*f*
- Alzheimer's disease, 86, 233, 246, 247
- androgen insensitivity syndrome, 219*t*
- aneuploidy, 226–227, 227*f*, 228
- aniridia, 167*f*, 219*t*
- Burkitt's lymphoma, 224
- camptodactyly, 219*t*
- chromosome number changes, 219*t*, 226–227, 226*f*, 227*f*
- chromosome structure changes, 219*t*
- chronic myelogenous leukemia, 219*t*
- cri-du-chat syndrome, 219*t*, 224
- cystic fibrosis, 201, 201*f*, 203*f*, 209, 213, 215, 215*f*, 219, 219*t*, 228, 244, 245, 246
- defined, 218–219
- Down syndrome (Trisomy 21), 219*t*, 226, 227*f*
- Duchenne muscular dystrophy, 223, 224
- Ellis–van Creveld syndrome, 218*f*, 219*t*
- epigenetic factors in, 219
- familial hypercholesterolemia, 219*t*
- Fanconi anemia, 219*t*
- fragile X syndrome, 219*t*
- galactosemia, 219*t*
- gene therapy for, 246, 246*f*
- hemophilia, 219*t*, 222*f*, 223, 223*f*, 228, 246
- hereditary hemochromatosis, 219*t*
- hereditary methemoglobinemia, 219*t*
- Huntington's disease, 218*f*, 219*t*, 220, 224, 245
- Hutchinson-Gilford progeria syndrome, 220–221, 220*f*
- incontinentia pigmenti, 219*t*
- Kartagener syndrome, 75, 75*f*
- Klinefelter syndrome, 219*t*, 227
- Marfan syndrome, 203*f*, 209, 209*f*, 219*t*
- multiple sclerosis, 219, 245
- muscular dystrophies, 219*t*, 222*f*, 223, 224, 228, 246
- neurofibromatosis, 219*t*
- number of known, 246
- Parkinson's disease, 246

persistence of, 219
- phenylketonuria, 219*t*, 228
- polygenetic factors in, 219
- progeria, 219*t*
- rare, cost of curing, 246
- research on, 218, 218*f*, 245, 246
- SCID-X1, 222*f*
- screening for, 228–229, 228*f*, 229*f*
- sex chromosome abnormalities, 226–227
- single-gene, 219–223, 219*t*
- Tay-Sachs disease, 203*f*, 219*t*, 221, 221*f*, 228
- Tetralogy of Fallot (TF), 131, 131*f*
- thalassemias, 158*f*, 159
- treatment of, 228
- Turner syndrome, 219*t*, 226–227
- X-linked, 219*t*, 222–223, 222*f*, 223*f*
- X-linked anhidrotic dysplasia, 219*t*
- XXX syndrome, 219*t*, 227
- XXY syndrome, 227
- XYY condition, 219*t*
- XYY syndrome, 227
Genetic diversity
- human, sources of, 217–225
- sexual reproduction and, 189
Genetic engineering, 242–247. *See also* Genetically modified organisms
- defined, 242
- ethical issues, 243, 244–245, 246–247
- history of, 244
- of humans, 246–247
- safety issues, 245
- uses of, 242
Genetic inheritance
- autosomal dominant, 218*f*, 219*t*, 220–221, 220*f*
- autosomal recessive, 218*f*, 219*t*, 221, 221*f*
- codominant alleles, 208, 208*f*
- dominant allele, defined, 203, 203*f*
- epigenetic, 172–173, 172*f*, 175, 175*f*
- in humans, 217–225
- incomplete dominance, 208, 208*f*
- karyotyping, 224
- law of independent assortment, 206–207, 206*f*, 207*f*
- law of segregation, 204–205
- Mendel's experiments on, 202–207, 202*f*, 203*f*
- overview of, 202–203
- patterns, analysis of, 218, 218*f*
- pleiotropy, 209
- recessive allele, defined, 203, 203*f*
- X-linked inheritance, 219*t*, 222–223, 222*f*, 223*f*
Genetic research
- comparative genome analysis, 240, 240*f*
- dihybrid experiments, 206, 206*f*
- on disorders and abnormalities, 218, 218*f*
- on DNA, 134–137, 147, 147*f*, 202, 233
- *Drosophila melanogaster* in, 166–167, 166*f*, 167*f*
- ethical issues, 247
- on gene expression, 244, 244*f*
- on genetic disorders, 218, 218*f*, 245, 246
- genomics, 240–241, 240*f*
- Human Genome Project, 239, 239*f*
- knockout experiments in, 166–167, 167*f*, 183, 240, 244
- Mendel's pea experiments, 202–207, 202*f*, 203*f*, 204*t*, 207*f*
- mice in, 183, 199, 199*f*, 240, 240*f*, 244, 244*f*, 247
- monohybrid crosses, 204–205, 205*f*

Index (continued)

safety issues, 245
SNP chips in, 240–241, 240f
on telomeres, 183
■ Genetic testing
DNA profiling, 240–241, 241f
individual, 240–241, 247
limitations of, 247
personal, 233, 233f
prenatal screening, 228–229, 228f, 229f
short tandem repeat profiles, 241, 241f
SNP testing, 233, 233f, 240–241, 240f
■ **Genetically modified organisms (GMOs)**, 242–247. *See also* Genetic engineering
Geneva Protocol, 149
Genome(s)
applications, 240, 240f
comparative analysis of, 240, 240f
definition of, 236
Human Genome Project, 239, 239f
human *vs.* other species, 240, 240f
isolation of genes in, 236–237, 236f, 237f
Genome library, 236
Genomics, 240–241
Genotype. *See also* Genetic inheritance
defined, 203, 203f
environmental factors. *See* Natural selection
karyotyping, 224
law of independent assortment, 206–207, 206f, 207f
linkage groups and, 207
Genus, 10, 10f
Germ cells, 191
Gey, George and Margaret, 177
■ **Global warming**
causes of, 113
impact of, 113
as scientific theory, 18t
Globin, 46, 47f, 158, 158f
Glucose
in aerobic respiration, 118f, 119, 124f–125f, 128, 128f
animal storage of, 129
■ blood levels, regulation of, 7, 129
breakdown of, 85, 85f. *See also* Glycolysis
in glycolysis, 120, 121f
phosphorylation of, 82, 82f, 92
plant production and storage, 42, 78, 81, 108–109, 109f
structure, 40f, 109
uses of, 42, 42f
Glucose-6-phosphate, 120, 121f, 129
Glucose transporter, 92, 92f, 120
Glutamic acid, 158–159, 159f
Gluten, 126
Glycerol, 128f, 129
Glycogen
animal storage of, 43, 43f, 129
Glycolipids, 88
Glycolysis, 118, 118f, 120, 121f, 124f–125f, 126–127, 126f, 127f, 128–129, 128f
Glycoproteins, 47, 88
Glyphosate, 243
Goat(s), genetically modified, 244, 244f
Golden Clone Giveaway, 133
Golden-mantled tree kangaroo, 16, 16f
Golden rice, 243
Golgi bodies, 60, 60f, 60t, 61f, 65, 65f, 95, 95f, 182
Gorilla (*Gorilla gorilla*), 225
Grass(es)
■ as biofuel source, 115, 115f

Griffith, Frederick, 134, 134f
Groucho gene, 166–167
Growth
as characteristic of life, 7
defined, 7
Growth factor, 184, 173, 220
Guanine (G), 136–137, 136f, 137f, 150, 150f, 151f, 152, 154, 154f, 172, 172f
Gunpowder, 81

H
Hair
color of, 218
proteins in, 40, 47, 47f, 68, 159
■ Hangovers, 96
Hare, coat color, 211, 211f
HDL. *See* High density lipoprotein
Heart
gap junctions in, 71
■ **Heart attack**
treatment, 244, 246
■ **Heart disease**
cause of, 219
Heat. *See also* Global warming
energy loss through, 78–79
Height, human, variation in, 212, 212f
HeLa cells, 177, 177f, 183, 186, 187, 187f
Helgen, Kris, 16f
Helicase, 140, 140f
Heliconius butterflies, 11f
Helium, atomic structure, 26, 26f, 27
Heme, 39, 39f, 46, 47f
as coenzyme, 86, 86f, 86t
in hemoglobin structure, 158, 158f
■ Hemochromatosis, hereditary (HH), 219t
Hemoglobin
genetic mutations in, 158–159, 158f, 159f
■ HbS (sickle), 158–159, 158f, 159f
molecular models, 39, 39f
oxygen binding and transport, 158, 158f
structure, 39, 39f, 46, 47, 47f, 158, 158f
■ Hemophilia, 219t, 222f, 223, 223f, 228, 246
■ Hepatitis, 77
■ Herbicides
resistance, 243
Hershey, Alfred, 135, 135f, 147, 147f
Heterotroph(s), 101
Heterozygous individuals
autosomal dominant inheritance, 220–221, 220f
autosomal recessive inheritance, 221, 221f
defined, 203, 203f
monohybrid crosses in, 204–205, 205f
Hexokinase, 82f, 92, 120, 121f
High density lipoprotein (HDL), 47, 47f, 51, 51f
Histones, 138, 138f, 164
■ **HIV (Human Immune Deficiency Virus)**
research on, 160
treatment, 246
Homeodomain, 166, 167f
Homeostasis
defined, 7
as defining feature of life, 7, 72
pH regulation, 33
tonicity, 90–91, 91f
water and, 30–31
Homeotic genes, 166–167, 167f, 168–169, 169f, 212

Homologous chromosomes, 179, 190, 190f, 192, 197f, 203, 204, 204f, 206–207
Homozygous individuals
autosomal dominant inheritance, 220–221, 220f
autosomal recessive inheritance, 221, 221f
defined, 203, 203f
homozygous dominant, 203, 203f, 204, 204f
homozygous recessive, 203, 203f, 204, 204f
Honeybees, honeycomb, 45
Hooke, Robert, 54
Hopf, Conner, 221f
Hops, 126
Hormone(s), animal
receptors
estrogen receptors, 173
progesterone receptors, 173
■ **Hormone(s), human**. *See* Hormone(s), animal
HPV. *See* Human papillomavirus
Human(s)
blood. *See* Blood
carbon in, 112
cells
division outside body, 177
research on, 145, 177, 177f
chromosome(s), 218–219
■ abnormalities, 177f, 226–227, 226f, 227f
vs. ape or monkey chromosomes, 225, 225f
number of, 138, 225, 226–227, 226f, 227f
number of nucleotides in, 158, 233, 239
sex chromosomes, 168–169
cloning of, 145
■ disease. *See* Autoimmune disorders; Disease, human; Genetic disorders; *specific diseases*
DNA. *See also* DNA
DNA fingerprinting, 241, 241f
genetic testing, 228–229, 228f, 229f, 233
genome, *vs.* other species, 240, 240f
Human Genome Project, 239, 239f
individual variations in, 233
number of nucleotides in, 159, 233, 239
early
evolution, 225, 225f
genetic screening of, 228–229, 229f
in vitro fertilization, 229, 229f
eye
color variations in, 203f, 212–213, 213f
fetus
development, 229f
prenatal screening of, 228–229, 228f, 229f
gastric fluid pH, 32, 32f
genetic diversity, sources of, 217–225
genetic engineering of, 246–247
hair
color of, 218
proteins in, 40, 47, 47f, 68
straightening and curling, 40
height, variation in, 212, 212f
impact on biosphere. *See* Biosphere
mitochondria, 66
pH
buffering system, 33
disorders, 33

gastric fluid, 32, 32f
internal environment, 33
■ psychiatric disorders, 211
reproductive system
female, 190f
male, 190f
sex determination in, 139, 168–169
sex hormones, 168–169
skeleton
sperm
■ disorders, 75, 75f
stomach,
gastric fluid pH, 32, 32f
stress, effects of, 210, 211
testes,
development, 168
location, 190f, 596f
tumors. *See* Tumor(s)
Human Genome Project, 239, 239f
Human Immune Deficiency Virus. *See* HIV
■ Human Microbiome Project, 329
■ Human papillomavirus (HPV), 184
■ Huntington's disease, 218f, 219t, 220, 224, 245
■ Hutchinson-Gilford progeria syndrome, 220–221, 220f
Hybrids
defined, 203
Hydrocarbons, 40
Hydrochloric acid (HCl), 32
Hydrogen
atomic structure, 26, 26f
isotopes of, 24
molecular, 28–29
and pH, 32–33, 32f
Hydrogen bond, 30–31, 30f
Hydrogen peroxide, 86
Hydrogen sulfide, 112
Hydrogenation, 37
Hydrogenosomes, 66
Hydrolysis, 41, 41f
Hydrophilic substances, 30
Hydrophobic substances, 30–31
Hydrothermal vents, 8f
Hydroxide ions, and pH, 32, 32f
Hydroxyl groups, 40–41, 40f, 40t, 41f, 150f
■ Hypercholesterolemia, familial, 219t
■ Hyperventilation, 33
Hypothesis, in scientific research process, 12, 13t, 15

I
■ Ice, polar, 113, 113f
IL2RG gene, 222f, 246
■ **Immune system**
autoimmune disorders
multiple sclerosis, 219, 245
blood type and, 208
cancer of, 224
CFTR protein and, 201, 213
immunization. *See* Vaccines
immunodeficiency. *See* HIV
organ transplants and, 245
respiratory tract and, 215
■ *In vitro* fertilization (IVF), 129, 229, 229f
Incomplete dominance, 208, 208f
■ Incontinentia pigmenti, 219t
Independent assortment
law of, 206–207, 206f, 207f
Independent variable, 13, 14
Induced-fit model, 82
Inductive reasoning, 12
■ Infectious disease. *See* Pathogen(s)
Inheritance
blending theory of, 202
defined, 7. *See also* Genetic inheritance

Insect(s),
polyploid species, 226
Insecticides. *See* Pesticides
Insertion mutation, 158, 158f, 159
Insulin, 242
Interleukin(s), 244
Intermediate filaments, 68, 68f
Internal environment. *See also*
Homeostasis
fluids in, 7
pH of, 33
Interphase, 178, 178f, 179, 179f, 180f,
181, 184
Intestines
bacteria in, 53, 53f, 96, 170–171
Introns, 153, 153f, 159,
164f–165f, 235
Inversion, chromosomal, 224, 224f
Invertebrate(s)
sex determination in, 139
Ion(s), 27, 27f, 28
Ionic bonds, 28, 28f
Ionizing radiation, genetic damage from,
142–143, 142f
Iris, 91f, 212–213, 213f, 226, 227f
Iron
in heme, 158, 158f
Irrigation, 242
Isotopes, 24–25
IVF. *See In vitro* fertilization

J
Jolie, Angelina, 233f

K
Kangaroo, 16, 16f
Kartagener syndrome, 75, 75f
Karyotype, 139, 139f, 228
Karyotyping, 139, 224
Ketone group, 40, 40t
Kinesins, 68f
Kinetic energy, 78
Kingdom, 11f
Klinefelter syndrome, 219t, 227
Knockout experiments, 166–167, 167f,
183, 240, 244
Krebs, Hans Adolf, 123
Krebs cycle (citric acid cycle), 118, 118f,
122–123, 122f, 123f, 124f–125f,
128f, 129
Kubicek, Mary, 177
Kuruvilla, Sarah, 131, 131f

L
Labrador retrievers, coat color in,
209, 209f
Lac operon, 170–171, 170f, 234f
Lacks, Henrietta, 177, 177f, 186
Lactase, 171, 758
Lactate, 127, 127f
Lactate fermentation, 126–127, 127f
Lactobacillus, 127
Lactose, 127
as bacterial nutrient, 170–171, 170f
intolerance, 171
structure, 42
Lactose (*lac*) operon, 170–171,
170f, 234f
Lamins, 68, 220–221
Law of independent assortment,
206–207, 206f, 207f
Law of nature, 18
Law of segregation, 204–205
LDL. *See* Low density lipoprotein
Leaf (leaves)
of C3 plant, 110, 111f
of C4 plant, 110–111, 111f
Learning
in autism, 249, 249f
brain DNA and, 211

Leeuwenhoek, Antoni van, 54
Length, units of, 56t
Leptin, 203f
Leucine, 154, 154f
Leukemia(s), 219t, 246
Leukocytes. *See* White blood cells
Levene, Phoebus, 136f
Life. *See also* Organic molecules
defining features of, 4, 6–7, 72
diversity of. *See* Biodiversity
emergent properties, 4
levels of organization, 4–5, 4f–5f
Life cycles
defined, 178
Light. *See also* Sunlight
as energy source, 102–103, 102f,
103f, 104, 104f, 105
properties, 102, 102f
Light-dependent reactions, in
photosynthesis, 106–107,
106f, 107f
cyclic pathway, 106f, 107, 108, 112
noncyclic pathway, 106–107, 106f,
107f, 108, 108f, 113
overview, 105
stomata and, 110
Light-harvesting complex, 106, 106f
Light-independent reactions, in
photosynthesis, 108–109, 109f
in C3 plants, 110–111, 111f
in C4 plants, 110–111, 111f
overview, 105
Light micrography, 56, 56f–57f
Lignin
in plant structures, 70
Linear metabolic pathway, 84, 84f
Linkage groups, 207
Linnaeus, Carolus, 10, 11
Lipid(s). *See also* Fat(s); Phospholipids;
Steroid(s); Waxes
defined, 44
metabolism of, 128f, 129
synthesis of, 64
types, 44–45, 44f, 45f
Lipid bilayer,
selective permeability, 54, 90, 90f
structure, 44f, 45, 54, 54f, 55f, 88,
88f–89f
Lipoprotein(s)
endocytosis of, 94, 94f
genes encoding, 233
synthesis of, 47
Liver, 694f
alcohol and, 77
diseases, 77, 96f, 97
functions, 77, 129
Locus, of gene, 203, 203f
Low density lipoprotein (LDL), 51,
51f, 94
Lutein, 103f
Lycopene, 103f
Lymphoblastic leukemia, 246
Lymphomas, 246
Lysosome(s), 61f, 64, 65, 95f
enzymes, 95, 95f
functions, 60t, 65, 65f
Lysosome-like vesicle, 61f
Lysozyme, 244

M
Mad cow disease. *See* Bovine
spongiform encephalopathy (BSE)
Magnetic field, animal's ability to
sense, 8f
Maize. *See* Corn
Male(s)
chromosomes, 139, 139f, 168
Maltose, 126
Mammal(s)
mosaic tissues in, 168, 168f

MAO-B, 25f
Marfan syndrome, 203f, 209,
209f, 219t
Marijuana (*Cannabis*), 10f
Markov, Georgi, 149
Marrow. *See* Bone(s)
Marsupials, 225f
Martin, Martine, 117f, 129
Martin, Tom, 117f
Mass number, 24–25
Mastectomy 163, 233f
Master genes, 166–167, 166f, 167f
and cancer, 173
homeotic, 166–167, 167f, 168–169,
169f, 212
in sex determination, 168–169
Mayr, Ernst, 11
Mazia, Daniel, 135
MC1R gene, 209, 218
McCarty, Maclyn, 134
Meiosis
abnormalities in, 199, 199f
and chromosome number,
190–191, 191f
chromosome segregation in,
194–195, 204, 204f,
206–207, 207f
comparison to mitosis, 196–197, 196f
crossing over in, 194, 194f
evolution of, 196–197
meiosis I, 190–191, 191f,
192–193, 192f
meiosis II, 190–191, 191f, 193, 193f,
196–197, 196f
nondisjunction in, 226, 226f
overview, 190–191
stages of, 190, 192–193, 192f–193f
and variation in traits, 194–195,
194f, 195f
Melanins
and eye color, 213
in feathers, 221
and fur color, 209, 209f,
210–211, 210f
and hair color, 218
in retina, 221
and skin color, 213, 217, 221
synthesis of, 217, 230
Melanoma, 185f
Melanosomes, 217
Membrane cycling, 95, 95f
Membrane trafficking, 94–95, 94f,
95f, 201
Mendel, Gregor, 202f
law of independent assortment,
206–207, 206f, 207f
law of segregation, 204–205
pea experiments, 202–207, 202f,
203f, 204f
Mendeleev, Dmitry, 24
Mercury
in fish, 23, 33
poisoning, 23
as pollutant, 23, 33, 35, 35f
Mesophyll, 110f, 111, 111f
Messenger RNA (mRNA)
in cDNA cloning, 235
and DNA libraries, 236
function, 150, 154–155, 156,
156f, 157f
and gene expression control,
164–165, 164f
guanine cap, 153, 153f
localization, 165, 166
poly-A tail, 153, 153f, 164–165
post-transcription modifications, 153,
153f, 164–165
riboswitches in, 171
stability of, 156, 165
structure, 153, 153f
translation, 150–151

MAO-B, 25f
Metabolic pathways
defined, 84
types, 84, 84f
Metabolic reactions, common,
41, 41f
Metabolism. *See also* Aerobic
respiration; ATP synthesis;
Digestion; Electron transfer chains;
Fermentation; Glycolysis; Protein
synthesis
of alcohol, 64, 77, 77f, 96–97
in cancer cells, 185
of carbohydrates, 42–43, 118–119,
118f. *See also specific pathways*
control of, 84–85, 84f, 85f
defined, 41, 41f, 84
as defining feature of life, 72
of drugs, 64, 65
endomembrane system and, 64–65
enzymes and, 82–83, 84–85, 84f. *See
also* Enzyme(s)
of lipids, 128f, 129
of polysaccharides, 128f, 129
of proteins, 128f, 129
Metaphase (mitosis), 178f, 180f, 181,
186, 196f
Metaphase I (meiosis), 190, 192,
192f, 195
Metaphase II (meiosis), 190, 193,
193f, 196f
Metastasis, 185, 185f
Methane
in early atmosphere, 112
structure, 40
Methemoglobinemia, hereditary, 219t
Methionine (met), 46f–47f, 154f, 155,
156, 157f
Methyl groups
attachment to DNA, 143, 172–173,
172f, 210, 211
attachment to histones, 164
defined, 40, 40t
Methylmercury, 23
Microfilaments, 61f, 68–69, 68f, 95
MicroRNAs, 150, 165
Microscopes and microscopy
history of, 54
types, 56–57, 56f–57f
Microtubules, 61f, 68–69, 68f, 69f,
177, 194, 195. *See also* Spindle
in cytokinesis, 182
defects in, 186
inhibition of, in cancer
treatment, 186
in meiosis, 191, 192–193, 192f–193f
in mitosis, 180f, 181, 181f
Miescher, Johann, 134
Milk
and lactose intolerance, 171
Millet (*Eleusine coracana*), 111f
Mineral(s)
as cofactors, 86
Mitochondrion (mitochondria), 117f
in aerobic respiration, 118, 118f, 122,
122f, 124, 124f–125f
ATP synthesis in, 117
diseases, 117, 117f, 129, 131, 131f
DNA, 66, 129
evolution of, 66
functions, 60t, 66
genetic code in, 155
in muscle, 127
structure, 66, 117, 122, 122f
Mitosis
in cell cycle, 178–179, 178f, 197
and chromosome number,
179, 179f
comparison to meiosis,
196–197, 196f
controls, loss of, 184–185, 184f

■ interference, in cancer
   treatment, 186
   stages, 178f, 180f, 181
   as term, 181
MLH1 protein, 197f
Model(s)
   of molecules, 38–39,
      38f, 39f
   in research, 13
Molecule(s). *See also* Organic
   molecules
   compounds, 28
   defined, 4, 4f, 28
   free energy, 80, 83
   and hydrophilia, 30
   as level of organization, 4, 4f
   models of, 38–39, 38f, 39f
   polarity of, 28, 29, 30
   representation methods, 29t
   size, 56f
Monkey(s)
   evolution, 225f
Monohybrid cross, 204–205, 205f
Monomers, defined, 41
Monosaccharides, 42
Monotremes, 225f
■ Mood disorders, 211
Morphological traits, 10–11
Mosaic tissues, X chromosome
   inactivation and, 168, 168f
Moth(s)
   sex determination in, 139
Motor proteins, 68–69, 68f, 94
Mouse (mice)
   in genetic research, 183, 199,
      199f, 240, 244, 244f, 247,
      249, 249f
   genome, *vs.* other species,
      240, 240f
mRNA. *See* Messenger RNA
MS. *See* Multiple sclerosis
Multiple allele systems, 208
■ Multiple sclerosis (MS), 219, 245
Muscle(s)
   fermentation in, 127, 127f
   human, muscle fibers, 127, 127f
Muscle cells
   contraction, 68
   size and shape of, 55
Muscle fiber, 127, 127f
■ Muscular dystrophies, 219t, 222f, 223,
   224, 228, 246
Mutation(s). *See also* Allele(s); Genetic
   abnormalities; Genetic disorders
   and alleles, 190
■ and cancer, 142, 143, 163, 173, 177f,
      184–185, 246
   causes of, 142–143, 142f,
      158–159, 159f
   chromosome number changes, 219t,
      226–227
   crossing over, 194, 194f, 206–207,
      224, 225, 225f
   deletions, 201, 224, 224f
   duplications, 224, 224f
   and evolution, 227, 227f
   frequency of, 142, 158
   inversions, 224, 224f
   protein synthesis and, 158–159,
      158f, 159f
   reintroduction of, 219
   severe, persistence of, 219
   single-nucleotide, 217, 218–223
   translocations, 224, 224f, 240
   types of, 158, 158f
Myoglobin, 127
Myosin
   in cytokinesis, 182, 182f
   as fibrous protein, 47
   in muscle contraction, 68

N
**NAD+ (nicotinamide adenine
   dinucleotide)/NADH**, 86t
   in aerobic respiration, 122, 122f, 123,
      123f, 124, 124f–125f, 128f
   in alcohol metabolism, 96
   as coenzyme, 86, 86t
   in fermentation, 126, 126f, 127
   in glycolysis, 120, 121f
NADP+/ NADPH, 86t, 105, 106f,
   107, 107f
   as coenzyme, 86t
   in photorespiration, 110, 110f
   in photosynthesis, 105, 106f, 107,
      107f, 108, 108f, 109, 109f
National Institutes of Health (NIH)
■ genetic engineering safety
   guidelines, 245
   and Human Genome Project, 239
Natural selection
   scientific research on, 14–15, 15f,
      15t, 21
Nature
   laws of, 18
   vs. nurture, 210–211, 210f, 211f
■ Neoplasms, 184–185, 185f
■ Nerve gas, 244
■ Neurodegenerative diseases, 87
■ Neurofibromatosis, 219t
Neutrons, in atomic structure, 24, 24f
■ New Guinea, cloud forests in, 3, 3f, 16,
   16f, 19
New Zealand mud snails, 189f
Newborn
   screening of, 228
■ Niacin (Vitamin B₃), 86
Nicotinamide adenine dinucleotide.
   See NAD+
NIH. *See* National Institutes of Health
Nitrocellulose (guncotton), 81
Nitrogen
   molecular, 29
   as pollutant, 33
Nomenclature, species, 10–11,
   10f, 11f
Noncyclic pathway, in photosynthesis,
   106–107, 106f, 107f, 108,
   108f, 113
Nondisjunction, 226, 226f
Nonpolar covalent bonds, defined,
   9, 29f
Norepinephrine
Notl enzyme, 234f
Nuclear envelope, 61f, 62–63, 62f,
   62t 63f
Nucleic acid(s), 49, 49f
Nucleic acid hybridization, 140,
   236, 236f
Nucleolus, 61f, 62f, 62t
Nucleosomes, 138
Nucleotide(s)
   in DNA, 49, 49f, 136–137, 136f, 137f,
      150, 150f, 151f
   in DNA replication, 140–141,
      140f, 141f
   in DNA sequencing, 238, 238f
   functional groups in, 40, 40t
   functions, 49
   and genetic code, 154–155, 154f,
      155f, 158
   methylation of, 172, 172f
   radiation damage to, 142–143
   in RNA, 150, 150f, 151f
   structure, 49, 49f
Nucleus, atomic, 24, 24f
Nucleus, cellular, 54f, 60f, 61f
   defined, 8
   functions, 60t, 62
   in mitosis, 178–179, 178f,
      180f, 181

in classification, 8
   structure, 55, 60, 60f, 61f, 62–63, 62f,
      62t, 63f
Nutrient(s)
   cycling of, in ecosystems, 6, 6f. *See
      also* Biogeochemical cycles
   defined, 6
Nutrition. *See* Autotrophs; Digestion;
   Heterotroph(s); Predators and
   predation
■ **Nutrition, human**. *See also* Diet
   essential fatty acids, 44f
   genetically modified foods and,
      242–243
   lipids, 37

O
■ **Obesity**
   causes of, 219
   genetic factors in, 203f
   health effects, 37
Obstetric sonography, 228, 228f
OCA2 gene, 229
Oil(s)
   in diet, 37, 45
Okazaki fragments, 141f
Oleic acid, 37f
Olestra®, 14, 14f, 17
Oligosaccharides, 42
Oliver, Paul, 3f
Omega-3 fatty acids, 44f
Omega-6 fatty acids, 44f
■ Oncogenes, 184, 184f
Operators, 170, 170f
Operons, 170–171, 170f
Opsins (visual pigments), 221, 222–223
Orangutan, 7f, 225
Orbitals, electron, 26–27, 26f, 27f
Organ(s)
   defined, 4–5, 4f
   as level of organization, 4–5, 4f
■ transplants, 97, 245
Organ systems
   defined, 4, 5, 5f
Organelles, 55
   in eukaryotic cells, 60, 60f, 60t, 61f,
      64–67
   prokaryotic, 58f, 59
Organic molecules
   cellular reactions and, 40–41, 41f
   defined, 38
   as defining feature of life, 38, 72
   functional group(s), 40, 40t
   models of, 38–39, 38f, 39f
   structure of, 38
Organism(s)
   characteristics of, 7
   classification of, 6f, 8, 8f, 9f
   defined, 5f
   as level of organization, 5f
Osmosis, 90, 90f, 92
Osmotic pressure, 91, 91f
Ovary(ies)
■ cancer of, 143, 173
   formation, 169
   functions, 169
   human, 169, 190f
   plant, 189f, 190f
■ Overweight. *See* Obesity
Oxaloacetate, 110, 110f, 122,
   123, 123f
Oxidation-reduction (redox) reactions,
   85, 85f, 128
   defined, 85
   in electron transfer chain, 106
■ Oxidative damage, 86, 117
Oxygen
   in aerobic respiration, 118, 118f, 119,
      124–125, 124f–125f
   atomic structure, 26f

chemical reactivity, 80–81, 84–85
■ in Earth's atmosphere
   early atmosphere, 112, 117
   and free radicals, 117
   molecular, 29
Oxygen cycle, 118, 118f

P
Paclitaxel, 186
■ Pandemics, 244f
■ Parkinson's disease, 246
Partially hydrogenated vegetable oil,
   37, 45
Passive transport, 92, 92f
■ Paternity disputes, DNA fingerprinting
   in, 241
■ **Pathogen(s)**
   bacteria as, 53, 53f
   prions as, 48, 48f
Pattern formation, 166, 166f
PAX6 gene, 167, 167f
PCR. *See* Polymerase chain reaction
Pea plant. *See* Garden pea
Peacock butterflies, 14–15, 15f, 15t,
   17f, 21
Pedigree, 218, 218f, 228
Pepsins, 83, 83f
■ Peptic ulcer, 71
Peptide(s), 46
Peptide bonds, 46, 46f–47f
Periodic table, 24, 24f
Permafrost
   remains preserved in, 147, 249,
Peroxisomes, 60t, 64, 86
■ **Pesticides**
   alternatives to, 242–243, 243f
   as pollutants, 242
■ PET (positron-emission tomography),
   25, 25f
Petals, flower, 169, 169f
PGA (phosphoglycerate), 109, 109f,
   110, 110f, 120, 121f
PGAL (phosphoglyceraldehyde), 109,
   109f, 120, 121f, 128f, 129
pH
■ of acid rain, 33
   defined, 32
   and denaturation of proteins, 48
   and enzyme action, 83, 83f
   functional groups and, 40, 40t
   homeostasis, 33
   human
      buffering system, 33
■   disorders, 33
      gastric fluid, 32, 32f
      internal environment, 33
   life processes and, 33
pH scale, 32, 32f
Phagocytosis, 94–95, 95f
Phenotype. *See also* Trait(s)
   autosomal dominant inheritance, 218f,
      219t, 220–221, 220f
   autosomal recessive inheritance, 218f,
      219t, 221, 221f
   codominant alleles, 208, 208f
   continuous variation,
      212–213, 212f
   defined, 203, 203f
   environmental factors, 210–211,
      210f, 211f
   epistasis, 209, 209f, 212
   incompletely dominant alleles,
      208, 208f
   pleiotropy, 209
   X-linked dominant inheritance, 219t
   X-linked recessive inheritance, 219t,
      222–223, 222f, 223f
Phenylalanine, 154, 154f,
   201, 228
■ Phenylketonuria (PKU), 219t, 228

Phosphate group(s)
in ATP, 49, 49f
defined, 40, 40t
in nucleotides, 49, 49f
in phospholipid, 44f, 45
transfers of. *See* Phosphorylation
Phosphoglyceraldehyde. *See* PGAL
Phosphoglycerate. *See* PGA
Phospholipids, 44f, 45, 54. *See also*
  Lipid bilayer
in cell membrane, 88, 88f–89f
Phosphorus
isotopes, as tracer, 135f
Phosphorylation
and ATP/ADP cycle, 87, 87f
defined, 87
electron transfer, 106–107, 118, 118f,
  124–125, 124f–125f,
  128f, 129
enzymes and, 82, 82f
of motor proteins, 68, 69f
in photosynthesis, 106–107, 107f
substrate-level, 120, 121f, 122, 123f
and translation, energy for, 157
in transport membranes, 92
Photolysis, 106, 118–119
Photon(s)
defined, 102, 102f
in photosynthesis, 103, 106
Photorespiration, 110, 110f, 111
Photosynthesis
adaptation to environment, 104
in C3 plants, 110–111, 110f, 111f
in C4 plants, 110–111, 111f
in CAM plants, 110, 111f
carbon cycle and, 112–113, 112f
and cycling of materials, 118, 118f
defined, 6
electron transfer chains in, 85,
  106–107, 107f, 108, 108f
energy flow in, 108, 108f
as energy source, 101
energy storage in, 81
equation for, 105
evolution of, 108, 112
light-dependent reactions, 105,
  106–107, 106f, 107f
  cyclic pathway, 106f, 107, 108, 112
  noncyclic pathway, 106–107, 106f,
    107f, 108, 108f, 113
  overview, 105, 105f
light-independent reactions,
  108–109, 109f
  in C3 plants, 110–111, 111f
  in C4 plants, 110–111, 111f
  overview, 105
light wavelengths for, 104, 104f
overview, 105
pigments in, 102–103, 103f
in prokaryotes, 58f, 59
research on, 104, 104f
Photosystem(s), 106–107, 107f,
  108, 108f
Phycobilins, 103f, 104
Pier, Gerald, 215
Pig(s)
  as human organ source, 245
  transgenic, 245
Pigment(s)
absorption spectrum of, 104, 104f
accessory, 102, 103, 106
chromoplasts and, 67, 67f
color of, 103, 103f
functions, 102–103
in iris of eye, 212–213, 213f
in photosynthesis, 102–103, 103f
in skin, 212–213, 217, 217f, 221, 230,
  231, 231f
structure of, 102–103, 103f
in thylakoid membrane, 106
visual (opsins), 221, 222–223

Pilus (pili)
sex, 59
structure and function, 58f, 59
Pinocchio frog, 3f
Pinocytosis, 94, 94f
Piven, Jeremy, 23
PKU. *See* Phenylketonuria
Placenta, 228–229
Plant(s). *See also* Crops (food);
  specific types
■ and carbon cycle, 112–113
characteristics, 8, 9f
defined, 8
diseases. *See* Disease, plant
as Eukarya, 8
gamete formation, 190, 191f,
  202, 202f
glucose production and storage in,
  42, 78, 81, 108–109, 109f
as kingdom, 11f
meiosis in, 190, 191f
as photoautotrophs, 101
pigments, 103f
polyploidy in, 226
reproduction, 189f, 190f
■ transgenic, 242–243, 243f
Plant cells
cell wall, 42, 70, 70f
cytokinesis in, 182, 182f
mitosis in, 180f, 181
structure and components, 54f,
  60f, 61f
turgor, 91, 91f
■ Plaque, dental, 8f, 59f
Plasma membrane, 54, 54f, 55f. *See
  also* Nuclear envelope
and cell size, 55
defined, 54
eukaryotic cells, 60, 61f
function, 54
membrane trafficking, 94–95,
  94f, 95f
prokaryotic cells, 58f, 59
Plasmid(s)
defined, 59
as vector, 234–235, 234f, 235f,
  242, 243f
Plasmodesma (plasmodesmata), 61f,
  71, 71f
Plastids, 67, 67f
Plate tectonics theory, 18t
Pleiotropy, 209
Polar molecules, 30, 30f
Polarity, of molecule, 30
defined, 28, 29f
functional groups and, 40, 40t
and hydrophilia, 30
■ Polio, 177
Pollen grain(s), 189f
Pollen tube, 189f
■ **Pollution**. *See also* Air pollution; Water
  pollution
fossil fuels and, 33
■ **Polydactyly**, 218f, 219t
Polygenic inheritance. *See* Epistasis
Polymer, defined, 41
Polymerase chain reaction (PCR),
  236–237, 237f, 241, 246
Polypeptides
defined, 46
synthesis of, 46, 46f–47f, 64–65,
  64f–65f, 150–151, 154–155, 154f,
  155f, 156–157, 156f, 157f
Polyploidy, 226, 226f
in animals, 226
defined, 226
in plants, 226
Polysaccharides, 42–43, 43f
Polysomes, 157, 157f
Polytene chromosomes, 164
Population(s)

defined, 5, 5f
dispersal research, genetic analysis in,
  229, 241
as level of organization, 5, 5f
Positron-emission tomography.
  See PET
Post-transcriptional modifications, 153,
  153f, 164–165
Post-translational modifications,
  164f, 165
Potential energy, 78, 79, 79f
■ **Precipitation**. *See also* Drought
  acid rain, 33
Predators and predation
predator-prey interaction
  models, 189
Prediction, in scientific research
  process, 12–13, 13t, 15
■ Preimplantation diagnosis, 229, 229f
■ Prenatal diagnosis, 228–229, 228f, 229f
Pressure
and diffusion rate, 90
osmotic, 91, 91f
turgor, 91, 91f
Primary plant cell wall, 70, 70f
Primers
in DNA replication, 140, 140f
in DNA sequencing, 238
in polymerase chain reaction,
  236–237, 237f
Prion disease, 48, 48f
Probability
in data analysis, 17
Probe, in nucleic acid hybridization,
  236, 236f
■ **Producers**
defined, 6
role in ecosystem, 6, 6f, 79, 79f
Product, of chemical reaction, 80, 80f
■ **Progeria**, 219t
Progesterone
receptors, BRCA proteins and, 173
Prokaryotes
in biofilms, 59, 59f
characteristics of, 8
defined, 8
DNA, 55, 58f, 59
gene expression, control of,
  170–171, 170f
habitats, 58f, 59, 59f
mRNA localization in, 165
as term, 58
transcription in, 152
Prokaryotic cells, 54f, 55, 58–59, 58f
Promoter(s), 152, 152f, 159, 164,
  164f–165f, 170–171, 170f, 172
Prophase (mitosis), 178f, 180f,
  181, 196f
Prophase I (meiosis), 190, 192, 192f,
  194–195, 197, 197f
Prophase II (meiosis), 190, 193,
  193f, 196f
Protein(s). *See also* Enzyme(s); Protein
  synthesis
adhesion, 88t, 89, 89f
cell membrane, 54, 55f, 63, 63f,
  88–89, 88f–89f, 88t, 92–93, 92f,
  93f, 95, 95f
defined, 46
denaturation of, 48
domain of, 46, 47f
fibrous, 47, 68, 70
as fuel for aerobic respiration, 128,
  129, 129f
functional groups in, 40, 40t
functions, 46–48
initiator, 140, 140f
■ medically important, production
  of, 244
metabolism of, 128f, 129
motor, 68–69, 68f, 94

post-translational modifications,
  164f, 165
receptor, 88t, 89, 89f, 94, 95
structure, 46–47, 46f–47f
  and function, 46–48
transport, 88t, 89, 89f
  active, 93, 93f, 201
  passive, 92, 92f
Protein synthesis, 46, 46f–47f,
  150–151. *See also* Transcription;
  Translation
ER and, 64–65, 64f–65f
■ genetic mutation and, 158–159,
  158f, 159f
regulation of, 157
ribosome-inactivating proteins
  and, 149
ribosomes in, 64
RNA and, 63
Protists
characteristics of, 8, 9f
defined, 8
as Eukarya, 8
fermentation in, 126
genetic code in, 155
as kingdom, 11f
pigments of, 103f
Proton(s)
in atomic structure, 24, 24f
charge on, 26, 26f
Pseudomonas, 215
Pseudopods, 69, 95, 95f
*Pst*I enzyme, 234f
■ Psychiatric disorders, 211
Pufferfish, 240f
Punnett squares, 204, 204f, 205, 206f
Purines, 136f
Pyrimidine dimers, 143
Pyrimidines, 136f
Pyruvate, 120, 121f, 122, 122f, 123,
  123f, 124f–125f, 126–127, 126f,
  127f, 128, 128f, 129

R
R group, 46, 46f
■ Rabbits, transgenic, 244
Radiation
bacteria and archaea resistance
  to, 58f
DNA damage from, 142, 143,
  184f, 224
■ Radical mastectomy, 163
Radioactive tracers, 24, 25, 25f, 56
Radioisotope(s), 25, 135f
Radioisotope dating, 113
■ **Rain forest**
in New Guinea, new species in, 19
Ranunculus, 142f
Rats (*Rattus*)
genome, *vs.* other species, 240f
Reactants, 80, 80f
Reaction. *See* Chemical reaction
Reasoning, types, 12
Receptor(s)
B cell, 89f
in cell membrane, 88t, 89f
hormone
  estrogen, 173
  progesterone, 173
Receptor proteins, 88t, 89, 89f,
  94, 95
Recessive alleles
autosomal recessive inheritance, 218f,
  219t, 221, 221f
defined, 203, 203f
X-linked recessive inheritance, 219t,
  222–223, 222f, 223f
Recombinant DNA, 234–235,
  234f, 235f

Red blood cells (erythrocytes). *See also* Sickle-cell anemia
  blood type and, 208, 628
  tonicity, 91*f*
Red-green color blindness, 219*t*, 222–223, 222*f*
Red Queen hypothesis, 189
Redox reactions. *See* Oxidation-reduction (redox) reactions
Reinbach, Gary, 96*f*
Replication fork, 140*f*, 141*f*
Repressor(s), 164, 170–171, 170*f*
Reproduction. *See also* Asexual reproduction; Life cycles; Sexual reproduction; *specific taxons*
  bacteria, 59
  as characteristic of life, 7
  defined, 7
  plants, 189*f*
Reproductive cloning, 144–145, 144*f*, 145*f*
Reproductive organs
  human, 190*f*
  plant, 190*f*
Reproductive system
  human
    development of, 168–169
    female, 190*f*
    male, 190*f*
Reptile(s)
  evolution of, 225*f*
Restriction enzymes, 234, 234*f*, 235*f*
Retina, 221
Retinal, 103*f*
Reverse transcriptase, 235
Rhea, 252, 252*f*
Ribonucleic acid. *See* RNA
Ribosomal RNA (rRNA), 150, 151, 154–155, 155*f*, 203*f*
Ribosome-inactivating protein (RIP), 149, 149*f*, 160, 161, 161*f*
Ribosomes
  eukaryotic cells, 61*f*, 62
  functions, 60*t*
  polysomes, 157, 157*f*
  prokaryotic cells, 58*f*, 59
  rough ER and, 64, 64*f*
  subunits, 155, 155*f*, 156–157, 156*f*, 157*f*
  in translation, 155, 155*f*, 156–157, 156*f*, 157*f*
Riboswitches, 171
Ribulose biphosphate (RuBP), 109, 109*f*
Rice (*Oryza sativa*)
  genetic engineering of, 242, 243
Ricin, 149, 149*f*, 160
RIP. *See* Ribosome-inactivating protein
RNA (ribonucleic acid). *See also specific types*
  compared to DNA, 150, 151*f*
  double-stranded, 165
  functions, 150, 151*f*
  genetic code in, 154–155, 154*f*, 158
  nucleotides, 150, 150*f*, 151*f*
  post-transcriptional modifications, 153, 153*f*, 164–165
  in protein synthesis, 63
  stability of, 157
  structure, 49, 49*f*, 150, 151*f*
  synthesis. *See* Transcription
    in translation, 156–157
  types, 150
RNA interference, 165
RNA polymerase, 152–153, 152*f*, 153*f*, 164, 171
Rose, 10, 10*f*
Rotifers (Rotifera), 197, 197*f*
Rough ER, 61*f*, 64, 64*f*–65*f*
rRNA. *See* Ribosomal RNA

Rubisco, 109, 109*f*, 110, 110*f*, 111, 111*f*
Runoff, 242

S
Saccharomyces cerevisiae (baking yeast), 126, 126*f*
Salamander, 181*f*
*Sal*I enzyme, 234*f*
Salmonella, 215, 215*f*
Salt(s)
  defined, 30
  denaturation of proteins, 48
Sampling error, 16–17, 16*f*, 218
Saturated fats, 44–45, 44*f*, 51*f*
Scanning electron microscopy, 57, 57*f*
Schizophrenia, 211
Schönbein, Christian, 81
SCIDs (severe combined immunodeficiencies), 222*f*, 246, 246*f*
Science, defined, 12
Scientific method, 12–13, 13*t*
Scientific theory, 18, 18*t*
SCNT. *See* Somatic cell nuclear transfer
Scrapie, 48, 48*f*
Second law of thermodynamics, 78, 79
Secondary sexual traits, 168–169, 227
Segmentation
  in fruit flies, control of, 166, 166*f*
Segregation
  law of, 204–205
  in meiosis, 194–195, 195*f*, 199*f*, 204, 204*f*, 206–207, 207*f*
Selective advantage. *See also* Adaptation, evolutionary
  of sexual reproduction, 189
Selective permeability, 90, 90*f*
SEM. *See* Scanning electron micrographs
Semiconservative replication of DNA, 141
Semipermeable membranes, 54, 90–91, 90*f*
Sex, determination
  factors affecting, 139, 210, 225, 225*f*
  in humans, 139, 168–169
Sex chromosomes, 139, 139*f*
  chromosome number, changes in, 219*t*, 226–227
  evolution of, 225, 225*f*
  human, 168–169
Sex hormones
  human, 168–169
Sex pilus, 59
Sexual reproduction. *See also* Meiosis; Reproductive system, human
  advantages and disadvantages of, 189
  vs. asexual reproduction, 189, 197
  evolution of, 196–197
Sharma, Ratna, 101*f*
Shell model
  of atom, 26–27, 26*f*, 27*f*
  of molecule, 28*f*, 29*t*
Shiga toxin, 149, 149*f*
Shigella dysenteriae, 149, 149*f*
Short tandem repeats, 212, 241, 241*f*
Shoulla, Robin, 163, 163*f*, 173
Sickle-cell anemia, 209, 219*t*, 228
  cause, 158–159, 158*f*, 159*f*, 190
  treatment of, 246
Silencers, 164, 169
Simple sugars. *See* Monosaccharides
Singh, Charlene, 48*f*
Single-nucleotide polymorphism (SNP), 233, 240–241, 240*f*, 247

Sister chromatids
  attachment of, 186
  in DNA replication, 138, 138*f*, 140
  in meiosis, 191, 191*f*, 192–193, 192*f*–193*f*, 195, 196–197, 196*f*, 207*f*
  in mitosis, 179, 180*f*, 181, 186, 196–197, 196*f*
Skeletal muscle
  contraction, 68
Skin
  cancers, 143, 185*f*, 221
  color variation in, 203*f*, 212, 217, 217*f*, 219, 221, 229, 231, 231*f*
  dermatological exams, 185
  UV radiation and, 143
  SLC24A5 gene, 217, 229
Smith, Hamilton, 234
Smoking
  and cancer, 185
  and DNA methylation, 210
  and epigenetic inheritance, 173
  genetic damage from, 143
  MAO-B activity in, 25*f*
Snapdragons, flower color in, 208, 208*f*
Snowshoe hare
  coat color, 211, 211*f*
SNP. *See* Single-nucleotide polymorphism
SNP chip analysis, 233, 233*f*, 240–241, 240*f*, 247
Sodium
  atomic structure, 26*f*, 27, 27*f*
Sodium chloride (NaCl; table salt), 28, 28*f*, 30
Sodium-potassium pumps, 93, 93*f*
Soil
  erosion of, 242
Solute, defined, 30
Solution
  defined, 30
  tonicity, 90–91, 91*f*
Solvent
  defined, 30
  water as, 30–31
Somatic cell nuclear transfer (SCNT), 144–145, 144*f*, 145*f*
Sonography, obstetric, 228, 228*f*
Sorghum, 243
Soybeans
  as biofuel, 101, 115, 115*f*
  genetically engineered, 243
Speciation
  chromosome changes and, 225, 225*f*
  genetic evidence for, 240, 240*f*
Species. *See also* Biodiversity
  classification of, 10–11, 10*f*, 11*f*
  new
    discovery of, 3, 3*f*, 16, 16*f*, 19, 19*f*
    naming of, 3*f*, 19*f*, 10
Sperm
  flagellum, 68–69, 69*f*
  formation, 190, 191*f*
  human
    disorders, 75
*Sph*I enzyme, 234*f*
Sphinx cats, 159, 159*f*
Spiders
  Aptostichus stephencolberti, 19*f*
  cuticle of, 70
Spindle, 180*f*, 181, 181*f*, 182, 182*f*, 186, 191, 192–193, 192*f*–193*f*, 194, 195, 196–197, 196*f*
Squamous cell carcinoma, 185*f*
SRY gene, 168–169, 225, 225*f*
Staining, in microscopy, 56, 57*f*
Stamen, 169, 169*f*
Starch
  as fuel for aerobic respiration, 129

  storage of in plants, 42, 43, 67, 109
  structure, 42–43, 43*f*
Statistical significance, 17
Stem cells
  adult, 183
  research on, 145
Steroid(s), 45, 45*f*
Steroid hormones, 45, 45*f*
  structure, 45, 45*f*
Stoma(ta)
  photosynthesis and, 110, 110*f*, 111
Stomach
  enzymes, 83, 83*f*
    gastric fluid pH, 32, 32*f*
  lining of, 71
  ulcers, 58*f*, 71
Stop codons, 155, 157, 157*f*, 158
Streptococcus, 127
Stress
  human health effects, 210, 211
Stroke
  risk factors, 86
Structural formula, of molecule, 28, 29*t*, 38, 38*f*
Substrate(s), 82–83, 82*f*
Substrate-level phosphorylation, 120, 121*f*, 122, 123*f*
Succulents, 111
Sucrose, 42, 42*f*
Sugar(s)
  as hydrophilic, 30
  metabolism of, 7
  simple. *See* Monosaccharides
  table, 42, 42*f*
Sugarcane, 101, 115
Sulfur
  isotopes, as tracer, 135*f*
  as pollutant, 33
Sunlight
  as energy source, 6*f*, 67, 78, 79, 79*f*, 81, 102–103, 102*f*, 104, 104*f*, 105. *See also* Photosynthesis
  skin, effects on, 185
Surface-to-volume ratio, and cell size and shape, 55, 55*f*
Surveys, 13*t*, 14, 16
Sweating
  cooling effects of, 31
Switchgrass
  as biofuel, 101, 101*f*
  as C4 plant, 110–111
Symington, James, 133, 133*f*
Syndrome, defined, 219

T
Tabuchi, Katsuhiko, 249, 249*f*
Taq polymerase, 237, 237*f*
Taxon (taxa), 10–11, 10*f*
Taxonomy
  six-kingdom system, 11*f*
  species, 10–11, 10*f*, 11*f*
  three-domain system, 11*f*, 11*t*, 332
Tay-Sachs disease, 203*f*, 219*t*, 221, 221*f*, 228
Team Trakr Foundation, 145
Telomeres, 183, 183*f*, 225, 225*f*
Telophase (mitosis), 178*f*, 180*f*, 181, 182, 196*f*
Telophase I (meiosis), 190, 192, 192*f*, 195, 195*f*
Telophase II (meiosis), 190, 193, 193*f*, 195, 195*f*, 196*f*
TEM. *See* Transmission electron micrographs
Temperature. *See also* Global warming
  defined, 31
  denaturation of proteins, 48
  and diffusion rate, 90
  and enzyme action, 83, 83*f*

Terminal bud (shoot tip), 169
Termites, 42
■ "Test-tube" babies, 229
Testcrosses, 204–205, 204f, 205f
Testis (testes)
  development, 168
  location, 190f
Testosterone
  action, 168–169
  as steroid hormone, 45f
■ Tetralogy of Fallot (TF), 131, 131f
■ Thalassemias, 158f, 159, 190
Theory. *See* Scientific theory
■ Therapeutic cloning, 145
Thermococcus gammatolerans, 58f
Thermodynamics, 18, 78, 79
Thermophiles, extreme, 8f
Thermus aquaticus, 83f,
  237, 330
Three-domain system, 11f, 11t
Threonine, 217
Through the Looking Glass
  (Carroll), 189
Thylakoid membrane, 105, 105f, 106,
  107, 107f
Thymine (T), 136–137, 136f, 137f,
  150, 151f
Thymine dimer, 143
Ti plasmid, 242, 243f
Tight junctions, 71, 71f
Tilman, David, 115
Tinman gene, 166
Tissue(s). *See also* specific types
  defined, 4, 4f
Toll gene, 167
Tonicity, 90–91, 91f
Tooth (teeth)
■ plaque, 8f
Topoisomerase, 140, 140f
Tracer, radioactive, 25, 25f, 56
  in DNA research, 135f
  in nucleic acid hybridization,
  236, 236f
Trait(s). *See also* Phenotype
  adaptive. *See* Adaptation,
    evolutionary; Selective advantage
  defined, 10
  sexual secondary, in humans,
    168–169, 227
  in species classification, 10–11
  variation within populations, 212–213,
    212f, 213f
Trans fats, 37, 37f, 44, 45, 49, 51f
Transcription
  control of, 164, 164f–165f, 168,
    170–171, 170f, 172
  vs. DNA replication, 152
  overview, 150, 151, 157f
  process, 152–153, 152f, 153f
Transcription factors, 164, 164f–165f,
  166, 169
Transfer RNA (tRNA)
  functions, 150, 151, 154–155, 155f,
    156–157, 156f, 157f
  structure, 155, 155f
Transformation, in prokaryotes, 327

■ Transfusions
  and transfusion reaction, 208
■ Transgenes, escape into environment,
  243, 245
■ Transgenic organisms, 242, 242f.
    See also Genetic Engineering;
    Genetically modified organisms
Translation
  control of, 164f, 165, 171
  genetic code and, 154–155, 154f,
    155f, 158
  overview, 150–151, 157f
  process, 156–157, 156f, 157f
  regulation of, 157
■ Translocation (of DNA), 224,
  224f, 240
Transmission electron microscopy
  (TEM), 56–57, 57f
■ Transplantation
  screening for, 97
  xenotransplantation, 245
Transport proteins, 88t, 89, 89f
  active, 93, 93f, 201
  passive, 92, 92f
  specificity of, 92
Transposable elements, 224
Triglycerides
■ blood levels, genetic factors in, 240
  energy stored in, 44
  structure, 44–45, 44f
■ Trisomy 21 (Down syndrome), 219t,
  226, 227f
tRNA. *See* Transfer RNA
Trypsin, 83, 83f
Tryptophan, 154f, 155
■ **Tuberculosis**
  body defenses against, 95f
Tubulin, 68, 68f, 181
Tumor(s)
  benign, 184
  blood vessels and, 185
  defined, 184
  malignant, 184–185
■ Tumor suppressor(s), 184
■ Tumor suppressor genes, 173
Turgor, 91
■ Turner syndrome, 219t, 226–227
Turtles
  sex determination in, 139
Twins
  artificial twinning, 144
  identical, 144
Typhoid fever, 213, 215, 215f, 219
Tyrosine, 155, 228
TYRP1 gene, 209

U

■ Ulcers, stomach, 58f, 71
■ Ultrasound, 228, 228f
Ultraviolet radiation. *See* UV (ultraviolet)
  radiation
United States Department of Agriculture
  (USDA)
  Animal and Plant Health Inspection
    Service (APHIS), 243
Units of length, 56t

Uracil (U), 150, 151f, 152, 154, 154f
USDA. *See* United States Department of
  Agriculture
UV (ultraviolet) radiation
  in electromagnetic spectrum, 102f
  and skin color, 217, 231, 231f
■ as threat to life, 103, 142, 143, 221

V

■ **Vaccines**, for polio, 177
Vacuoles, 57f, 60f, 60t, 65f
  central, 61f, 64
Valine, 46f–47f, 158–159, 159f
Van Helmont, Jan Baptista, 115
Variables, in experiment, 13, 14
■ Variant Creutzfeldt-Jakob disease
  (vCJD), 48, 48f
Vascular plants
  water movement through, 31
vCJD. *See* Variant Creutzfeldt-Jakob
  disease
Vegetable oils, 37, 45
Vendrely, Roger, 135
Venter, Craig, 239
Vesicle(s), 64–65
  endocytic, 94–95, 94f, 95f
  functions, 60t, 64–65, 64f–65f
  membrane trafficking, 94–95,
    94f, 95f
Victoria, queen of England, 223f
Virus(es)
  as DNA vectors, 246
  evolution of, 245
  size, 56f
Vision. *See also* Eye(s)
■ color blindness, 219t, 222–223, 222f
  pigment (opsins), 221, 222–223
■ **Vitamin(s)**
  as antioxidants, 117
  as cofactors, 86, 86t
Vitamin B_{12} (cobalamin), 171
Vitamin C (ascorbic acid), 42, 86t
Vitamin D, 45, 217

W

■ Warts, 184
■ Waste, environmental
  agricultural, as biofuel, 101, 101f
Water
  bonds, describing, 29t
  diffusion across membranes, 90–91,
    90f, 91f
  and enzyme action, 83
  evaporation, 31
  freezing, 31, 31f
  in metabolic reactions, 41, 41f
  and organic molecules, 40f
  properties, 30–31, 30f, 31f
  structure, 28, 28f, 29, 29f, 30, 30f
  and temperature homeostasis, 31
Water flea (*Daphnia pulex*), 211, 211f
■ **Water pollution**
  mercury as, 23, 33
Watson, James, 136–137, 137f, 143,
  239, 245
Waxes, 45

■ Wet acid deposition (acid rain), 33
Wexler, Nancy, 218f
Whale
  as air breather, 119f
Wheat (*Triticum*)
■ transgenic, 242
White blood cells (leukocytes). *See
  also* specific types
  in immune response, 95, 95f
  structure, 60f
Wilkins, Maurice, 143
Wilmut, Ian, 145
Wine, 126
Wingless gene, 166
Wood
  burning of, 80, 81f
Wood ducks (*Aix sponsa*), 45f
Work, 78
Wrinkled gene, 166

X

X chromosome
  evolution of, 225, 225f
  genes, 168, 222f
  inactivation of, 168, 168f, 173, 222
  and sex determination, 139, 168–169
■ X-linked anhidrotic dysplasia, 219t
■ X-linked disorders, 219t, 222–223,
  222f, 223f
X-ray crystallography, 136, 143
X-rays
  in electromagnetic spectrum, 102f
■ as threat to life, 142, 143
■ Xenotransplantation, 245
XIST gene, 168, 172, 173, 222f
■ XO individuals, 226–227
■ XXX syndrome, 219t, 227
■ XXY syndrome, 227
■ XYY condition, 219t
■ XYY syndrome, 227

Y

Y chromosome
  evolution of, 225, 225f
  genes on, 168
  and sex determination, 139, 168–169
Yarrow (*Achillea millefolium*),
  211, 211f
Yeast
  and alcoholic fermentation in foods,
    126, 126f
  baking (*Saccharomyces cerevisiae*),
    126, 126f
  mitochondria, 66
■ transgenic, 242
Yogurt, 127

Z

Zeaxanthin, 103f
Zebrafish, 240f, 244f
Zygote
  defined, 191
  formation, 191, 196